T0327444

Natural Catastrophe Risk
Management and Modelling

# Natural Catastrophe Risk Management and Modelling

A Practitioner's Guide

Kirsten Mitchell-Wallace
*SCOR, Zürich, Switzerland*

Matthew Jones
*Cat Risk Intelligence, UK*

John Hillier
*Loughborough University, Loughborough, UK*

Matthew Foote
*Argo Group International Holdings, London, UK*

This edition first published 2017
© 2017 John Wiley & Sons Ltd

The right of Kirsten Mitchell-Wallace, Matthew Jones, John Hillier and Matthew Foote to be identified as the authors of this work has been asserted in accordance with law.

*Registered Offices*
John Wiley & Sons Ltd, The Atrium, Southern Gate, Chichester, West Sussex, PO19 8SQ, UK

*Editorial Office*
111 River Street, Hoboken, NJ 07030, USA
9600 Garsington Road, Oxford, OX4 2DQ, UK
The Atrium, Southern Gate, Chichester, West Sussex, PO19 8SQ, UK

For details of our global editorial offices, customer services, and more information about Wiley products visit us at www.wiley.com.

Wiley also publishes its books in a variety of electronic formats and by print-on-demand. Some content that appears in standard print versions of this book may not be available in other formats.

*Library of Congress Cataloging-in Publication Data*

Names: Mitchell-Wallace, Kirsten, 1973- editor.
Title: Natural catastrophe risk management and modelling: a practitioner's guide / edited by Kirsten Mitchell-Wallace, Zurich, SZ, Matthew Jones, Southampton, UK,
    John Hillier, Loughborough, UK, Matthew Foote, London, UK.
Description: Hoboken, NJ : John Wiley and Sons, Inc., [2017] | Includes
    bibliographical references and index.
Identifiers: LCCN 2016053404| ISBN 9781118906040 (cloth) | ISBN 9781118906071 (epub)
    | ISBN 9781118906064 (Adobe PDF)
Subjects: LCSH: Emergency management. | Disasters–Risk assessment.
Classification: LCC HV551.2 .C38 2017 | DDC 363.34/2–dc23 LC record available
    at https://lccn.loc.gov/2016053404

Cover image: © cosmin4000/Gettyimages
Cover design by Wiley

Set in 10/12pt WarnockPro-Regular by Thomson Digital, Noida, India

10  9  8  7  6  5  4  3  2  1

# Contents

# List of Contributors and Acknowledgements

We, the main authors, wish to expressly thank the other authors and reviewers who have contributed to this book and made it possible. Catastrophe risk management and modelling is a discipline that draws upon a wide range of knowledge and expertise, so we could not have created a credible practitioner's guide without the widespread support that we have received from the catastrophe modelling community.

Each main author had oversight of one of the main chapters (i.e. 2–5), and sculpted it as appropriate; the structure of each topic and nature of the material required different approaches by chapter. This said, we have all re-written, polished and homogenized all the chapters of the book to produce what we hope is a coherent journey through the topic of catastrophe risk management and modelling.

We have chosen to accredit ourselves in reverse alphabetical order, selected with a degree of randomness befitting the subject matter of this guide.

As well as the four main authors, many contributors provided material for chapters, or sub-chapters, of the book; sometimes an author contributed in more than one place. These contributions are summarized in the table below, and, as appropriate, authors are listed at the start of sections that they co-wrote.

The contributions of the many external reviewers drawn from outside the editorial team were also invaluable; in particular, we are indebted to Alan Calder, George Cooper, Jeff Gall, Claire Crerar, Paul Nunn and Claire Souch who reviewed the entire book or a large part of it.

We thank Wiley's editors and team for giving us the opportunity to produce this guide and assisting us along the way.

## Main Authors

This guide has four main authors who also acted as editors. Here, we introduce ourselves in reverse alphabetical order.

***Kirsten Mitchell-Wallace*** is the Regional Cat Manager for Europe, Middle East and Africa at SCOR, leading cat teams based in Zurich and Paris in support of property treaty business written from this region, including US Cat. Before this, she was Head of Cat Pricing and Methodology and responsible for coordinating the development of SCOR's own view of risk, as well as managing the Zurich-based cat team. Prior to joining SCOR in 2009, she was a Senior Catastrophe Risk Analyst at Willis, London, for 5 years, working first on European, then Japanese business. She started her career with two years at Risk Management Solutions in London and two years at risk management consultancy, Risk Solutions, from 2001. Kirsten has a PhD in Atmospheric Physics from Imperial College, London, and a Masters in Meteorology from Reading University, as well as a degree in Chemistry from the University of Bristol. She also has an International Diploma in Risk Management from the Institute of Risk Management. Kirsten is passionate about catastrophe modelling.

*Acknowledgments*: I would first like to thank my husband, Richard, for keeping our family fed and Arthur clothed throughout this project. Without his incredible support, my contribution could never have been possible.

I would then also like to thank the many people outside the editorial team who reviewed my chapter and made valuable suggestions: Sibylle Steimen, Claire Crerar, Alan Calder, Paul Nunn, Claire Souch, Iakovos Barmpadimos, Roger Bordoy, George Cooper, Thomas Premier and Thomas Linford. Thanks to Parvez Chowdury and Tobias Hoffmann for their review of the actuarial matter. Special thanks to Tom for associated discussions and to Paul for his support of this endeavour.

I would like to thank my many co-authors for their contributions and for being, without exception, a pleasure to work with. I have learned a lot! Thank you also to my fellow editors for a real team effort. In loving memory of my father, Derrick G. Mitchell.

***Matthew Jones*** is the founding Director of Cat Risk Intelligence, a UK-based company providing catastrophe risk management consultancy to the (re)insurance industry.

In his previous role as Global Head of Catastrophe Management for Zurich Insurance Group, Matthew led the organizational change required to establish a global catastrophe management team with consistent processes to provide catastrophe knowledge, systems, models and services across Zurich's general insurance lines of business. He worked for Zurich for fourteen years, including various roles in the actuarial pricing and catastrophe risk management fields. Prior to this Matthew was a reinsurance pricing actuary for St Paul Re in London.

Matthew graduated from the University of Nottingham in 1993 with a degree in Physics. He then completed a PhD in Oceanography and Remote Sensing from University College London, while being based at the UK's National Oceanography Centre in Southampton. He is a Fellow of the Institute and Faculty of Actuaries.

*Acknowledgments*: I would like to thank my wife Zoe, and my children, Adam, Lauren and Becky, for their love, support and patience. I am extremely grateful to Jane Hayes for allowing me a sabbatical from Zurich to help write some of this book. I am also very thankful to Alan

Calder, Jeff Gall and Claire Souch for the time they put into reviewing and making improvements to Chapter 5; to Fortunat Kind for his input and improvements (including the 'Christmas tree' diagram!); to Gary Hemming for his review of some actuarial pricing aspects; to Federico Waisman for allowing many of his model comparison exhibits to be included; and to Ye Liu for some very helpful statistical discussions.

***John Hillier*** is currently a Senior Lecturer at Loughborough University, with research interests that include various geo-hazards (e.g. earthquake, landslide, flooding, extra-tropical cyclones), multi-hazard risk to property and infrastructure, use of impacts (e.g. loss data) to gain insights into physical processes, and inter-dependencies (i.e. 'links' or interactions) between perils.

He received B.A. (M.A. Cantab) and M.Sci. degrees in Natural Sciences specializing in Geology from the University of Cambridge, and a D.Phil degree in Marine Geophysics from the University of Oxford.

After studying he was awarded a research fellowship at St Catharine's College, Cambridge, and then worked for Zurich Insurance as a catastrophe modeller.

*Acknowledgments*: I am extremely grateful as ever for the love, tolerance and support of my wife, Katie; I do not say this enough, so I would like to say it explicitly here. My love and thanks also to Ben and Charlotte for putting up with Daddy working when you could clearly think of better things I should be doing.

Academically, I appreciate the stoic perseverance and professionalism of all 16 contributors on the perils in Chapter 3 in the face of requests for more detail in fewer words, or for correcting errors I inadvertently introduced when re-writing an initial submission to half its original length. The following reviewers did a great job of making sure we did a robust and thorough job, with numbers in brackets indicating sections: Rebecca Bell (Imperial College) [3.7], Thierry Corti (Swiss Re) [3.6], Mark Dixon (RMS) [3.3], Juergen Grieser (RMS) [3.4], Joanna Faure-Walker (UCL) [3.8], Greg Holland (NCAR) [3.2], Edmund Penning Rowsell [3.5], Tiziana Rossetto [3.9], Robin Spence [3.10]. This said, any remaining errors should be considered mine and mine alone.

***Matthew Foote*** is an exposure management specialist with over twenty years of experience in the (re)insurance industry, including roles with Willis and Guy Carpenter reinsurance brokers, Risk Management Solutions, Mitsui Sumitomo and most recently Argo Group. For seven years Matthew was the Research Director of the Willis Research Network, responsible for the coordination and development of industrial-academic collaborations involving over fifty research organisations and universities.

Matthew began his career as a cartographer and geographic information specialist, working primarily with earth observation (EO) data and other imagery-based data. He has a First Class degree in Geography from Birkbeck, University of London, is a Fellow of the Royal Geographical Society/IBG and a Chartered Geographer.

*Acknowledgments*: My thanks are short and simple, first, to Paula, my wife, for all she has done, and to my children Ciara and Michael for inspiring me. I would also like to particularly acknowledge the efforts of Barbara Page, Adam Podlaha, Shane Latchman, Rashmin Gunasekera and Claire Souch, and to express my gratitude to Stuart Lane for his advice in the very early stages of developing this book.

## Contributors

The following practitioners and academics contributed to this guide. For volunteering your time and expertise, we thank you.

| Contributor* | Section(s) | Biography |
| --- | --- | --- |
| Šárka Černá | 4.3 | *Šárka Černá* works in the business development team in Impact Forecasting doing product development and working with clients. She benefits from five years working as a catastrophe model developer, where she mainly focused on the mathematical and statistical aspects of flood models (e.g. Switzerland). She is involved in the development of the ELEMENTS platform, including testing and validation procedures for Solvency II. After joining Aon Benfield in 2010, Šárka was appointed their Chair of Research in 2014. Prior to 2010, she worked at the Institute of Thermomechanics of the Czech Academy of Sciences on statistical ultrasonic signal processing. |
| Arnab Chakrabarti | 2.15 | *Arnab Chakrabarti* is a Research Engineer at Nephila Advisors LLC, Larkspur, CA, USA. He works on developing methods and tools for catastrophe bond pricing, portfolio construction, and other problems that use a combination of math/stat/programming. He has been with Nephila since 2013. Prior to joining Nephila, Dr Chakrabarti worked in R&D at Qualcomm. He has a Masters in Financial Engineering from the Haas School of Business at UC-Berkeley. Earlier, he did his MS-PhD in Electrical and Computer Engineering from Rice University, and his B.Tech in Electronics and Electrical Communication Engineering from IIT-Kharagpur. |
| Ian Cook | 2.16 | *Ian Cook* is Chief Actuary and Managing Director at Willis Re, advising clients around the world on risk quantification and risk mitigation. He has represented Willis Re on EIOPA's Catastrophe Risk Subgroup, advising them on natural and man-made catastrophe risk and reinsurance. He has also presented and published papers on a number of actuarial and catastrophe-related topics. Prior to joining Willis Re in 2002, he spent over 10 years in actuarial consultancy helping clients on a wide range of actuarial matters. Ian is a Fellow of the Institute and Faculty of Actuaries, and holds a degree in mathematics from Cambridge University. |
| Claire Crerar | 2.7 | *Claire Crerar* has over 14 years experience working in catastrophe modelling, having started her career as a catastrophe risk analyst at Willis. After six years modelling for a wide range of global clients, Claire moved to Aspen where she led the London catastrophe modelling team, with responsibility for international treaty modelling and monitoring accumulations of catastrophe risk. After seven years in this role, Claire has recently shifted focus to concentrate on cat-related project work at Aspen. Claire has an MA in geography from Cambridge University and holds an Advanced Diploma in Insurance from the Chartered Insurance Institute. |
| Tom Dijkstra | 3.8 | *Tom Dijkstra* studied at Utrecht University, the Netherlands, taking Physical Geography with Geomorphology, Quaternary Geology and Soil Science as specializations. In 1989, he moved to the UK after being offered a research position as part of a large EU-funded project and this supported his PhD on landslide mechanisms in Chinese loess. He worked in academia, mainly in Civil Engineering Departments, focusing on geotechnical engineering and since 2012 has been an engineering geologist/geomorphologist at the British Geological Survey specializing in climate change effects on slope instability processes in |

*(Continued)*

| Contributor* | Section(s) | Biography |
| --- | --- | --- |
| | | natural and engineered (transport infrastructure) slopes, geohazards and landslide hazard forecasting. |
| Richard Dixon | 3.3 | ***Richard Dixon*** received a first-class honours degree in Meteorology followed by a PhD specializing in extratropical cyclones from University of Reading, UK. Since 2000, he has worked in the insurance industry building and researching catastrophe models at Risk Management Solutions, and for the past 10 years has worked in catastrophe model evaluation at Aon Benfield, Renaissance Reinsurance and Hiscox. He currently works as a consultant to the insurance industry in catastrophe model evaluation at CatInsight as well as carrying out meteorological research to aid in the understanding of catastrophe risk and catastrophe risk model development. |
| James Done | 3.2 | ***James Done*** received his BSc and PhD in meteorology from the University of Reading, UK. A post-doctoral position at the National Center for Atmospheric Research (NCAR), in the United States led to his current position of Willis Research Fellow and science lead of NCAR's Capacity Center for Climate and Weather Extremes. He works with stakeholders from the energy, water and insurance sectors to understand future weather and climate impacts. Examples of his recent work include assessing future hurricane impacts on the offshore energy industry and the electric power industry, and quantifying the benefit-cost ratio of hurricane building codes. |
| Radovan Drinka | 4.3 | ***Radovan Drinka*** is a catastrophe model developer at Impact Forecasting and a meteorologist involved in the European Windstorm project and atmospheric perils (e.g. hail, summer storm). He is also involved in an engineering approach to structure vulnerability to wind damage. He has two years experience as a catastrophe modeller after joining Aon Benfield in 2008. As a member of the Impact Forecasting team, Radovan has gained a wide knowledge of Impact Forecasting proprietary platform ELEMENTS, and has used it in his daily routine since 2010. Radovan earned a Masters degree in Physics, Meteorology and Climatology at the Comenius University, Bratislava, the Slovak Republic. |
| Matthew Eagle | 2.13 | ***Matthew Eagle*** is Managing Director and Head of GC Analytics for International at Guy Carpenter. In this role he is responsible for providing resource leadership across the EMEA and AsiaPac regions, with a specific focus on developing a market-leading catastrophe modelling and analytics proposition. He has over 20 years of experience in the reinsurance sector with a focus on catastrophe analytics. Previously he worked in statistical software and consulting. He has a MSc in Statistics from the University of Minnesota and is a Fellow of the Royal Statistical Society. |
| Juan England | 2.12 | ***Juan England*** is Managing Director of the Willis Re Latin America and Caribbean team, based in London. Previously he held various positions in the Analytics division as joint-head of Analytics for Willis Re, Deputy Managing Director of Willis Re's International Catastrophe Management team, and head of catastrophe modelling for Willis Re's Latin America and Caribbean. He joined Willis Re as a catastrophe analyst in 2006. Juan has a PhD in Structural Vulnerability from the University of Bristol, UK, and BSc in Civil Engineering with emphasis in earthquake risk assessment from the University of Los Andes in Bogota, Colombia. |

*(continued)*

(*Continued*)

| Contributor* | Section(s) | Biography |
|---|---|---|
| Chris Ewing | 4.3 | **Chris Ewing** is a member of Impact Forecasting's business development team, currently leading on the use of data for primary underwriting, and implementation of third party models on Aon's ELEMENTS loss calculation platform. Chris previously developed earthquake models (e.g. Greece) and worked on the global tsunami research project. Prior to 2011, Chris worked for a consultant engineering firm. Chris has a BSc in Geography from the University of Leeds and an MSc in GIS from Nottingham University. Chris is a Chartered Geographer (GIS), a Fellow of the Royal Geographical Society, and chairs AGI's Insurance and Risk special interest group. |
| Joanna Faure Walker | 3.7 | **Joanna P. Faure Walker** received BA (MA Cantab) and MSci. degrees in Natural Sciences, specializing in Geology, from the University of Cambridge, and a DPhil degree in Earth Sciences from University College London. Following her PhD, Faure Walker became an analyst for the catastrophe modelling firm, RMS. She is currently a Senior Lecturer at UCL IRDR (Institute for Risk and Disaster Reduction) with research interests that include earthquake geology, rates of fault deformation and interaction, seismic hazard, and the transitional phase of recovery. She lectures on natural hazards, vulnerability, risk and their integration into decision-making. |
| Guillermo Franco | 2.8 | **Guillermo Franco** is Managing Director and Global Head of Catastrophe Risk Research at Guy Carpenter. Prior to this, Guillermo was a manager and principal engineer at AIR Worldwide, where he participated in earthquake model development and headed the Decision Analytics practice. As a research fellow at Columbia, he studied the socioeconomic impact of natural hazards in developing countries. Guillermo is a structural engineer and holds MSc, MPhil and PhD degrees in civil engineering and engineering mechanics from Columbia University in New York and a BSc from the Technical University of Catalonia in Barcelona, Spain. |
| Yo Fukutani | 3.9 | **Yo Fukutani** received a BS degree in Physics from Tohoku University and an MSci degree in Natural Sciences, specializing in Meteorology, from the University of Tokyo, Japan. After studying, he worked for Tokio Marine & Nichido Risk Consulting Co., Ltd as a catastrophe modeller and surveyor, and then worked for Tohoku University as a research associate from 2012 to 2015. In 2016, he received a PhD in Civil Engineering from Tohoku University. He is currently a Senior Risk Analyst at Tokio Marine & Nichido Risk Consulting, with research interests that include uncertainty assessment and quantitative risk assessment especially in the coastal engineering field. |
| Peter Geissbuehler | 3.4 | **Peter Geissbühler** has worked at Tokio Millennium Re AG as the Head of the Actuarial and NatCat Department in Europe since 2011. Before he joined Tokio Millennium, Peter worked at RMS as a Director for European and International Product, mainly responsible for managing the European Winterstorm Model in 2011. Peter started his career at Converium as a cat modeller and pricing actuary where he built models for typhoons in South Korea. In addition, he was senior pricing actuary for cat business for many European countries. Peter holds a PhD in Climatology. |
| Alexandros Georgiadis | 4.3 | **Alexandros Georgiadis** joined Impact Forecasting (Aon Benfield) in 2009. His focus is the development of a probabilistic windstorm risk model for Europe, addressing the climate research-related aspects of the project in collaboration with the University of Cologne. His |

*(Continued)*

| Contributor* | Section(s) | Biography |
|---|---|---|
| | | background is in climatology, including: atmospheric dynamics and climate modelling, surface-atmosphere interactions, climate change, forest fire risk and remote sensing. Prior to 2009, he was a post-doc in the Space and Atmospheric Physics (SPAT) group at Imperial College London. Alexandros holds a PhD in climatology (Hurricane formation and modification in the south-east Caribbean) from the University of Sheffield, UK. |
| Rashmin Gunasekera† | 2.14, 3.10, 4.3, 6.2.4 | ***Rashmin Gunasekera*** is a disaster risk management specialist at the World Bank, focusing on disaster risk assessment and risk financing within the Latin American & Caribbean region. He has over 15 years experience extending to the public sector, re/insurance industry and academia. Prior to joining the World Bank in 2012, he was a Divisional Director of a global reinsurance intermediary and a coordinator of the Willis Research Network, the world's largest collaboration between public science and the financial sector. He has also been a research scientist for an EU project on volcanic risk, and holds an honorary lectureship at UCL. His PhD is in earthquake seismology. |
| Marc Hill | 4.3 | ***Marc Hill*** is a principal modeller in the model development group at Risk Management Solutions Ltd (RMS). He studied for his MSc in Concrete Structures at Imperial College London and received a PhD in Engineering at University College London. He joined RMS in 2009 and has participated in a variety of vulnerability development projects, including the climate peril of wind, as well as post-catastrophe damage surveys. |
| Michael Kunz | 3.4 | ***Michael Kunz*** studied Meteorology at the University of Karlsruhe (Dipl. Met.), where he also earned his PhD (Dr. rer. nat.) with his dissertation on orographic rain enhancement. In 2011, he completed his Habilitation at the Karlsruhe Institute of Technology (KIT) on the amplification of atmospheric processes over complex terrain. He is currently Senior Scientist and head of the working group 'Atmospheric Risks' at the Institute of Meteorology and Climate Research at KIT. His research focuses on extreme weather events (hail, heavy rainfall, and wind gusts), their probability, long-term variability, and related impacts. His working group has also developed various hazard and risk models in cooperation with insurance companies. |
| Rob Lamb | 3.5 | ***Rob Lamb*** studied at the Universities of Cambridge and Lancaster, where his doctoral research was in numerical modelling of river catchment hydrology and uncertainty analysis. He was a research scientist at CEH Wallingford for six years, before moving to JBA Consulting in 2002. Since then, he has worked on academic and applied research in hydrology, river hydraulics, flood risk management, climate change adaptation and systems-based infrastructure risk analysis. Since 2012, he has held dual roles as a Professor in the Lancaster Environment Centre, and Director of the JBA Trust, a charitable research and knowledge exchange foundation sponsored by the JBA Group. |
| Shane Latchman | 4.3, 4.5, 4.7 | ***Shane Latchman*** is Assistant Vice President in AIR's London office involved in some of AIR's Touchstone initiatives, such as the integration of third-party data and models, expanding AIR's capabilities in marine and energy, the Next Generation Financial Module, and the development of future multi-modelling/blending capabilities. He is a member of the catastrophe modelling and actuarial industry groups, and is heavily involved with rating agencies and regulators on topics, *(continued)* |

(*Continued*)

| Contributor* | Section(s) | Biography |
|---|---|---|
| | | such as Solvency II. After receiving a National Scholarship from Trinidad and Tobago, Shane studied Actuarial Science at City University and received a BSc with honours. His Masters in Mathematics is from the University of Cambridge, and he is a Certified Catastrophe Modeller. |
| Sue Loughlin | 3.10 | *Susan Loughlin* is the Head of Volcanology at the British Geological Survey and joint leader of the Global Volcano Model project. Her research interests include volcanic processes, hazards and risk, communication, social and environmental impacts of eruptions and the interaction of scientists and decision-makers. Dr. Loughlin spent several years at Montserrat Volcano Observatory and was Director there for two years. She has been an advisor to governments and communities during volcanic unrest and eruptions (e.g. Montserrat and Iceland/UK) and provided scientific evidence for longer-term planning. |
| Paul Nunn | 2.11 | *Paul Nunn* is Head of Catastrophe Risk Modelling at SCOR Global P&C, responsible for the management of natural hazard perils globally. A key aspect of the role is the provision of analytics and data for internal and external stakeholders, including SCOR's internal capital model, rating agencies, regulators and retrocessionaires. Before joining SCOR Global P&C, Paul was Head of Exposure Management at Lloyd's and previously worked for the cat modelling specialist firm, AIR Worldwide. He is also a Director of the non-profit Oasis Loss Modelling Framework company. |
| Brian Owens | 3.2 | *Brian Owens* is a meteorologist and specialist in catastrophe risk management. As a Senior Director at Risk Management Solutions (RMS), he is responsible for their global models and data market and release strategy. He has more than 20 years experience in insurance, catastrophe financing, and catastrophe risk management, and has written numerous scientific articles and blogs. Mr Owens holds a BS in computer science from the National University of Ireland, an MBA from the University of Pennsylvania (Wharton), and an MS in meteorology from the University of Miami. |
| Barbara Page | 4.3-4.6 | *Barbara Page* is a Senior Director in the model product management group at Risk Management Solutions Ltd (RMS). She obtained an MA in Natural Sciences from Cambridge University and a PhD in volcanology from Edinburgh University. Over the past 20 years, Dr Page has fulfilled a variety of roles at RMS in developing and managing catastrophe risk modelling products for the global private insurance market. |
| Adam Podlaha | 2.16, 4.3 | *Adam Podlaha* is the Head of Impact Forecasting at Aon Benfield, overseeing a team of 85+ catastrophe model developers responsible for over 100 models spanning 12 perils, plus the ELEMENTS loss calculation platform. In addition to managing the team, Adam collaborates with various governmental and non-governmental institutions to deliver the best quality models, finding ways to effectively quantify and visualise uncertainty and make catastrophe modelling platforms more open. He joined Aon Benfield in Prague as a flood model developer in 2003 and holds a PhD in Physical Geography from Charles University in Prague. |
| Petr Punčochář | 4.3 | *Petr Punčochář* is responsible for the Impact Forecasting flood model development team in the EMEA and APAC regions, implementing new workflows and methodologies. Additionally, he provides insights on |

*(Continued)*

| Contributor* | Section(s) | Biography |
|---|---|---|
| | | hydrology, hydraulics and geographical information systems. Prior to joining Aon Benfield in 2010, Petr was a research assistant at the Department of Hydraulics and Hydrology at the Czech Technical University. Petr is a member of the International Association of Headwater Control and Czech delegation of FAO committee for Mountain Watershed Management. Petr received his PhD in hydrology and open-channel hydraulics in 2011 from the Czech Technical University. |
| Gillaume Pousse | 3.6 | **Guillaume Pousse** gained a PhD degree in Earthquake Engineering in 2005 from the French Institute for Nuclear Safety. He is working for Guy Carpenter and is dedicated to delivering analytics value to clients. He previously worked at a reinsurer, pricing/monitoring inward catastrophe business in earthquake-prone or traditionally non-modelled countries. He also contributed to documenting the view of the risk in a continuous and prospective way. Prior to this, he developed probabilistic earthquake loss models in London for insurers and reinsurers. |
| Junaid Seria | 2.10, 2.11 | **Junaid Seria** works as a Solvency II Nat Cat Actuary at SCOR and is responsible for embedding Solvency II in the business activities of the Nat Cat team. This includes SII training, independent validation of cat model methods and results, and developing governance structures. Previously he worked as an Actuarial Executive at KPMG in London where he specialized in cat risk consulting. This included managing authorization engagements for two UK reinsurer start-ups, managing cat risk validation engagements for UK primaries. He was also the cat risk expert reviewer for internal model submissions to the Bermudian Monetary Authority. |
| Len Shaffrey | 3.3 | **Len Shaffrey** is a Senior Scientist in the National Centre for Atmospheric Science and a Professor in the Department of Meteorology at the University of Reading, UK. His research interests include understanding how extremes such as European Windstorms and extra-tropical cyclones have varied in the past and how they might respond to climate change. |
| Milan Simic | 4.3, 4.5, 4.7 | **Milan Simic** is Executive Vice President and Managing Director of International Operations for AIR Worldwide, and is responsible for business development, strategic growth initiatives, and client services. Milan has more than 25 years' experience in risk assessment, engineering consulting, teaching, and research. He is a Chartered Engineer of the UK Institution of Civil Engineers and has authored numerous papers. He is also a member of the OECD's High-Level Advisory Board on Financial Management of Large-Scale Catastrophes. He earned his MSc in Hydraulic Structures from the University of Belgrade and a PhD in Earthquake Engineering from the University of Bristol, UK. |
| Nilesh Shome | 4.3 | **Nilesh Shome** earned his PhD in Structural Engineering from Stanford University, USA. He is a Vice President of Risk Management Solutions (RMS) Global Earthquake and Terrorism products, leading the research and development works. Dr Shome joined RMS in 2009 and has more than 15 years of professional experience in modelling risk from hurricanes, earthquakes and other natural and man-made hazards, for RMS and other agencies including the World Bank, the Federal Emergency Management Agency (FEMA) and the Applied |

*(continued)*

(*Continued*)

| Contributor* | Section(s) | Biography |
|---|---|---|
| | | Technical Council (ATC). He authors, reviews and edits publications in international technical journals and refereed conferences. |
| Radek Solnický | 4.3 | ***Radek Solnický*** is a catastrophe model developer in the Impact Forecasting flood team. As a statistician, his domains are hydrological data processing, frequency analysis, stochastic event set generation, as well as client data-based vulnerability development. He also provides mathematical support. Apart from taking part in the Poland Flood model and Sweden Cloudburst model and flood model for Brazil, he was responsible for the Netherlands Flood and Storm Surge model and Hungary flood model development. Radek contributes to the implementation of flood models into ELEMENTS, Aon's own loss estimation software platform. |
| Claire Souch | 2.11, 4.8, 6.2.1 | ***Claire Souch*** has 15 years experience of catastrophe risk model development and usage across the global re/insurance industry. Claire is currently Head of Development at AgRisk, having held previous positions leading model development and evaluation at SCOR, and SVP of Model Strategy at RMS. She has advised on many aspects of catastrophe model usage and the impact of climate change on catastrophe risk for re/insurance companies across multiple markets. She has served on multiple industry task-forces on catastrophe risk and risk modelling, and is frequently invited to speak on topics such as catastrophe risk, development and the role of catastrophe risk insurance. Claire holds a BSc in Environmental Science and a PhD from Cranfield University, UK. |
| Anawat Suppasri | 3.9 | ***Anawat Suppasri*** is currently an Associate Professor at the International Research Institute of Disaster Science, Tohoku University, Japan, with research interests that include various topics on tsunami hazard and risk assessments. He received a BEng degree in Civil Engineering from Chulalongkorn University, an MEng degree in Water Engineering and Management from the Asian Institute of Technology, and a PhD Degree in Civil Engineering from Tohoku University. After studying, he was awarded a research fellowship at the Disaster Control Research Center, Tohoku University. |
| Rick Thomas | 6.2.2 | ***Rick Thomas***. After studying Natural Sciences at Cambridge and doing a PhD and postdoc in modelling volcanic eruptions, Rick started work at CARtograph building catastrophe models in 1995. He moved from CARtograph to PartnerRe in 1997 where he led the creation of PartnerRe's in-house model suite, starting with US Hurricane models, but expanding to Japanese Typhoon, European Windstorm and global quake models. Rick moved from modelling to underwriting in the early 2000s and took over as Head of the international property CAT book at PartnerRe in 2004. Subsequent to Partner Re, Rick worked as an advisor to a fund investing in ILS, and as Head of Model Development and Evaluation at Willis Re, where he was also responsible for the Willis Research Network. |
| Jane Toothill | 3.5 | ***Jane Toothill*** has over 20 years' experience in the modelling of natural catastrophes. Jane worked for the modelling company EQECAT and reinsurance broker Guy Carpenter, prior to joining JBA Group in 2008. She became one of the founding directors of JBA Risk Management in 2011, where she heads operations and catastrophe modelling. During her career Jane has also worked for the British Geological Survey and acted as an advisor on Zurich Financial Services' Natural Catastrophe |

*(Continued)*

| Contributor* | Section(s) | Biography |
|---|---|---|
| | | Advisory Council. She holds a degree in Geology from the University of Bristol and a PhD in Environmental Science from Lancaster University. |
| Goran Trendafiloski | 4.3 | ***Goran Trendafiloski*** is earthquake expert (e.g. seismic hazard assessment) and catastrophe model developer at Impact Forecasting (Aon Benfield). He develops damage and loss models to estimate the risk of properties and population due to earthquake shaking and tsunamis. He has published more than 100 scientific publications, books, papers and reports. Prior to 2010, he worked at the World Agency for Planetary Monitoring and Earthquake Risk Reduction, and the Institute of Engineering Mechanics, Harbin, China. Goran holds Dr.Tech.Sci. and M.Sc. degrees from the Institute of Earthquake Engineering and Engineering Seismology at the University St. Cyril and Methodius, Skopje, Macedonia, and postgraduate specialty degree from the University of Geneva, Switzerland. |
| Craig Verdon | 3.8 | ***Craig Verdon*** is currently the Head of Europe Catastrophe Modelling at Endurance Re in Zurich. He received BSc (Geology & Chemistry) and BSc (Hons) degrees from the University of Natal (Durban) and, after some time in industry, an MSc in Engineering Geology from Imperial College London. He worked in a variety of roles across the mining and engineering sectors before joining the catastrophe modelling industry as a product manager at RMS. |
| Renato Vitolo | 4.3 | ***Renato Vitolo*** is Head of Risk Modelling, Operational and Reputational Risk at Banca Monte dei Paschi di Siena, Italy, building on being a quantitative risk analyst there since 2011. Previously, he was a Willis Re Research Fellow based at the University of Exeter, UK, specializing in the clustering of extreme events (windstorms, hurricanes, typhoons, floods) and building event sets for tropical cyclone risk assessment using global climate model output. Renato has a PhD in Meteorology. |
| Dickie Whitaker | 6.3.3 | **Dickie Whitaker** has 30 years' experience in the (re)insurance business and for the last 20 years has specialized in risk and innovation and linking academia, government and finance. Dickie has written and presented extensively on these subjects and has operated globally, having worked in both London and New York. He co-founded and works for the Lighthill Risk Network, FiNexus Ltd, Oasis Palm Tree Ltd, and is chief executive of Oasis Loss Modelling Framework Ltd. |

*Disclaimer: The findings, interpretations, and conclusions expressed in this work are entirely those of the authors and should not be attributed in any manner to the organizations that they are currently or have been employed by.

†Disclaimer: The findings, interpretations, and conclusions expressed in this work are entirely those of the authors and should not be attributed in any manner to the World Bank, its Board of Executive Directors, or the governments they represent.

# Foreword

This is a ground breaking and essential book. For the last quarter of a century, some of us have witnessed the birth and development of a new, profound and integrating professional and scientific discipline called catastrophe risk modelling. It has transformed the quality of insurance protection upon which more than a billion people depend for their security, and reformed an industry from relative ruin in the early 1990s to relative resilience just two decades later, in spite of a sharp increase in natural hazard risk over that time.

And yet, despite this significance, the discipline is barely known outside its own circles. It is difficult to estimate, but perhaps just 25,000 people worldwide are active members of the 'cat modeling' community among insurers, reinsurers, regulators, modeling providers, and related communities across academia, engineering and civil protection.

I entered this domain in 2005 and, like everyone, learnt about this field through professional courses, experience, the wisdom of colleagues and reports and articles dealing with specific issues. There was no consolidated text that brought this world together. Meanwhile sensing the wider significance of these methods, there was nowhere to point the curious non-insurer when asked to recommend a comprehensive guide to the field. I am grateful that we now have one, and delighted that many of those who shone a light for me ten years ago have come together with others to share their knowledge and experience for all.

The authors and contributors are to be congratulated, not just for producing the first detailed reference on catastrophe modeling, but also for creating the framing of a new discipline. It is path-breaking and one can only imagine the length of internal debates on so many defining points and fundamental issues. This breakthrough will, I hope, spur many other books and publications to support the expansion and further deepening of the field.

Most of us work in specific areas and this text will be useful to experienced practitioners, to connect with wider aspects of the discipline, as well as essential reading for newer professionals. I am also excited that it will open up the field to many others across industry, science, NGOs and public policy, far beyond insurance, who can build on it, apply it and even criticize it as part of a wider, shared, endeavour to better understand, reduce and manage climate and natural hazard risks and protect the under-protected.

As we consider how to implement the UN Global Goals on climate risk and other natural hazards, catastrophe modelling will guide the changes needed to save millions of lives and livelihoods among exposed populations in the decades ahead. This book, and its future editions, will play a leading role in that journey.

*Rowan Douglas CBE*
CEO Capital Science & Policy Practice, Willis Towers Watson
Member, Prime Minister's Council for Science & Technology

# 1

# Fundamentals

*Matthew Jones, Kirsten Mitchell-Wallace, Matthew Foote, and John Hillier*

## 1.1  Overview

### 1.1.1   What Is Included

This chapter contains a broad overview of the topic of catastrophe modelling, including what catastrophe models are, why they are used, their overall structure and their output. Metrics used in catastrophe modelling are presented. Basic statistical concepts required for catastrophe modelling are also included for ease of reference.

### 1.1.2   What Is Not Included

Detailed information on every topic is not included. This is provided in the subsequent chapters.

### 1.1.3   Why Read This Chapter?

This chapter aims to give the reader an introductory background to catastrophe risk management and catastrophe modelling. It is targeted primarily at those new to the subject, but should also provide a refresher to those more familiar with the discipline. Reading this chapter together

*Natural Catastrophe Risk Management and Modelling: A Practitioner's Guide*, First Edition.
Edited by Kirsten Mitchell-Wallace, Matthew Jones, John Hillier and Matthew Foote.
© 2017 John Wiley & Sons Ltd. Published 2017 by John Wiley & Sons Ltd.

with any subsequent chapters should provide depth on the topic covered – be that the main uses of models, a discussion of the major perils, how to build a model, or how to develop a view of risk. Alternatively, this chapter can be read in isolation to provide an introduction to catastrophe risk management and modelling, from the basics of insurance to the elementary statistics required when using these models. The statistical basics are provided for completeness and reference; less mathematically-minded readers can avoid this section without compromising understanding of following chapters.

## 1.2 Catastrophes, Risk Management and Insurance

In its broadest sense, a **catastrophe** is something that exceeds the capability of those affected to cope with, or absorb, its effects; in the context of natural hazards the driver is an extreme event causing widespread and, usually sudden, damage or suffering. In the insurance industry, definitions of catastrophe are commonly based on an event exceeding one of a number of thresholds for loss (e.g. total economic losses, insured losses, loss of life – for an example by Swiss Re, see Table 1.1). Organizations may choose to define an event as a catastrophe if that company or the whole industry has large or unexpected losses or if significant media attention is expected. For example, the US Property Claims Service definition of a catastrophe is 'an event that causes 25 m USD or more in direct insured losses to property and affects a significant number of policyholders and insurers'.

The terms risk, peril, and hazard are often used interchangeably in conversation. However, in the context of this book, we use the following definitions:

- A **peril** is a potential cause of loss or damage such as an earthquake or windstorm.
- **Risk** is uncertainty leading to potential adverse outcomes. It is also used as shorthand for an insured object.
- **Hazard** is the danger from the peril.

Catastrophes are a risk to organizations and society. Managing this risk (**catastrophe risk management**) is the ongoing process of: (1) identifying the risk given the context of the organization or community, (2) quantifying the risk, (3) deciding what to do, given the level of risk and the **risk appetite** (i.e. how much risk an entity is willing to take) of the organization or community, and (4) monitoring the level of risk.

Table 1.1 Criteria used by Swiss Re in 2014 to determine if events were categorized as catastrophes and entered into their *Sigma* database.

| Threshold | Quantity |
| --- | --- |
| Insured loss, maritime disasters | US$19.6 m |
| Insured loss, aviation | US$39.3 m |
| Insured loss, other losses | US$48.8 m |
| Total economic loss | US$97.6 m |
| Casualties, dead or missing | 20 |
| Casualties, injured | 50 |
| Casualties, homeless | 2000 |

*Source:* Swiss Re, 2015.

The concept of **enterprise risk management (ERM)** (Sweeting, 2011) involves the preceding process, but on a holistic basis (i.e. assessing all risks together, allowing for diversifications and concentrations of risk, including risks that are easy to quantify and those that are not, e.g. reputational damage). The classic responses to risk are to reduce, avoid, transfer or retain (Sweeting, 2011). Insurance is one important mechanism to transfer risk.

Insurance is an arrangement whereby one party (the **insurer**) promises to pay another party (the **policyholder**) a sum of money in the case of a loss as a result of a specific cause. This obligation to provide compensation following a loss is called **indemnity**. A premium is charged to the policyholder to provide this service. Insurance companies provide products for individuals (**personal lines (PL)** insurance) and for corporations (**commercial lines** insurance). Insurance companies who provide money contingent on whether someone dies are called **life insurers**, whereas other insurance companies are called **non-life insurers** or **general insurers**. Companies that provide both life and general insurance are called **composite insurers**. The focus of this book is on general insurance, which can be categorized into different **lines of business**, depending on the type of assets that are being insured; for example:

- **Motor** (or **auto**) lines provide insurance for the physical car and sometimes also for the third-party liability.
- **Property** (or **Direct & Fac (D&F)**) lines provide insurance for properties, their contents and loss resulting from not being able to use the buildings because of an insured peril.
- **Marine** lines provide insurance for ships (**hull**) and the **cargo** they carry (often including goods not actually on a ship, but in transit and in warehouses; sometimes called **marine static risk**).
- **Aviation** lines provide insurance for aircraft, including third-party liability coverage.
- **Construction** (or **engineering**) lines provide insurance for building projects.
- **Liability** (or **casualty**) lines provide insurance to cover claims from third parties.

The precise origins of insurance are debated, but so long as there has been risk, people have tried to manage their individual exposure to it. The first record of insurance was the Babylonian King Hammarubi's code, an ancient tablet dating back to approximately 1750 BC. Insurance's origins were certainly in trade; Phoenicians and Greeks had similar schemes to minimize the impact of the potentially catastrophic (to any individual) loss of ships and cargo by forming a pool to spread the loss. Arrangements similar to modern marine insurance were in place by the mid-fourteenth century in Genoa. An explosion in trade in the seventeenth century led to a maritime information exchange in Lloyd's coffee shop in 1688, with the first recorded **underwriting** in 1757. After an insurance proposal was drafted, the participants all signed their names and participation on the risk underneath the proposal leading to the term 'underwriter' for those taking on the risk. Underwriting will be discussed together with insurance and **reinsurance** (the insurance of insurance companies) in Chapter 2.4. Meanwhile, in America, Benjamin Franklin founded the Philadelphia Contributorship in 1751.

The development of the (re)insurance industry, like modelling itself, has been driven by events. Large fires across Europe have been a driver for property insurance, notably the Great Fire of London in 1666. Before municipal fire-fighting facilities, insurance companies had their own services to protect the specifically marked properties of policyholders. The Hamburg Fire in 1842 precipitated the foundation of the first reinsurance company, Cologne Re. The foundation of Swiss Re has likewise been linked to the Glarus Fire of 1861. These severe events demonstrated the need for reinsurance, although many reinsurance companies were in fact founded to prevent the outflow of reinsurance premiums from local economies to foreign companies (Swiss Re, 2013). For details of the more recent evolution of insurance history, including the London Market Spiral in 1990s, see, for example, Thoyts (2010).

A fundamental concept of insurance is that **pooling risk** reduces the uncertainty in the **expected** (or average) **loss** (EL) over a specific time period. Put another way, the cost of losses, in any given year, from *a large number* of insured properties is *much more certain* than the loss cost from any *individual* property. An insurance company can, therefore, estimate the overall loss cost from a portfolio of insured properties with far more certainty than an individual could for one **policy**. In addition, for some loss scenarios (e.g. a property completely destroyed), the individual may be unable to cover the loss themselves. An insurance contract therefore provides the individual with much more certainty about the amount they will have to pay in any one year (the insurance premium, plus potentially an excess, see Section 1.9.2) as well as protection against an unaffordable loss. A more mathematical description of the rationale for insurance, in particular pooling of risk, is provided in Box 1.1.

---

**Box 1.1 How Pooling of Risk Works**

To illustrate how insurance works, consider an individual who owns a house worth £1 million, and that every year there is a 1% (0.01) chance that the house will be destroyed by an earthquake. Let us also assume that there are no other hazards and that the only outcomes are that the house is either completely intact or completely destroyed each year. The average losses expected each year, or expected loss, is relatively low compared to the value of the house (£1m × 0.01 = £10,000), however, the uncertainty in the amount of the loss is very high – either all or nothing. A measure of the uncertainty in an outcome is the **standard deviation ( SD )**; see Chapters 9 and 10 of Sweeting (2011) and Section 1.11 for some basic statistics. Quantitatively, the standard deviation of the annual loss for this scenario is £99,499 (£1m × $\sqrt{0.01 \times (1 - 0.01)}$, see Section 1.11). The **coefficient of variation** (the standard deviation divided by the expected loss cost) is 9.949.

Now consider a group of 1,000 such people with identical houses and an identical risk to earthquakes, but in different locations such that only one house can be affected by each earthquake, i.e. the risk to each house is completely independent. The expected loss per house remains at £10,000, so £10,000,000 in total across the 1,000 houses. The standard deviation for the group of independent houses is given by the square root of the sum of the individual variances and is equal to £3,146,427, giving a coefficient of variation of 0.3146, and a *standard deviation per person of £3,146*: much less than the £99,499 for the individual scenario. So although the expected loss cost per person stays the same, *the uncertainty in loss cost per person is much reduced by the pooling of risk*. This is also known as **diversification** benefit and is one fundamental reason why insurance makes sense as a concept.

An individual who wishes to avoid the potential financial ruin of an earthquake fully destroying their home can, therefore, choose to insure their house to take advantage of this pooling of risk. In doing this, they would have to pay an insurance premium that covers: (1) the expected loss, (2) a contribution to the expenses of running an insurance company, and (3) an amount to compensate the insurance company for retaining the (reduced because it is pooled) risk. This is the basis of insurance pricing discussed in Chapter 2.6.

The first point to note with the example above is that an assumption of independence of risk was introduced. Mathematically if, instead of the risks being independent, each risk was perfectly correlated so that all risks would be affected in the same way, the standard deviation of total loss would simply be the sum of the individual standard deviations; there would be no reduction in the coefficient of variation (or the SD per individual) and no advantage from pooling of risk.

Therefore, if risks are highly correlated, insurance is less economically attractive since the premium that the insurance company charges must be higher to cover the increased risk.

The second point to note is that, even when aggregating independent risks (i.e. considering their behaviour as a group), there is always some risk remaining which the insurance company must absorb. To do this, companies have **capital** (an excess of assets over their liabilities) to protect the insurance company against their *unexpected* losses, i.e. the chance that the losses in any given year could be more than expected (the average). Insurance companies will often buy insurance themselves (reinsurance) to help further protect themselves against large unexpected losses. Companies need to understand the risk they face in order to ensure they have sufficient capital or, put another way, insurance companies should only take on risk commensurate with the capital they have. Models help the companies understand their risk.

This book is about catastrophe risk management and modelling, and although in this context an earthquake completely destroying a house is quite realistic, the assumption of complete independence of risk used above is very unrealistic. In practice, the wide spatial scale of catastrophe events means that different properties can experience the same catastrophic event at the same time and so there is little independence between the risks within an event. Understanding just how independent (or correlated) the risks are, is a very important (and challenging) consideration for insurance companies, and is one of the main reasons that catastrophe models exist.

## 1.3 What Are Catastrophe Models?

**Catastrophe models** are models designed to estimate the potential loss from the extreme and wide-impact events that are termed catastrophes. The loss potential estimated by such models is usually financial.

Although various definitions of catastrophe models can be found in academic, developer, user or regulatory communities, the one used here is that provided by the United Kingdom Lloyd's Market Association (LMA, 2013):

> A catastrophe model is a computerized system that generates a robust set of simulated events and estimates the magnitude, intensity, and location of the event to determine the amount of damage and calculate the insured loss as a result of a catastrophic event such as a hurricane or an earthquake.

Catastrophe models, like all models, are abstractions of real-world processes. This determines their appropriateness and limits their use and interpretation. These models combine the science of natural hazards with engineering, socio-economic and financial processes (see Chapter 4.2). This amalgam is complex and depends on the evolving knowledge of the processes and the connections between them. The models themselves are products of specialist assumptions, which depend on scientific knowledge, often itself derived from other research or modelling sources. There are many complexities and uncertainties in such models.

However, catastrophe modelling allows the limited historical record for these losses to be extended beyond past events into the realm of what might plausibly occur and provides a framework to quantify the current risk from a catastrophe peril to a group of assets. Increasingly, catastrophe model users are required to understand the models' construction and justify each model's use in terms of appropriately representing the risk facing their particular organization (see Chapter 5).

## 1.4 Why Do We Need Catastrophe Models?

Understanding the risk associated with insurance is a fundamental aspect of managing an insurance or reinsurance company. This understanding ensures that the price charged is sufficient and that the capital and reinsurance protection of the company are adequate, given the risk faced.

For many types of risk, past insurance losses experienced by a company (its **claims experience**), are used to develop models to help the risk estimation. This is the mainstay of actuarial (i.e. mathematics applied to insurance) work, and is a good approach provided that: (1) sufficient claims data exist, and (2) there is a good means of bringing the historical claims to a level that reflects today's risk whether for the change in exposure or the different cost of paying the losses in today's environment (known as **on-levelling** the claims data). This is discussed more in Chapter 2.6.2.1.

For risk as a result of catastrophes, both of these criteria are problematic; **catastrophe risk**, by definition, must include the risk posed by large and unlikely events. Typically, a 10- or 20-year time series of insurance claims data is available, which is insufficient for the purpose of estimating extreme risk, such as the amount of a 1-in-a-200-year loss. However, even if a 200-year time series of losses did exist, the underlying trends (or cycles) within the data would likely make it unusable. Such trends include:

- changes in the type and location of underlying exposure, for example, the trend of increasing urbanization;
- changes in building standards, for example, wind-loading codes;
- changes in infrastructure, for example, flood defences;
- inflation, for example, increases in the cost to rebuild property;
- climate change, for example, increased sea levels or the multi-decadal cycle in North Atlantic hurricane activity.

These trends can be mitigated by using data that reflect the assets that are currently insured, together with a scientific representation of the risk posed by the peril today.

To provide this scientific representation of risk, an insurer could turn to academic work on the perils of interest and consider this, together with their insured assets, to form their risk assessment. However, this is extremely challenging and resource-intensive for two main reasons:

- *Catastrophe risk assessment involves multiple disciplines.* The insurer would need to bring together scientific experts in the perils (e.g. meteorology, seismology or hydrology, see Chapter 3) as well as engineers who can quantify the damage that these perils can cause. In addition, particular aspects of the discipline used for catastrophe risk assessment may fall outside the standard research areas of the scientific discipline or the interests and incentive structures that apply to the researchers (e.g. set by grant-awarding bodies).
- *Catastrophe risk assessment needs to include (re)insurance financial structures.* **Financial structures** within insurance policies are commonly used for the mitigation of risk (see Section 1.9.2). These can be complex and little understood outside of the (re)insurance industry. In order to do this, statisticians and actuaries are needed as well as software engineers to build a **platform** with associated data schema, so that the financial structures can be properly captured and applied.

Even with an established software platform, a high-quality catastrophe model, for a significant peril, can feasibly take around 50 person-years to build, and once built must be updated every few years. Because an insurance company will be exposed to multiple perils, a substantial team (multiple tens of people) is needed if the company is to construct their own models – even just for the main perils.

The resource-intensive nature of this process suggests that for most insurers there is a benefit in outsourcing model-building capabilities and sharing the development costs with other companies. Put another way, there is a market demand for companies that build catastrophe models. Significant catastrophe events help to crystallize this need, while the advent, and increasing prevalence, of computers have provided a mechanism for delivering models. So in the 1980s, the first catastrophe modelling companies (referred to as **vendors**) were formed.

## 1.5  History of Catastrophe Models

Traditionally, the beginning of catastrophe modelling is considered to be in the 1960s; pioneered by Don Friedman through research instigated by Travelers Insurance Company (Friedman, 1984). However, it was the late 1980s when the first commercially-produced model platforms were released. Grossi and Kunreuther (2005) provide a good background to the historical development of catastrophe models, and clearly highlight their evolution, identifying key developmental milestones across a range of technological, data, industrial, regulatory and disaster events, as well as the ongoing growth in model sophistication and peril coverage. New data, methods and scientific understanding have led to periods of significant model update since the late 1980s. Most recently, the advent of cloud computing and open architecture modelling has combined with the increased demand for quantitative risk models from the **disaster risk financing** (DRF) community (e.g. Ley-Borrás and Fox, 2015) to generate an upsurge in model provision, including a much wider community of model developers and users.

Figure 1.1 shows a simplified timeline of catastrophe model development milestones, including major catastrophe events, industry changes, key regulatory standardization, data and applications, commercial catastrophe modelling organizations, international academic and research initiatives, and technological milestones. Although only illustrating a small sample of activities and events, it shows how catastrophe modelling has evolved over its lifetime. The advent of computational analytics for weather forecasting and for seismic processes set the scene for the first computational models developed to estimate the risk from natural disasters. These provided the tools and data (illustrated by the 'i' symbol in Figure 1.1), particularly from government, academic and international organizations considering disaster safety and management. In addition, the development of engineering and insurance standards, including ATC in 1973 and the CRESTA organization in 1977, provided the framework around which models could integrate scientific models of hazards with engineering vulnerability data. A number of large catastrophic events (illustrated by the 'lightning' symbol in Figure 1.1), including Hurricane Hugo, and the Loma Prieta earthquake (both in 1989), highlighted the need for the insurance industry to better understand its risk to large-scale, infrequent loss events. The convergence of the capability to augment historical actuarial data with scientific hazard models and a commercial demand for quantitative risk assessment from the insurance industry, especially in the United States, led to the first commercial models from AIR and RMS in the late 1980s. In addition, the digital spatial frameworks around which hazard, exposure and vulnerability could be integrated were being developed in the **Geographic Information Systems** (GIS) community.

The cluster of large catastrophic events in the 1990s which affected the United States, Japan and Europe (Hurricane Andrew in 1992, the Kobe earthquake in 1995 and wind storms 87J and Daria in 1990) intensified the demand for additional and more sophisticated computational modelling for those territories and perils. Catastrophe modelling enabled more quantitative assessment of insurance risk management and reinsurance transactions and led to the rise of

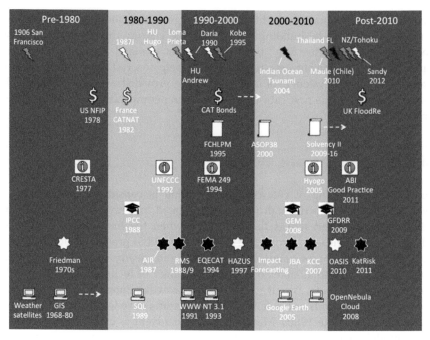

Figure 1.1 History of catastrophe modelling. Time runs left to right. From top to bottom, symbols representing activities and events are: Lightning bolt = natural hazard event; $ = financial idea or mechanism; scroll = regulation; i = data or information; star = modelling company launched; mortar board = academic contributions; computer = computational development. For acronyms, see the Glossary.

new insurance products that took advantage of the catastrophe model as a means to define both risk and price; particularly catastrophe bonds which rely on models to define the triggers and loss terms under which both the insureds and investors could operate with confidence (see Chapter 2.15). A third significant commercial modelling vendor, EQECAT, was formed in 1994, and the US FEMA organization's HAZUS multi-peril platform was first released in 1997. At this time data collection by the insurance industry began to increase in quality, granularity and coverage, for example, with the use of postcodes to collect and report on accumulations of exposure at relatively high levels of detail. The creation of high resolution digital boundaries for **CRESTA** zones, postcode and other datasets for key regions, as well as the use of **geocoding** technologies drove the increased use of electronic accumulation systems, especially for US and European insurance companies. The increase in computational power in the 1990s enabled the widespread use by insurers of database management systems such as Oracle, SQL and desktop spreadsheet systems such as Lotus and Excel. This gradually led to the shift of catastrophe models from the back office mainframe computer environment towards the actuarial and underwriting business support areas in both insurance and reinsurance. Higher resolution and more sophisticated models were made possible by increases in computational power, and the availability of higher resolution data, for example, **digital elevation models** (DEMs). This enabled other perils, such as floods, to be included for the first time in catastrophe modelling platforms.

The 2000s were characterized by the growth of the Internet for data and scientific research, and this enabled catastrophe modelling companies to further widen their engagement with insurance companies. In particular, Google Earth in 2005 (a spatial visualization platform combining satellite and aerial imagery) significantly widened the use of spatial data and visualization by insurers at the desktop. This was a major step in the development of 'point of underwriting' quantitative risk assessment and also drove the development of higher resolution and more complete data for model construction, validation and calibration.

The internationalization of climate change concerns, most particularly in relation to the Intergovernmental Panel on Climate Change (IPCC) Third and Fourth Assessment Reports (in 2001 and 2007 respectively) also heightened insurance and reinsurance concerns and demand for catastrophe modelling, as did other events, such as the Indian Ocean tsunami in 2004 and Hurricane Katrina in 2005. However, perhaps the most important driver of catastrophe modelling development in the 2000s was the increased regulatory demand for quantitative risk management and capital rigour, including the US Actuarial Standards Board (2000), which defined actuarial standards for catastrophe risk assessment and model use, and **Solvency II** in Europe, which led to the 2011 publication of the Association of British Insurers' 'Good Practice' guidelines for catastrophe modelling (ABI, 2011).

The first half of the 2010s has seen significant changes in the catastrophe modelling community, again driven by a combination of large-scale events (in Chile, the 'Maule' earthquake (2010), the New Zealand earthquakes (2010 and 2011), the Thailand flooding (2011), the Tohoku 'Great East Japan' earthquake (2011) and Hurricane Sandy (2012)). These events led, in nearly all cases, to a reassessment of the scientific and methodological bases of the models, and to large-scale revisions of the models and their loss estimates, or the construction of new models. In some cases, these new models have been produced by organizations other than the original commercial modelling companies, including regional specialists as well as reinsurance intermediaries and governmental/academic organizations. These are illustrated in Figure 1.1 as a star (blue star denotes a commercial organization, and a white star denotes a not-for-profit organization). The role of international disaster risk management and finance groups, such as the World Bank, in driving catastrophe model use and development has also grown. This has been in parallel with the advent of the cloud as a potential computational framework for the next generation of high performance and in some cases, open access catastrophe modelling platforms, including the Oasis loss modelling framework (see Chapter 6).

The insurance industry has also evolved over the 2000s and 2010s both in terms of its requirements (regulatory and business) and in how catastrophe modelling is integrated within the business function. Increasingly, catastrophe modelling results are being used at the point of underwriting to inform decisions on price, return on capital and capacity (see Chapter 2.6). This in turn is driving advances in data quality, and in the visualization of information for effective decision making. The need for 'model completeness' and validation of the use of models to represent the view of risk (see Chapter 5) will continue to drive demand for increased model resolution and coverage, including the widening of model perils and regions across non-traditional areas.

In addition, there will continue to be growth in commercial model coverage, both geographically, and in terms of perils, in response to market demand and new scientific advances. The Key Past Events sections in Chapter 3 provide an overview of how events have driven major changes to model availability.

It is, perhaps, not surprising that major updates to models have had some impact on their interpretation and use in the (re)insurance industry. Updates have, in some cases, led to considerable challenges, including revising business plans and capital requirements. These have been amplified by the increased complexity of the models being developed and the increased choices available to users. In particular, changes in US hurricane models over the last decade, which have reflected major updates in fundamental assumptions and modelling approaches, and the provision of multiple viewpoints of event frequency, have resulted in a more careful and considered approach to the use and application of catastrophe models.

## 1.6 Who Provides and Uses Catastrophe Models?

Catastrophe models are used across multiple segments of the (re)insurance industry as well as, increasingly, in the disaster-risk financing communities. In (re)insurance, these models are used in businesses which take on risk themselves (insurers, reinsurers, insurance linked security investors) as well as those who advise them ((re)insurance brokers and consultants) and assess their activity (regulators and rating agencies).

Model provision and use can be explored using the concept of model agents, all of whom have specific influence in, and requirements of, the design and function of a model. Four main roles are considered: contributors, developers, analysts and assessors/overseers. Individuals may, of course, assume more than one agent role during the modelling process.

- Catastrophe model contributors are the creators of the constituent scientific, mathematical and statistical theories and methods used within the models or their components. They include primary source data providers, software and model process designers and suppliers. Contributors may not have designed or produced the components specifically for use within a catastrophe model, and instead may have developed them for other purposes (e.g. flood-risk maps).
- Catastrophe **model developers** derive model components, integrate, calibrate and validate them while creating loss modelling systems or platforms. Developers may be collaborative academic, industrial or commercial organizations producing either generic or bespoke models.
- The **model vendors**, the specialist companies who develop and license catastrophe models, are model developers; these include AIR Worldwide, Ambiental, Catrisk Solutions, Corelogic (formerly EQECAT), ERN, Impact Forecasting, JBA, KatRisk, KCC, and RMS.

   The evolution of **plug and play** open architecture model platforms and formats (i.e. the idea that a new model, or model component can be 'plugged in' to an existing model platform and quickly and easily used), including Oasis, expands the roles of developers and contributors, and future model construction and design are likely to reflect an evolution in contributor and developer roles.
- **Catastrophe risk analysts** are responsible for data entry, model selection, model operation, analysis, outputs and reporting. Catastrophe risk analysts can be considered the *primary users* of catastrophe models. The analyst is responsible for ensuring that the characteristics of the model system are tuned to be as representative as possible of the objects (hereafter referred to as **risks**) being assessed. In particular, the analyst will often be responsible for the input of **exposure data** (see Section 1.9), which are the data captured by the organization, representing the location, characteristics and value of the assets being insured. Catastrophe risk analysts often have a background in science, engineering or mathematics and may have a specialization in a hazard or engineering discipline. These backgrounds are helpful in interrogating the model components and developing a view of the suitability of the model for use within the organization (see Chapter 5). Depending on the structure of the company, there may be an overlap between the activity of the catastrophe risk analyst with the traditional roles of both the underwriting and actuarial departments. For example, catastrophe pricing can be carried out by a catastrophe risk analyst or by an underwriter or actuary. In some territories (particularly London), organizations differentiate between exposure management and catastrophe modelling, with catastrophe modelling being a subset of the broader field of exposure management. In other territories, catastrophe modelling, catastrophe management and exposure management teams often do the same thing.
- Other potential direct users of catastrophe models within a (re)insurance company may include **actuaries** who are responsible for many aspects of statistical analysis and pricing, **underwriters** who select and price risk, and capital modellers who determine the relative solvency of the (re)insurance entities. However, it is more likely that people in these roles will

use the model output rather than using the models directly. Senior management (e.g. chief risk officers, chief underwriting officers) will not usually use the models directly but will rely on **risk metrics** from these models to inform fundamental business decisions related to catastrophe risk.

Catastrophe modelling is a cross-functional activity and therefore can be viewed either as part of risk and exposure management or part of underwriting support. The organizational structure and reporting line of the catastrophe modelling, or exposure management, team can provide insight into how the team is viewed by the organization's management and how well it is integrated into the day-to-day business activities.

Catastrophe model assessors and overseers are those with responsibility to ensure best practice for the operation and use of the model, for example, catastrophe risk managers, exposure managers, risk managers and regulators. Increasingly, the growth in regulator-driven standards and associated external assessment of model design, operation and validation, particularly around the incorporation of catastrophe models into organizational **capital models** and **enterprise risk management** (ERM) systems, is developing a wider community of specialists throughout organizations requiring catastrophe model literacy. An example of this is **Solvency II** (a pan-European regulatory regime for insurers, see Chapter 2.11.4), or the US ASOP 38 (Actuarial Standards Board, 2000), both of which require model oversight and robust governance structures to understand model construction, assumptions and methods.

Models will certainly continue to be created by communities of specialists across a range of disciplines, and whether based on proprietary platforms, or open architecture systems, the developer role will continue to determine model character and applicability. Equally, all model agents, whether contributors, developers, operators or assessors, require insight into the decision processes and trade-offs necessary in model construction to enable effective and appropriate use. Put simply, all model agents must 'get inside the mind' of the model developer, to achieve successful and appropriate use of the model.

## 1.7   What Are Catastrophe Models Used For?

Chapter 2 provides details on the applications of catastrophe modelling. The most common use of catastrophe models is to quantify risk, allowing its transfer between parties in the insurance and reinsurance industry. Catastrophe modelling is now used in many aspects of the daily operations of those insurers, reinsurers, funds, broking houses, consultants and regulatory bodies concerned with catastrophe risk.

As much of the (re)insurance industry has a specialist vocabulary, we will not attempt to define the model uses in detail in this chapter, but rather explore them together with a discussion of the business purposes in Chapter 2. However, common questions that catastrophe model output is used to address include:

- How much should we charge for a (re)insurance policy?
- Which new business should we add?
- How profitable is it likely to be?
- How much could we typically lose and with what likelihood?
- What might a specific event cost us?
- What are the potential causes of loss, considering our business model?
- How can we best mitigate these?
- Do we have enough money set aside for specific eventualities?
- Are we operating within the constraints set by our board, our regulators and supervisory agents?

There are many different stakeholders within the (re)insurance company at multiple levels, from those choosing which policies to add to the portfolio up to the board level. Catastrophe models in the societal role of reinsurance are also briefly explored when looking at government pools (Chapter 2.13). As models are increasingly being used in the disaster-risk financing communities, some discussion of this topic is included in Chapter 2.14. Models have not been developed for all perils, but must adequately represent those that are of most concern for their users. A description of, and considerations for, modelling these perils are given in Chapter 3.

Understanding that a catastrophe model is a series of interconnected sub-models from various sources is critical to understanding how those models have been produced, should be used, and should be interpreted with respect to the specific questions they may be used to tackle. This topic is covered in more detail in Chapter 4.

Models are constructed so that they should be tested, rebuilt against the lessons learnt, and retested. On that basis, there will never be a final version of a model, simply various improved versions. This raises a doubt in the mind of the user – how wrong can a model be before it is useless, or worse still, misleading? Also, how can a continually evolving and uncertain model be used effectively for assessment and decision-making? These questions would be difficult enough if the users of models were always the same people who built them. In the financial industry, including insurance, this is generally not the case. These questions can be partially answered by developing a company-specific view of the risk; this process is described in Chapter 5.

## 1.8  Anatomy of a Catastrophe Model

Conceptually, most catastrophe models tend to follow a similar modular structure, reflecting the integration of the multidisciplinary geophysical, engineering and financial components, which each contribute to the overall estimate of risk. Figure 1.2 shows one representation of such a structure.

The components **hazard**, **vulnerability**, **exposure** and loss (or **financial calculation**) require validation both separately and collectively as well as integration within a computational workflow and data management framework (the platform). The following sections describe the features of each component as well as what constitutes a platform.

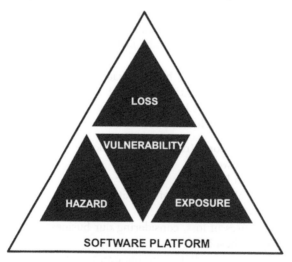

Figure 1.2  Structure of a catastrophe model showing the five key components: hazard, exposure, vulnerability, loss and a platform. Courtesy of Impact Forecasting (http://www.aon.com/impactforecasting/impact-forecasting.jsp)

### 1.8.1 Hazard

The hazard component reflects the extent and intensity of a peril as defined by a specific hazard metric. This often represents the hazard intensity variation across a pre-defined geospatial framework, either in a regular **raster** (grid cell-based), or in an irregular vector structure (e.g. postcode zones or points). Either way, each **event footprint** reflects the relative intensity of the hazard over the defined time period of the event. Examples include **peak ground acceleration** (PGA) for an earthquake, flood depths across a floodplain from a river breach, or **peak gust wind speeds** across a storm track. An example of such a footprint is given in Figure 1.3.

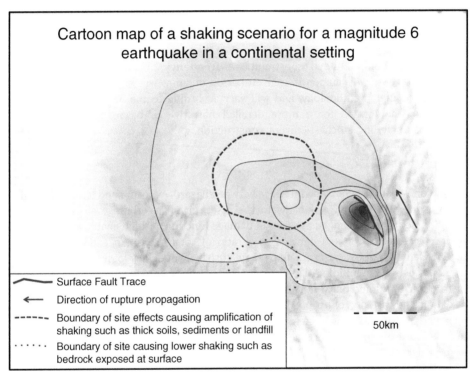

| Perceived Shaking | Not felt | Weak | Light | Moderate | Strong | Very strong | Severe | Violent | Extreme |
|---|---|---|---|---|---|---|---|---|---|
| Potential Damage | None | None | None | Very Light | Light | Moderate | Moderate /Heavy | Heavy | Very Heavy |
| Peak Acceleration (%g) | <0.17 | 0.17-1.4 | 1.4-3.9 | 3.9-9.2 | 9.2-18 | 18-34 | 34-65 | 65-124 | >124 |
| Peak Velocity (cm/s) | <0.1 | 0.1-1.1 | 1.1-3.4 | 3.4-8.1 | 8.1-16 | 16-31 | 31-60 | 60-116 | >116 |
| Instrumental Intensity | I | III-III | IV | V | VI | VII | VIII | IX | X+ |

**Figure 1.3** The spatial footprint of shaking due to an earthquake, coloured according to various common and equivalent measures of intensity. Red line represents surface rupture. Note that in general the most intense shaking occurs where there is the greatest deformation, and decreases away from the fault. However, surface conditions affect the shaking, creating a non-uniform decrease in intensity away from the fault, such as due to thick soils west of the fault in this illustration. The scale is used in USGS ShakeMap (Figure 2.5). *Source:* USGS, 2006, adapted from Wald *et al.* (1999).

The footprints are produced via a model development process (see Figure 4.1 in Chapter 4.2) that reflects the geophysical processes operating to create the hazard as accurately as possible. The footprints are based on the particular data, assumptions and computational approaches used and taken by the model development team.

The choice of hazard metric (see Chapter 4.3.2) will also be a critical element of the hazard model, and will often be based on a generally accepted approach, although this may not be fully representative of all damage caused, for example, flood depth may not be the only causal factor for flood damage, when duration of inundation, velocity of flow or water pollution may all have an effect on damage. Instead the chosen metrics may simply be the most effectively modelled within the geophysical modelling framework. In general, most hazard-intensity factors used will be reasonable proxies, after calibration of the model.

In addition, for probabilistic models, **stochastic** event sets will also be developed, which are a collection of individual event footprints that could happen in some synthetic history (see Chapter 4.3.1). The number and range of stochastic scenarios in the overall **catalogue** of events will be designed to accurately represent the scope of the hazard in the particular model region, while recognizing computational constraints. A **frequency** or **rate of event occurrence** (i.e. how many events will happen in a given period of time) will be assigned to each event, based on a particular methodology, and will vary according to the hazard and region being modelled. See Chapter 4.3 for a more detailed narrative of hazard model construction considerations, including stochastic event construction.

### 1.8.2 Vulnerability

The vulnerability component is the interface between hazard, exposure and loss, and provides a means to estimate the relative damage to the asset, given a certain level of hazard. Most vulnerability models are arranged as series of damage functions, which enable look-up between hazard intensity and estimated damage as a ratio of total value. An example is shown in Figure 1.4. Damage ratios are estimated repair cost (i.e. modelled loss) as a fraction of the replacement cost of the building (i.e. insurance exposure or **sum insured**).

Construction of vulnerability functions will be dependent on available data, for example, claims data in the case of (re)insurance. For the more extreme and rare events there is usually insufficient data to construct vulnerability functions using data alone. In these cases much reliance is placed on engineering studies. There is clearly much uncertainty, especially at the higher hazard levels, and so often vulnerability functions are constructed including an estimate of this kind of uncertainty. See Chapter 4.5 for more details of the model vulnerability development process.

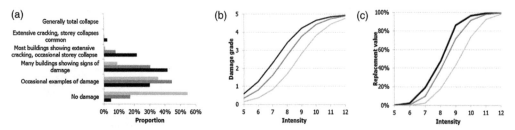

Figure 1.4 Illustrative vulnerability functions, derived according to Risk-UE methodology (Mouroux, 2006): (a) shows the distribution of damage states for a particular intensity level in a particular area for three different buildings; (b) displays the average damage status per intensity value for three different buildings in a particular area; and (c) displays the resulting replacement cost as a percentage of insured value, which can be computed once the damage grade is known.

### 1.8.3  Exposure

Exposure is used in two distinct ways in catastrophe models. First, exposure data for the specific objects being modelled are entered by the user into the model. This is described in Section 1.9.1. Second, a representation of industry exposure for the region covered by the model is also used in the process of building a catastrophe model. This takes the form of a database of exposure values split by area and by type of object being modelled (the primary modifiers; see Section 1.9.1.3). The values will typically be either insured or economic (i.e. insured plus non-insured) values, and may be split between buildings values, contents values and other aspects, such as business interruption values (see Section 1.9.1.2). The database will also usually contain information on insured financial structures such as deductibles and limits (see Section 1.9.2). Such a database is called an **industry exposure database** (**IED**).

The main uses of an IED are as follows:

- To enable calibration of the model, the industry exposure database can be run through the model to calculate the modelled industry losses; these can be compared to historical observations for model calibration purposes.
- To enable assumptions to be made when input exposure characteristics are missing. As discussed in Section 1.9.1.3, it is fairly common for primary modifiers to be absent in the input data. If this the case, then an aggregated or composite vulnerability curve can be formed by weighting together the individual vulnerability curves; using the IED to provide the weights.
- To enable disaggregation of coarse resolution, aggregated input data. In some cases, where limited address information is provided on the input data, the best geocoding resolution (see Section 1.9.1.1) may be too coarse for the modelled peril. The IED can be used to disaggregate the coarsely geocoded data in proportion to the geographical distribution of values contained in the IED.

As most catastrophe models have traditionally been focused on static property asset risk assessment, this tends to influence the form and scope of parameters used to represent asset inventories, although some variation is applied for specific sub-classes of assets, such as agriculture, population and vehicles. More details on the use of exposure data as part of the model development process are given in Chapter 4.4.

### 1.8.4  Loss and Financial Perspectives

The output from the vulnerability module of a catastrophe model is the total loss, i.e. before the application of any insurance or reinsurance financial structures. This is usually termed the **ground up loss (GUL)**. This is calculated for each location and **coverage** (buildings, contents, business interruption, see Section 1.9.1.2) for each event in the model. Whether this granularity (i.e. level of detail) of loss information is actually saved into the results database depends upon the settings when the model is run, but it will be calculated for all detailed models. Most current models do not just calculate a single loss amount for each event, location and coverage, but rather a distribution of likely loss, reflecting some of the uncertainty in the loss estimate. This is often represented by a mean and standard deviation of loss (the maximum loss is, by definition, the coverage sum insured), with some newer models also providing a full distribution of loss uncertainty (for more details, see the discussion on uncertainty in Chapter 2.16.1).

For the output to be useful to the insurer or reinsurer, the model needs to perform several further functions, specifically:

- It must use the data describing insurance financial structures (see Section 1.9.2) to *share* (or *partition*) the ground up loss between the various parties involved in the risk transfer: the insured and the insurer at a minimum, and potentially other insurers and reinsurers.

- It needs to *aggregate*, or *combine*, the loss statistics at different output resolutions. For example, location level, policy level, event level, and portfolio level (see Section 1.9.3).
- If different financial structures apply at different resolutions, the model could *back allocate* the impact of the structures applied further up in the hierarchy (e.g. portfolio or policy) down to the more detailed level (e.g. location) to allow for additivity in subsequent analysis of the detailed level losses.

These are all difficult tasks, because the range and complexity of financial structures are huge. For example, financial structures can do the following:

- Operate at different levels, from coverage level to location level to policy level to the level of the whole portfolio (for reinsurance structures).
- Be nominal (or flat) amounts or percentages of loss with minimums and maximums.
- Apply to individual events or across multiple events.
- Cover a single peril or a range of perils.
- Be interdependent: one level of financial structure can impact the next.

At the time of writing, there are no catastrophe models in production which can cope with the full range of financial structures, however, some do better than others, and this can be a distinguishing feature in model selection. The different kinds of financial structures (deductibles, limits, coinsurance) are described in Section 1.9.2. Reinsurance is discussed in Chapter 2.4.

The partitioning of modelled loss, according to the financial structures, gives rise to the need for a terminology describing how the loss is shared among different parties, or **loss perspectives** (or financial perspectives). Commonly used loss perspectives are:

- Ground up loss: The entire loss with no financial structures applied.
- **Retained** or **client loss**: The loss to the insured. Sometimes this is defined as the loss due to deductibles, more often it also includes the loss exceeding any limits.
- **Gross loss**: The loss to the insurer after limits and deductibles and co-insurance are applied, but before any forms of reinsurance.
- **Net Pre-Cat**: The gross loss with facultative and per risk reinsurance applied (see Chapter 2.4.2), but not catastrophe treaties.
- **Net Post-Cat**: The net pre-cat loss with catastrophe treaties applied.

These terms are used in most vendor models, and will often mean the same thing; however, it is important to note that the precise definition may differ from model to model, so it is always worth checking the model documentation carefully.

In practice, the most commonly used perspectives are *ground up*, *gross* and *net pre-cat*. Ground up losses are often used in model validation, and for comparison to the gross losses to quantify the modelled effect of the financial structures, particularly deductibles and limits. The net pre-cat losses are often the key output taken from the catastrophe model and used within subsequent analyses. Reinsurance that applies to a whole group of policies (treaty reinsurance; see Chapter 2.4.2) is often applied outside the catastrophe model in Asset Liability Modelling (ALM; see Chapter 2.10.6) software such as Remetrica[1] or Igloo.[2] This is because these packages offer more flexibility in applying reinsurance structures and because they are often already used for the capital modelling of a (re)insurance company, which is an important use of catastrophe model output covered in Chapter 2.10.

Whether models accurately represent the impact of financial structures depends on how well models represent uncertainty and propagate it through the respective financial structures and aggregation levels. Different vendors have different mathematical approaches to this. Some use closed form integration of a parametric distribution (such as the **Beta** distribution, see Section 1.11.2.1) fitted to the mean and standard deviation of the ground up loss at the relevant aggregation level to partition the loss. Some use numerical convolution to aggregate and apply financial structures. More recent models often use **Monte Carlo simulation**. All models will contain an assumption about the level of uncertainty **correlation** between coverages for a given location, and between locations for a given peril. Often an assumption is made of 100% uncertainty correlation between coverages. The treatment of location uncertainty correlation varies from platform to platform: sometimes the default value is zero, sometimes it is a non-zero value that varies by peril. In some platforms the level of location correlation is a user-supplied input, in others it is 'hard-wired' into the model.

### 1.8.5   Platform

The four model components of hazard, exposure, vulnerability and loss are implemented and integrated together in a piece of software called a platform. In addition to integrating these components, a platform usually also provides:

- a mechanism for the user to input and validate exposure items;
- a way of converting address data to latitude and longitude (geocoding), although not all platforms provide this;
- an interface to allow the user to select run-time model options and output settings, and to initiate the model runs and monitor progress of the runs;
- a structured way of storing the input exposure and output results data;
- a method of running reports and analysing and visualizing the exposure data and output results, although not all platforms provide this.

The way a platform is constructed, the methodologies and features within a platform, and the performance of the platform are linked to the database structures that the platform is based on and the computational power available.

#### 1.8.5.1   Computational Power and Catastrophe Models

The computational framework for a catastrophe model requires considerable data management, storage and calculation-processing capability. Catastrophe models have benefitted from a continuous evolution in hardware and systems architecture and application developments. This has enabled the granularity and sophistication of the calculations within a catastrophe model platform to steadily improve. For example, several current platforms now use Monte Carlo simulation to propagate uncertainty through the financial calculations, whereas a decade ago this would have been impossible in a business setting due to computational limitations. These limitations have been a major design consideration in all catastrophe models, affecting all elements of model construction, analytical resolution, model coverage and their use. In particular, computational limitations to 'run-time' calculations have driven many of the key design trade-offs necessary to enable effective use of the models in business environments, including statistical parameterizations, re-sampling and other optimizations, and the uncertainty inherent in these practical decisions must always be considered (see Chapter 2.16.1).

Since the development of the first commercially available catastrophe models in the late 1980s, the hardware and operating systems available to host the models have evolved continuously, determining to a large part the model versioning histories of each commercial system. Increases in processor power have driven the underlying development of increasingly sophisticated computational engines, and enabled improved data management. Early platforms were designed to operate on single processor PCs, then progressed to multiple processor/PC architectures, applying process queuing methods to enable multiple model calculations to be managed in a run-time environment.

The advent of distributed computational architectures, employing groups of machines, as well as multiple-core processors, for business applications in the early 2000s provided a means for greater computational efficiency, with catastrophe model redesigns taking this approach in the mid-2000s. In the late 2000s and early 2010s, Microsoft High Performance Computing (HPC) clustered computational products enabled commercial vendor companies to develop bespoke server architectures to optimize their data and analytical management processes even further, with the enhanced power and computational efficiencies enabling the development of high resolution, regional flood catastrophe models, and improved model run-times for large and detailed calculations such as model sensitivity analyses (see Chapter 5.4.2).

The latest platform designs harness cloud-based systems, reducing the need for business-hosted hardware and allowing the development of models with greater user application and control. This includes the much-vaunted model transparency arising from reducing model parameterization using greater processing power. These advances have also enabled a new 'plug and play' community of model developers to grow, through the creation of 'open architecture' platforms, such as the Oasis Loss Model Framework.[3] Open architecture has the potential to enable model users to select individual model components, created around common architectural standards, to better represent their own risk, although this does result in an increased responsibility on the model user to understand and justify their own choice of model components and approaches (see Chapter 5).

### 1.8.5.2 Platform Database Management and Data Structures

Catastrophe model analytical functionality has been remarkably stable for most of the history of catastrophe models, with the core stochastic 'exceedance probability' calculation being applied in most models, in one form or other. This approach has therefore tended to drive most computational requirements. Around this core model requirement, the other primary computational requirement has been in the database management approach used to import, store and manage the increasing sizes of exposure, hazard and loss data produced in the computation.

Early platforms employed a range of relational database management systems (RDBMS), both bespoke, but also those produced by specialist database vendors, including Oracle, Microsoft and Bentley. Most commonly, Microsoft SQL Server has been used as the database management system underlying the commercial vendor platforms. There are limitations to the computational efficiencies possible using traditional RDBMS approaches, and this has led to the development of other data warehouse approaches, including 'data lakes' such as Microsoft Azure, and operating methods such as Hadoop, to further enhance data management for catastrophe modelling, particularly in parallel with cloud-based, open data architectures.

Additional functionality, most particularly in respect to geospatial analytics, geocoding and geo-referencing, informatics and visualization, has also evolved over time with increased demand for detailed model outputs, sensitivity analyses and other risk metrics (see, e.g. Slingsby *et al.*, 2010). The most recent platforms enable seamless end-to-end data management, input, modelling and visualization, and widen the end-user community from a core of specialist catastrophe analysts to include underwriters, risk managers and others across organizations.

### 1.8.5.3 Exposure and Result Database Structures

All models require a standard approach to data structuring, for exposure data import (see Section 1.9) and for results reporting. These all include field structures, and data dictionaries.

As noted previously, the commercial model providers have developed their own proprietary data formats around the database management system used and their particular data structures. These provide, in some cases, a pseudo-standard for specific models, but the costs and errors associated with translation between proprietary standards, originally considered to be worthy of Intellectual Property Right (IPR) protection, have been seen as a major challenge to improving the quality and confidence of catastrophe modelling. In particular, the lack of a common and transferable data format standard has been considered a major limitation in the desired evolution towards greater model compatibility. Although initiatives such as ACORD,[4] have driven initial attempts to create model agnostic data standards, these have only been partially successful in their aim to standardize data structures, not least due to the wide range of insurance policies, covers and data sources. Recent work by the Lloyd's Market Association to derive consistent structure frameworks for exposure data (the Exposure Data Design Project, or EDP), has considered the need for a combined process and platform approach to successfully develop a model agnostic data structure across all classes. This considered the harnessing of data formats as well as technological solutions to data integration, management and manipulation, such as data lakes, while looking to other data standards organizations, including ACORD and the Open Geospatial Consortium (OGC), for similar standards that could potentially be adopted or modified for exposure requirements.

## 1.9  Model Input

Catastrophe models are **exposure**-based **models** and as such provide an estimate of the risk without using the historic claims data or claims experience from the specific locations or policies being modelled. The catastrophe risk analyst should enter the following information into the models:

- Exposure details: Location information (Section 1.9.1.1), exposure values such as sum insured (Section 1.9.1.2), exposure characteristics (primary and secondary modifiers – Section 1.9.1.3), and user-defined information for classification and reporting purposes.
- Financial structure information, such as deductibles, limits and reinsurance (Section 1.9.2).
- Information about how the locations are grouped or categorized (by legal contract) into policies (Section 1.9.3).

The preceding list seems fairly tractable. However, collecting, transforming, cleaning and assessing the quality of such data always constitutes a difficult, time-consuming and costly process for a (re)insurer (see Chapter 2.5.4).

In addition to entering the preceding information, the catastrophe risk analyst will also need to decide which 'switches' to turn on in the model and which output options to select. Some of the most common switches and options are:

- whether **demand surge** is used (see Chapter 5.4.2.10);
- whether **secondary perils** are used (e.g. **storm surge**; see Chapters 2.6.3, 3.2.5 and 4.3.7);
- whether **secondary uncertainty** is used (see Chapters 2.16.1 and 4.5.1.1);
- which event set is used (e.g. long-term or warm sea surface temperature for North Atlantic hurricane, see Chapter 3.2);
- how many samples to use (for those models using Monte Carlo simulation, see Chapter 4.7.3).

The specified output options can have a significant impact on the model run-time and data storage requirements for the output data. Most models allow the user to specify options regarding which financial perspectives (Section 1.8.4) are used, what type of statistics are output (see Section 1.10), and the granularity of output statistics. For example, whether result statistics are summarized across locations, or output for each individual location. The main concern from a data space perspective is the granularity at which event loss tables or year loss tables (see Section 1.10 for a definition of these) are output, as each such table can contain tens of thousands of records, and if these are output at each location for a large number of locations (1 million would not be uncommon), the space requirements are substantial.

### 1.9.1 Exposure

The main exposure characteristics input to models are summarized below and then defined in detail in the subsequent sections.

- Location: Address information or coordinates.
- Sum insured: The value of the exposure.
- Primary modifiers: 'Modifiers' is the catastrophe modelling terminology for exposure characteristics that can differentiate the potential damage to the exposure from the catastrophic event (otherwise known as rating factors). Primary modifiers are those which are useful predictors of damage for most perils.
- Secondary modifiers: These are modifiers whose importance is usually peril-specific.
- User-defined information for reporting purposes: A (re)insurance company will have many reporting needs and so information must be attached to the exposures and entered into the model to ensure that these needs can be met. This includes information such as legal entity, business unit, underwriter, class of business, etc.

#### 1.9.1.1 Location

Because natural catastrophe models are designed to evaluate the geographically correlated risk from a set of locations, it is important that information about the location of each input exposure is provided. The location of assets is the primary link between hazard, exposure and vulnerability in the model.

Most models either accept coordinate information directly (and will determine a computationally applied location via a standard coordinate and geodetic system; commonly World Geodetic System (WGS)-84[5]) or will use geocoding toolsets (geocoders) to create coordinates from supplied address information. The level of spatial (or geocoding) resolution is determined by the quality of the address information supplied and the geographical granularity of the geocoder. Coordinates will often be stored as decimal degree, bounded by +/− 90 degrees latitude and +/− 180 degrees longitude depending on the hemisphere of the location, for example:

$$38.889931, -77.009003;$$

This is the decimal degree latitude and longitude to six decimal places of the Senate Building in Washington, DC. A precision of six decimal places represents a location to an accuracy of approximately 6–10 cm in the real world, depending on the latitude of the position. Usually, a precision of three to five decimal places is an appropriate accuracy to locate an individual property within a catastrophe model given the resolution of hazard data used (typically tens of metres).

In some cases, locations are supplied with latitude-longitude coordinates, often via **GPS** or other surveyed information. Typically, however, locations are determined from address information. The address information required will be based on a political, postal or other administrative geographic framework. These may be hierarchically structured, for example, varying levels of postal code, or adopted from other administrative systems, such as CRESTA[6] zones. The level of address granularity achieved will determine the overall modelling resolution, and will also, in some models, influence the calculation of model uncertainty. The precision of such information can be high within data systems, often shown as a six-digit decimal degree, but care should be taken when interpreting the address and location data as precision does not necessarily imply an equal level of accuracy. In particular, coordinates interpreted from address data by catastrophe models, or via free-standing geocoding tools, should be carefully validated, as the assumptions made in the geocoding algorithms may introduce significant error, for example, by selecting the wrong coordinate position for ambiguous locations with multiple potential locations.

Geocoding resolution indicates the best granularity that the geocoder believes it has been able to achieve with the address information supplied. Commonly used resolutions, in descending order of spatial accuracy are:

- Building/parcel
- Street/address
- ZIP/postcode
- City/town name
- District/parish
- CRESTA zone
- State/municipality
- Country
- Unknown

The precise resolutions returned often depend upon the country and geocoder in question; however, 'street-level' geocoding is a commonly used term that can be ambiguous. For example, street-level geocoding can mean:

- the geocode of the exact building on the street (i.e. equivalent to building level);
- an estimate of the location of the building by knowing the street and interpolating based on the street number (this is the usual definition of street level geocoding);
- the midpoint of the street, with no adjustment for building number.

It is important to clarify the precise meaning of geocoding levels, such as 'street' geocoding within the geocoder or catastrophe model that is being used.

Similarly, 'postcode'-level geocoding can have different meanings in different countries depending on the size of the postcodes. For example, in the UK, a full postcode will typically narrow the geocode down to one of 15–20 residential houses (often equivalent to 'street'-level geocoding in other countries, depending on the definition) or for a large commercial building will narrow the geocode down to the exact building. In most other countries 'postcode' resolution will not be this accurate. Further considerations related to geocoding are provided in Chapter 5.4.2.6.

### 1.9.1.2 Exposure Value

Capturing the correct 'value' of the exposure being assessed is a critical component of risk analysis, and notoriously difficult to achieve in a consistent and accurate way. In general terms, most exposure models will apply a financial approach to representing value. This will generally be the *100%*, or *ground up* total value, most often interpreted as the 'total rebuild' or 'reinstatement' value in monetary terms determined for calculation of loss against a vulnerability function. For other, non-property assets, value can be defined in other ways, for example, the total yield for a given crop, or population; but in most cases translation to a monetary value will be undertaken regardless of the type of asset.

In most models exposure value is required in three coverage categories: buildings, contents and **business interruption** (BI). In some regions, for example, the United States, 'other structures' such as sheds and outbuildings are entered as a fourth category of sum insured. In some models **Additional Living Expenses** (ALE) values are entered instead of BI values for personal lines exposures. Each category of exposure value will be specific to a given type of vulnerability.

The *buildings* value usually represents the total structural rebuild value, the *contents* value represents material items or other assets within but not part of the main structure, and *business interruption* (BI) represents the estimated loss of profits resulting from closure or non-operation of the asset due to direct damage.

Some exposure models will apply predefined splits between buildings and contents, and business interruption, determined from the underlying exposure parameters pre-set by the model development process.

Care is needed to ensure the correct amounts are entered into the sum insured fields, as although this seems straightforward, there are pragmatic issues that can complicate this, as discussed below.

### Building Sum Insured

In all models this should be the estimated rebuilding cost. Complications occur in personal lines insurance where the insured party (i.e. householder) can mistake the market value for the rebuilding cost, which often leads to an overstatement of sum insured. Another complication is where sum insured amounts are calculated by the insurer (**notional** sum insured) to avoid the issue of customers supplying incorrect values, and the insured is provided with a policy limit that is much higher than the estimated rebuilding cost so that the insured is comfortable that the policy meets their needs. If this limit is entered into the model as the building sum insured, an overstatement of likely modelled loss will occur; the notional sum insured is likely to be a more accurate representation of the risk.

For commercial insurance, the most likely concern is that of **under-insurance**, or **insurance-to-value** (**ITV**) issues (see Chapter 2.6.3 for more details). A further consideration is whether or not a **day-one** sum insured is used (i.e. a value with no explicit allowance for inflation during the policy and rebuilding period, but with a day-one uplift provision), or whether a **full reinstatement** sum insured is used (i.e. with an allowance for inflation already within the sum insured). This is explained and discussed further in Chapter 5.4.2.7.

### Contents Sum Insured

Contents sum insured for both commercial and personal lines may be subject to underinsurance concerns. For personal lines it is common for the insurer to estimate the *notional* value required by the customer (in a similar way to the estimation of rebuilding cost) and provide higher limits than those actually required. As with buildings limits, these will provide an upper bound on the policy, not a realistic estimation of values and so using the limit figures directly as contents sum insured will overstate results. Using the notional value for modelling should prove more representative of the true risk.

### Business Interruption Sum Insured

The main issues here arise around the interaction between the sum insured and the **period of indemnity** (i.e. the period of time for which loss of profit or revenue can be calculated for a business interruption loss). Some models require an annual sum insured to be input for business interruption. In some models a period of indemnity field is provided for information only and does not change the results in any way. In practice, the BI sum insured on a policy will be the maximum amount recoverable in a specified indemnity period. This indemnity period can be less than or greater than a year, depending on the specific terms of the contract. The question then arises as to how this non-annual sum insured should best be represented in a model requiring an annual sum insured. This again needs careful discussion with the model vendor to ensure the input assumptions reflect the assumptions inherent in the model. A method sometimes used is to pro-rate or scale the sum insured in order that it represents an annual amount. For example, if the policy period of indemnity is two years, the BI sum insured would be halved. In the case of increasing the modelled BI sum insured due to an actual period of indemnity that is less than a year (e.g. doubling the BI sum insured because the BI period of indemnity is six months), then a BI limit should also be added to the model input to ensure that the actual modelled loss is never greater than the actual BI sum insured.

For personal lines policies, some models require the ALE sum insured to be entered in the BI field. These are policy specific and will usually either be a specified fixed maximum amount or some percentage of the building or contents sum insured.

### 1.9.1.3 Exposure Characteristics: Primary and Secondary Modifiers

The four main primary modifiers are usually:

- **Occupancy class:** The purpose for which the insured building is being used. At a high level this is usually residential, commercial, industrial or agricultural. At a more granular level the *residential* classification splits into single and multiple-occupancy, and *commercial* splits into industry trade codes (e.g. retail, manufacturing, and so on).
- **Construction type:** The material and method of construction. For example, whether the building is made from wood, masonry, reinforced concrete or steel.
- Year built: The year that the property was built.
- Building height: The height of the building; usually specified as number of floors.

Other modifiers that are sometimes classed as primary are *number of buildings* (which is commonly used where data is aggregated and sometimes used within a campus-type scenario), and *square footage* (commonly used in the United States).

The impact of primary modifiers varies by peril, but these modifiers are useful information to gather for most, if not all, perils. In most models, different values (e.g. year built is 1950 rather than 2000) of primary modifiers will cause the model to select different vulnerability curves (see Section 1.8.2). This means that the impact of a primary modifier can vary with the intensity of the hazard.

Secondary modifiers are peril-specific and very much depend upon the model being used. For example, for the wind peril, these will allow extra information to be specified about roof type or roof anchoring. For flood, these may contain modifiers around secondary defences or resilience of a property's contents. For earthquake, these may contain modifiers related to retro-fitting. Unlike primary modifiers, these do not normally have separate vulnerability curves assigned, but rather usually act as percentage scalings (i.e. multipliers) applied to the loss cost calculated by using the primary modifiers; thus, the size of the percentage scaling does not vary with the intensity of the hazard.

The extent to which exposure data is available for these modifiers depends mainly upon the peril-region in question and whether a risk engineering report is available for a specific location. In catastrophe-prone regions in the United States, it is common to obtain all the primary modifiers and several secondary modifiers. In the rest of the world it is common to obtain one or two primary modifiers (usually occupancy) and no secondary modifiers. If a risk engineering report is available (normally only for high value facilities), this will usually contain all primary and many secondary modifiers. The challenge in these cases is systemizing the process to extract the right information from the risk engineering report in an efficient and appropriate way.

Some notes on the four main primary modifiers are as follows:

- The usage, or *occupancy*, of the asset is a key modifier influencing the choice of vulnerability function within the model. This will commonly be coded using either model specific codes such as UNICEDE[7] or standard industry codes such as SIC,[8] NAICS[9] or ATC.
- Standardized *construction* classifications, such as ATC-13, originally developed for regional-level aggregate Californian earthquake risk assessment purposes, are often applied across many other territories, and at the site location level. Equally, individual catastrophe modelling

organizations will often create proprietary classifications based on their own vulnerability models, such as UNICEDE[7] or RMS construction codes.

- *Year built* or *building age* is assumed to reflect the type of building standards and regulations likely to have been applied as well as wear and tear on the building structure. Pragmatically, similar ages are often grouped together within a model (e.g. properties built in 1950–1959 may be assigned the same vulnerability curve).
- The *building height* is used by the model to determine the response of the structure to a given hazard impact, such as shaking resonance, wind speed, or flood depth. Like year built, height is often banded within the model and entered as number of floors. Care must be taken not to confuse the specific floors occupied with the total number of floors; the latter should always be used to represent the building height.
- For most territories and perils it is unusual to have insurance exposure data populated with all primary modifiers (US catastrophe-prone areas are an exception to this). Very often occupancy is present, and other primary modifiers are not. In the case of missing primary modifiers the model should contain assumptions about how to treat the exposure data, i.e. which vulnerability curve to assign. This is one reason why models will contain details of industry exposure data, as described in Section 1.8.3.

### 1.9.2 Financial Structure

Insurance financial structures (also commonly referred to as **policy terms**) are features designed to modify the loss payments. Financial structures can be illustrated by considering a simple buildings insurance policy (see Figure 1.5). The box represents the rebuild cost of the property, also known as its replacement cost. It is unusual for insurers to pay very small claims since they wish insurance to be priced reasonably and including many small claims with associated handling costs would make the cost of the policy uneconomical. Therefore, the most common insurance structure is the **deductible**, the amount the policyholder has to bear before they can reclaim from the policy. It can also be known as an **excess** since payments are in excess of this value or a **retention** since the losses are retained by the policyholder. Deductibles can be set as a proportion of the original policy value, a fixed monetary amount or a proportion of the loss, sometimes with a minimum and maximum value. Deductibles may vary by peril. It is common that household policies will be subject to a deductible but cover the entire rebuild cost of the home in excess of this

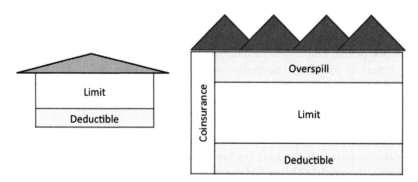

**Figure 1.5** Simple representation of insurance financial structures or policy terms. See text for details. Loss is imagined as increasing upwards from the bottom of the box, i.e. small losses fall within the deductible layer. The smaller box represents a typical situation for a lower-value property. The larger box, with the blue roof, represents a common situation for a higher-value property. White boxes represent areas covered by insurance, while grey areas remain the liability of the policyholder.

value. For higher value properties, the insurer is unlikely to cover the entire rebuild cost of the property but instead only cover a specified **limit** (their limit of liability). Limits generally take the same form as deductibles. If the sum of the deductible and the limit is less than the rebuild cost, the difference will again fall to the policyholder. This difference is known as the **overspill**. There may also be insurance shared between multiple parties, known as coinsurance.

These basic structures can be applied in a variety of ways: they may differ by coverage (see Section 1.9.1.2) and also may apply at either an individual location as in the example above or across multiple locations when the structures are applied at a policy level. For instance, a chain of shops may have multiple locations with deductibles and limits at each location, but then may also have a master policy that has an overall maximum deductible so that if multiple properties are affected by the same event, the amount to be retained by the policyholder is capped. On a similar basis, there will be an overall limit to cap the payment by the insurer. This limit may only apply to particular locations, for instance, a geographical sub-set of the total, or for a specific peril, e.g. flood, and in this case it is known as a **sub-limit**.

There are variations on this theme: a **franchise deductible** is a deductible that prevents a pay-out to the policyholder until the loss reaches the level of the franchise, but then vanishes as a deductible once that level is reached. A **step policy** is a policy that pays out only a set number of specific amounts, for example, five possible levels of pay-out, each pay-out responding to a range of assessed building damage. These are common in Japan.

Policy terms vary significantly by country, and catastrophe risk analysts will need to thoroughly understand the type of policies they are modelling, by close examination of the wordings and discussion with the underwriting teams. The translation of these terms into model input depends very much on the coding schema of particular model vendors, so careful reference to the vendor documentation is also required.

Policy structures can apply to insurance and reinsurance. More detail on reinsurance policies can be found in Chapter 2.4.2.

### 1.9.3 Portfolio Hierarchy

In catastrophe modelling, the order of loss calculation is important, particularly when location and policy characteristics can be complex; for example, where locations may be affected by more than one set of peril-specific policy conditions, or where there are multi-territorial limits.

In general terms, the calculation hierarchy will be reflected in the structure of the data within the model database. A generic hierarchy is shown in Figure 1.6. As can be seen, location (a single asset entity, ideally with an associated location reference which can have insured value and risk characteristics (modifiers) applied to it) is the basic and most granular unit of exposure. Locations tend to aggregate to the policy (or sometimes termed **account**) level. In addition, locations will usually aggregate at varying levels of geographic hierarchy, such as ZIP/postcode, state or country. Policies may apply conditions across all locations in a single country, or across different geographic regions. In many cases, for example, policies may apply across various countries, or may exclude particular countries. In general, policies will then accumulate to the **portfolio** level. This is often defined at the insurance **business unit** level, for instance a particular property line or product may be considered as a portfolio.

**Accumulation**, **roll-up**, or **aggregation** are terms employed to describe the overall combination of multiple portfolios into a time-specific snapshot of exposure and modelled results, for example at a quarterly or monthly intervals. This involves the calculation of loss after each set of financial conditions are applied at each hierarchical level. This is a complex process (see Section 1.8.4) and in some models, this can create difficulties in interpretation of losses, given the challenges in calculating across multiple geographical and policy levels. Accumulation is discussed in more detail in Chapter 2.7.

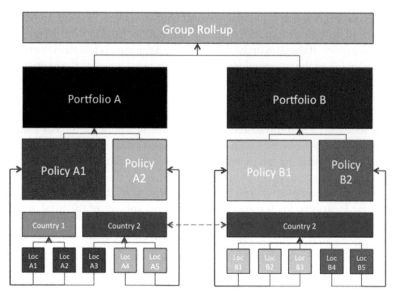

Figure 1.6 A generic insurance hierarchy. 'Loc B2' is location B2. Locations are collected together or 'aggregated' into policies, here shown in the same colour. Policies are grouped into portfolios. An insurer usually has a number of portfolios that it periodically wishes to assess in a 'roll-up'.

For some types of insurance exposure, time-varying levels of insured value, and the use of limits rather than actual sums insured can complicate this process. In these cases, it is often necessary to determine an agreed, but sub-optimal approach to the estimation of insured value at each level and apply it consistently, even though this is likely to be inaccurate.

## 1.10 Model Output: Metrics and Risk Measures

Catastrophe models are designed to quantify catastrophe risk, and so the fundamental output from such models is a probability distribution of loss at the appropriate aggregation level, for the relevant financial perspective. Summary statistics such as the mean and standard deviation of loss are also calculated.

As the catastrophe modelling industry has evolved, a terminology specific to this industry has also emerged and is described in this section. The distribution of the maximum loss in a year is called the **Occurrence Exceedance Probability** (OEP) distribution. The distribution of the *sum* of losses in a year is termed the **Aggregate Exceedance Probability** (AEP) distribution. These are the two main distributions obtained from catastrophe models. Catastrophe models were initially designed to answer questions such as 'What level of reinsurance should I buy?' – requiring an estimated distribution of maximum loss, and 'How much capital do I need?' – requiring an estimated distribution of the sum of losses in each year. Hence the OEP and AEP curves evolved as a standard output of most catastrophe models. These distributions and other metrics commonly output and used are described below.

### 1.10.1 Common Metrics

There are many metrics that are either calculated by the catastrophe modelling platform or can be calculated from the underlying model output. Some common metrics are described below. These can be calculated for each financial perspective output from the catastrophe model.

### 1.10.1.1 Annual Average Loss (AAL) or Average Annual Loss

The annual expected loss. Sometimes called **Annual Mean Loss** (AML) or **pure premium**.

### 1.10.1.2 Standard deviation (SD) around the AAL

A measure of the volatility of loss around the AAL. It is often used in conjunction with the AAL as an input to forming technical price (see Chapter 2.6.2). However, this usually does not represent the full uncertainty in the AAL. The largest component of uncertainty missing is usually the uncertainty associated with the hazard or event generation process (the primary epistemic uncertainty, see Chapter 2.16.1 for more details).

### 1.10.1.3 Exceedance frequency (EF)

The annual frequency of events expected to give losses greater than a given amount. This is not usually output directly by catastrophe models, but is a useful metric and one that is readily calculable from ELTs or YLTs (see Section 1.10.5.4). Note that since this is a frequency (i.e. a count divided by a time period), not a probability, the EF can be greater than 1.0. The **return period** of a particular size of event is the reciprocal of its exceedance frequency.

### 1.10.1.4 Occurrence Exceedance Probability (OEP)

OEP is the probability that the *maximum event loss* in a year exceeds a given level. The occurrence return period is the reciprocal of the OEP. Another equivalent way of thinking about this is that it is the probability that *at least one* event in a year exceeds a given level. While this may not at first be obvious, it is always the case that if at least one event in a year exceeds the given level, the maximum event in that year must also have exceeded the same given level. This equality is used later on when calculating the OEP.

### 1.10.1.5 Aggregate Exceedance Probability (AEP)

AEP is the probability that the *sum of event losses* in a year exceeds a given level. The aggregate return period is the reciprocal of the AEP. The area under the AEP is equal to the AAL.

### 1.10.1.6 Value at risk (VaR)

VaR is the loss value at a specific quantile of the relevant loss distribution. For example, a 99.5% VaR based on the aggregate loss distribution would be the value at the 0.5% level (1-in-a-200-year return period) on the AEP curve.

### 1.10.2 Exceedance probability curve characteristics

A schematic of an AEP curve is shown in Figure 1.7. The area under the AEP curve is equivalent to the AAL.

Although the VaR from AEP and OEP curves is a widely used metric (e.g. Solvency II specifies the 99.5% VaR as the regulatory capital requirement) it should be noted that VaR is not a **coherent risk measure** as defined in Artzner *et al.* (1999); see Section 1.11.3. In particular, there is no guarantee that adding two EP curves together gives a combined EP curve that is less than or equal to the individual curves (i.e. sub-additivity is not a property of VaR). In other words, diversification is not always properly reflected when using VaR, as demonstrated in Woo (2002). Following Woo (2002), this can be illustrated by using an example of two independent portfolios (A and B), each of which contains just two events with the same event frequency (0.5%), as shown in Table 1.2 and Table 1.3.

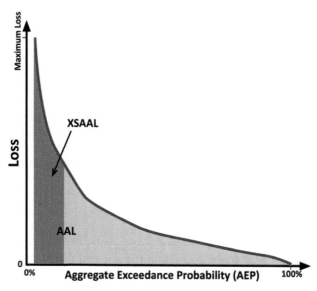

Figure 1.7 An AEP curve. The area under the whole curve (i.e. dark and light blue) is the average annual loss (AAL), and the area beyond some threshold or *excess* (dark blue) is the excess annual average loss (XSAAL). In this case, the x-axis is plotted in terms of probability, but it is also common to create AEP curves with return period in units of years (i.e. 1/probability) on the x-axis.

Table 1.2   Event loss table for a simple idealized portfolio: Portfolio A.

| Loss | Rate (%) | Exceedance frequency (%) | Return Period |
|------|----------|--------------------------|---------------|
| 10,000 | 0.5 | 0.5 | 200 years |
| 1,000 | 0.5 | 1.0 | 100 years |

Table 1.3   Event loss table for a simple idealized portfolio: Portfolio B.

| Loss | Rate (%) | Exceedance frequency (%) | Return period |
|------|----------|--------------------------|---------------|
| 20,000 | 0.5 | 0.5 | 200 years |
| 2,000 | 0.5 | 1.0 | 100 years |

The combined portfolio event loss table is shown in Table 1.4 and it is clear that the 99% VaR metric is 1,000 for Portfolio A, 2,000 for Portfolio B, but *10,000* for the combined portfolio. Despite the fact the portfolios are independent, combining them has not resulted in a reduced risk measure at 1 in 100 year VaR; this is counterintuitive as one would expect diversification to reduce the risk. Irrespective of this, VaR measures are commonly used in setting overall risk limits and thresholds, largely because they are easy to explain and understand. However when *allocating* such limits (e.g. capacity or capital allocation, as discussed in Chapter 2.10), it is even more important that a coherent risk measure is used. Metrics such as TVaR, and XSAAL are often used and these are described below.

Table 1.4 Event loss table for a portfolio that combines portfolios A and B (from Tables 1.1 and 1.2, respectively).

| Loss | Rate (%) | Exceedance frequency (%) | Return period |
| --- | --- | --- | --- |
| 20,000 | 0.5 | 0.5 | 200 years |
| 10,000 | 0.5 | 1.0 | 100 years |
| 2,000 | 0.5 | 1.5 | 67 years |
| 1,000 | 0.5 | 2.0 | 50 years |

### 1.10.3   More Advanced Metrics

#### 1.10.3.1   Tail Value at Risk (TVaR) or Tail Conditional Expectation (TCE)

TVaR or TCE is the expected value of loss above a specific quantile of the relevant loss distribution, given that the specific quantile has been exceeded. It is, therefore, a conditional metric (unlike XSAAL, see later). It is equivalent to the area under the curve above the specified quantile divided by 1 – quantile. The value of TVaR is always greater than the VaR value at the same quantile.

#### 1.10.3.2   *Excess VaR* (xVaR) and Excess TVaR (xTVaR)

'x' here means excess of the mean so these are the VaR and TVaR metrics, but with the mean value (AAL if the AEP distribution is being used) subtracted.

#### 1.10.3.3   Excess Average Annual Loss (XSAAL)

XSAAL is the expected loss cost above a certain threshold (or *excess* point). It is equivalent to the area under the AEP curve above the excess point. It can be thought of as that portion of the overall AAL from events greater than the threshold. Unlike the TVaR metric, the area under the curve above the threshold is not re-normalized by dividing by (1 – threshold quantile), so the XSAAL will always be smaller than the AAL. In other words, XSAAL is *unconditional* whereas TVaR is *conditional*.

### 1.10.4   Event Loss Tables and Year Loss Tables

**Event Loss Tables** (ELTs) or **Year Loss Tables** (YLTs) (see Sections 1.10.5 and 1.10.6 for a description of these) are the fundamental outputs from catastrophe models used to calculate the metrics above. These tables are widely used as input to capital and pricing models, so it is important the practitioner is familiar with ELTs and YLTs. Models will either output an ELT or a YLT, but not usually both.

ELTs and YLTs have a significant impact on the size of the output dataset. For example, an event set could contain 40,000 events, each containing a mean and standard deviation of loss. If this is output at location level, even a relatively small portfolio of 10,000 locations can generate a large amount of output data (multiple GB). The resolution at which event set level output is turned on should be checked carefully against database size constraints and/ or disk space. These tables, some methods for calculating basic metrics from them, and the advantages and disadvantages of each approach are described in Sections 1.10.5 and 1.10.6.

### 1.10.5   Event Loss Table (ELT)

An ELT contains the loss statistics, resulting from the input exposure, for every event in the catastrophe model (for a given peril); typically tens of thousands of events. These events may

Table 1.5  An example of a typical ELT with fictitious numbers.

| Event ID (*i*) | Rate (*r*ᵢ) | Mean loss (*L*ᵢ) | Standard deviation (independent) *SDind*ᵢ | Standard deviation (correlated) *SDcor*ᵢ | Exposure impacted (*E*ᵢ) |
|---|---|---|---|---|---|
| 1 | 0.04 | 850,000 | 1,500,000 | 1,000,000 | 200,000,000 |
| 2 | 0.02 | 700,000 | 1,600,000 | 1,300,000 | 50,500,000 |
| 3 | 0.01 | 1,000,000 | 2,000,000 | 1,500,000 | 100,000,000 |
| 4 | 0.03 | 800,000 | 1,500,000 | 900,000 | 60,000,000 |
| 5 | 0.01 | 650,000 | 1,000,000 | 800,000 | 150,000,000 |

*Note:* See text for details of the columns.

well be a 'boiled-down' set from a larger catalogue (perhaps hundreds of thousands); the reduced number of events in the production model are chosen to preserve the important overall statistics while reducing the model run-time to acceptable levels.

The loss statistics reflect the selected output resolution and financial perspective. So, for example, an event loss table could represent the ground up loss for every event for an individual location, or the gross loss for each event summarized over a portfolio of 1 million locations. In the example shown in Table 1.5, the ELT contains just five events. In practice, the ELT will contain many thousands, or even tens of thousands, of events.

Going through each field in in turn:

- *Event ID*: This is a number or reference uniquely identifying the event. It is often used when combining multiple ELTs to ensure that the correlation between losses from the same events is preserved; in other words, the losses should be added for events with the same event ID.
- *Rate*: This is the annual frequency of occurrence of the specific event. It is not a probability, as in theory this could be greater than 1.0 (although this would be very unusual). For a given model version and Event ID, this is always the same; it does not vary according to the input exposure or financial perspective.
- *Mean loss*: This is the accumulated mean loss for the given event. It varies according to the input exposure, the accumulation level, and the financial perspective.
- *Standard deviation, independent and correlated*: The standard deviation in an ELT represents the **secondary uncertainty**, which is the uncertainty present, given that there has been an event (more details are provided in Chapters 2.16.1 and 4.5.1.1). A model will contain an assumption as to how correlated the uncertainty is, both between coverages for a given location and between locations for a given event. In order to combine standard deviations from loss distributions (e.g. when aggregating from a lower level to a higher level), it is important to know whether the distributions are correlated or not. If the distributions are 100% correlated, the standard deviations can simply be added. If the distributions are completely uncorrelated, the standard deviations are combined by taking the square root of the sum of squares. Typically the *coverage* correlation is assumed to be 100%, and so the coverage standard deviations are usually added to obtain the location standard deviation. The *location* correlation is often a non-zero quantity, and is used within a model to partition the standard deviation into a correlated part (= total location standard deviation × correlation) and an independent part (= total location standard deviation × (1-correlation)). The overall standard deviation for a given event is found by summing the independent and correlated standard deviations.

- *Exposure impacted*: Exposure impacted is the sum of the sum insured for any location potentially exposed to the event in question. It is typically used together with the mean and standard deviation of loss to parameterize a statistical distribution (e.g. Beta) in order to reflect uncertainty in subsequent calculations.

In addition to the fields mentioned above, an ELT often has a description field capturing details about each event.

The subsequent sections show how statistics can be calculated from ELTs, using the sample ELT in Table 1.5 as an example.

### 1.10.5.1 Limitations and Benefits of ELTs

The main form of output from some models is an ELT. Other models provide a YLT form of output instead, which is discussed in Section 1.10.6. The main drawback of an ELT approach is that it can be difficult to implement a non-parametric frequency assumption, whereas in a YLT a frequency distribution is an inherent part of the output and so parametric and non-parametric distributions (e.g. the output of a climate model) can be incorporated. Some benefits of an ELT approach are that the number of years simulated from the ELT can be changed to match user requirements, and annual rates can easily be modified if a user wishes to implement a revised view of risk.

### 1.10.5.2 Calculating the Mean Loss Across All Events (the AAL)

The AAL is calculated as the sum over all events of the product of the rate and mean loss. Each entry, $i$, in an ELT can be thought of as a **compound distribution** (a distribution resulting from combining the individual distributions of the number of losses $N_i$ and the size of losses $X_i$ (e.g. Kaas *et al.*, 2009) of aggregate loss in a year, $S_i$.

Using $E[\ ]$ to denote the expectation (or mean value) of a quantity, the mean number of events for a given entry in an ELT, $E[N_i]$, is $r_i$ (the event rate). The mean event loss $E[X_i]$ is $L_i$ (see Table 1.5). The mean of a compound distribution is $E[N_i] \times E[X_i]$, which is equivalent to $r_i \times L_i$ for each ELT entry. The AAL is found by summing over all entries in the ELT table:

$$AAL = \sum_i r_i \times L_i \tag{1.1}$$

### 1.10.5.3 Calculating the SD of Loss for One Event and Across All Events

The variance (denoted $Var[\ ]$) of a compound distribution, $S_i$, with frequency and severity assumed independent, is given by (e.g. Kaas *et al.*, 2009; Klugman, Panjer and Willmot, 1988):

$$Var[S_i] = E[N_i]Var[X_i] + Var(N_i)E[X_i]^2 \tag{1.2}$$

If a **Poisson** assumption is made, then both $E[N_i]$ and $Var[N_i] = r_i$ (see Section 1.11.1.1).

Referring to the columns in Table 1.5: $E[X_i]^2 = L_i^2$, and $Var[X_i] = (SDind_i + SDcor_i)^2$. This gives the following formula for the variance of an individual ELT entry $i$:

$$Var[S_i] = r_i(SDind_i + SDcor_i)^2 + r_iL_i^2 \tag{1.3}$$

Assuming events are independent, we can sum the variance across all ELT entries and then square root the resulting sum to obtain the standard deviation ($SD$) around the AAL:

$$SD = \sqrt{\sum_i r_i(SDind_i + SDcor_i)^2 + r_iL_i^2} \tag{1.4}$$

**Table 1.6** Sample ELT, ordered, and with EF calculated.

| Event ID ($i$) | Rate ($r_i$) | Exceedance Frequency ($EF_i$) | Mean Loss ($L_i$) | Standard Deviation (Independent) $SDind_i$ | Standard Deviation (Correlated) $SDcor_i$ | Exposure Impacted ($E_i$) |
|---|---|---|---|---|---|---|
| 3 | 0.01 | 0.01 | 1,000,000 | 2,000,000 | 1,500,000 | 100,000,000 |
| 1 | 0.04 | 0.05 | 850,000 | 1,500,000 | 1,000,000 | 200,000,000 |
| 4 | 0.03 | 0.08 | 800,000 | 1,500,000 | 900,000 | 60,000,000 |
| 2 | 0.02 | 0.10 | 700,000 | 1,600,000 | 1,300,000 | 50,500,000 |
| 5 | 0.01 | 0.11 | 650,000 | 1,000,000 | 800,000 | 150,000,000 |

#### 1.10.5.4 Calculating the Exceedance Frequency (EF) Without Secondary Uncertainty

The exceedance frequency for a given level of loss is simply the annual frequency of events greater than or equal to this level of loss. It is not a statistic normally output by catastrophe models, but it is nonetheless useful. It answers questions that OEP and AEPs cannot answer (e.g. how many losses above a certain amount can I expect in a year?), so can be useful when comparing and contrasting models with different frequency distribution or clustering algorithms in order to gain insight into the underlying model differences.

The EF is calculated by sorting the ELT in descending order of mean loss, $L_i$. The event rate, $r_i$, for the largest ELT entry is the exceedance frequency at this level of loss. The rates can then be summed, from each level up to the largest level of loss, in order to calculate the exceedance frequency for each level. This is shown in Table 1.6.

So the EF of a 700,000 loss is 0.1. Another equivalent way of expressing this is that a 700,000 event has a return period of 10 years.

#### 1.10.5.5 Calculating the Exceedance Frequency with Secondary Uncertainty

The calculation of EF above ignored secondary uncertainty. In order to take this into account, the uncertainty distribution must be known. It is becoming more common for models to output this distribution for each event (as well as the mean event loss). However, for many models currently used, a standard deviation (perhaps separated into independent and correlated parts) as well as the maximum exposure impacted is all the uncertainty information that is provided for each event. An assumption must therefore be used about the size of loss distribution. It is fairly common for a Beta distribution to be used, to represent the size of loss normalized by the maximum exposure (or exposure impacted). The steps in calculating the EF for a particular amount of loss (A) are as follows:

1) For each ELT entry calculate the probability of the loss exceeding the amount A, using either the distribution provided by the model or a parametric assumption based on the mean and standard deviation (and potentially maximum exposure) of loss for that ELT entry. A description of the Beta distribution is given in Section 1.11.2.1.
2) For each ELT entry, multiply the probability of loss exceeding the amount A by the event rate.
3) Sum the quantity produced in (2) across all events. This is the EF with secondary uncertainty for loss amount A.

To obtain the EF for other loss exceedance levels, steps 1–3 need to be repeated for the different levels.

Table 1.7 Sample ELT with OEP calculated.

| Event ID ($i$) | Rate ($r_i$) | Exceedance Frequency ($EF_i$) | Occurrence Exceedance Probability ($OEP_i$), % | Mean Loss ($L_i$) | Standard Deviation (independent) $SDind_i$ | Standard Deviation (correlated) $SDcor_i$ | Exposure Impacted ($E_i$) |
|---|---|---|---|---|---|---|---|
| 3 | 0.01 | 0.01 | 0.995 | 1,000,000 | 2,000,000 | 1,500,000 | 100,000,000 |
| 1 | 0.04 | 0.05 | 4.877 | 850,000 | 1,500,000 | 1,000,000 | 200,000,000 |
| 4 | 0.03 | 0.08 | 7.688 | 800,000 | 1,500,000 | 900,000 | 60,000,000 |
| 2 | 0.02 | 0.10 | 9.516 | 700,000 | 1,600,000 | 1,300,000 | 50,500,000 |
| 5 | 0.01 | 0.11 | 10.417 | 650,000 | 1,000,000 | 800,000 | 150,000,000 |

#### 1.10.5.6 Calculating the OEP

The OEP (without secondary uncertainty) can be calculated from the ELT entries, provided an event frequency distribution is known or assumed. The steps are as follows:

1) Sort the event losses in descending order and calculate the EF as shown above.
2) Use the EF at each level in the ELT, together with a frequency distribution, to calculate the probability of there being at least one event greater than the level in the ELT (which equals 1 minus the probability that there are no events). This is the OEP.

For example, if a Poisson distribution is assumed, the probability of there being at least one event greater than level $i$ is $1 - e^{-EF_i}$ (see Section 1.11.1.1). The OEP values using the Poisson assumption are shown in Table 1.7.

The OEP figures above ignore secondary uncertainty. To calculate the OEP with secondary uncertainty, the EF with secondary uncertainty would first be calculated using the process described in Section 1.10.5.5, and the OEP would then be calculated in the manner described above, but using the EF with secondary uncertainty instead.

#### 1.10.5.7 Calculating the AEP

Calculating the AEP is not possible in the same way as OEP, since the AEP represents the distribution of the sum of losses within a year. In order to evaluate this, the distributions from the individual ELT entries must be convolved (added together). In practice, this can either be done through Monte Carlo simulation or through the use of a technique involving discretization of the distributions, such as Panjer or FFT (e.g. Embrechts and Frei, 2009). The use of simulation is discussed in Section 1.10.5.9; the use of Panjer or FFT is beyond the scope of this book.

#### 1.10.5.8 Correlation between ELTS

A natural question to ask is the extent of correlation between locations, policies or portfolios. If an ELT for the respective locations, policies or portfolios is available and the variance for each ELT and for the grouped (combined) ELT is known, the correlation can be calculated. Consider two ELTs, A and B, with standard deviation $SD_A$ and $SD_B$ respectively (calculated using Equation (1.4)). The combined ELT has a standard deviation of $SD_{A+B}$.

The general formula for the variance of the sum of distributions is given by:

$$Var[A + B] = Var[A] + Var[B] + 2Cov[A, B] \tag{1.5}$$

where $Cov[A, B]$ is the covariance of A and B.

The Pearson (or linear) correlation coefficient between $A$ and $B$ is defined by:

$$\rho_{A,B} = \frac{Cov[A,B]}{\sqrt{Var[A]Var[B]}} \tag{1.6}$$

This leads to the following formula for correlated variables, expressed in terms of standard deviations instead of variances:

$$SD_{A+B}^2 = SD_A^2 + SD_B^2 + 2\rho_{A,B}SD_A SD_B \tag{1.7}$$

Rearranging, we can see that the correlation between two ELTs, A and B, can therefore be calculated as:

$$\rho_{A,B} = \frac{SD_{A+B}^2 - SD_A^2 - SD_B^2}{2SD_A SD_B} \tag{1.8}$$

### 1.10.5.9 Use in Simulation: Generating YLTs

Monte Carlo simulation (e.g. Klugman, Panjer, and Willmot, 1988) can be used with ELTs to perform further calculations in order to generate statistics such as the AEP, or to calculate the impact of financial structures that operate at the level of the ELT (e.g. reinsurance on a portfolio or group of portfolios). Commercial tools may be used to do this (see Chapter 2.10.6); the process is as follows:

1) Calculate the overall ELT frequency.
2) Use this frequency, together with any other information known about the frequency distribution, to simulate the number of events each year.
3) Given the number of events in a specific year, sample the events that occur such that the chance of them occurring is in proportion to the relative frequency of the events within the ELT.
4) For each event, sample from the secondary uncertainty distribution to obtain the realization of the loss from that event, including uncertainty.
5) The event loss figures from (4) can then be used in order to calculate further statistics.

Further information on each step is as follows. In step 1, the overall ELT frequency is simply calculated as the sum of the events rates in an ELT. In step 2, if a Poisson distribution (Section 1.11.1.1) is assumed, then the only parameter that is needed is the sum of event rates. If a different distribution is used, then more information may be needed, for example a **negative binomial** will need an assumption about the standard deviation of event rate. More sophisticated algorithms can also be used, for example, a specific clustering algorithm could be implemented through a simulation framework; perhaps including a dependency between the number of events and the size of each event.

In step 3, once a number of events have been sampled from a frequency distribution, the same number of specific event entries needs to be selected from the ELT. This can be done by first normalizing the exceedance frequency for each event in the ELT by ordering the events in the ELT in descending order, calculating exceedance frequency for each event, and then dividing each event exceedance frequency by the overall sum of the event rates. For example, the normalized exceedance frequency from the sample ELT in Table 1.7 is calculated and shown in Table 1.8.

In the simulation framework, for each event sampled in a given year, a uniform distribution ranging between zero and 1 (denoted Uniform(0,1)) can be sampled, and the realization of this is used to look up the ELT entry through the normalized exceedance frequency. For example, say the frequency distribution sampling results in two events occurring in a given year, and the Uniform(0,1) distribution sampling results in 0.05 and 0.62 being sampled for each event. This

Table 1.8    Sample event loss table with normalized exceedance frequency calculated.

| Event ID ($i$) | Rate ($r_i$) | Exceedance Frequency ($EF_i$) | Occurrence Exceedance Probability ($OEP_i$), % | Normalized Exceedance Frequency, % | Mean Loss ($L_i$) | Exposure Impacted ($E_i$) |
|---|---|---|---|---|---|---|
| 3 | 0.01 | 0.01 | 0.995 | 0.01 / 0.11 = 9.1 | 1,000,000 | 100,000,000 |
| 1 | 0.04 | 0.05 | 4.877 | 0.04 / 0.11 = 45.5 | 850,000 | 200,000,000 |
| 4 | 0.03 | 0.08 | 7.688 | 0.03 / 0.11 = 72.7 | 800,000 | 60,000,000 |
| 2 | 0.02 | 0.10 | 9.516 | 0.02 / 0.11 = 90.9 | 700,000 | 50,500,000 |
| 5 | 0.01 | 0.11 | 10.417 | 0.01 / 0.11 = 100.0 | 650,000 | 150,000,000 |

would result in Event ID 3 (0.05 < 9.1%) and Event ID 4 (45.5% < 0.62 < 72.7%). This ensures that the events are selected in proportion to their relative frequency.

The final step is to sample each realization of every event selected allowing for secondary uncertainty. If a parametric distribution (such as a Beta distribution, see Section 1.11.2.1) is used, then this can be done directly within the functionality of most simulation tools. If an **empirical distribution** (i.e. one that has not got a particular parametric form) is provided by the model, then a similar approach to the normalized exceedance frequency approach provided can be used to obtain the realization of the event.

The incorporation of uncertainty is straightforward for one ELT, but if multiple ELTs are being simulated (for a given peril), as is often the case, then the question is how to properly reflect the uncertainty correlation between ELTs. A cautious approach sometimes used (which will bias the results high) is to assume full correlation in uncertainty, and use the same sampled value of the Uniform(0,1) distribution for each ELT entry for a given realization of an event. For example, if a value of 0.75 is sampled from the uniform distribution and used to calculate the appropriate value from the Beta distribution representing the uncertainty for Event 1 in ELT A, then the same 0.75 figure is used to calculate the value from the Beta distribution for Event 1 in ELT B, and so on. This will not preserve the correlation in the same way that the grouping procedure described in Section 1.10.5.10 does. However, more sophisticated approaches are available that do preserve the correlation in uncertainty maintained by using the combination techniques described in Section 1.10.5.10. They are, however, proprietary, and the interested reader is directed to either their model vendor or the vendor of their simulation software.

The end result of the simulation process is a set of event realizations for every year, at the granularity and for the financial perspective of the ELT. These figures can then be used to derive further metrics in a very flexible way. For example, the AEP can be calculated by calculating the sum of event losses in each year. The ordering of the number of years, together with the probability of each year occurring being 1/(Number of Simulation Years) gives the AEP probability distribution. For example, if 100,000 simulation years are sampled, and ordered from high to low, then the loss value of the 1000th year represents the 100 year (1000 × (1/100,000) = 100) AEP loss. The OEP can also be calculated in a similar way, but instead of calculating the sum of losses for each year, the maximum event loss for each year is used instead.

Financial structures can also be applied within a simulation framework, so long as they work at the same level (or a higher level) than the original ELT (a policy-level ELT could not be used to evaluate location-level financial structures). The financial structures are simply applied to the event losses for each year and the losses net of the financial structures evaluated and captured. The flexibility of Monte Carlo simulation often means that more complex structures can be evaluated than can readily be done using either closed form integration or numerical convolution techniques.

Historically, insurance and reinsurance companies tend to take the net loss pre-cat ELT tables out of the model and use a simulation framework to calculate the losses for other perspectives. In many catastrophe models the convolution of loss distributions to different levels, and the application of financial structures are performed using either closed form integration or a discretized convolution technique. However, models are emerging that use Monte Carlo simulation as the basis for aggregating and applying financial structures from location-coverage level upwards. We anticipate this trend continuing as computing power increases.

### 1.10.5.10 Combining ELTs

It is often necessary to combine ELTs in order to calculate metrics at a more aggregated level. For example, ELTs for individual locations could be combined in order to calculate the metrics from a multi-location policy. Alternatively ELTs from separate policies may need to be combined to give a portfolio view. In order to combine ELTs, and preserve the correlation provided by the catastrophe model, the ELTs must be of the same version of the model (or at least a version with consistent event IDs). Although the model itself will contain a consistent set of events, each ELT may contain a different number of events, as events that do not generate any loss are sometimes excluded from the model output.

The first step in combining two ELTs is to compare the event IDs; events that are not in both ELTs can be added to the combined ELT. For events that are in both ELTs, the following steps are necessary:

1) The event ID and rate should be the same in both ELTs, so these can be bought across without change.
2) The mean loss, the correlated standard deviation, and the exposure impacted, for the same event ID, should be added.
3) The independent standard deviation for the same event in each ELT should be squared, summed and square-rooted.

The resultant combined ELT can then be used to calculate the metrics (e.g. AAL, SD) described above.

### 1.10.6 Year Loss Table (YLT)

YLTs can be divided into two different types:

- *Type 1*: A YLT where each loss entry loss represents the realization of an uncertainty distribution – NOT the mean loss. There is no uncertainty distribution provided in such a YLT – it is inherent within the loss figures.
- *Type 2*: A YLT where each loss entry represents the mean event loss, and a standard deviation, or distribution of uncertainty, is provided together with the mean loss.

A typical YLT structure is shown in Table 1.9. The standard deviation ($SD_i$) of each event, $i$, is only present for Type 2 YLTs.

### 1.10.6.1 Limitations and Benefits of YLTs

A disadvantage of YLTs is that making adjustments to the frequency is not as straightforward as for ELTs as the events have essentially already been drawn from the frequency distribution.

A second disadvantage is that the number of years in the YLT may not be the same number of years needed for the simulation framework that the insurance company is using. For example, perhaps an internal model framework is set up that uses 10,000 years and the YLT contains

Table 1.9 Sample year loss table.

| Year (k) | Loss Number in Year (j) | Event ID (i) | Loss (X_i) | Exposure Impacted (E_i) | Standard Deviation (SD_i) |
|---|---|---|---|---|---|
| 1 | 1 | 8 | 850,000 | 200,000,000 | 2,150,000 |
| 1 | 2 | 6 | 700,000 | 50,500,000 | 1,950,000 |
| 2 | 1 | 9 | 1,000,000 | 100,000,000 | 2,300,000 |
| 3 | 1 | 11 | 800,000 | 60,000,000 | 2,100,000 |
| 3 | 2 | 7 | 650,000 | 150,000,000 | 2,050,000 |

20,000 years. In this case, stratified sampling (e.g. Calder, Couper and Lo, 2012) can be used to reduce the number of years while minimizing the impact on the simulation error. If an increase in the number of years is required, then years can either be re-sampled from the YLT, however, the increased number of years will not reduce the simulation error for a type 1 YLT and will only partly reduce the simulation error for a type 2 YLT.

An advantage of YLTs is that any dependency between events in a year can be explicitly incorporated (e.g. clustering). There is no requirement that the distribution of events in a year fit a particular parametric distribution; for example, they could be the output of a global climate model, or the output of a complex clustering scheme where there is some dependency between the frequency and severity of events. A dependency between years and seasonality could also be incorporated in a YLT (although this is unusual).

A practical advantage of type 1 YLTs is that they can be used 'as is', without recourse to a simulation engine, to calculate metrics; including the impact of financial structures that depend on multiple events in a year such as reinstatements and aggregate event limits. A type 2 YLT still needs the uncertainty around the mean event losses to be simulated.

An advantage of a type 2 YLT is that the uncertainty is explicitly represented, and so increased sampling from the uncertainty distribution within a simulation framework can result in a more robust propagation of the uncertainty. Since the mean and the standard deviation of event size are known, a type 2 YLT can also be used to construct an ELT: each event having rate $1/N$ (assuming no repeated events). Although the event rate can be calculated, and the mean and standard deviation of loss populated for each event, the precise frequency distribution for the ELT may not be known, given the flexibility of frequency distributions that can be represented in a YLT.

### 1.10.6.2 Calculating Metrics from YLTs

For either kind of YLT, the AAL is calculated in the same way: by summing the losses (mean losses or realizations of loss depending on the type of YLT) across all years (k) and dividing by the number of years (N), i.e.

$$AAL = \frac{1}{N}\sum_{k=1}^{N}\sum_{j=1}^{m_k} X_{j,k} = \frac{1}{N}\sum_{k=1}^{N} S_k = \frac{Z}{N} \tag{1.9}$$

where $m_k$ = number of events in year k, $X_{j,k}$ = realization of loss or mean loss for the $j^{th}$ event in year $k$, $S_k$ = the sum of event losses within year $k$, and $Z$ = the sum of event losses for the whole YLT.

The method of calculation of the standard deviation of a YLT depends upon the type of YLT. If the YLT is type 1 (i.e. represents the realization of each event loss), the uncertainty is inherent within the YLT realization and so the variance must be calculated explicitly from the data using the formula for sample variance:

$$Var[S] = \frac{1}{N-1} \sum_{k=1}^{N} (S_k - AAL)^2 \tag{1.10}$$

where:

$$S_k = \sum_{j=1}^{m_k} X_{j,k} \tag{1.11}$$

$$SD = \sqrt{Var[S]} \tag{1.12}$$

If, however, the YLT is of the second type (i.e. the loss represents the mean and the SD of each event is provided), then using the formula above would understate the uncertainty as it would not take into account the SD in each event loss. Instead, the variance and standard deviation of this type of YLT can be calculated as (courtesy of Ye Liu):

$$Var[S] = \frac{1}{N} \sum_{k=1}^{N} \sum_{j=1}^{m_k} (SD_{j,k})^2 + \frac{1}{N} \sum_{k=1}^{N} (S_k - AAL)^2 \tag{1.13}$$

$$SD = \sqrt{Var[S]} \tag{1.14}$$

As a practical note, in some cases, the number of years in a YLT may be less than $N$, perhaps because for some years there are no event losses generated for the portfolio. This might be particularly the case for low frequency perils (such as an earthquake in Austria). However, even if this is the case, the full value of N must still be used in calculating the AAL and SD; N must not be reduced even if there are years with no events.

The calculation of the OEP and AEP from a type 1 YLT is described in Section 1.10.5.9. YLTs can be combined in the same way as ELTs; the years and events are matched between YLTs and the losses and SDs combined in the same way as for ELTS. The correlation between two YLTs can be calculated in the same way as for two ELTs.

The calculation of the OEP and AEP from a type 2 YLT is, however, more complex. An approximate AEP and OEP can be calculated by treating the mean losses for each event as realizations of loss and following the process described in Section 1.10.5.9. However, this approach would only provide an approximation and would not use the uncertainty information provided in a type 2 YLT. A more comprehensive approach would involve simulating a number of type 1 YLTs by sampling the uncertainty distributions around each event in the type 2 YLT. For each generated type 1 YLT, an AEP curve can be calculated (as described in Section 1.10.5.9). At each probability level the distribution of losses can be averaged to calculate a final AEP. This, however, is considerably more complex and computationally intensive than the approximate approach. Research would be needed to establish under which conditions the approximation is valid, and when the more complete but complex approach should be used.

## 1.11 Statistical Basics for Catastrophe Modelling

This section introduces some of the basic distributions used within catastrophe modelling. This is a catastrophe modelling and management textbook, not an actuarial or statistical primer, and so many statistical topics will be beyond the scope of this book. The purpose of this section is to enable the reader to have some statistics for relevant distributions 'to hand' rather than provide a complete primer in statistics for a full range of distributions. Readers are referred to Forbes *et al.* (2011) for more details of many distributions, and Chapters 9 and 10 of Sweeting (2011) for some basic statistics.

Throughout this section, $X$ is a random variable with $x$ a specific value of this random variable. For the distributions described below, the following statistics are provided.

- Probability function $Pr(x)$ (for discrete distributions): the probability that each discrete value occurs.
- Probability density function (pdf) $f(x)$ (for continuous distributions): a function that gives the relative chance of a value occurring, such that the integral over the entire function is equal to one, but the probability of each precise value occurring is zero.
- Cumulative distribution function $F(x)$ (for continuous distributions): the probability that the random variable is less than or equal to a given value. It is equal to the integral of the pdf from the lowest possible value of the distribution to the given value.
- Mean or expectation of the random variable $E[X]$
- Variance of the random variable $Var[X]$.

The formulas for estimating the mean and variance from a sample of data of $N$ observations (rather than the full but unobservable population) are as follows:

$$Sample\ Mean = \overline{E[X]} = \frac{1}{N}\sum_{n=1}^{N} X_n \tag{1.15}$$

$$Sample\ Variance = \overline{Var[X]} = \frac{1}{N-1}\sum_{n=1}^{N}(X_n - \overline{E[X]})^2 \tag{1.16}$$

Also, note that the standard deviation (SD) is the square root of the variance, and the coefficient of variation (CV) is the standard deviation divided by the mean.

In some cases it can be useful to estimate parameters of the distributions from the observed (or sample) mean and variance. Where possible, these estimators are also given for the readers' convenience. Note that this 'method of moments' type approach is often less robust than other methods of parameter estimation; in particular, maximum likelihood estimation (e.g. Forbes *et al.*, 2011) is the preferred technique. However, for ease of application, very often the method of moments approach is still used.

The concept of a *compound distribution* is used in this chapter; the sum of a random number of independent and identically distributed random variables is a compound distribution. Namely, if $S = X_1 + X_2 + X_3 + \cdots + X_N$, where the $X$ variables are independent and identically distributed random variables and $N$ is also a random variable that is independent from the $X$ variables, then $S$ has a compound distribution. The expectation and variance of a compound distribution (e.g. Kaas *et al.*, 2009; Klugman *et al.*, 1988) are as follows:

$$E[S] = E[X]E[N] \tag{1.17}$$

$$Var[S] = E[N]Var[X] + Var[N](E[X])^2 \tag{1.18}$$

As an example of the application of this, consider the example of a property worth £1,000,000 with a 1% annual frequency of being destroyed:

$E[N] = 1\% = 0.01$
$Var[N] = 0.01 \times (1.00 - 0.01) = 0.0099$ (assuming a Bernoulli distribution which is appropriate for this scenario)
$E[X] = £1,000,000$

$Var[X] = 0$ (there is no uncertainty in the size of loss as the property is assumed completely destroyed if an event happens)

$$E[S] = E[N]E[X] = 0.01 \times 1,000,000 = £10,000$$

$$Var[S] = 0.01 \times 0 + 0.0099 \times (1,000,000)^2 = £9,900,000000$$

Which gives a standard deviation of £99,499.

### 1.11.1 Discrete Distributions

Discrete distributions are often used in catastrophe modelling to model the frequency of events occurring within a particular year for a particular peril-region. The main distributions used are the Poisson and the negative binomial distributions, and these are described below.

#### 1.11.1.1 Poisson

The Poisson distribution is commonly used as the distribution for event numbers in a catastrophe model. It gives the probability of a number of independent events occurring in a specified time. The assumption of independence and the fact that the mean equals the variance indicates that it may not be suitable as a frequency distribution for some perils; for example, for some European windstorm regions (see Chapter 3.3.1) or for North Atlantic hurricane numbers (see Chapter 3.2.1) where the variance in numbers is known to be greater than the mean (over-dispersion/clustering).

For $x = 0, 1, 2, \ldots$ and parameter $\lambda > 0$:

$$\Pr(x) = \frac{e^{-\lambda}\lambda^x}{x!} \tag{1.19}$$

$$E[X] = Var[X] = \lambda \tag{1.20}$$

A probability function for a Poisson with $\lambda = 2.0$ is shown in Figure 1.8.

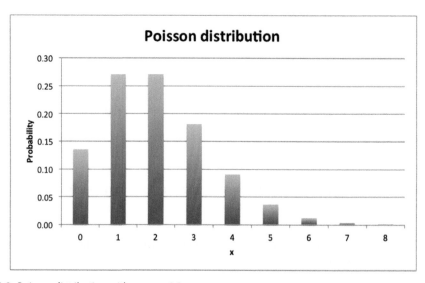

Figure 1.8 Poisson distribution with mean = 2.0

### 1.11.1.2 Negative Binomial

The negative binomial distribution is sometimes used within catastrophe modelling when a parametric frequency distribution is required, but there is some known dependence between events. The variance is greater than the mean for a negative binomial distribution. This distribution has been used to model numbers of European windstorms or North Atlantic hurricanes. There are many different ways of defining the negative binomial distribution. The definition below utilises the gamma function ($\Gamma$), which allows the parameter $r$ to take non-integer values.

For $x = 0,1,2, \ldots$ and parameters $r \geq 1$, $m > 0$

$$\Pr(x) = \frac{\Gamma(x+r)}{x!\Gamma(r)} \left(\frac{m}{r+m}\right)^{x} \left(\frac{r}{r+m}\right)^{r} \tag{1.21}$$

$$E[X] = m \tag{1.22}$$

$$Var[X] = m + \frac{m^2}{r} \tag{1.23}$$

If the mean and variance are known then $m$ and $r$ can be estimated from:

$$m = E[X] \tag{1.24}$$

$$r = \frac{E[X]^2}{Var[X] - E[X]} \tag{1.25}$$

A negative binomial distribution for $r = 2$ and $m = 2.0$ (mean $= 2.0$) is shown in Figure 1.9.

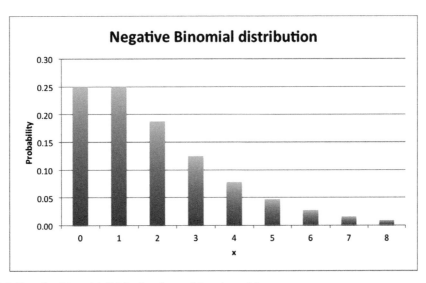

Figure 1.9 Negative binomial distribution for r = 2.0 and m = 2.0

### 1.11.2 Continuous Distributions

There are many parametric continuous distributions available. In this section we describe only two, the Beta distribution and the single parameter Pareto distribution, because of their common use in catastrophe model building and catastrophe pricing respectively. For information on other distributions, refer to Forbes *et al.* (2011).

#### 1.11.2.1 Beta

The Beta distribution is commonly used within the vulnerability component of catastrophe modelling to describe the uncertainty around the mean damage ratio (the expected percentage of sum insured lost given a certain level of hazard). It is well suited to this, since it describes values in the range zero to one, which encompasses all possible values of damage ratio. This distribution is also used within the financial component of some catastrophe models to parameterize the aggregated uncertainty of event losses in order to quantify the impact of financial structures at that particular level of aggregation; in this usage the random variable $X$ is the mean event loss divided by the exposure within the footprint of the event.

Before describing the Beta distribution, we must introduce the incomplete Beta function and the Beta function. The incomplete Beta function:

$$B(x; a, b) = \int_0^x t^{a-1}(1 - t)^{b-1} dt \tag{1.26}$$

The Beta function is the incomplete Beta function where $x = 1$:

$$B(a, b) = \int_0^1 t^{a-1}(1 - t)^{b-1} dt \tag{1.27}$$

The Beta distribution is then given as follows:

For $0 < x < 1$ and parameters $a > 0$, $b > 0$:

$$f(x) = \frac{x^{a-1}(1 - x)^{b-1}}{B(a, b)} \tag{1.28}$$

$$F(x) = \frac{B(x; a, b)}{B(a, b)} \tag{1.29}$$

$$E[X] = \frac{a}{a + b} \tag{1.30}$$

$$Var[X] = \frac{E[X](1 - E[X])}{a + b + 1} = \frac{ab}{(a + b)^2(a + b + 1)} \tag{1.31}$$

If the mean and variance are known, the parameters $a$ and $b$ can be estimated as follows:

$$CV = \frac{SD}{E[X]} \tag{1.32}$$

$$a = \frac{E[X]^2(1 - E[X])}{Var[X]} - E[X] = \frac{(1 - E[X])}{CV^2} - E[X] \tag{1.33}$$

$$b = \frac{a(1 - E[X])}{E[X]} \tag{1.34}$$

The probability density function for a Beta distribution with $a = 4.0\ and\ b = 3.0$ is shown in Figure 1.10.

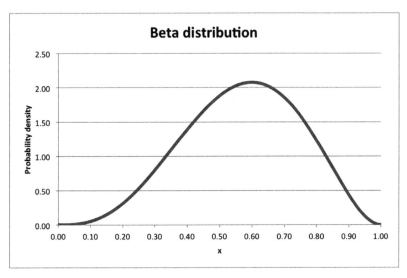

Figure 1.10  Beta distribution with a = 4.0 and b = 3.0

### 1.11.2.2  Pareto (One Parameter)

There are many different types of Pareto distribution. These distributions are 'heavy-tailed' distributions which means that there is an increased probability of extreme losses compared to many other distributions – which makes them appropriate for modelling catastrophe losses. The distribution described here is usually described as the one parameter Pareto distribution. Although it has two parameters, one is prescribed up-front by the threshold beyond which the loss distribution applies, so it is not a free parameter. This distribution is not normally used within the building of exposure-based catastrophe models, but is often used within catastrophe pricing to describe the severity of event losses. Particular values of the shape parameter, $\alpha$, are commonly used as starting points to parameterize the distribution for different types of peril (see Table 2.3 in Chapter 2.6.4.2). It can therefore also be a useful distribution when trying to represent non-modelled risk using an actuarial approach (see Chapter 5.4.7).

Other types of Pareto distribution commonly used include the two parameter Pareto and the Generalized Pareto distributions. Description of these is beyond the scope of this book, but references can be found in standard actuarial and statistical texts (e.g. Klugman, Panjer and Willmot, 1988).

The one parameter Pareto distribution is given as follows:

For threshold $t > 0$, $x > t$ and parameter $\alpha > 0$:

$$f(x) = \frac{\alpha t^{\alpha}}{x^{\alpha+1}} \tag{1.35}$$

$$F(x) = 1 - \left(\frac{t}{x}\right)^{\alpha} \tag{1.36}$$

$$E[X] = \frac{\alpha t}{\alpha - 1} \tag{1.37}$$

$$Var[X] = \frac{\alpha t^2}{(\alpha - 1)^2(\alpha - 2)} \tag{1.38}$$

If the mean is known, then $\alpha$ can be calculated from:

$$\alpha = \frac{E[X]}{E[X] - t} \tag{1.39}$$

The probability density function of a Pareto distribution with $t = 10.0$ *and* $\alpha = 2.5$ is shown in Figure 1.11.

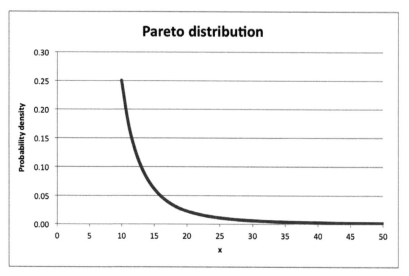

Figure 1.11  Pareto distribution with t $=$ 10.0 and $\alpha = 2.5$

### 1.11.3  Coherent Risk Measures

This section describes coherent risk measures. Understanding of this term may useful to the catastrophe risk analyst when considering appropriate metrics to use for reporting and risk tolerance purposes (see Chapter 2.10.1). A coherent risk measure (e.g. Artzner *et al.*, 1999; Sweeting, 2011) is one which satisfies the criteria below, where $F(\ )$ is a function which implements a particular risk measure:

**Sub-additivity**: $F(X + Y) \leq F(X) + F(Y)$

Portfolio diversification should always lead to lower risk. In other words, combining two risks should not create any additional risk. The VaR metric does not always satisfy this criterion.

**Positive Homogeneity**: $F(cX) = cF(X)$, where $c$ is a constant.

Multiplying the losses by a constant factor (e.g. inflation or a currency conversion – or even aggregating identical risks) should result in the risk measure changing by the same factor. Variance does not satisfy this criterion: $Var(cX) = c^2\, Var(X)$.

**Monotonicity**: $F(X) > F(Y)\, if\, X > Y$

If each entry in an Event Loss Table for portfolio X is greater than that for portfolio Y, the risk measure for X should be greater than the risk measure for Y. In other words, if the losses increase, the risk measure should increase.

**Translation Invariance**: $F(X + c) = F(X) + c$, where $c$ is a constant.

If the amount of loss is increased by a fixed amount, then the risk measure increases by the same amount.

## Notes

1. http://www.aon.com/reinsurance/analytics/remetrica.jsp
2. https://www.towerswatson.com/en/Services/Tools/igloo
3. http://www.oasislmf.org/

4. https://www.acord.org/webfiles/acord_knowledge/20130313_CatExp.pdf
5. See, for example, the US Department of Defense standard: http://earth-info.nga.mil/GandG/publications/tr8350.2/tr8350_2.html
6. https://www.cresta.org/
7. http://www.unicede.com/
8. http://www.ehso.com/siccodes.php
9. http://www.ehso.com/naics.php

## References

ABI (2011) *Industry Good Practice for Catastrophe Modelling: A Guide to Managing Catastrophe Models as Part of an Internal Model under Solvency II.* Association of British Insurers, London.

Actuarial Standards Board (2000) *ASOP No. 38: Using Models Outside the Actuary's Area of Expertise (Property and Casualty)* Actuarial Standards Board, London.

Artzner, P., Delbaen, F., Eber, J-M. and Heath, D. (1999) Coherent measures of risk. *Mathematical Finance*, **9** (3), 203–228.

Calder, A., Couper, A. and Lo, J. (2012) *Catastrophe model blending: techniques and governance. In General Insurance Convention (GIRO)* UK Actuarial Profession, Brussels.

Embrechts, P. and Frei, M. (2009) Panjer recursion versus FFT for compound distributions. *Mathematical Methods of Operations Research*, **69** (3), 497–508.

Forbes, C., Evans, M., Hastings, N. and Peacock, B. (2011) *Statistical Distributions*. John Wiley & Sons, Inc., Hoboken, NJ.

Friedman, D. (1984) Natural hazard risk assessment for an insurance program. *Geneva Papers on Risk and Insurance*, **9** (30), 57–128.

Grossi, P. and Kunreuther, H. (2005) *Catastrophe Modelling: A New Approach to Managing Risk.* Springer, New York.

Kaas, R., Goovaerts, M., Dhaene, J. and Denuit, M. (2009) *Modern Actuarial Risk Theory Using R.* Springer, Berlin.

Klugman, S., Panjer, H. and Willmot, G. (1988) *Loss Models: From Data to Decisions*. John Wiley & Sons, Inc., New York.

Ley-Borrás, R. and Fox, B. (2015) Using probabilistic models to appraise and decide on sovereign disaster risk financing and insurance. *Financing and Insurance Policy Research Working Papers.* World Bank Group, Washington, DC.

LMA (2013) *Catastrophe Modelling: Guidance for Non-Catastrophe Modellers.* Lloyd's Market Association, London.

Mouroux, P. and Le Brun, B. (2006) Presentation of RISK-UE Project. *Bulletin of Earthquake Engineering*, **4** (4), 323–339.

Slingsby, A., Wood, J., Dykes, J., Clouston, D. and Foote, M. (2010) Visual analysis of sensitivity in CAT models: interactive visualisation for CAT model sensitivity analsis. Paper presented at Accuracy 2010 Symposium, Leicester.

Sweeting, P. (2011) *Financial Enterprise Risk Management*. Cambridge University Press, New York.

Swiss Re (2013) *A History of Insurance.* http://media.swissre.com/documents/150_history_of_insurance.pdf

Swiss Re (2015) Natural catastrophes and man-made disasters in 2014. *Sigma*, 2/2015.

Thoyts, R. (2010) *Insurance Theory and Practice*. Routledge, London.

Wald, D.J., Quitoriano, V., Heaton, T.H. and Kanamori, H. (1999) Relationship between peak ground acceleration, peak ground velocity, and modified Mercalli intensity in California. *Earthquake Spectra*, **15** (3), 557–564.

Woo, G. (2002) Natural catastrophe probable maximum loss. *British Actuarial Journal*, **8** (V), 943–959.

# 2

# Applications of Catastrophe Modelling

*Kirsten Mitchell-Wallace*

## 2.1 Overview

### 2.1.1   What Is Included

This chapter describes the uses and users of catastrophe modelling, focusing on its wide variety of applications and their relationships. It is primarily aimed at catastrophe model developers, catastrophe risk analysts, scientists wishing to develop catastrophe models and those interested in the wider applications of catastrophe modelling and catastrophe risk management, i.e. anyone who builds or uses catastrophe models. As a primer, it aims to reference literature wherever possible and to provide insight into the catastrophe modelling aspects of the wider topics rather than provide particular depth on the specific topics. Since catastrophe models are mostly used in the (re)insurance industry, many of the sections relate specifically to these areas, but there are

*Natural Catastrophe Risk Management and Modelling: A Practitioner's Guide*, First Edition.
Edited by Kirsten Mitchell-Wallace, Matthew Jones, John Hillier and Matthew Foote.
© 2017 John Wiley & Sons Ltd. Published 2017 by John Wiley & Sons Ltd.

also sections on their use in the public sector. The chapter is intended to be independent of any particular modelling philosophy, but for simplicity and as an illustration, it is restricted primarily to discussion of property business.

### 2.1.2 What Is Not Included

This chapter does not attempt to cover topics in detail, but to provide a main pointer to themes. It deliberately does not cover actuarial topics in depth as much material is available elsewhere as part of actuarial study courses.

It specifically excludes:

- detailed description of the mechanisms of insurance and reinsurance;
- detailed insight into the actuarial aspects, proofs or formulae;
- detail of underwriting principles and practice;
- capital allocation methodologies;
- mathematical exploration of portfolio optimization techniques;
- liability catastrophe modelling as a consequence of natural hazards, e.g. workers' compensation or group life policies;
- terrorism or other non-natural catastrophe losses.

### 2.1.3 Why Read This Chapter?

To understand the application of catastrophe models, some understanding of the commercial context is required. This chapter therefore aims to provide an overview of the broad use of catastrophe models in catastrophe risk management; primarily this use is for financial risk transfer but uses are emerging in disaster risk reduction (DRR).

## 2.2 Introduction

As discussed in Chapter 1.5, since catastrophe models evolved over 25 years ago they have become the basis for quantifying and transferring the risk associated with catastrophes across the insurance industry. This chapter focuses on the practical uses of catastrophe modelling for different user groups. The use of catastrophe models and their outputs varies significantly in different segments of the (re)insurance industry depending on the significance to, and integration with, the business process. For instance, if the organization takes risk itself, as in the case of an insurer or a reinsurer, the integration of the models into the workflow will necessarily be different than that of a regulator or a broker who have no requirement to hold capital (see Chapter 1.2) or reimburse policyholders in the event of a loss. While business purpose shapes the application of the models, application is also determined by the goal of the model users. The same output can inform risk selection (i.e. which policies to underwrite), to set limits on how much can be written in a particular area (exposure management or accumulation) or to determine that the company has is enough money to back the loss potential.

Although this chapter touches on the rapidly developing area of catastrophe models in the public sector, its main focus is their established use for financial risk transfer in the private (re)insurance industry. This chapter begins with a description of risk transfer, continues with an introduction to insurance and reinsurance, before explaining the role of catastrophe risk management as a function in this industry. The main model-related (re)insurance business activities are then considered: underwriting, pricing, accumulation control, portfolio management, portfolio optimization and event response. The use of catastrophe models in capital modelling forms an introduction to the subject of regulation in catastrophe modelling. Following this, the chapter considers the wider applications of catastrophe modelling in governmental schemes, the public sector and insurance linked securities. The chapter concludes with a discussion of uncertainty and considerations for effective decision-making using catastrophe models.

## 2.3 Risk Transfer, the Structure of the (Re)insurance Industry and Catastrophe Modelling

**Risk transfer** is the process of shifting risk from one party, such as an individual or company, to another. One can think of the risk of loss flowing from the party who has purchased insurance, the **policyholder**, to the insurance company via individual insurance policies. A group of a large number of similar policies (known as a portfolio, see Chapter 1.9.3) can then be covered by reinsurance, which transfers the risk from the insurer to the reinsurer in the same way as from the policyholder to the insurance company. This transfer is known as **cession** of the risk and cover is often attained via a reinsurance broker. Some of this risk may then be transferred further via **retrocession**, so that a portion is kept by the reinsurer and some further transferred to the retrocessionaire, who is often another reinsurer. Instead of reinsurance or retrocession, catastrophe risk may also be transferred to the financial markets by either insurers or reinsurers. As described in Chapter 1.2, for most sources of risk, such as fire or theft, pooling is an effective method of dealing with the risk. However, natural catastrophe events tend to be large and severe, i.e. impacting a great many properties (or whatever is being insured) at the same time and to a highly damaging degree. It is this correlation of risk that has created the need for specialized catastrophe risk transfer mechanisms, and the models to enable it.

The transfer of specific insurable risks for a given premium from policyholders to insurers and from insurers to reinsurance companies is critical in ensuring financial stability. This risk transfer mechanism enables better functioning of the economy as it helps preserve the continuity of businesses that are subject to unexpected shocks. Companies can use the insurance market to transfer some of these unexpected shocks to a reinsurance company, so that the capital that would have been allocated to pay for losses arising from catastrophe events can be deployed more efficiently.

The transfer of the risk due to natural catastrophes within the insurance industry is illustrated in Figure 2.1; this example is an approximate calculation based on risk sharing of losses from major natural catastrophes in 2011, a record year at the time of writing, scaling up the declared losses from reinsurers' annual reports and assumptions based on the retrocession premium triggered (Von Dahlen and Von Peter, 2012). Although these figures will vary from year to year, Figure 2.1 illustrates the value of losses transferred and the number of members in the pool available to share the risk in different parts the risk transfer chain. Globally, the reinsurance market is dominated by the United States, as shown in Figure 2.2.

Throughout the risk transfer process, adequate risk quantification is necessary to allow this transfer to be fair and equitable and ensure that the appropriate premium is charged for the risk. It is important that the price reflects the risk itself (known as being 'actuarially sound' or 'risk-reflective') rather than simply market considerations such as demand and supply. For low frequency, high severity catastrophe events where there are few, if any, data points from

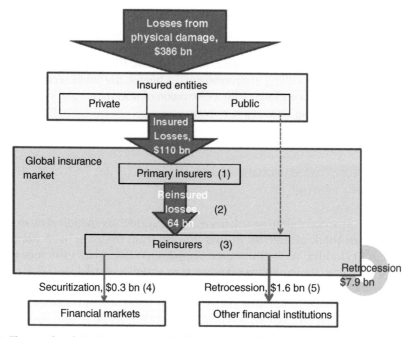

**Figure 2.1** The transfer of risk that is associated with the physical damage caused by natural catastrophes, with values in US$ from 2011, highlighting the major entities involved. See Chapter 1.2 for a description of the entities, and main text for details of the transfers. *Source*: Adapted from Von Dahlen and Von Peter (2012). Reproduced with permission of Bank for International Settlements.

experience, catastrophe models are essential tools, particularly for the most extreme and rare events. This contrasts with some other types of insurance, e.g. auto insurance, for which actuarial models based on claims data are adequate. Catastrophe models are used in determining both the amount of risk to be transferred and the price for this.

We can examine these processes, at a high level, by considering an example that illustrates some of the catastrophe modelling tasks which will happen in the different organizations

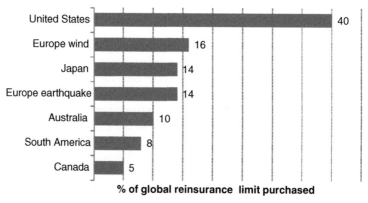

**Figure 2.2** Proportion of global reinsurance limit purchased by region, based on data from Guy Carpenter (2014). Note the peril of wind damage in Europe 'Europe wind' and earthquake risk in 'Europe Earthquake' cover slightly different geographic regions.

within the typical reinsurance purchase chain described above. In this example, we can again consider the parties shown in Figure 2.1. The first is the *insurer* who underwrites the original policies and is the organization that the policyholder (e.g. homeowner) will interact with. Next, we have a *reinsurance broker* who works on behalf of the insurer to transfer risk to one or more *reinsurers* via reinsurance policies. The broker will often provide analytical services as part of this service, and this will include catastrophe modelling for all but the smallest reinsurance brokers. An insurer may decide not to directly license specific catastrophe models from third party vendors nor develop their own internal models, but instead rely on outputs provided to them as part of a brokers' service; these outputs can originate either from vendor catastrophe models or those developed by the broker. Reinsurance companies also want to quantify the risk that they take on and will again likely use the models to do this. If they then find that they have more risk than they want to retain, they will transfer it further, either into the financial markets (via securitization with an ILS (Insurance Linked Securities) fund, described in Section 2.15) or share it with other reinsurers (known as retrocessionaires). There are many activities undertaken, and different parties may perform these depending on the circumstances, e.g. some activities of the reinsurance broker will be undertaken by the insurer or the reinsurer if no reinsurance broker is used. This list is illustrative, rather than exhaustive and gives a flavour of what will follow in this chapter. The position of an entity in the flow in Figure 2.1 is given in brackets, e.g. (1).

At an insurer (1), the catastrophe model may be used to:

- inform the risk selection, i.e. which types of properties or geographical areas to include in the insurer's portfolio;
- inform the setting of price (i.e. the premium) charged to the policyholder. Catastrophe risk is usually a small part of the original premium, e.g. 10% for windstorm for a typical UK homeowners' policy. However, larger proportions are possible; US hurricane may typically make up 60% of a Florida homeowners' policy premium;
- allocate and monitor the maximum amount of business that can be written for individual business units (i.e. their capacity);
- ensure that the business has enough capital to meet solvency regulation;
- monitor adherence to stated risk guidelines and risk policy for key peril regions and business units;
- allocate reinsurance costs to different business segments, e.g. underwriting business units, branches, or even individual risks.

At a reinsurance broker (2), catastrophe models may be used to:

- provide analysis of the client insurer's portfolio, perhaps using multiple, alternative models;
- design reinsurance or retrocession structures;
- price reinsurance or retrocession structures;
- peer review client exposure and loss profile versus other insurers;
- provide insight into the strengths and limitations of different models.

At a reinsurer (3), catastrophe models may be used to:

- design and price reinsurance structures;
- allocate and monitor the maximum amount of business that can be written for individual business units;
- target the most profitable combination of different contracts, given the company's constraints;
- ensure that there is enough money held to pay back claims and keep the company solvent.

At an insurance linked securities fund (4) and retrocessionaire (5), catastrophe models may be used to:

- price reinsurance structures;
- define and monitor variables used in determining payment factors for catastrophe bonds.

The further from the original risk in the process, the less detailed information will typically be available, so considering Figure 2.1, typically a reinsurer will have less information about the original risks than the insurance company who will, in turn, know less than the policyholder themselves.

The activities and duties of a catastrophe risk management team are explored in more detail when discussing the role of a catastrophe risk management function in Section 2.5.

## 2.4  Insurance and Reinsurance

### 2.4.1  What Is Insurance?

Insurance is an arrangement whereby one party (the insurer) promises to pay another party (the policyholder) a sum of money in the event of a loss due to a specific cause (see Chapter 1.2). A premium is charged to the policyholder in order to provide this service. Only certain risks can be insured: these risks must be measurable in monetary terms; they must be accidental from the point of view of the insured; they must be 'pure' risks (i.e. there can only be a loss and there is no possibility of a gain); and there should be a large number of similar exposures which are reasonably independent (see e.g. Diacon and Carter, 2005; Thoyts, 2010, for discussions of insurability). Additionally, the loss potential must be reasonably large so it is worthwhile for the policyholder to purchase the insurance in the first place. One of the principles of insurance is that the risk bearer (the policyholder) retains some of the risk themselves, and thus has an incentive to minimize their risk, e.g. by taking basic preventative measures (e.g. installing fire alarms). This is typically via an 'excess' or 'deductible' on the policy up to a certain amount; see Chapter 1.9.2 for the basics of these financial structures, e.g. Figure 1.5.

There are two major types of insurance company: **mutual companies** and **stock companies** (e.g. Randall, 1994). Stock companies have shareholders who have purchased stock whereas a mutual company is owned by its policyholders. A mutual company exists to provide cover to its insureds rather than to make profit. Stock companies have shareholders who expect a return and therefore the stock companies may be subject to more efficient use of capital and are more concerned with profit. Many insurers are geographically constrained in the assets they cover (e.g. regional) and may specialize in particular types of insurance (marine, property, life); this latter type is known as **mono-line insurance**. They may, particularly in the case of mutual companies, also be associated with particular affinity groups (e.g. farmers, driving instructors, teachers). Global insurers are insurers that have a multi-territory presence (e.g. Axa, Allianz, Generali, Zurich) and tend to offer multiple lines of business (**multi-line insurance**). There may be more emphasis on catastrophe modelling in stock companies and larger insurers who may have their own catastrophe management teams, although as described in Section 2.3, catastrophe modelling services are often accessed by smaller insurers through a broker if a broker is used for reinsurance purchase. It is comparatively expensive to license catastrophe models and to have appropriately skilled staff to interpret their output.

**Takaful insurance** is a particular form of insurance which is used to overcome cultural and religious objections to insurance in the Islamic world (see e.g. Thoyts, 2010). From a catastrophe modelling perspective, it is no different from any other insurance.

### 2.4.2    What Is Reinsurance?

**Reinsurance** is the insurance of insurers. Reinsurance has lots of jargon which makes it relatively inaccessible to the uninitiated, so some of the most common terms are provided here. Reinsurance is purchased as protection by the **cedant** who may have a number of motivations, including risk transfer, reducing volatility in income by preventing unexpected or unpredictable financial events (financial shocks), ensuring that no particular type of business is too dominant (balancing the risk profile) or allowing the company to write more business by increasing **solvency**. Reinsurance is generally classified as either treaty or facultative. **Facultative reinsurance** is the specific reinsurance of individual risks (e.g. large factories or other high value locations, or risks which do not fit in to the target profile of the insurance company) where the reinsurance is negotiated on a case-by-case basis (e.g. Swiss Re, 2007). In contrast, **treaty reinsurance** is an overall umbrella-type agreement that all risks written in a given year or period of a certain type (e.g. property, marine, risks in France, homeowners', earthquake, etc.) are reinsured and the treaty will automatically cover all the risks within a particular portfolio (the **book**), e.g. a homeowners' portfolio, for that time period, usually annual. Reinsurance can be further divided into **proportional** and **non-proportional** reinsurance.

In a proportional reinsurance contract, the risk is shared between the insurer and the reinsurer in proportion to the premiums, whereas in non-proportional reinsurance, it is not. **Excess of loss reinsurance** can function on a **per risk** or **per event** basis, the former is designed either to protect specific locations against possible losses and the latter is designed to protect against the accumulation of multiple losses caused by a catastrophe event. While this **catastrophe excess of loss (CAT XL)** business has the most immediate and obvious connection with catastrophe modelling, modelling is also used to quantify the catastrophe component (**catastrophe load**) for stop loss, per risk, surplus and quota share treaties. The modelling of per risk and surplus share treaties is dependent on detailed exposure data since these treaties apply at specific locations, aggregate data cannot be used. The same is true of facultative modelling, which additionally will require representation of the facultative structure (proportional or non-proportional).

In reinsurance, there are many different policy types; the most fundamental ones are presented here. Reinsurance is a broad subject, so only the major points of relevance to catastrophe modelling are discussed in this section – details of any reinsurance contract clauses are included only where directly relevant and financial accounting aspects are excluded completely. Basic introductions to reinsurance can be found in insurance books (e.g. Diacon and Carter, 2005; Thoyts, 2010) or in specialist texts (Kiln and Kiln, 2001; Bellerose, 2003; Riley, 2013), as well as industry training material (Charted Insurance Institute (CII), or Chartered Property Casualty Underwriter (CPCU), e.g. Harrison, 2010), Associate in Reinsurance (A Re) or publications by Swiss and Munich Re (e.g. Munich Re, 2004; Swiss Re, 1997; 2002; 2003; 2014).

The discussion of treaty types will follow the structure and hierarchy of the boxes in Figure 2.3 from top to bottom and left to right, i.e. proportional including quota share and surplus, followed by non-proportional including per risk per event, then aggregate excess of loss and stop loss.

### 2.4.2.1    Proportional Treaties

Proportional treaties are also known as **pro rata** treaties. For proportional business, the premium is shared, or 'pro rated' between the insurer and reinsurer in proportion to the risk. However, since the insurer has extra costs when acquiring and maintaining the original business, the reinsurer will compensate the insurer for these on top. This compensation is known as **ceding commission**. Additionally, there may be another commission returning money to the insurer if the loss results of the insurer are low and the treaty is profitable for the reinsurer. This is known as **profit commission**. Alternatively, the ceding commission may be adjusted to reflect the actual profitability of the treaty (**sliding scale commission**); profitability

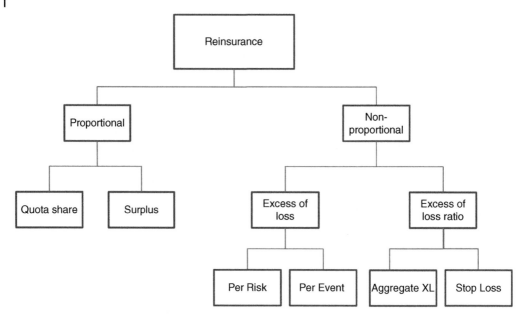

**Figure 2.3** Types of treaty reinsurance, in particular, illustrating the split between proportional and non-proportional reinsurance detailed in the text. Aggregate XL and Stop Loss are identical except that Aggregate XL is expressed in absolute terms.

is covered in Section 2.6.2.3. Proportional treaties tend to be long-term partnerships between the reinsured and the reinsurer.

A **risk** in the context of (re)insurance often is used to mean an insured object (see Chapter 1.2 for insurance basics), i.e. a policy which may apply at one or many locations.

### 2.4.2.1.1 Quota Share

In a **quota share** treaty, all business is shared in a fixed proportion of the original risk, i.e. a ratio set on the original sum insured (liability) which is applied to each and every risk in the entire portfolio. The reinsurer receives a percentage of the original premiums and is liable for the same percentage of the losses. The sharing of liability is illustrated in Figure 2.4, which shows that 75% of each risk is retained and 25% ceded. This is known as a 25% quota share.

It is comparatively rare for quota share treaties (QS) to cover natural perils alone but there are instances of this (e.g. Earthquake QS in Slovenia, Storm QS in Germany, wind perils in Florida). However, in general, QS treaties will contain an element of catastrophe cover as they reimburse the cedant for a proportion of *all* losses incurred, whatever the cause. While QS treaties can therefore provide unlimited catastrophe protection, an **event limit** is usually imposed to cap the potential loss from one catastrophe event. For example in 2005 in the United States, some reinsurers had very large losses from quota shares where no event limit had been applied. The very large, uncapped, individual losses from each event which were then even further aggregated across the season as a result of multiple landfalling hurricanes that season. Quota share treaties work well where the risks are reasonably homogeneous, i.e. similar in nature and loss potential, so that the insurer is not paying for unnecessary cover where it could easily retain the risk itself. For example, if there was a mixture of small, low-value and very large, high-value properties, the proportion of the low-value properties would be a lower absolute amount compared to the high-value ones and so might be more easily retained by the insurer. In this case, and for this reason, a quota share would not be the best solution.

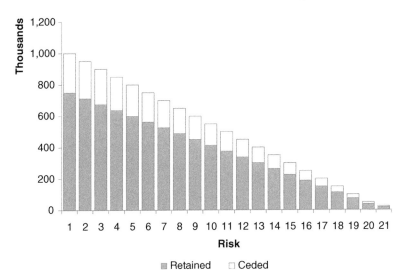

Figure 2.4  A 25% quota share treaty. The x-axis shows the insured object or risk number, the y-axis shows the sum insured, the retained part is grey and the transferred (ceded) part is white.

Quota share treaties tend to be used for new companies or new products (lines of business) or are commonly a method used to access additional capital and so underwrite more business. The sharing of premium and losses encourages knowledge transfer between the reinsurer and the cedant, e.g. if the reinsurer has particular insight into the risk in that region, they may share that with the cedant as an additional benefit. This provision of this insight, taken together with the ceding commission, explains the popularity of quota share treaties for new business.

### 2.4.2.1.2 *Surplus*

In a surplus treaty, also known as a surplus share treaty, the insurer chooses a set amount to retain on every risk, called a fixed retention. This allows reinsurance (or cession) of variable proportions of the original risks, depending on their original sum insured (or **liability**). The same fixed amount will be a larger proportion of smaller sums insured and a smaller proportion of larger sums insured and it is the ratio of the value of retention to the value of the sum insured that defines the cover on each risk. The amount of cover (known as the **capacity** of the treaty) is generally expressed as a number of **lines** where one line is the amount of the retention. For example, if the insured decides that 150,000 is the maximum it would like to retain on every risk then a three-line treaty offering 450,000 coverage (capacity) may be chosen (Figure 2.5). The choice of the number of lines depends on the number of risks of particular liability, since again the insurer wants to make sure it has enough cover, but is not paying for too much, for instance, by paying for cover for only a few risks.

In the case of a surplus, the reinsurer has more liability on the larger risks, which tends to make surplus treaties more volatile from a reinsurer's perspective compared to QS. The surplus treaty helps the insurance company by both decreasing its liabilities and increasing the value of its assets via the additional income from the commission. Surplus reinsurance should not be confused with surplus lines insurance; surplus lines insurance is US-specific terminology for insurance purchased from an insurer not regulated by the State where the policyholder resides.

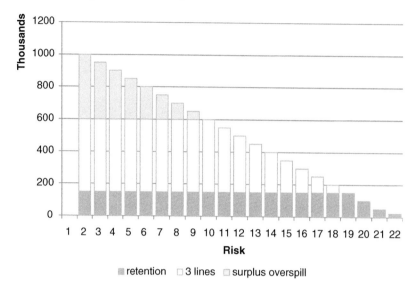

**Figure 2.5** A three-line surplus treaty with line size of $150,000. The x-axis shows the individual property and the y-axis the sum insured. The retention is taken by the cedant, the three lines form the reinsurance coverage where losses are taken by the reinsurer, and the overspill also falls to the cedant. The retention is grey and the ceded part is white.

### 2.4.2.2 Non-Proportional Treaties

In **non-proportional** treaties, there is no sharing of premium and loss. Instead, the reinsurer assumes the losses above the retention of the treaty in return for a premium. Although this structure may look like a surplus treaty, no fixed proportion of loss is paid. Unlike in a surplus treaty where the relationship of the retention to the sums insured determines the payment, the payment is determined *solely by the size of the loss*. Excess of loss treaties may be subject to reinstatements which means that after a loss the cover can be reset ('reinstated') to the original amount, usually in exchange for additional premium. This allows the cedant protection for more than one loss incident. When a loss is paid, the amount of cover that remains is the original amount less the paid amount. However, if the contract features a **reinstatement**, the cover is automatically reset (or reinstated) to its original level via the payment of this **reinstatement premium** which is usually calculated using the relationship between the amount of cover needed and the original limit (known as pro-rata by amount).

A non-proportional cover will be structured in layers which relate to the loss amount. The layers will be expressed in the form *limit xs attachment* point i.e. 'b-a xs a', where the b-a is the limit, for example, a cover which starts at 10 and offers a limit of 10 will be known as a 10 xs 10 (see Figure 2.6). The loss value at which the programme starts is known as the **attachment point** (a) and the point at which the cover ends is known as the **exhaustion point** (d). The area below the attachment point is the **retention** (deductible) paid by the cedant and the area above the exhaustion (known as the **overspill**) is also retained by the cedant. **Coinsurance** is the sharing of the reinsurance between multiple reinsurers or between the insurer and the reinsurer. In this example, it refers to the sharing between insurer and reinsurer. Coinsurance between insurer and reinsurer encourages good risk management on behalf of the insurer who will also have a significant share in any loss. Markets such as Lloyd's exist so that insurance shared can be shared among many insurers via coinsurance.

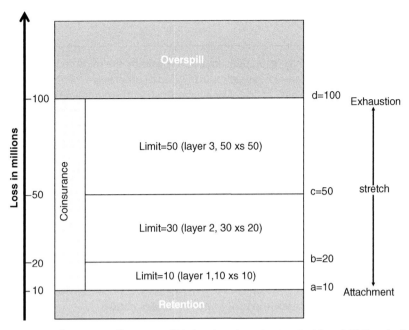

**Figure 2.6** Structure of an excess of loss treaty. This is a three-layer treaty, attaching at 10 M and exhausting at 100 M, with a stretch of 90 M.

### 2.4.2.2.1 Per Risk Excess of Loss

In this type of treaty, each loss is treated separately on a per risk basis, i.e. the retention and limit as shown in Figure 2.6 are applied to each individual risk/policy. This treaty is designed to cover individual locations against losses, however they occur, e.g. fire, explosion or flood. Multiple risks can only be considered up to an aggregate limit defined by the number of reinstatements (e.g. if there are nine reinstatements, this would cover ten risks). This cover is designed to minimize the number of large losses an insurer would experience. Decreasing the loss frequency allows stabilization of results while allowing the insurer to take on larger risks than would otherwise be possible.

### 2.4.2.2.2 Per Event or Per Occurrence Excess of Loss, i.e., Catastrophe Excess of Loss

Catastrophe excess of loss (CAT XL) is also known as per event or per occurrence excess of loss. As the name implies, this treaty is designed to protect against an aggregation of losses in a particular event – a natural or man-made catastrophe – where multiple risks are affected at once. This is usually across a wide area known as the event footprint (see Chapter 1.8.1) for an event which is unusually severe. All of the claims from the event are added together and if this total loss is more than the retention, a payment will be made. If the loss is above the exhaustion point, a payment of the full amount of the programme is made up to the value of the limit. Particularly with CAT XL, the reinsured is encouraged to retain some of the programme by either significant retention underneath the attachment point or coinsurance – either method leads to the insurer participating in losses which is thought to encourage better risk management and selection.

As with the per risk excess of loss, the treaty is divided into layers in order to create smaller packages (or quanta) of risk, which can be more easily transferred between reinsurer and client. The catastrophe excess of loss contract reduces volatility in the insurer's results. The retention and limit as shown in Figure 2.6 are applied to the sum of the applicable claims across the portfolio for that event; this sum of applicable claims is known as the **Ultimate Net Loss**. Ultimate Net Loss has a particular form, detailed in Box 2.1. In simple terms, it usually is the money lost by the insurer reduced for any gains they have. Losses include the amount of the catastrophe loss itself and certain associated expenses, whereas gains could be money that

---

**Box 2.1  Ultimate Net Loss**

The term Ultimate Net Loss shall mean the sum actually paid by the reinsured in respect of any loss occurrence including expenses of litigation, if any, and all other loss expense of the reinsured (excluding, however, office expenses and salaries of officials of the company), but salvages and recoveries, including recoveries from other reinsurances, shall first be deducted from such loss to arrive at the amount of liability, if any, attaching hereunder.

All salvages, recoveries or payments recovered or received subsequent to any loss settlement hereunder shall be applied as if recovered and received prior to the aforesaid settlement, and all necessary adjustments shall be made to the parties hereto.

Nothing in this clause shall be construed as meaning that a recovery cannot be made hereunder until the reinsured's ultimate net loss has been ascertained (Riley, 2013, p. 15).

---

the insurer receives from another reinsurance contract which applies before the one considered or income from salvage of property (e.g. the sale of partially damaged goods).

Two important features of a catastrophe excess of loss contract relate to the catastrophe event definition. These are the **two risk warranty** and the **hours clause**.

The **two risk warranty** is a condition within the reinsurance contract that says that at least two distinct risks must be affected in order for the catastrophe cover to be activated. This ensures that the reinsurance covers an aggregation of risk, that is multiple risks being affected at once rather than being used to cover individual risks, which is the function of the per risk treaty. The **hours clause** or, more properly, the **loss occurrence clause** is also an important contract clause relevant to catastrophe modelling. This clause provides the definition of the duration and extent of the event and can be found in the contract in the definition of loss occurrence. This allows the cedant to define a time window over which it can aggregate the claims. Time windows of 72 hours, 96 hours, 168 or 504 hours are common intervals chosen, although these vary by market and by the nature of the peril (e.g. UK flood tends to be 504 hours, whereas US hurricane may be 96 hours). If the event lasts longer than the specified window, the cedant will choose to aggregate their claims over the set of specified hours which give the best recovery from the reinsurance treaty considering its structure and retentions; this depends on the amounts of loss in the different intervals compared to the layers and attachment point of the treaty. A separate retention will apply to each event's loss (see Figure 2.6). Any difference between geophysical definitions assumed within the model and event definition within the reinsurance contract itself should be understood and accounted for in further use of the modelling output. For example, peak gust windspeeds are commonly integrated across the entire lifecycle of a storm in wind models (see e.g. Chapter 3.2.4) and the hours clause will probably specify a different time period from this storm lifetime in the contract; this may, or may not, be significant, given the geographical scope of the contract, type of storm and the contract hours compared to the likely duration of the event considering the underlying exposure. Some newer modelling software will allow the specification of the contractual hours clause within the software and some older software may provide timestamps on events. Assumptions and definitions may vary by vendor model, so care is required. Hours clauses may also be present in primary insurance policies covering multiple locations.

Catastrophe excess of loss contracts may contain an additional feature related to catastrophe modelling. Catastrophe modelling output can be used in adjusting the amount paid by the cedant for the catastrophe coverage when it is related to the level of risk (known as **risk-adjusted premium**). A minimum premium is charged at the start of the year and the final balance is adjusted to reflect the amount of exposure over the duration of the contract. Often this adjustment is on the basis of the amount of premium written by the cedant, but the adjustment can sometimes also be measured on the basis of changes in the modelled output, e.g. a rate applied to the modelled losses. For instance, taking a reinsurance programme which attaches at around the 1-in-25-year return period and exhausts at the 1-in-100-year return period, the average of the OEP losses at the 1-in-100-year and 1-in-25-year levels might be used. The cedant may wish to further average these losses over two models, to decrease the risk of sensitivities in one model affecting the result.

### 2.4.2.2.3 Aggregate Excess of Loss

An aggregate excess of loss treaty protects against an aggregation of losses in a period, usually a year, and stabilizes the insurer's results. The retention and limit apply to the total loss across the year rather than to individual losses. This treaty therefore provides the potential to protect against one catastrophe, but also multiple catastrophes which could occur in the year.

### 2.4.2.2.4 Stop Loss

This type of treaty is an excess of **loss ratio** cover and again protects against an aggregation of events in a year leading to a fluctuation in the cedant's results. **Loss ratio** is the loss per unit premium and is a common metric used in describing insurance company results. The cover is almost identical to an aggregate excess of loss but is expressed in terms of loss ratio, where the reinsurer will provide cover if the insurer's results fall between a lower and upper bound. For example, the reinsurer may take 80% of the claims between a loss ratio of 90% and 120%.

### 2.4.2.2.5 Other Treaty Types

There are various other cover types which may be relevant to the catastrophe risk analyst such as those where the nature of the treaty can change, e.g. drop down covers where the retention (attachment) decreases when the cover is reinstated after an event, top and drop covers which can act either as a top layer or become a reinstatement back up for a lower layer. Umbrella covers which protect against the accumulation of retentions or overspills are also common. Unfortunately, these more complex covers are not covered in most reinsurance primers and readers are directed to the internet or discussions with underwriting or broking colleagues, if possible.

### 2.4.2.3 Reinsurance Programmes

As illustrated above, different types of reinsurance are used to achieve different ends, e.g. catastrophe insurance is to protect against the aggregation of multiple losses whereas per risk excess of loss is used to protect against individual large losses. An insurer is therefore likely to have more than one type of reinsurance, and have a combined set of contracts known as a **reinsurance programme** (see Swiss Re, 1999). Particular contract types are more dominant in different markets. Catastrophe excess of loss tends to be found in more established insurance markets and less often in developing markets: here insurers are fond of proportional reinsurance which provides capacity. However, in these markets there is often also a CAT XL protecting the retention of the proportional programme.

In a reinsurance programme, there is an order in which the reinsurance treaties apply. A treaty which applies before another treaty is said to 'inure to the benefit' of that treaty. Facultative treaties tend to apply first, followed by proportional treaties (surplus then quota share) followed by per risk XL treaties then CAT XL treaties. Stop loss applies last of all. However, the order may be different from this and must be understood. It is important to know this order, especially as in most cases the reinsurance programmes are shared by different reinsurers. A programme does not have to have all of these kinds of treaties, although many do. In the example in Figure 2.7, the 40% QS applies first, i.e. it insures to the benefit of the CAT XL. The CAT XL applies to the 60% of the exposure that is left.

The ability of catastrophe models to represent these different treaty types is variable. Catastrophe models are known to be particularly poor in representing per risk treaties because of limitations in their financial models (see Chapter 4), historically caused by the need to introduce simplifying assumptions in order to constrain computer run times. Many practitioners are also sceptical of the use of models at individual locations, as required for modelling of per

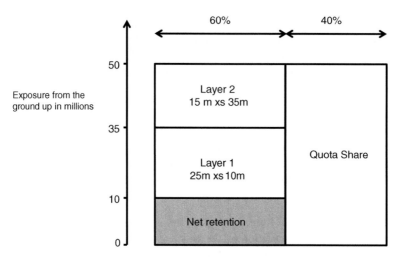

**Figure 2.7** A simplified reinsurance programme showing a 40% quota share and a two-layer catastrophe excess of loss programme. This is known as an XL on retention, since it covers the retention of the quota share.

risk and surplus treaties. New catastrophe models are promising to tackle these issues by taking advantage of parallel processing and cloud computing to provide simulation at a location level. Treaties where aggregation across multiple perils is required (e.g. stop loss, aggregates or quota share) can also be problematic as sometimes only a subset of the perils will be modelled and modelling must be combined with actuarial approaches. It is essential that the catastrophe risk analyst is able to understand the flow of loss through the different financial perspectives of the catastrophe model (see Chapter 1.8.4), i.e. how the ground up, or pure loss is modified by the application of financial terms (deductibles, limits, coinsurance), to the application of underlying insurances and through to the treaty loss itself.

## 2.5 Catastrophe Risk Management and Catastrophe Modelling

*Kirsten Mitchell-Wallace and Matthew Foote*

### 2.5.1 What Are Catastrophe Risk Management and Exposure Management?

There are many definitions of **catastrophe risk management** and **exposure management**. What is called 'exposure management' by some may be called 'catastrophe management' or 'catastrophe risk management' by others, and often the terms are used to describe the same job or function. The term 'exposure management' is more commonly used in London and the Lloyd's market than elsewhere in the world where the term 'catastrophe risk management' is more usual. Although, as will be described, these functions cover more than catastrophe modelling, catastrophe models are critical tools to help manage the exposure and risk from natural catastrophes within a (re)insurance organization. From this point on, the term catastrophe risk management will be used to include exposure management unless specifically referring to Lloyd's.

As (re)insurers have moved from an empirical, experience-based claims pricing culture to a more statistically-based quantitative risk framework, the role of the catastrophe risk manager has grown to encompass many activities, including:

- underwriting decision support – providing analytical and quantitative information to support underwriting decision processes, including pricing and return on capital (see Section 2.6, in particular 2.6.2.3) and monitoring of underwriting limits (see Section 2.6.1);

- exposure data accumulation and management – identifying, monitoring and communicating patterns of exposure distribution and aggregation, whether spatially (i.e. exposure accumulations by defined cluster or hazard zones, ZIP or postcode, region or territory), or across classes (i.e. correlations of exposure across the business), at varying scales and business focus points (see Section 2.7);
- post-event loss estimation, claims analysis and model adjustment (see Section 2.9);
- contributing to reinsurance purchase and other risk transfer strategies (see Section 2.12);
- integration into the organization's capital requirement via internal capital models (see Section 2.10);
- exposure governance, including regulatory submissions (see Section 2.11);
- supporting and advising on strategic planning for the organization's growth and sustainable business, including underwriting strategies, and reinsurance optimization (see Section 2.8);
- assessment of risk materiality and the identification/quantification of key emerging and non-modelled risks (see Chapter 5.4.7);
- adjustment of models and analytics to be as representative of the business as possible, including modifications and calibrations or adjustments to models to the specifics of the insured risks (see Chapter 5.5);
- validation and evidencing of quantitative and analytical approaches taken by the organization to manage exposure (see Chapter 5.3.1);
- raising awareness of model limitations and proposing strategies to mitigate these (see Chapter 5);
- model selection and licensing decisions (see Chapters 5.3.4 and 5.4.4);
- liaison with model vendors and IT to ensure operational testing of models and platforms (see Chapter 1.8.5);
- broader hazard and risk research to support organizational risk awareness.

Many catastrophe risk management techniques are structured around requirements defined by market regulators (see Section 2.11). For example, Lloyd's define 'minimum standards' for exposure management (Lloyd's, 2015a), which state the requirements for all entities operating in the Lloyd's market (termed managing agents and syndicates). Entities are expected not only to *ensure effective systems and controls for managing exposure for all classes of business are in place* but also to show clear *evidence of their application and governance*.

Although there is a quantitative basis to catastrophe risk management, there is also significant uncertainty, 'fuzziness' of information, imprecision and qualitative assessment. This is discussed in more detail in Section 2.16.1. A catastrophe risk manager must deal with, and express, this uncertainty effectively to those responsible for risk decision-making within the business. Equally, catastrophe risk management must reflect the aims of the business and its underwriting decision processes. As noted previously, the role is also closely allied to other disciplines and functions found within most insurance organizations, including: actuarial and capital modelling, risk management, information and data operations, regulatory and compliance, auditing and risk transfer management.

### 2.5.2 Catastrophe Risk Management Metrics

Particular metrics are used to quantify and communicate the level of risk that an insurance entity is exposed to. Metrics associated with catastrophe modelling are described in Chapter 1.10.1.

A term which often appears in catastrophe risk management is **Probable Maximum Loss (PML)**. Traditionally, this has been defined as the value of the largest loss that is considered *likely* to result from an event. In this context, it is typically the loss assuming the normal functioning of protective features (e.g. firewalls, static flood defences) including the proper functioning of most, but perhaps not all, active responses (e.g. sprinklers, temporary flood barriers erected, timely flood alerts). However, since probabilistic catastrophe models have been in use, Probable Maximum

Loss (PML) is also commonly used to refer to exceedance probability losses at particular thresholds e.g. often the 1-in-100 OEP VaR or TVaR loss for wind perils or the 1-in 250 OEP VaR or TVaR loss for earthquake perils. However, confusingly it can be also used for be any exceedance probability loss e.g. the "1-in-200 or 1-in-50 PML". The term is used inconsistently, but most often as shorthand to represent a metric for catastrophe exposure monitoring, especially in the underwriting community. Care should be taken when using the term PML to ensure the definition in any particular context is clearly understood. In this chapter, PML is used to mean an exceedance probability loss at a key, defined and communicated, threshold used for catastrophe management.

A catastrophe risk management risk metric can be viewed as a form of **Key Risk Indicator** (KRI) as applied in broader Operational Risk (Lam, 2014) or other risk management environments, with the same three desirable criteria:

- The metric should be quantifiable.
- It should allow comparability, particularly over time to enable trend analysis.
- It must be readily reportable and representative of the risk in question, enabling and supporting informed decisions.

Metrics should represent the form and scale of risk, allow prediction or forecasting and account for the uncertainties inherent in the derivation of the metric. The metrics should also reflect the business considerations discussed later in this chapter.

These metrics form ready summaries of the company's performance for stakeholders, such as customers, regulators, shareholders, or reinsurers. Increasingly, these metrics are considered not only in terms of absolute value, but rather with respect to the uncertainties associated with the analytical estimates produced; as yet, however, there is no standardization of uncertainty measures.

### 2.5.3 Catastrophe Risk Management Data

Catastrophe risk management consumes and creates information to represent risk as effectively as possible. *Data quality* (covered by Actuarial Standards Board, 2004, ASOP 23) is critical since the appropriateness of data used to represent the risks together with accurate interpretation of the policy coverage determine the confidence in results. In particular, the feedback between underwriter, catastrophe risk analyst and risk management function determines the validity and appropriateness of the data used. Coverage, completeness and representativeness of the data at any time are critical and form a large part of the day-to-day considerations of catastrophe risk professionals.

The lack of universally adopted data standards together with varying approaches to modelling almost always results in sub-optimal data quality. The role of the catastrophe management function is, therefore, to clean and validate the raw data provided to ensure it is fit for purpose. Sometimes the data may also be augmented, i.e. filled in using other data sets, to estimate missing parameters.

Increasingly, holistic risk assessment and management of exposure across the organization require catastrophe risk managers to have processes in place which enable consideration of the significance (**materiality**) of each source of risk, and an ability to quantify that risk, whether via available models or based on best estimates.

In particular, the consideration of **non-modelled risk (NMR)** (see Chapter 5.4.7) and the interface between **emerging risk** and modelled risk is becoming important, with recent guidance from the Association of British Insurers (2014) highlighting points in relation to catastrophe risk, including non-modelled perils, regions, and secondary impacts and potential correlations not explicitly covered by existing models.

Treatment of non-modelled risk is part of the role of the exposure or catastrophe manager, this is covered in detail in Chapter 5.4.7 which outlines the definition, identification and quantification of non-modelled risk.

### 2.5.4 Exposure Data

Exposure data is the primary input to catastrophe models and an element over which the user has influence. The need for, and benefit of, better data capture (where 'better' means more timely, complete, accurate and appropriate) have been an ongoing theme in the industry since the inception of catastrophe modelling and gained momentum after Hurricane Katrina in 2005 (see also Section 2.11.1.2). Before the advent of catastrophe models, premium information was the portfolio information of interest to insurers since it represented income and could be used as a proxy for exposure. Sums insured were not systematically collected since they were not useful. Although the risk characteristics and locations of large individual risks may have been known at the time of underwriting, these were also not recorded in a referenced system, not available nor accessible.

#### 2.5.4.1 Data Quality and Exposure Analysis

The concept of 'garbage in, garbage out' is easily communicated and well understood when considering data quality. The desire for representative modelling outputs has prompted heavy investment in data capture and recording systems, particularly in large multinational insurers but also in markets where catastrophe modelling was adopted comparatively early (e.g. in the United States and the United Kingdom). Regulatory pressure and an emphasis on data quality reinforce this trend (see also Section 2.11).

The quality of data recorded varies greatly in different parts of the world and tends to be lower in areas where catastrophe modelling is less mature. For instance, in some reinsurance submissions, only aggregates per zone will be received with no detail even on whether the exposure is residential or commercial. In developing markets, it is also important to benchmark data against premiums to see that the values make sense and are consistent. Of course, this must be done with care since the majority of the insurance premium is for the fire peril, but consistency and patterns should be evident within a market. Databases of building stock can be useful in benchmarking assumptions about sums insured, occupancy and construction, but again care must be taken when considering the difference between the insured building stock and the total building stock, given that the insured building stock will tend to be of a higher quality (or at least more expensive, which may imply better building standards). How the data are best coded will also depend on the view of the default assumptions within the model (see Chapter 4.5). If, for example, the model assumptions applied for 'unknown' construction are found to be unrepresentative of the insured building stock, different assumptions which better reflect the views of catastrophe risk analyst or underwriter will be taken. Some more basic models may not include 'unknown' vulnerability functions.

Considerable insight can be obtained from interrogation of exposure data; a profile of exposed limit by physical risk characteristics and geographical area will provide a summary picture of the shape of a portfolio. This can be compared with expectations and a quick visual check can ensure adherence to any relationship to guidelines restricting the addition of business in a particular area. Such profiles allow easier management of the sums insured (**aggregates**) see also Section 2.7.1.1. For a reinsurer or broker, comparison of the limit profiles year-on-year allows insight into changes in underlying exposures, how the model results might be expected to change, and provides an opportunity to sense-check the development of the portfolio and whether the client has stuck to the communicated business plan (e.g. the client communicated that they wished to grow in Texas for small commercial business. Has this happened?). Year-on-year comparison also provides a check on the data as it provides insight into the variability of data, with implications for its quality and reliability. Data quality algorithms can, for instance, be developed on the basis of exposure data to facilitate sense-checking, e.g. it is unlikely that in a reasonably stable portfolio in terms of sum insured that in one year it will mostly contain one-storey properties and in the following year mostly two-storeys ones. This type of change should raise flags for discussion. Such summaries of key characteristics (e.g. construction, year built) also allow the opportunity to benchmark the exposures against a set of identified peers and against modelling companies' available industry exposure databases. Exposure market shares can be calculated to compare with historical and modelled loss market shares. Visualization of these data is a powerful communication tool. Such summaries also allow identification of **bulk coding**, i.e. where a set of standard assumptions has been used for a large proportion of the portfolio. If, for instance, 85% of the commercial exposure has been coded as 1970, four-storey, reinforced concrete, either the company is unusually specific in what they write or are bulk coding as a default option. Some catastrophe models provide sophisticated exposure analysis in their catastrophe modelling tools. For others, these tools may have been developed in a standard template outside the models. The form that these tools take is less important than the opportunity for discussion and insight from examining the summary data.

### 2.5.4.2  Practicalities of Exposure Data Transformation

Data need to be transformed from the data held within insurers' systems into data that can be digested by the catastrophe model. Given that different entries for occupancy, construction, age and height may result in the use of different vulnerability functions (see Chapter 1.8.2 and Chapter 4.5), it is important that the original data describing the risk are mapped into the model input using a consistent approach. For a large organization, it will be important to have standard mapping tables which describes how the attributes of the original data are transformed into inputs for each model used. Different models recognize different characteristics. Although attempts have been made to share data schemas between certain model vendors, time and effort are often still required when transforming exposure in one model's format into data for input into another model.

Even where insurers have sophisticated policy management systems, the exposure data (often known as the **schedule of risks**) must be carefully cleaned for transformation into an exposure file. The records must be examined in detail to look for duplicate data (this may be repeated rows or two records where almost identical data are entered), erroneous data (e.g. negative or zero sums insured) or nonsensical data (e.g. data where the deductible is larger than the sum insured). Average sums insured can be benchmarked against other portfolios or rebuild cost databases.

In some cases, especially for global or larger regional insurers, this data transformation will be conducted in-house. However, especially for smaller insurance companies, this data transformation will be a service either outsourced to a third party supplier, or offered by the reinsurance broker and related to the placement of the reinsurance contract. Often where data quality is poor (e.g. less developed or developing markets), the aggregates will often be provided direct to the reinsurer for modelling by them.

Conversion of raw exposure data into catastrophe model input can be difficult even for standard exposures. As well as assumptions about which are the most appropriate codes to represent the primary risk characteristics (occupancy, construction, year built, number of stories, see Chapter 1.9.1.3), the sums insured and policy terms must be applied with care. As a principle, all definitions within the model should be compared with those used within a specific organization to identify and minimize differences; this is also true for exposure definitions.

There may be multiple, valid ways to code the same underlying data describing the physical risks, particularly where the data are not comprehensive. This may give significantly different results because although most interpretations are credible, different vulnerability functions are used, for instance, using reinforced concrete as the predominant construction type rather than using the modeller's default for unknown construction (see Chapter 4.5).

Equally valid ways to code the underlying data may exist be because the categories used in the policy system of record do not exactly match up with those in the model (i.e. different classifications) or because certain data are unknown at the time of underwriting. If the occupancies have existing coding (e.g. SIC or NAICS in the United States: http://www.ehso. com/siccodes.php; http://www.ehso.com/naics.php), analysts should be aware of any differences for codes which have been developed for fire rating and those for catastrophe perils (e.g. codes with significant fire resistance differences may not be that different for wind). There may be risk mitigation measures which can also be coded (sometimes known as secondary modifiers, see Chapter 1.9.1.3).

For an insurer, especially of large complex commercial exposures, reading the original policy forms is an important part of understanding the best way of coding the exposure. As described in Chapter 1.9.2, deductibles may apply at policy, location or coverage, have minimum or maximum values, be set as a percentage of the loss or a fixed value, as well as varying by peril. Limits and sub-limits face similar complexities and workarounds may be needed to enter the original policy terms in the model. Decisions such as whether to apply deductibles and limits on a location (i.e. at a particular site) or policy basis will lead to differences in results. Insurance policy terms, such as debris removal and guaranteed replacement value, require additional loading since these will not be modelled as part of the standard approach. Ultimately the scope of what is modelled must match the scope of the coverage offered under the original policy. This requires collaboration between the catastrophe risk analysts and their underwriting colleagues.

**Geocoding** (see Chapter 1.9.1.1) can also make a very large difference to modelled loss estimates. Sometimes data will be ungeocoded; the geographical attributes provided do not allow a georeference to be found. In these cases, assumptions must be made to include this exposure. Depending on the portfolio modelled, an analyst might choose to distribute the ungeocoded exposure in the same proportions as the geocoded exposure (e.g. for a relatively well-spread portfolio), to allocate the exposure to the most usual location (e.g. for a less well-spread portfolio of larger risks), to allocate the exposure to the area of highest risk (e.g. if extreme conservatism is regarded as an appropriate penalty for insufficient data quality), or in some cases to allow the model to disaggregate the exposure or disaggregate it in proportion to the overall industry data (e.g. if it is a 'standard' portfolio for a particular type). Some models include geocoding software, but sometimes geocoding will happen externally to the model; in these cases, care must be taken over falsely precise geocoding. For example, lower resolution data at CRESTA centroid may appear to be at latitude-longitude level accuracy as it has been pre-geocoded, giving the impression that more is known about the location of the risks than actually is, and 'better' geocoding may be rewarded with a lower uncertainty. Alternatively, pre-geocoded data may be geocoded to an improper location, e.g. the middle of a bay. GIS can be used to sense-check the data; it will quickly be visually evident if many points lie on top of each other or the latitudes and longitudes have been reversed. Free software can be very useful for sense-checking the location and risk attributes of the exposure with another source. Increasingly, vendors are integrating GIS functionality into their platforms. This is particularly recommended for high value risks, for which a change in location can have a major impact on the modelled loss.

### 2.5.4.3 Non-Standard Exposures

Modelling and coding of non-standard exposures are much more challenging than for small, simplistic, static, land-based properties, such as houses or offices. However, once exposure can be recognized by the catastrophe model, the event set framework can be used to consider losses across different classes of exposure (see Section 2.7.1.2). However, if the workarounds used to put non-standard exposures into this framework are very large, the results must be treated carefully as the results may be misleading, depending on their use.

Table 2.1 compares what might be considered more standard exposures (i.e. the main types of residential, commercial and industrial property) with other classes of exposure, which may require specific consideration by the catastrophe risk analyst.

Even though industrial is considered reasonably 'standard', larger industrial facilities have unique attributes which can be difficult to capture, e.g. they may cover a large area of many square kilometres with multiple buildings with different exposure characteristics, including

Table 2.1  Summary of exposure characteristics for different exposure types.

| Class of business | Data availability and quality | Model availability | Exposure characteristics | Key assumptions and generalizations | Strengths | Weaknesses |
|---|---|---|---|---|---|---|
| Residential/ Commercial/ Industrial Property | Variable by territory and class – high quality in developed markets, poorer in emerging markets | Variable, depending on region and peril, most common in highly developed markets | Mainly static assets generally split into buildings/ infrastructure, contents and business interruption | Risks may not be accurately characterized by available model attribute options Insured values may be inaccurate Business interruption risk is difficult to quantify at the location level | Residential and commercial exposure is relatively well captured for larger developed markets | Data quality in emerging markets (all sub-classes), and for industrial classes |
| Power/ Onshore Energy | High, typically high resolution, attribute rich for main plants, less accurate for transmission | Variable, depending on region and peril | Static assets, but variable characteristics and location quality (e.g. pipelines) | Power station construction matches model vulnerability Business interruption is difficult to quantify | Engineering basis for risk, high resolution georeferencing for main sites | Variability of data quality between generators, transmission, distribution |
| Marine Cargo | Variable | Poor, largely reliant on onshore models | Highly mobile, wide variability of asset type and vulnerability, varying length contracts | Property model vulnerability classes applied to cargo, poor calibration of loss to historical | US cargo exposures, main ports | Commodity and trading risks, limit based insured values |
| Marine Specie | Highly variable | Poor, largely reliant on onshore models | Wide variability of asset type and vulnerability, varying length contracts | Property model vulnerability classes applied | US, European model coverage | Lack of detailed vulnerability data |
| Marine Hull | Very poor – some for hull construction | Very poor - dependent on onshore models suitability | Highly variable – time and space | No locational capture of oceanic shipping locations | US Gulf of Mexico | Sparse data coverage, inability to represent transit |

| | | | | | |
|---|---|---|---|---|---|
| Offshore Energy | Good (international standards applied) | Limited to particular regions and perils, such as Gulf of Mexico | Complex, additional risk characteristics i.e. control of well, debris removal, liability | Risk modifiers are captured, accumulation control | US Gulf of Mexico | Limited coverage of models elsewhere |
| Aviation (hull) | Limited | Limited | Non-static, variable | Global coverage | International aviation flight cycle data | Time-varying location specific data |
| Crop/ agricultural | Variable- depends on insurance scheme and country | Limited geographical availability (US, China, India) | Time/space variant, crop/livestock type, growing season, historical yields | Time varying (planting, cropping and growth cycle) | National/international crop yield/livestock & land use data, climate modelling | Pest risks, aquaculture, irrigation |
| Auto/Motor | Variable-vehicle years, numbers, location difficult | Good | Time-varying, age of driver, distance, usage | Time-varying location | Comprehensive data collection (vehicle statistics at national scale | Time-varying location specific |

office blocks, manufacturing areas and warehouses. In an ideal situation, the different buildings types would be entered individually so as to reflect their relative rebuild costs, differing occupancy and construction types and also be geocoded individually. In some models, there are also specific detailed vulnerability functions which differentiate the risk for specific types of engineered structures. The mapping of these and best practice around their use should be studied carefully for each model as uninformed use can potentially have large unintended consequences of outputs.

### Static Versus Non-Static Exposure

Most catastrophe models assume that exposure can be characterized at a static location, whether that location is a point on the ground or an area such as a postcode or CRESTA zone. However, for many classes of business, this is not the case. Furthermore, even when structures can be mapped to a permanent position, the assets within them, and so the insured value, may change considerably over time. For example, warehouses are likely to include variable levels and types of stock contents over time. Construction policies, e.g. CAR (Construction All Risks) or EAR (Erection All Risk), also tend to have varying levels of insured asset value over the project period, and this may mean that for any given policy period, the insured value may be different from the project maximum insured value, depending on the status of the project.

Considering construction risks, in many cases, an 'S-curve' approach is taken to estimate the 'build-up' factor that describes how the amount of exposure changes throughout the life-cycle of a project. This approach assumes that the build-up of insured values increases first slowly then more steeply in the middle of the project and more slowly again towards the end, like an S on its side. The model must also describe how the vulnerability of a partially built structure varies at different points in its lifespan (e.g. foundation, erection of the frame, building envelope through to completion) and attempt to differentiate between these. There are specialist Engineering/ Builders' Risk available for limited geographies, but these currently require detailed input data, so are suitable for insurance rather than reinsurance applications. Many construction policies have maintenance periods at the end of the policy period where catastrophe cover is not provided, so it is also important this is recognized where it applies.

For many classes, such as Marine Hull, Aviation, Auto/Motor, the insured assets are likely to be highly mobile. This makes the capture and quantification of exposure data, and hence effective monitoring, even more difficult. Catastrophe risk analysts will often develop bespoke analysis to estimate these temporally and spatially variable 'non-modelled' components of risk, including probabilistic models based on assumptions of varying insured value levels and trending.

Marine Cargo is particularly problematic from a catastrophe management perspective. Policies tend to be variable in length, asset accumulations in each place vary significantly over time and data may provide an estimate of total limit *rather than* insured value. The range of assets being transported is also extremely wide, and makes characterization of risk harder, given the range of vulnerabilities to hazards this can create. When losses occur, they also tend to be highly variable, with little apparent correlation between simplistic measures of hazard intensity such as wind speed, and the losses incurred. Post-event claims calibration of marine cargo loss tends to be particularly difficult.

High value exposures and accumulations, such as automobile hard standing at ports, can also exhibit wide variations in exposure levels at any time. Equally, yacht and marina accumulations can also be significant drivers of potential loss from natural catastrophe events.

In particular, commodity trading account cargo policies, where exposure values often reflect the total value of the transaction rather than the physical assets at any given location and time, are notoriously difficult to assess in terms of value. Storage durations may vary depending on the business conditions that the traders experience at a particular point, so any exposure assessment is difficult, with large ranges of possible values.

From a geographic perspective, catastrophe management for cargo has tended to concentrate on analysis of exposures at major points of risk accumulation such as ports, airports, key warehousing and industrial centres and other significant freight nodes. Difficulty arises when information is not available to adequately calibrate the values at risk at any given time. Albeit not a natural catastrophe, the Tianjin explosion in 2015 illustrated some of the issues around accumulation and assessment of loss in a port; even 6 months after the event, the total loss was not well understood.

Exposure analysis of Fine Art and Specie insurance business (covering art and various valuables such as bullion, bankers' bonds and cash) is also challenging. It is characterized by high value assets held in variable locations, often in museums or retail, but also in private dwellings or on loan to others, or held in vaults. **Cash in transit** risks are also difficult to assess, given the nature of the assets and the movement and storage of them.

Motor exposure brings its own challenges. Cars are mobile, although car lots and dealerships, of course, can lead to high concentrations of value. The time of day can have a large influence on losses, e.g. if a hail loss occurs during rush hour (e.g. see Chapter 3.4.6).

There are also issues with so-called **linear exposures**, e.g. pipelines, transmission lines, power lines, railway lines, which are not easily represented in models. Best practice is to divide the asset into segments and geocode the centre of each segment, smaller segments will be required where the hazard changes more with distance.

### 2.5.4.4  Understanding the Impacts of Exposure Data Choices

Advanced users of catastrophe models will have a detailed understanding of the impacts of different physical risk characteristics on model output. Some catastrophe analysts will do significant work testing all of the input-output combinations, not only as an input into model validation, evaluation and own view of risk (see also Chapter 5.4) but also to ensure the most appropriate representation of the risk for the modelling purpose. For instance, in the case where two possible codings are equally valid, the more conservative of these might be chosen for an insurer's internal risk management and a less conservative approach provided to external parties. Although not unusual market practice, some argue that this approach is not really in the spirit of equitable risk transfer. It certainly becomes difficult to defend when the modelling output is used for regulated processes.

Exposure data may also be manipulated to represent an organization's view of risk (see Chapter 5.5.3.1). If, for example, an insurance company has conducted an internal study and discovered that its performance in a particular area (say, municipal lines of business in the United Kingdom) can be shown to demonstrate better performance than implied by the modelled vulnerability, the insurance company will want that outcome to be reflected in the modelling results. In the absence of more sophisticated options to implement this finding, the insurer may adjust some characteristics which affect the vulnerability. This might include attributes such as the building size or artificially filling the 'percentage complete' input option to effectively scale the sum insured. In other cases, for instance in the United States, insurance companies may wish to completely exclude storm surge losses by coding building elevations to 1000 ft. above sea level; this may be valid if the insurer excludes flood and is confident that their loss adjustment practices and claims control mean that no claims for flood-related damage 'leaks' through into the wind policy claims. Practices like these decrease transparency and decrease the integrity of the exposure data. Where these cannot be avoided, it is important that such changes are documented and ideally a fully representative, unadjusted set of exposures should also be provided.

Changes in the modelling results should be compared with changes in the exposure characteristics to check that the change in results is consistent with changes in the exposure profiles, i.e. exposure summarized by location and risk characteristics. If the behaviour of the model is understood, verification of the changes in results can be achieved by examination of the changes in exposures.

Analysts should also develop 'portfolio benchmarks' derived from e.g. the ratio of AAL and key return period losses (see Chapter 1.10.1) to exposure and be able to explain these at a high level to ensure that the portfolio results behave as expected. As with exposure, peer comparison and market share analyses should be undertaken.

In some cases, multiple model runs will be required to track down changes or identify potential bugs. When interrogating apparent inconsistencies between model input and output, an alternative view using the principle of Occam's razor can be formulated. Years of catastrophe analysis experience show that rather than thinking of complicated explanations for changes in model results, the simplest explanation is often the best; in other words, if it looks like a mistake, then it probably is.

### 2.5.5  Common Tools Used in Catastrophe Risk Management

Catastrophe models are the main tool used in catastrophe risk management, but analysis is, of course, limited to the scope and coverage of the models available. In addition to the catastrophe models, many other tools are required, including data feed tools, analytics, and reporting tools. The following list provides some of the most commonly used:

- policy management systems to manage the details of the policies written and store data and documentation (including all the contractual documents which determine the exposure of the insurer to each policyholder);
- accumulation management and exposure databases which are centralized databases to store and report location-based exposure by line of business, subclass, geography and business-specific classification at any given point in time;
- Geographic Information Systems (GIS) are used both in terms of geographic visualization (geo-visualization) of exposure and defining risk outputs, but increasingly as a core component of the wider analysis framework especially for non-modelled risk (see Chapter 5.4.7);
- underwriting decision support tools, which provide information to help inform the decisions required when selecting and pricing risk in the underwriting process (described in Section 2.6.1). These are often tools which combine functionality and data from other sources including GIS, catastrophe Models and accumulation systems (see Section 2.7.1 for a description of accumulation).
- internal capital models, as discussed in Section 2.10, catastrophe risk management is an input into these models but also receives additional modelling information for calibration and assessment purposes;
- various statistical, modelling and analytical tools to enable bespoke risk assessment e.g. R (https://www.r-project.org), @Risk (http://www.palisade.com/risk/) and SaS (https://www.sas.com).

## 2.6  Underwriting and Pricing

*Kirsten Mitchell-Wallace and Matthew Jones*

### 2.6.1  What Is Underwriting?

**Underwriting** can be defined as 'the process of defining (1) which risks are acceptable, (2) determining the premium to be charged and the terms and conditions of the insurance contract and (3) monitoring each of these decisions' (Randall, 1994). This activity lies at the heart of insurance and reinsurance. The majority of insurance and reinsurance organizations exist to make a profit, certainly those that are shareholder-owned. This profit can be from investment income, from underwriting, or a combination of both. Underwriting profit is the earned premium once the losses have been paid and expenses have been deducted. To generate profit, the right risks must be selected and the appropriate premium charged. This is where catastrophe models become useful: both in risk selection and in ensuring adequate premiums are charged by establishing the expected losses (and their volatility) from these infrequent and severe events. Underwriting decision support is therefore an essential application of catastrophe modelling.

The complexity of underwriting varies with the type of exposure. Simple, homogeneous business such as personal lines may often be primarily rule-based and potentially automated so that underwriting for individual contracts is an exception, whereas large commercial, or reinsurance underwriting will require more analysis and judgement.

Underwriters work within a framework where there are specific rules governing the limits of their **underwriting authority**, i.e. the maximum amount and the type of business that they can underwrite. Decisions outside of this authority need to be referred, either to more experienced underwriters or to underwriting management. There are usually many internal guidelines governing what constitutes a referral and the number and thresholds for these. What the underwriter is able to write is related to the risk appetite (see Section 2.10.1) of the entity, and this is translated from the global level into a limit known as the capacity per underwriting business unit. The monitoring of this capacity is often undertaken or augmented by use of a catastrophe model and may fall under the remit of the catastrophe risk management function (Section 2.5.1), and is usually done at least partially using a catastrophe model loss accumulation (see Section 2.7.1.2). The purchase of reinsurance or retrocession (see Section 2.12), which allows more capacity to support the underwriting, is also related to this.

The underwriter will receive a **submission**, which is an application for cover and the associated supporting information. From this information, they must decide whether to accept the risk, whether to reject it or whether it can be made acceptable by modifying the application in some way: the risk, price, coverage or retention (the amount of loss paid by the insured). The underwriting decision-making process can be thought of as a number of steps as illustrated in Figure 2.8.

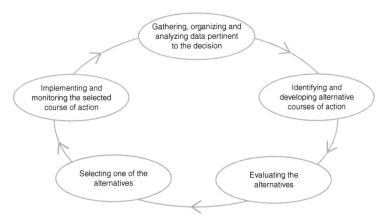

**Figure 2.8** Underwriting decision-making process. *Source*: Randall (1994). Reproduced with permission of The Institutes.

We can consider how the role of catastrophe modelling fits in this process by looking at these steps:

- gathering, organizing and analysing data pertinent to the decision;
- identifying and developing alternative courses of action;

- evaluating the alternatives;
- selecting one of the alternatives;
- implementing and monitoring the selected course of action.

First, through quantification of the risk, catastrophe modelling provides pertinent data, allows evaluation of the alternatives and allows monitoring of the course of action. In this context, catastrophe modelling is a decision-support tool. Although the model can help inform and enhance the underwriting process, it should never replace it. The underwriting process should also consider factors beyond those provided to, or produced by, the models, including financial and mitigation information. Close examination of supplementary risk mitigation information is particularly important when deciding whether to accept commercial policies for a large commercial insurance underwriter or a facultative reinsurance underwriter.

The role of an underwriter encompasses aspects such as interaction with accounting, claims and legal departments (see e.g. Kiln and Kiln, 1996), but here we limit our examination of the role to their interaction with and context of catastrophe models.

(Re)insurance underwriters have different styles, with varying degrees in the belief and applicability of catastrophe models. One reinsurance treaty underwriter (Linford, 2015, pers. comm.) ranked the factors which drive his decision whether to underwrite a treaty as follows:

- client selection;
- wordings acceptability;
- portfolio fit;
- price.

Catastrophe modelling can play a role in three of these steps, as we will see below.

Client selection refers to the choice of client over its peers and whether the underwriter believes that this is a 'good' client. The catastrophe model framework can help form this view. The modelling results will give some indication of this, in conjunction with their historical loss performance (out- or under-performance in a loss compared to their market share). However, any conclusion based on modelled output also heavily depends on judgemental factors. Similarly, the completeness and quality exposure data provided as model input may also indicate the rigour with which catastrophe risk management is approached in the organization. The monitoring of the client's growth against their communicated plan can be assessed with the help of catastrophe exposure data analysis, as discussed in Section 2.5.4. All of these factors which use the model may influence client selection.

Wordings acceptability refers to whether the contractual terms are acceptable – not only within the (re)insurer's company's underwriting guidelines but also reasonable with respect to market standards. This analysis is completely separate from modelling activity.

Portfolio fit is how well this new piece of business combines with the rest of the existing business. This fit has two main aspects: first, whether the addition leads to an optimal portfolio in terms of the metrics of risk and return (see Section 2.8.2); here there is a clear role for the model. The second is whether this new business fits with the strategy of the portfolio, e.g. there may be stated preference for a particular outcome such as less volatility, or a preference for more predictable business such as homeowners' lines or a desire not to have an outsize loss compared to the market. Some of these preferences can be measured using the model.

Last of all, is the pricing which although critical in terms of commerciality is the last step in this decision. It distils all the above elements and acts as a signal of the acceptability of the risk. Pricing and the role of catastrophe models are discussed in the rest of this section. The relative weighting of different steps varies between different underwriters, but this discussion shows that modelling is a tool to aid the underwriting process, rather than a replacement for underwriting. Some people are concerned that blind application of modelling could lead to 'underwriting by numbers' resulting in of a lack of consideration for the nuances of decision-making. It is, therefore, critical that the catastrophe analysts, actuaries and underwriters work together: each applying particular expertise to come to the best mutual view. In the end, underwriters must decide how to incorporate modelled information into their judgements.

### 2.6.2 What Is Pricing?

The aim of pricing any product is to determine how much to charge to cover the cost of the product and generate sufficient profit. For most (non-insurance) products the cost of the product is well defined and known at the point of sale. However, in (re)insurance this is not the case: *the cost of a policy (or treaty) is unknown at the point of sale*; the policy may incur no claims over the insurance period, or it may incur a very severe claim, or several claims. Statistical models must therefore be used to estimate the expected loss for the policy (see Chapter 1.2). Since, for any policy, the actual loss will differ from the expected loss, there is uncertainty (or risk) present, so capital (see Chapter 1.2) must be set aside to cover this uncertainty. Actuaries calculate how much capital is needed: capital for the company as a whole as well as at the level of individual policies. For stock companies, shareholders will also require a return or profit on this capital. The price charged must also cover the expenses associated with the policy, including the overheads of running a (re)insurance company and any reinsurance, or retrocession, costs.

Simplistically, the price consists of three elements: the expected loss, expenses and a profit loading (which should vary with the risk that policy brings to the company) (Figure 2.9).

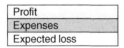

**Figure 2.9** Components of technical price.

The term 'price' needs amore precise definition: technical, market, target and hurdle price are all commonly used. The **technical price** is the price required to cover the loss cost, expenses and the company's profit requirement over some medium term. The price the customer pays may be different to this, and is known as the **market price**. In any given year, for any particular tranche of products, the company may target above or below the technical price, either to take advantage of a beneficial market or to retain its market share in anticipation of an improving market. This can lead to the concept of a target price. Sometimes the concept of a minimum (or **hurdle**) **price** is used, below which the (re)insurance product will not be written, irrespective of the market. The definition of these terms varies by organization, so careful communication is required to ensure meanings are clear.

Insurance and reinsurance have been subject to market cycles where the prices and conditions fluctuate. The reinsurance cycle depends on the supply of, and demand for, reinsurance. In a soft market, there is generally lenient risk selection combined with low prices, whereas in a hard market there are both restrictive selection standards and high prices (see e.g. Randall, 1994; Thoyts, 2010). For instance, a soft market in the late 1980s hardened significantly in the wake of a series of storms in the 1990s, which combined with market problems led to a significant shortage of capacity (Kiln and Kiln, 1996); the market also hardened significantly in the wake of the World Trade Center attacks in 2001. Availability of capacity, which can be related to the recent event history, is often cited as one major causes of the market cycle. In the past, insurance and reinsurance markets have followed different cycles which may be in opposite phases (insurance is hard when reinsurance is soft and vice versa). Technical pricing should be independent of this market cycle. Indeed, it has been argued that improved analytical tools, such as catastrophe models, should provide a 'floor' to the reinsurance market softening by providing a clear limit to the market price, although it remains to be seen if this will occur – especially given the uncertainties in cat models.

As well as the position in the market cycle, market pricing will also be influenced by other commercial factors, including what clients are willing to pay (given competition), the strength and value of the client relationships, the availability of capital, economic factors and opportunities for cross-selling. These commercial factors are decided within the underwriting framework, but should not influence the technical pricing: technical pricing should remain reasonably consistent over time.

The remainder of this section will focus on the technical price, examining the three main elements of expected loss, expenses and profit in turn, before focusing on how catastrophe model output is used in technical pricing.

### 2.6.2.1 Expected Loss

The expected loss represents the arithmetic average (i.e. the mean) loss over the period of insurance for the policy in question. It is also sometimes called the pure premium although this terminology can vary from company to company. The catastrophe modelling terminology for this is average annual loss (AAL), but in this section we will use the term expected loss (EL, denoted as E[L]) as this is used more commonly in pricing.

There are two main types of (re)insurance pricing approach: **experience rating** and **exposure rating.** These are alternative ways of deriving the expected loss. Experience rating is based on the policy-specific historical losses, whereas exposure rating is based on the characteristics of the underlying exposure. Within exposure rating there are several different techniques. For example, exposure curves (also known as **first loss scales**) are used particularly in pricing per risk treaties, Generalized Linear Models (GLMs) are often used in personal lines pricing, as well as scenario-based models which include probabilistic catastrophe models.

#### 2.6.2.1.1 Experience Rating

Although catastrophe models were developed to overcome some of the limitations and challenges of experience pricing, experience rating can still provide a useful supplement to catastrophe modelling when considering the pricing of catastrophe perils. This is particularly true for perils and regions where the catastrophe events, although severe, are also reasonably frequent, for example, US convective storm, or where there have been few socioeconomic changes over the duration of the reported losses. Experience pricing is only possible where credible, applicable claims experience is available and losses are sufficiently frequent. In these cases, if used with care, it can provide an alternative or a means of cross-checking the modelled output.

The first, and most challenging, task in experience pricing is adjusting or **on-levelling** historical loss experience so that it reflects current conditions. Experience pricing works best for stable portfolios, where the character and mix of business, underlying policy conditions and legal environment have not changed over the time period considered. Two major types of adjustment are required: volume adjustment and loss inflation adjustments. Volume adjustment accounts for the difference between today's portfolio volume (i.e. the amount of business in the portfolio) and the amount at the time of the loss. Indicators of volume include the number of risks, the sum insured (although this will also include an inflationary factor) and the premium. Since the premium received by the insurer depends on the original rates charged to the insured, this measure is not a pure reflection of the portfolio volume. For instance, the total premium received may have decreased because the insurer had an aggressive growth plan and decided to undercut the competition by decreasing the rates to have lower prices per insured, whereas in fact the exposure volume actually increased. Historical premiums must be adjusted to remove bias introduced by changing rates. Indices can be used to adjust for inflation of loss payments, for instance, the relative value of a property rebuilding cost index, or the consumer price index today compared to the time of the loss. Loss inflation may not always be completely accounted for using these indexes and so consulting an in-house specialist (e.g. a reserving actuary, or economic research department), who has good knowledge of how well historic loss inflation compares to published indexes, is advisable and a treatment of trends may also be included.

The second task in experience pricing is to use this adjusted loss data (known as 'on-levelled', 'as if' or 'trended' data) within an experience analysis. One approach is to compare the adjusted loss to the attachment and exhaustion of the layer, or policy, being priced to obtain an expected loss from experience. Losses will be counted if they are above the attachment point and below the exhaustion point. The losses to the layer will then be averaged over the number of years considered. This is sometimes termed a **burning cost** approach. This could be appropriate for low layers with high frequency perils where there is sufficient claims experience to yield a stable estimate of the expected loss cost.

However, for most catastrophe pricing analyses, a better approach is to fit distributions for both frequency and severity to the on-levelled claims data, and use these distributions to smooth and extrapolate the experience.

The frequency distribution is the distribution of the number of loss events in a year. This is often assumed to be a Poisson distribution (see Chapter 1.11.1.1). However, the associated assumption of independence does not hold for some perils. For these perils, the frequency may be better represented using a negative binomial distribution or even a clustering algorithm which attempts to represent the clustering of the events within a year temporally and in some cases, spatially based on geophysical factors. The choice will be related to the dispersion characteristics of the frequency distribution (see Chapter 1.11.1.1).

The severity distribution is the distribution of the size of the loss, given that a loss has occurred. Common choices of distribution to parameterize severity of losses include the Pareto, generalized Pareto and Lognormal distribution (see e.g. Klugman *et al.*, 1998; Parodi, 2015) choices. A range of standard statistical approaches may be used to estimate the distribution parameters (e.g. method of moments, maximum likelihood or method of least squares). Details of such fitting methods can be found in Klugman *et al.* (1998). The fitting method may often be decided within a tool which has been developed especially for the pricing processes.

The selection of the length of the time window, i.e. the number of years to be included in experience pricing is also important. This is influenced not just by the claims history, but also by what is reasonable from an on-levelling perspective, given the requirement for portfolio stability and homogeneity: 10 years is often chosen as a reasonable time period. For low-frequency catastrophe perils, recent history may or may not contain an extreme loss. If so, this loss therefore needs to be assigned an appropriate return period, so that it does not overly bias the expected loss. Assigning return periods to these large losses is heavily judgement-based. It depends on experience and market knowledge, as well as scientific input. However, care must be taken that estimates of market return period do not become a self-fulfilling prophecy. Often after an event there will be many discussions, e.g. what is the return period of the 2011 aggregate US convective storm loss or the 2011 Danish cloudburst loss? If industry loss data are available, fits can be used to derive the return period of losses. Additionally, hazard data can be used to derive local hazard return periods, however, these need to be used carefully: hazard exceedance probabilities are not the same as loss exceedance probabilities since the most severe hazard intensity within a time period may not cause the largest loss.

An alternative approach to adjusting the return period of a large loss is to use a blended approach – capping the experience losses at a certain level and above this level using catastrophe model output. Such an approach is described in Section 2.6.4.1.

Integrating new information can also be challenging. In principle, an additional loss data point should not change the underlying distribution of loss. However, since these time series are generally short and not fully representative, a further loss often does result in significant changes in the parameters and resulting expected losses. The distribution will also change if the peril is **non-stationary** over time (see Chapter 4.3.3.1). The catastrophe pricing analyst must therefore also decide if the original loss distribution is valid when assigning a return period to an event.

### 2.6.2.1.2 Exposure Rating

Exposure rating techniques use the characteristics of the exposure for the policy in question, to determine or derive an expected loss. Losses are usually used in building an exposure model; these will be losses from a wider pool of experience than just from the specific policy being priced. For example, Generalized Linear Models (GLMs) are an exposure rating technique widely used in personal lines insurance pricing. GLMs use the company-wide or market-wide loss experience for a particular line of business to build models which estimate the expected loss cost (usually split into frequency and severity components), given a set of exposure characteristics (known as rating factors) for a specific policy. Rating curves were originally derived from event loss data (e.g. Swiss Re, 2004).

Catastrophe models use industry-wide loss experience in their validation, as discussed in Chapter 4. They are also a form of exposure model.

### 2.6.2.2  Expenses

Expense costs will vary from organization to organization, and will differ substantially between reinsurers and insurers. The main elements of expense are as follows:

- *Commission*: For primary insurers, this term is interchangeable with brokerage – it is the amount that an insurer pays a broker (or partner such as a bank) in exchange for acquiring and performing some of the administration associated with the insurance business. From a reinsurer's perspective, it is the amount a reinsurer pays an insurer, usually as part of a proportional treaty agreement to cover the insurer's administration and acquisition costs. As described in Section 2.4.2.1, some contracts will also specify a profit commission where the amount of commission varies according to the profitability of the contract.
- *Brokerage*: The amount a reinsurer pays a reinsurance broker, or a primary insurer pays a retail broker.
- *Fees and levies*: This covers costs such as fees for risk engineering activity that may be needed to provide a more detailed risk assessment of key sites within a policy. It also covers levies that may be payable in certain territories, for example, Flood Re will impose a levy on all UK primary insurers in order to help fund the UK flood pool scheme (see Section 2.13.6). Another typical charge might be a fire brigade charge.
- *Taxes*: Any local premium taxes, e.g. insurance premium tax (IPT) in the United Kingdom, and corporation tax (tax on profits) need to be included in the technical price. Note that corporation tax is usually allowed for in the profit requirement, and IPT is added on to the actual price charged at the end of the process, not in the technical price formula.
- *Loss adjustment expenses*: allocated and unallocated. These are the expenses associated with handling claims. Allocated loss adjustment expenses (ALAE) are the expenses that can be directly associated with a particular claim. For example, the expenses in employing loss adjustors to investigate claims, or the legal costs associated with a particular claim. Unallocated loss adjustment expenses are those expenses associated with a claims department which cannot be associated with an individual claim, for example, claims management overheads. Since ALAE are directly associated with a claim, this is usually allowed for as a percentage of loss cost. Catastrophe models do not include this in their estimates of loss.
- *Reinsurance and retrocession costs*: these are the premiums associated with of the company's reinsurance purchase.
- *Internal expenses*: These are the expenses of running a (re)insurance company, besides those already mentioned.

Capital costs are also important but are treated separately in Section 2.6.2.5.

Expenses, particularly internal expenses, are often categorized into variable (which vary over some short term, e.g. one year) or fixed (which do not). For example, a company's premises' cost may be considered fixed since they do not vary, over the short term, in proportion to the amount of business or amount of claims. The classification of fixed and variable depends upon individual companies' definition of short term: in the long term, everything is variable.

The expense items discussed must be incorporated in the calculation of technical price, as they are additional to the expected loss provided by the model. They will, therefore, be expressed as either a flat amount per policy, as a proportion of the overall premium, as a flat amount per expected number of claims, or as a proportion of the incurred claims cost.

### 2.6.2.3 Profit (and Risk Loading)

As already discussed, it is a fundamental principle of insurance that pooling of risk reduces uncertainty due to diversification. However, some risk always remains, and a (re)insurance company must have capital (excess of assets over liabilities) in order to cope with this uncertainty. Shareholders require a return (or profit) on this capital, and will require a higher return if more risk is taken. Techniques for calculating and allocating capital are discussed in Section 2.6.2.5, but the important point is that the higher the risk, the higher the required capital and the higher the required profit.

In terms of pricing, this means that the higher the risk a policy brings to the company, the higher the price; in this case, 'risk' is not the expected loss cost, but rather the potential for loss above and beyond expectations. There are several approaches to incorporating risk in the price. Some just consider the risk that the individual policy brings. Some consider the incremental impact to the company as a whole. Some make an assumption that the risk is proportional to the expected loss (or premium). The choice of technique will depend on the homogeneity of the business, the size of the risk an individual policy brings to the company, and the sophistication of the company's pricing.

One approach is to represent profit is as a loading on the standard deviation of the losses:

$$profit = \alpha\sigma \tag{2.1}$$

where $\alpha$ is the risk loading calibrated to ensure the overall target return is met for the company, and $\sigma$ is the standard deviation of the loss to the policy.

When using a catastrophe model, the standard deviation of loss can be calculated directly from an ELT (see Chapters 1.10.5 and 1.10.6).

However, this standard deviation pricing approach is less useful where the premiums depend on the losses (e.g. as in reinsurance where there are reinstatements). Depending on the sophistication of the variation in $\alpha$, it often does not account for the portfolio effect, i.e. whether the policy diversifies (lower risk) or concentrates (higher risk) the company's portfolio, the cost of capital, or the payout patterns (are the losses likely to take a long time (higher risk) or short time (lower risk) to pay out?).

Often the profit load is calculated as a percentage of the technical price with reference to the required risk-based capital (see Section 2.6.2.5, and Swiss Re, 1999b, for a simple introduction to concepts, although outdated with regard to regulation), overall targeted return on equity, and the risk-free return being earned on the company's capital. Return on Equity (ROE) (i.e. the targeted company return on risk-based capital), can be calculated as follows:

$$RoE = RoC + \frac{RoP}{CIR} \tag{2.2}$$

where $RoC$ is risk-free return on risk-based capital, $RoP$ is Return on technical premium (i.e. the profit margin), and $CIR$ is Capital Intensity Ratio (i.e. the risk-based capital/technical premium).

Equation (2.2) is easily rearranged to:

$$RoP = (RoE - RoC) \times CIR \tag{2.3}$$

For example, if the company's targeted return on risk-based capital (RoE) is 10% and the risk-free return is 3% (RoC), the return needed from writing the premium is 7% of risk-based capital (RoE-RoC). Further, if the Capital Intensity Ratio for a particular line of business is 50%, then the company must price for a 3.5% profit margin (i.e. the RoP). Using these example values in Equation (2.3):

$$RoP = (10 - 3) \times 0.5 \tag{2.4}$$

While the RoE and RoC will normally remain the same across the company as these are set by the overall company aspiration and the prevailing risk-free interest rates, the CIR will vary according to the riskiness of the business. For a homogeneous class of business where there is little difference between the type and value of the insured risks, such as personal lines, the CIR may only vary by broad type or segment of business (e.g. Personal Lines UK Property). For more complex business the CIR could have a more variable structure and could even change according to the individual characteristics of the policy, perhaps factoring in the policy standard deviation as per the discussion earlier in this section and Equation (2.1). This will largely depend upon the sophistication of the RBC allocation and of the pricing tools in use within the company. RBC calculation and allocation will be discussed later in Section 2.6.2.5, however, first we discuss how the different components of price (expenses, loss and profit) can be brought together in a technical price formula.

There are other, more complex, measures of profitability including RARoC (Risk Adjusted Return on Capital) (Goldfarb, 2010) and RoRAC (Return on Risk Adjusted Capital). The latter is more common in banking. The differences between these can be found in literature such as Froot and Stein (1998), but are well beyond the scope of this text.

### 2.6.2.4 Technical Price from Profit, Expenses and Expected Loss

Technical price is defined as the sum of Loss, Expenses and Profit; see Section 2.6.2. However, it is usual for a company to calculate these components as a ratio of the most relevant metric, rather than flat amounts. For example:

- profit is often expressed as a percentage of premium: $ROP\%$;
- some expenses will be best represented as a flat monetary amount per policy (e.g. policy printing costs): $FE$;
- some expenses will best represented as a percentage of premium (e.g. commission): $VE\%$;
- some expenses will be best represented as a percentage of expected loss (e.g. $ALAE\%$).

For example, if $P$ is the technical price, and E[L] is the expected loss:

$$P = \mathrm{E}[L] + ALAE\% \times \mathrm{E}[L] + \mathrm{FE} + VE\% \times P + RoP\% \times P \tag{2.5}$$

Rearranging Equation (2.5) provides the following technical price:

$$P = \frac{FE + \mathrm{E}[L] \times (1 + ALAE\%)}{1 - VE\% - RoP\%} \tag{2.6}$$

This is, in fact, a simplified example, but serves the purpose of illustrating the likely form that is used in practice. The following factors would also be accounted for in any technical price formula:

- Expected losses could be developed for all perils, and are often split between attritional, large and catastrophic claims. Each company will have their own exact definition of these. However, generally, attritional losses are those that happen with a frequency and regularity such that the average loss to the insured book in any year can be bound within a narrow range with high probability. This makes it easier to estimate capital requirements and thus to insure attritional losses. Large losses are infrequent and larger than attritional, for example, a large fire, but they are not catastrophic. They are more difficult to estimate but techniques exist to parameterize these losses. Losses due to catastrophes are of course rare, but can be very large, as discussed throughout this text;

- tax would be taken into account in the calculation of RoP%;
- RoP% could vary with the CIR (as discussed in Section 2.5).

Moreover, the technical price formula also needs to be adjusted for the **time value of money**, i.e. the way that the value of money changes through time. Assuming interest rates are positive, $1 now is worth more than $1 in a year's time. This is because $1 now could be invested and in a year's time would be worth $1 \times (1+r)$, where r is the interest rate, i.e. $1 a year from now is worth $1/(1+r)$ now. More generally, if t denotes the time (in years) and r is the annual interest rate, then the value now (the **present value**) of an amount X received at some point, t, in the future is:

$$Present\ Value = X \times \frac{1}{(1+r)^t} \tag{2.7}$$

The component $\frac{1}{(1+r)^t}$ is termed a discount factor, and is made up of the risk-free interest rate and the average time to payment. The risk-free interest rate is theoretical rate of return of an investment with no risk of financial loss. In reality, the situation is more complex than shown, as there is likely to be a stream of payments rather than one payment at one point in time. This is the remit of any standard actuarial textbook, and is beyond the scope of this book.

Considering the elements of technical price, each component has associated discount factors, used to allow for the time value of money at the time of the inception of the policy.

### 2.6.2.5   Risk-Based Capital: Calculation and Allocation in Profitability Calculations

The actual share capital that a company has is the difference between the company's assets (including retained earnings) and its liabilities. This can be calculated directly from the balance sheet of the company.

Where there is good governance, a company's management will want to understand the risk that is being borne by the company and to compare this risk to the capital of the company. This results in estimation of the required capital, the capital that the company thinks it needs, given its risk profile and risk tolerance, which reflect what it underwrites and its limits (see Section 2.10.1). This required capital is called the **economic capital**, or the **risk-based capital** (RBC). The regulator also has an interest in ensuring the company remains solvent and so they will insist that an estimate of the required capital is calculated using the regulator's rules. This is called **regulatory capital** or in Solvency II parlance, the **Solvency Capital Requirement** (SCR) (see Section 2.11.4). Comparing the actual capital of a company to the SCR will help the regulator to assess the financial condition of a company. Comparing the economic capital to the actual capital will help the company's management ensure capital is at a sufficient level.

The amount of capital a company requires will vary enormously, depending on the size of the company, the type of business being written, the diversification of the company (geographically, by line of business, by risk type, and so on), the amount of reinsurance purchased, and many other factors.

There are a number of risk types that will be considered when calculating risk-based capital: These include: underwriting risk (including catastrophe risk), reserving risk, operational risk, investment (or asset) risk and credit risk.

---

**Box 2.2 Case Study: A Simple Approach for Risk-Based Capital**

This simplistic approach is commonly used to calculate risk-based capital (RBC):

$$RBC = VaR_q[L] - E[L] \tag{2.8}$$

where $VaR_q[L]$ is the loss amount at the specified percentile of the loss distribution q (e.g. q is 99.5% for Solvency II), and $E[L]$ is the expected loss. The rationale for subtracting the mean loss is that this amount of loss is expected, and capital is only needed for the unexpected loss (i.e. the amount above the mean).

There are many issues with a VaR approach for capital. The main two issues arr: (1) it is only one point on the probability distribution; and (2) it is not a coherent risk measure (see Chapter 1.11.3). To expand the first issue, two companies could have very different distributions of loss, but if the distributions have the same mean and coincide at the percentile chosen for RBC calculation, then they would show the same RBC. Considering the second issue, it is desirable to be able to allocate the risk-based capital to parts of an organization and $VaR_q$s calculated at a more granular resolution do not sum up to the $VaR_q$ at the overall company level (because of diversification). Nonetheless, $VaR_q$ is widely used to calculate overall RBC, perhaps inappropriately carried over from banking and the Basel II Accords.

---

There are a number of approaches to calculate the RBC, but the overall level is usually set according to the company's risk tolerance (see Chapter 2.10.1). It is beyond the scope of this book to discuss the different methods available for risk-based capital calculation, however, risk measures that can be used as a basis for setting capital are usually based on the distribution of aggregate loss (or Present Value) and include metrics such as standard deviation (or variance), VaR, TVaR, XTVaR (see Chapter 1.10).

After the overall RBC has been calculated, it needs to be allocated to smaller units within the company. There are two main reasons for this: (1) to enable the performance of different entities within a company to be assessed relative to their risk-based capital (entities with higher RBC should generate higher returns); and (2) to enable the RBC to be used to calculate CIRs (or risk loads) for use in the technical price formula.

This process of taking the overall capital and disaggregating it to a finer level, whether that is business unit or even policy, is known as **capital allocation**. It is a complex subject and it is beyond the scope of this text to give anything but a basic overview. However, it should be noted that differences in these methods can lead to very different results (e.g. Robbin, 2013).

There are a variety of risk measures used to allocate capital, such as those already mentioned for the calculation of overall capital (and defined in Chapter 1.10). There are also various methodologies for allocating capital, for example, Myers-Read (Myers and Read, 2001), Rhum-Mango-Kreps (Rhum and Mango, 2003; Mango, 2003; Rhum *et al.*, 2003; Kreps, 2005; Bodoff, 2007).

Approaches to capital allocation can be proportional or marginal. Proportional approaches allocate capital to an entity in proportion to the entity's risk measure compared to the sum of all the other entities' risk measures. Marginal approaches allocate capital to an entity in proportion to the impact on the overall company risk metric of adding (or subtracting) the entity from the company. Vaughn (2007), Cordier and Lacoss (2011) and Venter *et al.* (2006) provide a comparison of methods. The concept of suitable allocation can also be considered (e.g. Tasche, 2000), and marginal allocation meets these criteria (Venter, 2010).

An example of a marginal approach using catastrophe models is to create a planned 'ideal' or reference portfolio which is used to determine the impact on the catastrophe risk of adding a particular treaty. This is known as 'last in' marginal capital allocation (Venter *et al.*, 2006), and shown in Equation (2.9).

$$\text{Marginal contribution} = \text{risk measure for portfolio with new treaty}$$
$$- \text{risk measure of original portfolio} \tag{2.9}$$

The amount of risk, and therefore capital, a (re)insurance company has, are related to how well the company is diversified. This means that the degree of dependence between business entities (e.g. portfolios) needs to be taken into account. Dependence between portfolios can be modelled using copulas, which are generalized dependence structures (see e.g. Gorge, 2013); these are used instead of standard linear correlation (Upton and Cook, 1996) because standard linear correlation tends to independence in the tails of distributions. However, the geographic aspects of dependence between events and portfolios are already considered by catastrophe models and therefore instead of using copula-based dependence structures to split down the capital into individual treaties, capital can be allocated to individual treaties using risk metrics derived from catastrophe models. This approach requires the modelling to be part of the pricing framework and the choice of a particular model per region-peril (e.g. for European wind or North Atlantic Hurricane), as well as an appropriate risk measure. Here TVaR is a common choice because of its coherence, as discussed in Chapter 1.11.3.

The assumption of independence between peril-regions may need to be interrogated (see Chapter 6.2.2) and is a subject for further research, although broadly there is some evidence to suggest that the major basins are not dependent, or are anti-correlated. Dependence in the tail under a changing climate is a situation that could affect the need for capital in extreme circumstances.

Different capital allocation methods are one of the main reasons for differences in technical prices between competing reinsurers participating on the same business (known as the **reinsurance panel**), although internal costs can also play a role.

### 2.6.3 Practicalities of Using Catastrophe Model Output for Pricing

The previous sub-sections have introduced (re)insurance pricing and capital calculation and allocation in order that the practitioner has an understanding of this important use of catastrophe models. This section focuses on the practicalities of using catastrophe models, and catastrophe model output, for pricing purposes.

It is important to realize that the catastrophe model output alone will almost never lead directly to the technical price, even once the risk and expense loadings are added. First, usually other perils are covered within a policy. These may be high frequency perils for which there is a large wealth of data such as fire or escape of water (e.g. washing machine flooding) for which standard actuarial approaches are likely to have been developed (e.g. based on GLMs; Parodi, 2015). Second, there may be other catastrophe perils for which no model exists and assumptions must be made (non-modelled risks, see Chapter 5.4.7). Even if the policy only covers the specific peril mentioned, the model output will still need to be adjusted for any limitations in the model.

There are a number of considerations when adjusting catastrophe model loss estimates to be appropriate for pricing purposes:

*Exposure Development*

The exposure data used for catastrophe modelling are a snapshot of the insured's portfolio (see Chapter 1.9) at a point in time (the **'as at' date**). As the year progresses, the insurance company may add more, and different, locations to its portfolio, possibly growing or reducing in particular zones depending on their business plan. The exposure used to price should be representative of the exposure at the mid-point of the risk period covered. Ideally, this adjustment would be undertaken by modification of the exposure data and the provision of a pro-forma exposure set based on expected growth of the portfolio. This would provide the insured's best estimate of exposure, and would be especially useful if inuring treaties must be accounted for. However, in practice, this adjustment is often made to the losses.

### Insurance to Value

In some legal environments (e.g. the United Kingdom and the Caribbean) the concept of **average** applies i.e. that underinsurance (insurance purchased for a sum which is lower than the value of the insured property) will lead to a proportional scaling down of the loss paid (see e.g. Diacon and Carter, 2005). However, where the average does not apply, the concept of **insurance to value** (ITV), i.e. the reported insured value is less than the actual value, may be important. Although insurance policies usually contain provision to scale down if there is a large discrepancy between insured value and the value of the property (e.g. if the insurance to value is less than 80%, the policy will work on an **Actual Cash Value (ACV)** basis where the payment is the replacement cost less depreciation and obsolescence), in practice, this often depends on the legal environment. Under-reporting of the replacement costs leads to incorrect original sums insured as the basis for damage ratio application and therefore the calculated loss estimates will be too low. Insurance companies spend a large time validating their replacement costs using external databases or comparing claims to reported values.

### Allocated Loss Adjustment Expenses

Allocated loss adjustment expenses (ALAE) are the insurer's expenses directly linked to a claim (e.g. lawyer's fees, loss adjustment fees). These are not usually included in the catastrophe model and so must be explicitly added. It is common practice to add a flat percentage load across the different return periods, although this is unlikely to be a very realistic assumption since the percentage of ALAE is likely to differ between a very large loss and a small loss. Loads for enhanced legal fees may be considered in especially litigious jurisdictions.

### Secondary Perils or Non-Modelled Aspects

Secondary perils are hazards which are caused by the original peril but have a different physical mechanism for causing loss and (see Chapter 3). Some models include multiple secondary perils as analysis options. However, more basic models may not give the full range of perils which need to be considered in a particular territory. Consequently, there may be non-modelled secondary perils such as fire following earthquake, earthquake sprinkler leakage, storm surge or tsunami (see Chapter 3 and Chapter 4.3.7).

There are also aspects that may not be modelled, such as the increase in the cost of paying a loss after a large event due to shortages of material and labour (demand surge or post-event loss amplification which also includes additional impacts for very big catastrophe events such as non-linear effects when high-density concentrates of exprosure such as major cited are impacted and loss escalations that occur as a result of mandatory evacuations).

Non-modelled elements and secondary perils should be accounted for by a factor, which may vary by the severity of the loss or the return period of that loss.

### Underwriting Judgement

Whether, and how, underwriting judgement should feed into the technical price (as opposed to a commercial price) is a subject of debate. Underwriting judgement is predominantly a factor in large commercial insurance pricing or reinsurance pricing, since personal lines insurance pricing is generally quite prescriptive.

There may be attempts to quantify underwriting judgement in a systematic fashion weighting factors, such as management strength and experience, robustness of the contractual wording, financial strength, quality of the business concept, or reinsurance buying behaviour. The factors are blended to give a client quality score to capture the softer factors which contextualize the modelled output. The ranking of these clients may qualify them for a discount or a load in pricing. In other organizations, underwriting judgement may be applied in a more arbitrary fashion.

*View of Risk*

An organization may have derived a particular view of the risk for the peril-region being modelled; for instance, the company view may be that the frequency of earthquakes is incorrect according to the historical record and so they may consider it appropriate to apply an adjustment on the basis of that. This may also include the framework related to either model blending or taking account of an alternate model view within the pricing. This is discussed in more detail in Chapter 5.5.

### 2.6.3.1 Calculating Pricing Metrics from ELTs and YLTs

Several metrics have been used in the preceding sections on pricing and capital calculation. Many of these metrics can be obtained directly from a catastrophe model interface, however, it is very likely that a (re)insurer will have its own pricing tool or pricing model, and these metrics will instead be calculated from an event loss table (ELT) or year loss table (YLT). Chapter 1 specifies how risk metrics ($E[L]$, $\sigma$, VaR, TVaR, XTVaR) are calculated from ELT and YLTs.

Ideally, a YLT is required to price an annual contract since this provides the losses for the period covered and allows easy application of financial structures (especially aggregate features, such as an aggregate deductible or limit). An ELT can be used directly in some limited cases such as in the calculation of AAL for insurance. This YLT can come from simulation of an ELT (see Chapter 1.10.4.9), or directly from the model itself if the model is configured to do this. Most reinsurers will have in-house pricing systems, although sometimes pricing software packages available from brokers will be used. Some newer models also provide loss tables for non-annual periods (PLTs).

### 2.6.4 Pricing Specifics for Insurance and Reinsurance

Insurance pricing relates to a product which will be offered to many customers. Many more claims data are usually available to a primary insurer for pricing insurance as compared to a reinsurer pricing reinsurance. More statistical analysis of claims data is possible for insurance, and the expected value of the distribution is important. In contrast, treaty reinsurance pricing relates to pricing a cover for a portfolio, and there tends to be less information available. Since non-proportional reinsurance is purchased for more extreme events, the tail of the distribution is more important in this case. Catastrophe pricing considerations specific to insurance and reinsurance are discussed in Sections 2.6.4.1 and 2.6.4.2.

### 2.6.4.1 Insurance Pricing Specifics

The use of catastrophe modelling in primary insurance pricing depends on the type of business, as well as the organization's philosophy and resources. This setting of the premium is sometimes called 'rating' or 'rate-making' (e.g. Walters, 2007).

When considering original insurance policies, the catastrophe component is often only one element of the price. Most types of policies, fire, escape of water, theft, and other non-catastrophe perils, will have a large contribution to the premium in addition to the contribution from any catastrophe perils. The pricing approach used depends on the type of business;

business can be classified as either high-volume business or low-volume business, with 'volume' dictated by the number of policies typically written. To be profitable, high-volume business needs a rapid quotation and renewal process as each individual policy is relatively low value; this is possible as the exposures are fairly homogeneous, typically with one location per policy, and have simple financial structures. In low-volume business there are fewer, higher-value policies; this comprises a more heterogeneous mix of exposure with more complex policies, often with multiple locations and of larger value, and so more time is needed to underwrite and price the risk. High-volume business includes personal lines and small commercial business. Low-volume business is everything else including global corporate and industrial business, as well as middle market, which lies between the very large and complex and the high volume. The differentiation between these different types of business will vary from company to company, but is usually related to the size of the sum insured or premium, the occupancy class (see Chapter 1.9.1.3) of the risk and the geographical spread of risk.

The pricing of high-volume business is usually performed using a multiplicative rating table approach, which can easily be incorporated in an automated system. A base premium is set, which is typically expressed as a premium per unit of sum insured. This is then modified through the application of multiplicative terms, which depend on the level of each rating factor (e.g. building age). If the base premium were 1000, a factor for the house being new (2001–2010) is 0.9 and that for it being a terrace is 1.2, the premium would be $1000 \times 0.9 \times 1.2 = 1080$. The full table relating to this example is shown in Figure 2.10. There may also be a discount on the rate for higher sums insured and/or a load for lower sums insured. For this kind of business, the catastrophe model will be used to develop the rating table, as discussed below, but not used to model each new policy as it is underwritten.

The different approaches to constructing and presenting the final premium tables largely depend on the sophistication of the insurer's pricing tools and on the distribution channel. However, underlying each set of premium tables will be a statistical analysis of the frequency and severity of each major peril that the policies are exposed to. A GLM analysis is a common statistical method used in this field, and proprietary tools such as Emblem (https://www .towerswatson.com/en/Services/Tools/emblem) have evolved to enable insurers to perform these analyses in an efficient way, although they can also be undertaken in common statistical packages such as R (https://www.r-project.org) and SaS (https://www.sas.com). The outputs of these analyses are usually a frequency and severity model for each peril. Frequency here is normally defined as the expected number of claims per policy year, and severity is normally defined as the expected incurred amount of claims divided by the number of claims. These specific definitions of frequency and severity are commonly used in personal lines GLM analysis, but frequency and severity can be defined in other ways depending on the context. The frequency and severity of claims for each peril are usually expressed in terms of how they vary from a base level by multiplicative rating factors.

The pricing of low volume policies is more involved as they will likely have multiple locations per policy as well as a more complex policy structure. For example, there may be limits and sub-limits by peril, and varying deductibles and the exposures may span multiple countries. For this type of business a rating table approach will usually not be able to reflect the complex characteristics of the policy, and it is more common for the catastrophe component of the pricing to be set by running each policy through a catastrophe model in order to better resolve this complexity. Global Corporate and Industrial business is usually sufficiently complex and high value to warrant this approach.

In some cases, a hybrid approach is adopted. In a hybrid approach, multiplicative rating tables are used to construct an expected loss, and then a more sophisticated approach is used to apply the impact of expenses and policy financial structures. Commercial middle market business often uses this hybrid approach.

**Input data**

| | |
|---|---|
| Storm Base Rate Per Mille | 0.05 |
| Base Year Built | 1985 |
| Base Rating Area | 10 |
| Base Property Type | Semi Detached |
| Base Sum Insured | 300'000 |

| Modifier | Data | Relativity |
|---|---|---|
| Risk Year Built | 1905 | 1.500 |
| Risk Postcode | AB1 1 | |
| Risk Rating Area | 2 | 0.450 |
| Risk Property Type | Detached | 0.800 |
| Risk Sum Insured | 310'000 | 0.980 |

| Storm loss cost | 8.2026 |
|---|---|

**Age**

| Year Built | Relativity |
|---|---|
| Pre 1900 | 1.600 |
| 1900 - 1910 | 1.500 |
| 1911 - 1920 | 1.400 |
| 1921 - 1930 | 1.300 |
| 1931 - 1940 | 1.200 |
| 1941 - 1950 | 1.150 |
| 1951 - 1960 | 1.100 |
| 1961 - 1970 | 1.050 |
| 1971 - 1980 | 1.025 |
| 1981 - 1990 | 1.000 |
| 1991 - 2000 | 0.950 |
| 2001 - 2010 | 0.900 |
| 2011 - | 0.800 |

**Location**

| Rating Area | Relativity |
|---|---|
| 1 | 0.400 |
| 2 | 0.450 |
| 3 | 0.500 |
| 4 | 0.550 |
| 5 | 0.600 |
| 6 | 0.650 |
| 7 | 0.700 |
| 8 | 0.800 |
| 9 | 0.900 |
| 10 | 1.000 |
| 11 | 1.050 |
| 12 | 1.100 |
| 13 | 1.200 |
| 14 | 1.300 |
| 15 | 1.400 |
| 16 | 1.500 |
| 17 | 1.750 |
| 18 | 2.000 |
| 19 | 2.500 |
| 20 | 3.000 |

**Type**

| Property Type | Relativity |
|---|---|
| Terraced | 1.200 |
| Semi-Detached | 1.000 |
| Detached | 0.800 |
| Bungalow | 1.500 |
| Flat | 0.300 |

**Sum insured**

| Sum insured low | Sum insured high | Relativity |
|---|---|---|
| below | 99999 | 1.42 |
| 100'000 | 109999 | 1.40 |
| 110'000 | 119999 | 1.38 |
| 120'000 | 129999 | 1.36 |
| 130'000 | 139999 | 1.34 |
| 140'000 | 149999 | 1.32 |
| 150'000 | 159999 | 1.30 |
| 160'000 | 169999 | 1.28 |
| 170'000 | 179999 | 1.26 |
| 180'000 | 189999 | 1.24 |
| 190'000 | 199999 | 1.22 |
| 200'000 | 209999 | 1.20 |
| 210'000 | 219999 | 1.18 |
| 220'000 | 229999 | 1.16 |
| 230'000 | 239999 | 1.14 |
| 240'000 | 249999 | 1.12 |
| 250'000 | 259999 | 1.10 |
| 260'000 | 269999 | 1.08 |
| 270'000 | 279999 | 1.06 |
| 280'000 | 289999 | 1.04 |
| 290'000 | 299999 | 1.02 |
| 300'000 | 309999 | 1.00 |
| 310'000 | 319999 | 0.98 |

**Geocode**

| | Rating Area | |
|---|---|---|
| AB1 1 | | 2 |
| AB1 2 | | 10 |
| etc | | |

Figure 2.10 Example of an insurance rating table showing rating factors such as age, location, type, sum insured. The multiplicative terms are denoted 'relativity'.

Catastrophe modelling can assist with primary insurance pricing in three main ways:

- helping to establish the long-term level of expected loss for catastrophe perils;
- providing a source of external input when constructing rating area tables, in particular, the variation of expected loss cost by geography;
- helping the insurer understand the variation in capital required due to catastrophe perils.

Taking these in reverse order:

Capital allocation has already been discussed in Section 2.6.2.5. For high volume business, the ratio of the capital required per unit of technical premium (the **capital intensity ratio**) is usually established for broad sections of this business and this ratio will often not vary significantly (if at all) from policy to policy. For low volume business, individual policy characteristics such as the standard deviation of loss and the extent to which a policy concentrates or diversifies a portfolio can be used to vary the capital requirement on a policy-by-policy basis.

A rating area table classifies each geographical area (e.g. a postcode) into a specific level of frequency, severity, loss cost or premium. Insurers often develop these rating area tables using their own claims experience combined with external data, e.g. flood maps. Catastrophe model output can assist these analyses by providing an external data source. For example, a table containing AAL, for a standard risk, for each postcode would be a useful form of input to such an analysis, to help an insurer form their rating areas. Often the pricing teams and catastrophe modelling teams within an insurer are quite separate, and the pricing actuary may not realize that such useful information is readily available from a catastrophe model.

Although GLMs are widely used within insurance pricing, they are mainly useful for establishing the rating relativities. In other words, which rating factors are significant and how the metric (frequency, severity, loss or premium) changes with rating factor. Typically up to 10 years of data would be used for such an analysis, although more is ideal. The insurer must also understand the long-term level of loss – or at least the level of loss over whatever future period over which the technical price is designed to be stable (often three to five years to fit in with the typical business planning period). A 10-year time series is not sufficient to establish a long-term expected loss cost level for catastrophe perils; the time series may contain no catastrophe events and so likely understates the long-term level, or it may contain an extreme event and overstates the level. One of the reasons catastrophe models exist is to estimate the long-term level of loss.

Catastrophe model output allows the time series to be extended. Small, high frequency (i.e. attritional) losses form a significant part of the insurance loss. For example, the storm and flood perils will often include high frequency losses from very small and localized events. In contrast, this is less often the case for the earthquake peril as shaking intensity and spatial scale of impact are strongly associated, so for large impacts it is usually a very low frequency peril. Catastrophe models were originally created and designed to model large, rare events. As they have evolved, they have increasingly sought to represent the entire spectrum of events for a given peril, but it is generally accepted that they tend to be weaker in the area of high frequency, small claims. Conversely this is where the insurer will have much information. Thus, in this situation, for perils with a high frequency component where the insurer has claims experience, an approach blending catastrophe model output with insurers' experience is recommended. This is described in Box 2.3.

Blending combines the insurer's experience where it is most credible with the catastrophe model output in the frequency range where models are most useful. However, there are some practical considerations when applying such an approach:

- the definition of event from the claims time series needs careful work. Even if a fixed period window (72 hours, i.e. three days for storm) is chosen, the window must be placed carefully to ensure events are properly represented. For example a time window that cuts an event in half (e.g. a four-day event where two days are in one time window and two in another) would

---

**Box 2.3 Blending Catastrophe Model Output with Loss Experience for a Primary Insurer**

1) Define events from daily time series of claims. An insurer must use their own judgement here as there is no strict definition of an event. Typically a 72-hour period will be used for windstorm and a longer period (perhaps 7–14 days expressed in hours) for flood. Claim amounts within these periods (e.g. consecutive 3-day periods for the entire time series) are summed to form 'events'.

2) Select a level at which to cap event losses and calculate the capped expected loss from the time series. The aim is to prevent individual events (which may have extreme return periods) contributing too much to the expected loss.

This capped expected loss is given by:

$$E[L]_{cap} = \frac{\sum_i Min(L_i, cap)}{N} \tag{2.10}$$

where $L_i$ is the Loss from the event $i$ in the time series (defined in (a)) and $cap$ is the capping level, $N$ is the number of years that the time series represents.

From an ELT or YLT (Chapter 1.10.4), calculate the excess loss, $E[L]_{XS}$, the loss above the level chosen for capping the experience. The excess expected loss, calculated from an ELT, is given by:

$$E[L]_{XS} = \sum_i Max(L_i - cap, 0) \times Rate_i \tag{2.11}$$

Where $Rate_i$ is the rate for event $i$ from the ELT.

From a YLT the excess expected loss cost is given by:

$$E[L]_{XS} = \frac{\sum_i Max(L_i - cap, 0)}{N} \tag{2.12}$$

where $N$ is the number of years represented by the YLT.

Add the capped experience loss to the excess catastrophe model loss to obtain an estimate of the overall loss cost:

$$E[L] = E[L]_{cap} + E[L]_{XS} \tag{2.13}$$

---

understate the event loss. One approach to mitigate this is to run through the time series and pick out the peak dates (e.g. dates of maximum incurred loss above some materiality threshold compared to the days either side) to ensure the time windows are centred on these dates for the important events;

- the on-levelling considerations described in Section 2.6.2.1 are required to ensure loss data and modelled exposure data are comparable so that they can be combined;
- the definition of an event by cause of loss (known as **head of claim**) in a primary pricing analysis can often be different from the definition of peril in a catastrophe model, even though they use the same name. For example, storm can sometimes be defined by a primary insurer to include pluvial flood, lightning, hail, tornado, and so on. A catastrophe model may only represent one component of this 'storm' definition. A primary insurer may define flood as fluvial and coastal, whereas often the coastal component of flood is in the catastrophe vendor's storm model. The claims and the catastrophe model output must be consistent in terms of definition.

- some catastrophe models will 'bundle' high frequency events into individual model events which are within the model to ensure the overall AAL is correct. Although referred to as 'events', these do not represent any individual event or have any physical basis. If robust experience from loss statistics is available, this should be used in preference to these pseudo-events;
- the capping level can be varied, and the sensitivity of the choice of capping level to the calculated overall loss can be plotted to investigate whether the choice will materially alter the results for a range of levels. Typically, experience shows the capping level to be approximately commensurate with a 1-in-2 to 1-in-5-year return period loss.

### 2.6.4.2 Reinsurance Pricing Specifics

Reinsurance pricing approaches for Natural Catastrophe can be grouped into three major types (Qiu *et al.*, 2012):

- market approaches based on catastrophe modelling output;
- market approaches based on non-modelled factors;
- academic approaches based on financial and actuarial theories.

In daily reinsurance practice, the first two of these are used; academic approaches such as Capital Asset Pricing Methods are not generally used in reinsurance (see e.g. Venter et al., 2006). The approaches described here are a mixture of the first two; we emphasize the role of catastrophe modelling and the potential for adjustments to modelling output within the pricing process.

In reinsurance, there is a difference between the pricing of facultative and treaty business. As described in Section 2.4.2, facultative reinsurance relates to individual policies so the pricing and underwriting methodology is comparable to that for large and complex primary insurance contracts. The specific locations and risk characteristics are particularly important, as is the correct application of the financial terms. Most of the considerations which differentiate facultative pricing are therefore modelling considerations such as adequate representation of the physical risk characteristics and policy conditions rather than pricing considerations. This discussion therefore focuses on treaty reinsurance pricing.

The first step in a pricing analysis is retrieving the data. Reinsurers receive data in the form of a submission package. This should contain:

- an underwriting narrative including a description of expected portfolio development (growth or shrinkage) in the context of the business plan;
- information on changes to policy form or financial structures (e.g. flood deductibles) or basis of coverage;
- portfolio information such as total sum insured (TSI), risk count and premium;
- losses for historical events, including the cedant's 'as if' estimation of loss for these (e.g. including impact of new deductibles and limits);
- losses should be provided in excess of a threshold which is small enough that the losses are still useful to price the structure after indexing considerations and for a sufficient length of time;
- historical portfolio development (TSI, risk count and premium), ideally per line of business and region. This should go back at least as far as information on the historical events.
- modelling input and/or results with a modelling narrative, including client-specific assumptions made, data checks undertaken and specific adjustments relating to their view of risk.

All of the elements listed above are relevant to the pricing task, which reinforces the point that there is more to the pricing task than simply applying the modelling results directly. The catastrophe pricing analyst will need to consider the loss history and its development to benchmark the modelling output. As discussed in Section 2.6.3 and more detail below, the modelled loss estimates are almost always adjusted before they are used for pricing. Guidelines related to pricing should be developed for each and every market priced. There should be consistency between pricing and accumulation control in order to ensure consistency of decision-making in the context of the use test of Solvency II (Article 120, see Table 2.7).

Further adjustments to modelling output for reinsurance, such as exposure development, ITV (insurance to value, see also Chapter 1.9.1.2), ALAE, non-modelled perils and underwriting judgement are applied to insurance and reinsurance pricing. In addition to these factors described in Section 2.6.3, there are also specific adjustments relevant to a reinsurer. These may include the following.

### Cedant Specifics – Exposure Peculiarities

The specific portfolio may be contain unusual types of risks or structures for which few data are available for validation of the model's vulnerability functions. Thus, the particular cedant may not 'model well' when compared to appropriate historical benchmarks. There may also be factors not included within the model that reduce the likely losses such as localized flood defences or initiatives limiting the payout of the loss compared to the historical losses used to validate the vulnerability functions.

### Cedant Specifics – Business Model

Additionally, the cedant's loss experience may no longer be representative; the insurer may have started selling a different product, or there may have been material change to an existing product even if the amount of exposure is similar. Care is always required when using loss experience since assessing the representativeness of that experience is critical. Although catastrophe models are able to capture some of these aspects (e.g. financial structures), they cannot always capture them all.

### Wordings Issues

As described in Section 2.6.1, part of the role of the underwriter is to analyse the contract and to check that it is acceptable in terms of any guidelines and underwriting approach. The terms of the contract are known as the wording. There may, however, be aspects, such as an hours clause variation, which may not be possible to model within the available software; many models will look at the time integrated footprint of a physically-defined event. If this is the case, the pricing analyst should work with the underwriter to evaluate the potential impact of these contractual issues on the price.

Adjustments can be grouped broadly into a wider framework which is generally applicable (see Chapter 5.4.7):

- missing exposure (increases the loss);
- missing sub-peril (increases the loss);
- hazard or vulnerability adjustment model adjustment (can increase or decrease the loss).

Table 2.2 illustrates the types of adjustment that can be made. The individual adjustments are sometimes considerable and can have a significant impact on the price; the impact of such adjustments can be larger than the choice of event set for example. The typical scale of some adjustments is shown in Figure 2.11. In this example, we see the maximum and minimum adjustments at program mid-point across a nationwide US book. Of course, these will vary considerably by region and cedant.

Figure 2.11 shows a simple example of linear adjustment, i.e. a constant multiplier. Adjustments are also often made non-linearly for instance as a function of return period. More

**Table 2.2** Types of adjustments to model output.

| Adjustment | Purpose | Type |
|---|---|---|
| Growth | To bring exposure to mid-point of contract | Missing exposure |
| ALAE | To add allocated loss adjustment expenses | Missing exposure |
| Wind Pool | Approximation that represents the cedant's exposure for participation in wind pools (although these can, of course be modelled explicitly) | Missing exposure |
| APD | Auto Physical Damage exposure which is included in the submission totals but not in the modelling data | Missing exposure |
| Storm surge | To account for storm surge where they have not been provided in the modelling | Missing sub-peril |
| CQI | Client Quality | Soft underwriting factors |
| Model fit | How well does the model represent these exposures | Hazard/vulnerability |
| Adjustment | Overall adjustment | The sum of the adjustments. It may be assumed that adjustments are independent when they are not in reality. There may be double counting in any overall adjustment level, e.g. if a significant model fit adjustment is entered this may be in lieu of more specific factors such as ALAE or APD that are missing in the raw results |

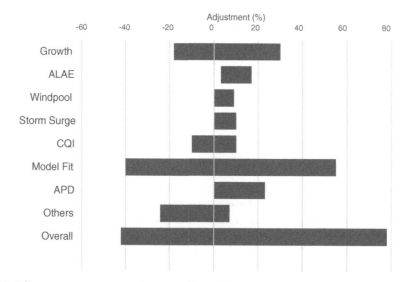

**Figure 2.11** Adjustments on a nationwide US wind book, illustrating the possible scale of such adjustments. *Source*: (SCOR, 2015b).

technically, multipliers may also be based on some underlying hazard or vulnerability characteristic, for example, for earthquakes of a particular magnitude to exposures of a type where the vulnerability functions are deemed inappropriate. As discussed above, these are tied closely with model evaluation, validation and the in-house view of risk (see Chapter 5). Depending on a company's in-house catastrophe pricing philosophy, there may be flexibility in individual catastrophe pricing judgement or a more prescriptive approach.

For multi-national reinsurance programmes, all exposures which are capable of triggering the treaty should be included in the scope, even if they are not provided in the modelling. This can easily be evaluated by applying a possible, indicative damage ratio to the exposure to derive a possible loss and comparing this loss to the attachment point of a program. These damage ratios can be derived from claims experience, industry losses or model benchmarking. The same applies to perils, modelled or not. The catastrophe analyst should have an overview of the potential perils (see Chapter 3) and their magnitude in each territory, perhaps by reference to an in-house grid of perils by territory or a peril map.

Adjustments should be made before the application of inuring reinsurance if possible so that they can flow through the financial perspectives correctly.

### 2.6.4.2.2 *Validating the Catastrophe Pricing*

Figure 2.12 shows a comparison of expected losses calculated using different catastrophe pricing approaches for a fictitious three-layer reinsurance programme, 100 xs 100, 200 xs 200 and 400 xs 400 M. As described above, the burning cost is the pure loss to the programme from 'as if' historical losses. In this example, the burning cost is compared to three alternative approaches. The first is experience pricing, i.e. an actuarial frequency-severity model built from indexed historical losses (see Section 2.6.2.1). The second is the unadjusted catastrophe model output, and the third is adjusted model output. In this case the model output has been adjusted for a perceived understatement of event frequency in the model. Figure 2.12 shows some common issues encountered when pricing catastrophe perils.

The burning cost only provides losses to the lowest layer (100 xs 100); this is because the length of historical record used does not capture losses which are severe enough to penetrate the second layer. For this first layer, the modelled losses are lower than the burning cost. The pricing analyst must judge if this is because the record contains an unusual event, or whether it is likely that the model understates the loss for some reason. This judgement will be underpinned by model evaluation and detailed knowledge of the peril behaviour, as well as close examination of

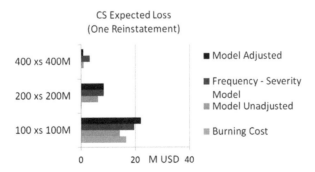

**Figure 2.12** Example of comparison of burning cost, a frequency-severity model, unadjusted and adjusted model results. The three-layer programme structure is given on the left, the loss amounts using the different methods are shown in US dollars. *Source*: (SCOR, 2015a).

the loss history. The expected losses from the experience pricing are, of course, sensitive to the frequency and severity distributions chosen; however, a common feature of comparing an experience pricing to catastrophe model output is that the losses to the upper layer are overstated by experience pricing as compared to a physically-constrained catastrophe model. This comparison of methods is provided for illustration as one would not usually consider an experience pricing for catastrophe perils as a credible approach, except for frequency perils and low layers. However, pricing with loss history can be useful to test the sensitivity of assumptions such as the over-dispersion (see Chapter 1.11.1.1) if, for instance, there is evidence of clustering in the loss history that is not seen in the model.

Some checks to consider when pricing treaties are presented below. None of these can, however, give a complete picture and the derivation of the price is therefore heavily judgement-based and depends on the benchmarks used. Each of these methods has drawbacks. The catastrophe pricing analyst is using models with known issues, and only has experience to judge them by, for example, a particular model may have incorrect individual hazard footprints, vulnerability issues, geographical skews). The issues of representativeness of experience are, of course, the very reason that catastrophe models were built in the first place (see Chapter 1.4). So the catastrophe pricing analyst is left with a conundrum – how to judge a model with known issues using experience which is by its nature not representative of the probabilistic spectrum. The relative weight of the judgements therefore depends on both model evaluation and close discussion with the underwriters.

Examining this in more detail, several questions can be considered when establishing the potential validity of a catastrophe pricing.

### How Does the Burning Cost Compare with the Modelled Expected Loss?

This check compares the 'as if' history to the modelled expected loss. Burning cost is dependent on the quantity and representativeness of the claims experience and the robustness of the indexing approach chosen. It is not especially suitable for low frequency, high severity perils; perils such as hurricane or earthquake should not generally be treated with a burning cost, or experience, analysis. The burning cost will be dependent on the length of the time window chosen and will be biased by major events appearing within this window. Ten years is a period commonly chosen since this is usually available and can usually be on-levelled reasonably well. If the losses from these large events are simply treated with their empirical frequency (i.e. a large loss that occurred last year but realistically might be considered to be a 1-in-35, but is considered a 1-in-10 because 10 years of loss history are used), the expected loss will be higher in their immediate aftermath of an event, thus the reinsurance cycle will be 'implicitly' priced when using a burning cost; this is not the aim for technical pricing. Burning cost is, however, a commonly used benchmark since it shows the recent performance of a treaty.

### How Does an Experience Pricing Compare with the Modelled Expected Loss?

This question examines how the expected loss to the programme calculated by fitting frequency and severity distributions fits the claims history, i.e. an experience pricing, compared to that obtained by using the catastrophe model. This check is similar to the one above, but uses distributions rather than the claims directly. Again, this method is dependent not only on the quantity and representativeness of the claims experience and the robustness of the indexing approach chosen, but also on the goodness of fit of the distributions. As well as this, the distributions should be evaluated by considering the return periods of the losses of historical events. Indeed, judging these return periods may be a part of the experience pricing.

It may not be possible to make this comparison using actual experience for catastrophe events, especially where there are no or very few historical losses. However, a benchmark 'experience' pricing can also be completed using synthetic 'as if' historical losses to benchmark the modelled output; these 'as if' scenarios use event footprints derived from a model's recreation of actual events. This may be desirable if, for instance, the frequency assumptions differ from that assumed by the model. This may be needed if significant clustering is seen in the record but not assumed by the model.

### How Do the Client Losses Compare to the Modelled 'As If' Footprint Losses?

Here the loss experienced by the client are compared to the 'as if' losses generated by using a modelled representation of the event (see also Chapter 5.4.3). Fewer losses are required for this approach than for an experience pricing, one or two losses may suffice. Again, the methods of indexing the client loss are critical. This approach is also very sensitive to the footprint of an individual event and its accuracy in representation of the hazard footprint and the vulnerability at the hazard values experienced in that event. The relationship between these, of course only illustrates a difference at one or two points in the OEP curve, but may be a reasonable indicator that some model adjustment is required for a stable book where there is a known issue with exposures (e.g. in France where the original sums insured are not recorded) or vulnerability. Such analysis can be more difficult when there are multiple losses within a season for historical events, e.g. 2004 in the United States. In such cases, although the modelled total loss across the season may be correct, the proportion of loss attributed to each geophysical event in the model may be different from industry practice where clients will likely assign losses to maximize recovery from their reinsurance (this is related to the discussion of hours clause in Section 2.4.2.2.2).

### Is The Return Period of the Actual Event Loss in a Credible Range on the Modelled EP Curve?

Subject to the same caveats re. indexation above, here the historical loss is checked against the EP curve for credibility; namely, the return period of the 'as if' indexed loss is calculated from the modelled EP curve. The return period of loss depends on the characteristics of the exposure, so for very regional portfolios, this might not be a good comparison, e.g. a client with little exposure in the path of the storm will have different event losses and EP curve, and of course a different return period from the industry as a whole. However, for regionally spread portfolios, expectations around the return period of that event can be pre-defined and used as a benchmark. For example, a market loss may be expected to have a specific return period; Anatol in Denmark may be expected to have a return period of 1-in-80–100 years, and deviations from this may merit further investigation or event adjustment. A modelled return period of greater than 1-in-5000 years suggests an issue with the modelled representation of the risk; this was true for some models post the Christchurch earthquake or after the Alabama tornadoes of 2011.

### Does the Probability of Loss to Layer Make Sense?

This is a simple comparison where the frequency of indexed historical loss to a CAT XL layer is compared to the modelled loss to that layer. Knowing what we do about recent history, does the modelled loss frequency make sense? This is more useful check for the first and second layers.

### Do the Attachment and Exhaustion Points Make Sense Compared to What a Cedant Might Buy for the Main Peril?

A 'typical' wind reinsurance programme in the United States or Europe might attach around the 1-in-20 and exhaust around the 1-in-100, but this type of information should be benchmarked by market and across peers with similar buying behaviour. Further, different cedants will have different levels of conservatism depending on their company culture. If, for instance, a programme to primarily protect against US hurricane appears to be offering cover to the 1-in-500-year loss, either the client has an extremely conservative reinsurance buying strategy, the version of modelling being examined is very different from the one used to inform the buying decision, or there is something wrong with the modelling results, exposure or the model itself. Regulatory requirements may also affect the exhaustion point in many markets.

### Is the Weight of Perils Consistent with Expectations?

Are the relative contributions from different perils consistent with expectations for the particular region, compared with other similar programmes and your understanding of the relative risks? Are these consistent and reasonable? If not, are they unreasonable. This might be an indication that adjustment is required.

### How Stable Is the Expected Loss Under Different Assumptions?

How does the expected loss change, given alternative credible pricing assumptions? These could include differing model input assumptions, different adjustments on outputs or different assumptions with respect to on-levelling for an experience pricing. The ranges when conservative and optimistic assumptions are taken should be examined to give an indication of the stability of the result and the sensitivity to changes.

### How Does the Pricing Compare to the Approach Taken Last Year (if a Renewal)?

The explanation of differences in results and a summary of causes is critical to communicate with stakeholders, whether underwriters, cedants or brokers.

#### 2.6.4.2.3 Incorporating Non-Modelled Perils

The price must include all of the perils which expose the programme (see Chapters 5.4.7 and 5.5). Although the global suite of perils is growing, the catastrophe pricing analyst will likely meet perils for which no catastrophe models exist and must include them in the pricing.

A simple and popular approach for reinsurance catastrophe excess of loss pricing is so-called 'Pareto pricing' where the severity distribution is represented by a Pareto distribution with one shape parameter and a threshold (see Chapter 1.11.2.2 and Schmutz and Doerr, 1998). It is based on the peak over threshold approach of extreme value theory (e.g. Coles, 2001). This distribution is favoured because (1) it is easy to work with, (2) a long-tailed distribution, and (3) standard benchmarks exist for the severity of particular perils. Alpha, the exponent, is the shape parameter and can be considered to vary by peril as shown in Table 2.3. Low values of alpha represent a heavy tail. This is illustrated in Table 2.3 and Figure 2.13.

Since predicted losses using this statistical model may exceed the sum insured, sometimes an upper cap must be introduced to the distribution. If no claims data are available, assumptions can be made about the frequency at threshold and a standard alpha used to provide a simple first approximation of the loss distribution. This can be especially useful for non-modelled perils. It can also provide a little insight when benchmarking the output of models where losses are frequent enough, although this is a rather crude approach.

An alternative for non-modelled perils is the development of a market curve, e.g. a set of damage ratios by return period that are applied to a representative exposure to derive a simplistic pricing approach.

**Table 2.3** Suggested values of the Pareto distribution's alpha parameter, i.e. shape parameter, by peril.

| Peril | 'Standard' Alpha (shape parameter) |
|---|---|
| Earthquake | 0.6–1.0 |
| Windstorm | 0.8–1.3 |
| Flood | 1–1.5 |
| Convective storm | 1.8–2.2 |
| Wildfire | 1.5–1.8 |

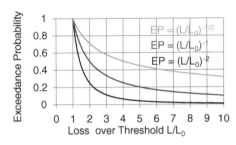

**Figure 2.13** Effect of different Pareto alpha values on the shape of the distribution. L is Loss. $L_0$ is the threshold, which must be above zero, and the exponent is alpha. As the alpha decreases, the tail of the distribution becomes heavier.

#### 2.6.4.2.4 Reinsurance Pricing Metrics

Some metrics commonly encountered by catastrophe risk analysts are included here. In reinsurance pricing the expected loss is commonly normalized by the reinsurance layer limit to allow easy comparisons between different programmes. Since, as we saw earlier, the limit was called the 'line' for certain types of treaty, the normalization of the expected loss by unit of limit is known as the **loss on line** (LoL). LoL is sometimes also referred to as the 'pure premium rate on line'. The LoL gives the expected loss per unit limit and is useful as it is equivalent to the frequency of total layer loss.

$$LoL = \frac{E[L]}{limit} \tag{2.14}$$

The premium divided by the limit is called the **rate on line** (RoL). Technical rate on line (technical price divided by limit) and market rate on line (market price divided by limit) are also used.

The concept of payback is also sometimes used in reinsurance pricing and used to benchmark the catastrophe model output.

$$Payback = limit/premium \tag{2.15}$$

The **payback period** is the length of time in years it would take to pay for the limit with the current premium, i.e. limit divided by premium. This metric may be used as a benchmark in pricing and varies from region to region and for different perils. Payback can also be used more generally to mean reimbursement of a loss after an event.

Another commonly used metric in both insurance and reinsurance is the underwriting ratio (UWR).

$$UWR = \frac{E[L] + external\ expenses}{premium} \tag{2.16}$$

In addition to calculating the price required, the underwriter also wants to be able to benchmark the new pieces of business against each other. This is done to understand how much of the allocated capacity each uses and decide whether a new treaty is a good addition to the portfolio.

### 2.6.4.2.5 Pricing Curves

To check the integrity and validity of catastrophe excess of loss pricing, so-called **market pricing curves** are frequently used (see Figure 2.14). These show the market rate on line (RoL) on the y-axis against the layer mid-point normalized by some measure of exposure volume on the x-axis. These plots are used to show the relationship between layers bought by different cedants in the market, show price development year on year and benchmark pricing. The measure of volume used for normalization can be sums insured or limit or PML if models are used. Two other measures can also be used: PZA (**Peak Zone Aggregate**), which is the sum insured in a major zone considered for aggregation (e.g. CRESTA 3 for Chile, CRESTA 1 for Colombia or CRESTA 1 to 5 for Peru); subject premium, which is a measure of the business covered by the treaty may also be used. The mid-point used can be the arithmetic or geometric mid-point although sometimes a third-point is also used. The third point is simply the point a third of the way between the attachment and exhaustion point of the layer. The geometric midpoint is often preferred as since the loss distribution is not symmetrical through the layer, the geometric mid-point is closer to the 'centre of gravity' of the layer loss, i.e. where most of the loss to layer happens (compare with Figure 2.24). Pricing curve outliers can be identified and the reasons for these interrogated. Although it is very commonly used in practice, discussions of the weaknesses of this approach can be found in Morel (2013).

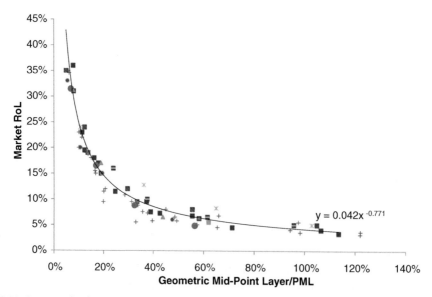

**Figure 2.14** An example of a market pricing curve, different symbols are used for different cedants. Catastrophe modelling output can be used in normalization of the x-axis. Courtesy of SCOR.

## 2.7   Accumulation, Roll-Up and Capacity Monitoring

*Claire Crerar and Kirsten Mitchell-Wallace*

### 2.7.1   What Is Accumulation?

Accumulation is the process by which (re)insurance organizations add up the individual risks underwritten to understand the total level of risk to which the organization is exposed. This understanding is vital for both underwriting and risk management decision-making, and, as such, accumulation is one of the most important and heavily scrutinized functions of a catastrophe management team within a (re)insurance organization.

The accumulation process, which is also known as aggregation or roll-up, can be applied to either exposure (total sum insured, exposed limit) or losses. This section will discuss these two types of accumulation along with the most common uses for accumulation output. This section will also discuss the practical considerations associated with the accumulation process.

#### 2.7.1.1   Exposure Accumulation

Exposure accumulation involves selecting a measure of exposure, such as total sum insured or exposed limit, and aggregating this exposure for all risks underwritten within specified **accumulation zones**. The accumulation zones are geographical regions, which typically correspond to a pre-existing administrative definition such as country, city or postcode. The exposure measure may also be broken down further by additional criteria which the organization considers useful. The criteria may relate to the characteristics of the risk itself (e.g. property type or construction types) or the organization (e.g. office or business unit underwriting the risk). The output from exposure accumulation is often described as **aggregate exposure** or **aggregates**. An example is shown in Table 2.4.

The key considerations for exposure accumulation are the definition of the zone to be used for accumulation and the choice of the exposure metric. The definition of the accumulation zone is often based on pre-existing administrative boundaries such as country, CRESTA zone or postcode. The resolution chosen will depend on the intended use of the output; for example, lower resolution aggregates such as by country or CRESTA zone are useful for high level reporting, but for individual risk selection decisions, it is beneficial to have smaller geographical zones such as postcode.

Accumulation zones may also be defined according to criteria designed to reflect the varying levels of risk. For example, distance to coast (e.g. within 200 m of the coastline) may be used to

Table 2.4   An example of sum insured by product and country, illustrating accumulation zones.

| Country | Personal lines sum insured (m EUR) | Commercial lines sum insured (m EUR) | Total sum insured (m EUR) |
| --- | --- | --- | --- |
| France | 400 | 40 | 440 |
| Germany | 250 | 250 | 500 |
| The Netherlands | 0 | 50 | 50 |
| Italy | 100 | 300 | 400 |
| Total | 750 | 640 | 1390 |

define accumulation zones in areas at risk from tropical storms as this determines where the hazard is greatest.

The exposure metric chosen will depend largely on the type of business written. For insurance companies writing relatively simple policies such as personal lines, total sum insured is a representative measure of the risk they are exposed to. However, for those writing policies where financial structures are in place to restrict the amount payable in the event of a loss (e.g. commercial and industrial policies), exposed limit is likely to be a more appropriate measure. For reinsurance companies, the exposure metric should consider the net position once terms such as attachment points, limits and coinsurance have been applied.

The basis of the exposure metric has significant implications for the loss potential of that exposure. For example, exposure reported on a total sum insured basis is more likely to generate losses in a specific event than an equal amount of exposure reported on a net (net of financial structures) basis, assuming all other characteristics of the exposure are the same. This is illustrated in Box 2.4.

It is worth noting that while catastrophe models are typically used to calculate potential losses, many commercially available catastrophe models also have the ability to carry out exposure accumulations. The accumulation zones used can be based on geographically defined zones, event footprints or defined dynamically based on the areas of highest exposure concentrations within a portfolio. This is particularly useful for non-modelled perils and man-made catastrophe risks such as terrorism. This functionality is possible within commercially available catastrophe models because the insurance exposure data required for loss modelling and exposure accumulation are very similar.

The main uses of exposure accumulation are aggregate reporting and capacity monitoring. These are discussed in more detail in Section 2.7.2.

### 2.7.1.2 Loss Accumulation

Catastrophe modelling will often be undertaken for individual risks or groups of risks within a business unit of an insurance company, often termed a portfolio within the catastrophe modelling context (see Chapter 1.9.3). Loss accumulation is the process by which the modelled loss estimates for individual risks and portfolios are combined to produce a modelled loss estimate for the organization as a whole.

For probabilistic catastrophe models, the modelled loss estimates for each portfolio are accumulated to find the total loss associated with each event in an event or year loss table (ELT or YLT). This new loss table represents the combined loss distribution for the portfolios included in the accumulation, and summary metrics can be reported from it as per any other loss table output (see Chapter 1.10). The combined loss distribution will vary from the underlying loss distributions due to the differences in composition and distribution of each portfolio. This is clear in the example shown in Table 2.5 where event 10001 produces the

Table 2.5 Example of a loss table showing accumulation across different portfolios. P1 is portfolio 1, and so on.

| Year | Event | Loss P1 | Loss P2 | Loss P3 | Total loss |
|------|-------|---------|---------|---------|------------|
| 1 | 10001 | 500 | 0 | 300 | 800 |
| 1 | 10002 | 200 | 5 | 150 | 355 |
| 2 | 10028 | 25 | 100 | 400 | 525 |
| 2 | 10036 | 100 | 600 | 1000 | 1700 |

**Box 2.4  The Importance of Exposure Metric in Determining Loss Potential**

Insurance company A insures two factories in Europe. Each factory is worth $10 million and is insured for the full value without any financial conditions. Insurance company A reports its aggregate exposure on a total sum insured basis giving a total of $20 million for Europe.

Insurance company B also insures two factories in Europe. Each factory is worth $20 million but is insured with a deductible of $10 million for each factory. Insurance company B reports its aggregate exposure on a net basis giving a total of $20 million for Europe ($20 million − $10 million deductible for each of the two factories).

On first appearance, insurance company A and B have the same amount of exposure in Europe: $20 million, and hence are equally at risk of incurring losses. In practice, company A is more likely to experience greater losses from a specific event. If there was a windstorm that caused damage of $5 million at each of the four factories in this example, the insurance companies would pay the following losses:

Insurance company A: $2 \times \$5$ million $= \$10$ million insured losses
Insurance company B: Each $5 million loss falls within the $10 million deductible
    $= \$0$ insured losses.

An additional challenge with reporting exposure on a net basis is ensuring that financial structures are not double- or triple-counted within the accumulation zones. This can arise where policies cover locations in multiple accumulation zones but the financial structures apply across all locations irrespective of accumulation zone. This is illustrated in the example below:

Insurance company C provides insurance to a multinational company with a factory in France worth $50 million, a factory in Germany worth $50 million and a factory in Belgium worth $50 million. The insurance policy includes a limit which states that the maximum possible claim in any one event from all locations is $30 million.

Insurance company C wishes to report its aggregate exposure on a net basis for Europe by accumulation zone, where each country is a separate accumulation zone. It produces the following report of aggregate exposure on a net basis:

France $= \$30$ million
Germany $= \$30$ million
Belgium $= \$30$ million

This report implies that the total net exposure for Europe is $90 million. However, in practice, the net exposure for Europe as a whole is $30 million because the policy limit does not apply to each individual factory or country; it applies to the losses from all locations.

In this example, it is obvious that the solution to this issue is to calculate the aggregate exposure for each level in the accumulation zone hierarchy separately (i.e. calculate totals for each country and Europe-wide separately rather than calculating for each country and adding the totals together).

However, there may be instances where it is necessary to report the exposure by zone, so, when added together, it reflects the overall total without double counting. This mainly arises where accumulation zones are small, and to count the full policy limit in every zone would imply the exposure is much greater than it really is.

In these instances, the policy limit can be allocated to the different accumulation zones. There are many different allocation methods with common examples including allocating the limit equally to each zone, splitting the limit in proportion to the total sum insured in each zone or assigning the full limit to the zone with the largest sum insured.

largest loss for portfolio 1, whereas for the organization as a whole, event 10036 is more significant.

The main requirement when accumulating probabilistic model output is a consistent loss table across all the portfolios to be accumulated. The same events must be present in the output for each portfolio in order for the individual losses to be combined into a total loss. While this requirement is relatively simple, it presents considerable challenges for catastrophe management teams who are using multiple models or adjusted model output. This is discussed in more detail in Section 2.7.3.

An important benefit of accumulating losses in this way is that by adding losses event-by-event to create a new ELT, it is possible to provide the full range of exceedance probability metrics and consider the whole loss distribution rather than simply looking at individual points or scenarios. Accumulating at event level will also naturally account for the diversification benefit associated with adding new portfolios (see Chapter 1.2).

There are also disadvantages to this approach. Without a loss table, it can be very difficult to account for non-modelled portfolios (such as marine cargo) or non-modelled perils. Additionally, exceedance probability metrics can be difficult to clearly communicate to non-specialist decision-makers who are often key users of accumulation output.

To address this, many organizations also use scenario-based accumulations to complement, support and benchmark the full probabilistic accumulation. In this case, one or more event scenarios are chosen and the loss estimates for each portfolio affected are accumulated to provide an overall loss estimate. The scenarios chosen are either from the historical events or defined as a credible event which could happen in an area of interest, for example, a hurricane, typhoon or earthquake. The choice of scenarios will depend on where the companies' exposure is highest, or may be driven by management interest based on other factors, such as recent natural catastrophe events or research. In some cases, a set of standard scenarios must be used and reported either within a company or to a regulator, rating agency or pseudo-regulatory body. The most widely used set of industry standard scenarios are Lloyd's Realistic Disaster Scenarios (see case study in Section 2.7.1.3).

The loss estimate for each portfolio affected in a given scenario can be produced using a catastrophe model, or by other means such as a market share analysis or underwriter estimate of loss. This makes it possible to assess the impact across all areas in the organization including non-modelled areas such as assets and life insurance. This type of exercise is therefore useful in assessing the **clash potential**, i.e. the cases where multiple different portfolios are affected at once.

Scenarios can also provide a framework for communication and a method for visualizing the consequences which can be particularly powerful in opening the topic to non-specialists.

A more recent alternative accumulation method is that using 'characteristic events' (KCC, 2012). These are events which are created for defined probabilities, usually those of most interest to the (re)insurance industry, e.g. the 1-in-100 and 1-in-250-year return periods. These baseline characteristic events will vary by region. Taking the example of hurricane, be the 1-in-100-year event for Florida is moving cat 5 hurricane, whereas for the North-east of the United States is a cat 3 hurricane. The location of the characteristic event is then changed by moving the track to obtain different loss scenarios for the same event probability. This approach is considered by some to form an intuitive complement to both traditional probabilistic models and scenario modelling, since it contains elements of both: a distribution of losses for a fixed probability. Although primarily an accumulation approach, its proponents also advocate use in pricing, underwriting and portfolio management.

### 2.7.1.3 Scenario Modelling Case Study: Lloyd's Realistic Disaster Scenarios

Lloyd's have employed a framework of **Realistic Disaster Scenarios** (RDS) since 1995. Syndicates are required to calculate and report their estimated exposures to a range of defined catastrophe events, including man-made disaster scenarios.

The natural catastrophe scenarios have defined footprints and estimated industry losses. A complete list of the specifications can be found on the Lloyd's website (see Lloyd's, 2015b). Examples include a hurricane making landfall in Florida and a major earthquake in Japan. RDS deliberately does not take account of probability in the catastrophe modelling sense; they are intended to be a test of exposure management and disaster planning for a set of severe but realistic events. Syndicates can use catastrophe models to produce loss estimates for the events, or for a simpler approach, apply the damage factor included within the event specification provided by Lloyd's.

The RDS approach offers three distinct advantages. First, the scenarios are tangible and defined in a way which is easy for non-experts to understand. It is easy to visualize a category 5 hurricane making landfall in Miami, compared with exceedance probability analytics, which are more difficult to understand conceptually.

Second, the RDS have defined footprints and this makes it simpler to consider non-modelled risks. For instance, potential accumulations of cargo can be considered for the specific ports and airports within the footprint of the event.

Third, the scenarios are consistent and static; this means that they can be compared over time, across portfolios and different businesses. They are also additive, which makes it easy for Lloyd's and others to calculate total exposure from all syndicates (or portfolios in the case of other organizations besides Lloyd's) by simply summing the loss estimate for a specific scenario. Probability distributions are not additive in this way, and the variety of approaches to catastrophe model usage among syndicates means that obtaining a consistent loss tables (ELT and YLT) to perform a probabilistic accumulation is impractical. As RDS do not consider a range of outcomes at various levels of probability, they are a weak tool for pricing (see Section 2.6.2) and capital management (see Section 2.10) but their relative simplicity means that they are used as robust benchmarks to compare with the probability metrics.

### 2.7.2 Use in Underwriting and Risk Management

The most important use of accumulation output is the reporting of the overall catastrophe risk that a (re)insurance organization is exposed to, along with the relative contribution that each portfolio makes to the total risk. An understanding of this is crucial to the successful running of a (re)insurance organization and consequently there are many stakeholders to whom this information must be communicated and explained. The stakeholders in an organization range from underwriters through to actuaries, risk management teams, senior management and the board of directors. Interested parties external to the organization include the organization's reinsurers, investors, regulators and rating agencies.

This presents a considerable challenge to catastrophe risk analysts responsible for reporting accumulation output. Each stakeholder will have differing requirements with regard to the frequency and granularity of reports required. The frequency of reporting may range from a single annual snapshot, through quarterly, monthly, daily or even dynamic reports updated as each new risk is underwritten. Granularity may vary from a single metric measuring risk for the whole organization for a given region and peril (such as US hurricane or Japan earthquake) to a breakdown by line of business (e.g. commercial lines, personal lines) or other organizational criteria (e.g. office underwriting the risk, business unit). Additionally, the stakeholders will have differing levels of understanding, so the catastrophe risk analyst will need to ensure that the metrics chosen for the report are appropriate and clearly explained. This may involve making use of scenarios (as described in the case study in Section 2.7.1.3) or visualizations such as maps and event footprints to assist in the clear communication of the output.

Examples of the types of report that might be requested, and the stakeholders who might be interested include:

- aggregate exposure within a given hazard zone which dynamically updates as each risk is underwritten, e.g. of exposure within 100 m of the US coastline. Stakeholders: underwriters operating in the hazard zone;
- annual snapshot of the 1-in-100-year and 1-in-250 return period loss estimate for major peril regions such as US hurricane or Japan earthquake. Stakeholders: rating agencies;
- examples of natural catastrophe scenarios causing extreme loss to an organization taking into account the loss from all portfolios. Stakeholders: risk management team conducting a stress testing exercise;
- TVaR metrics describing the risk in the tail of the distribution broken down by line of business and region. Stakeholders: senior management making strategic decisions;
- the relative contribution of individual portfolios to the total average annual loss for a given region and peril. Stakeholders would be finance department allocating reinsurance costs;
- total sum insured by postcode. Stakeholders: reinsurance purchase team intending to include information in reinsurance submissions;
- accumulated YLT broken down by region, peril and portfolio: Stakeholders: actuarial team undertaking organization-wide portfolio optimization exercise.

Table 2.6  Uses of accumulation output.

| Use | Exposure accumulation | Loss accumulation | Topic |
|---|---|---|---|
| Capacity allocation and monitoring | Used to monitor aggregate exposure in defined zones to ensure it does not exceed the threshold determined by the organization's risk appetite. | Used for monitoring purposes where the organization's risk appetite has been defined using modelled loss metrics. This is often a VaR or TVaR metric. | Section 2.7.2.1 |
| Portfolio optimization | Not used | Used for understanding the total catastrophe risk and the relative contribution of each component (this could be a portfolio or individual risk) in order to provide a starting point for portfolio optimization. The accumulated ELT/YLT is often used for further data analysis. | Section 2.8 |
| Marginal pricing | Not used | Used to provide a baseline to which the new risk can be compared to assess its marginal contribution to the overall portfolio risk. In practice, this involves combining the ELT/YLT of the new risk with the accumulated ELT/YLT of the portfolio. | Section 2.6 |
| Reinsurance design, purchase and cost allocation | Aggregate exposure information is often included within reinsurance submissions | Understanding of the total accumulated risk, and the relative contribution of each portfolio is crucial to designing effective reinsurance programmes. Accumulated ELT/YLTs are often provided in reinsurance submissions. | Section 2.12 |
| Capital Modelling | Not used | Accumulated ELT/YLTs are the main way in which data regarding catastrophe risk are fed into the capital model. | Section 2.10 |

In practice, most catastrophe risk management teams responsible for reporting accumulation output will develop a standard set of reports that they produce regularly, meeting the main requirements of their organization, with any other reports needed being produced on an *ad hoc* basis. The types of reports produced regularly will vary greatly from organization to organization depending on both the reporting need and the resources (e.g. tools and staff) available. For example, a large complex organization may need more granularity on a regular basis. To help facilitate this, where the accumulation process is largely automated, it may be possible to produce reports more frequently than in organizations where accumulation is done manually.

While the reporting of catastrophe risk is an important end use of accumulation output, there are also a number of other applications where accumulation output is used as an input into other processes; this is particularly true for accumulated loss data in the form of loss tables (i.e. ELT and YLT). Table 2.6 provides an overview of some of the most common uses of accumulation data as an input; this also shows where applications are described elsewhere in this chapter, with capacity allocation and monitoring in Section 2.7.2.1.

### 2.7.2.1   Capacity Allocation and Monitoring

All (re)insurance organizations will have a maximum amount of risk that they are prepared to take on. This risk tolerance, or capacity, can either be defined in terms of a maximum aggregate exposure allowed within a given geographical zone, or expressed in terms of VaR or TVaR metrics for a given region and peril. It is typically set at a corporate level (see Section 2.10.1) and then cascaded down through the organization to individual portfolios and underwriters. Accumulation is necessary to monitor the amount of risk taken on for each portfolio to ensure the capacity allocated to that portfolio is not exceeded. Additionally, real-time monitoring of the current capacity allows underwriters to identify the remaining capacity available for deployment and identify potential opportunities to use it.

Where capacity is defined in terms of maximum aggregate exposures in a given zone, careful consideration must be given to the zone definition. The orientation and size of these zones will depend on the peril being considered and should relate to the hazard gradient of the peril. For instance, in the United States many insurers may place a restriction of the total value that can be underwritten in Tier 1 counties; these are the counties closest to the coast. However, for simplicity, some organizations choose an existing administrative definition such as CRESTA zones or counties and set a limit for the 'key' zone in terms of either exposure or hazard, or a combination of the two. For example, for Chilean earthquake, CRESTA zone 3 (i.e. Santiago) is considered to be the 'key' zone as it represents the peak concentration of exposure in the country.

A more sophisticated approach is to define zones in terms of hazard footprints. These can be events or probabilistic hazard maps. Events could include reproduction of historical events of importance to the particular region (e.g. the footprint of windstorm 87J in Europe) or 'spiced' versions of relevant historical events (e.g. Lloyd's' realistic disaster scenario 'Super Daria'). The probabilistic hazard maps could show the 1-in-50-year hydrological flood extent or the 1-in-100-year wind speed. This approach usually requires specialist knowledge to implement successfully, as Geographical Information Systems (GIS) are typically needed to define the zones and overlay the exposure data so the aggregate exposure can be accumulated. Consequently, this function is often carried out in catastrophe and exposure management teams, even though it does not necessarily require the use of catastrophe models.

Aggregate exposure approaches to capacity allocation and monitoring are particularly useful where no models exist or there are concerns about the quality of the catastrophe models available for that region and peril. However, care must be taken to ensure that concentrations of risk do not build up in areas outside of the defined zone, particularly where hazard footprints are used as the basis of the zone. For example, monitoring and restricting exposure based on a 1-in-50-year hydrological flood extent could inadvertently lead to a build-up of exposure just outside this zone, which is still likely to carry some flood risk, albeit at a lower level.

To address this concern, many organizations choose to use VaR or TVaR measures in regions and perils where reliable catastrophe models are available to them. By using the probabilistic model output, the full spectrum of possible scenarios will be taken into account. In these cases, loss accumulation is necessary to monitor the capacity used, both on an individual portfolio level and overall, because it is not possible to add these metrics across portfolios due to the impact of diversification (see Chapter 1.2).

These are the two most common approaches to capacity allocation. While there are a number of others such as monitoring ratios of premium to modelled loss, or using risk based capital measures, these are beyond the scope of this chapter.

### 2.7.3 Practicalities of Accumulation

Accumulation is, in theory, a straightforward process: the simple aggregation of exposure or losses. However, for many organizations the reality is significantly more challenging due to the large and varied amount of data which needs to be collated.

For exposure accumulations, detailed information about the risks underwritten is required including the sum insured, location information (to determine which accumulation zone it falls into) and any relevant policy terms (to calculate the net exposure after terms such as limits have been applied). This can be challenging for insurance organizations that need to aggregate across different geographical regions or different products (such as personal and commercial lines), if varied and potentially incompatible IT systems are in place for capturing and storing those data. Reinsurance organizations face an additional challenge, in that it is common in many regions for the data on location and sum insured that they receive to already be aggregated. In these cases, it may be impossible to carry out some types of exposure accumulation, particularly those using bespoke accumulation zones, due to the lack of detail in the data.

For loss accumulations, a consistent loss table (ELT/YLT) is required for all the risks to be accumulated. This can be particularly challenging in organizations using multiple catastrophe models for a given region and peril. Use of multiple models will require multiple sets of loss tables to be generated and combined (also see Chapter 5.5.4). This may be an arduous process, starting with the creation of analogous input files through choosing coherent options for modelling in different software. Also, a set of appropriate adjustments will need to be applied to the accumulation to ensure that it captures the entire risk (i.e. including non-modelled exposures and perils, see Chapter 5.4.7). These adjustments are likely to be variable by model to account for the differing limitations. It may also be necessary to apply an adjustment to account for differences in data capture between the differing models used. This step is vital to ensure that accumulations using multiple models are directly comparable and that differences in results are genuine rather than being caused by differences in the underlying basis of the results.

From a reinsurance perspective, accumulation of the net position (losses after terms have been applied) across the different cedants' portfolios is required to take into account the effect of the reinsurance structures, as well as different geographies and product lines. Gathering this information, and applying it to the model output for each risk prior to accumulation, can be a huge undertaking to ensure the correct terms are applied appropriately and consistently in each model used. Some companies have been able to automate this process, but in many organizations (particularly smaller companies), this often remains a time-consuming manual task.

For large or complex organizations, the ELT/YLTs being accumulated can become very large and thus require stable and powerful IT infrastructure. The major catastrophe modelling software companies do provide platforms for loss accumulation. However, many companies do not use these and prefer to export loss tables, combining them in systems such as Remetrica (http://www.aon.com/reinsurance/analytics/remetrica.jsp), Igloo (https://www.towerswatson.com/en/Services/Tools/igloo) or their own bespoke accumulation systems. These often offer greater ease and flexibility to apply reinsurance terms at multiple levels, and with an improved audit trail compared to the platforms available from modelling companies.

The frequency of accumulation varies considerably between organizations and is dependent on a number of factors. Where accumulation is an integral part of underwriting support, it will often take place daily (usually only in large or sophisticated organizations), but where it is a purely risk management and reporting function it may be as infrequent as monthly or quarterly. The IT infrastructure available for the exposure and catastrophe management, the resource in the catastrophe management area, and the materiality and propensity to change of the catastrophe risk will also impact the frequency of the accumulation.

## 2.8 Portfolio Management and Optimization

*Kirsten Mitchell-Wallace and Guillermo Franco*

### 2.8.1 What Is Portfolio Management?

Portfolio management is the creation and maintenance of a portfolio (or book of business) that produces the best overall combination of the business available within the commercial constraints, both of the company and market. In the example shown in Figure 2.15, the Chief Underwriting Officer must balance aims: to maximize return on the capital deployed; to obtain diversification between perils and geographical regions to earn the most premium for the catastrophe capacity deployed (in this case measured by a PML metric; e.g. the 1-in-100 OEP

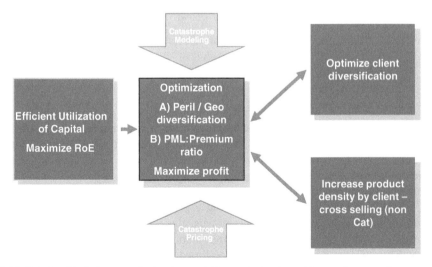

Figure 2.15 Relationship between catastrophe modelling, catastrophe pricing and treaty reinsurance underwriting strategy. *Source*: (SCOR, 2013a).

loss) and to develop a portfolio that allows the opportunity to sell a wider range of products (cross-selling) beyond property catastrophe reinsurance, for instance liability or motor insurance. As Figure 2.15 shows, portfolio management is closely related to catastrophe modelling operations such as accumulation and catastrophe pricing. The exposure needs to be interrogated and understood, and the accumulation process is used to monitor the capacity consumption which should be minimized with a well-diversified portfolio. This balancing of different aims and constraints in order to build the best possible portfolio is true to a greater or lesser extent in all insurance and reinsurance portfolios, and it is inherent in the underwriting process. Catastrophe models provide the risk metrics that are relevant to this process; thus, interaction between the catastrophe risk analysts and underwriters is required not just on an operational daily basis, but also to provide analysis to help inform tactics and strategy for the book as a whole.

The term 'optimization' is often used loosely in the industry and does not always mean optimization in the true mathematical sense covered in the next section. Often it is used to refer to the ability to investigate the optimal portfolio structure by considering different underwriting strategies, e.g. how does the diversification behaviour and hence Return on Equity (RoE) change with respect to premium income under different business scenarios? Typical investigations involve ascertaining how the key risk metrics would change if:

- a specific new large treaty was written?
- business was concentrated on a new geographical region?
- a new product type was added?
- a different **signing profile** was taken for catastrophe excess of loss reinsurance?

The term signing profile describes how the reinsurers' participation on layers changes with the layer height. For instance, the reinsurer may prefer premium-rich but loss-heavy bottom layers. Alternatively, they may prefer to write equal shares on every layer, called **writing across the board**. Whether the reinsurer has a preference for top or bottom layers will significantly change both their premium and risk metrics.

This optimization exercise will require various scenarios to be developed and built. This scenario testing is often also very closely related to the underwriting planning process and evaluating the impact of different strategies for growth.

It is also important to analyse the portfolio to better understand its composition and provide material for discussion around opportunities for growth and diversification. For instance,

- which are the key accounts? How much of the tail risk to they make up?
- are any of these contributing disproportionately?
- what is the composition of the portfolio PML? (VaR or TVaR?) (see Figure 2.16).
- what is the profitability of these accounts in terms of their normalized capacity consumption?
- how do different segments (e.g. lines of business or layers or geographies) compare?

Figure 2.16 shows an example of the types of events that can be found in the range of the PML (in this case, 1-in-100 OEP VaR). This helps to visualize the events in this range and where the concentrations of risk for this portfolio are.

Evaluation of growth is part of the planning process. The planning process is the process of deciding how the business should develop over the next period and projecting the performance during this period, for instance, by constructing a projecting balance sheet and examining income and expenditure. The period chosen is usually a year, but for a strategic plan for communication to investors may be three to five years.

As part of this, a projected portfolio will be developed. This portfolio will need to incorporate information on where the company plans to grow, or shrink and in which products or lines of business. The process will require significant engagement with the underwriters in order to

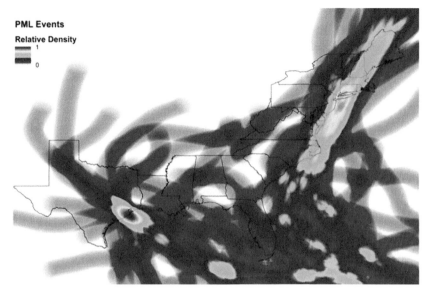

**PML Events**

Relative Density

Figure 2.16 Composition of VaR PML for US hurricane, by choosing the events above and below threshold and considering their spatial (i.e. geographic) density. *Source*: SCOR.

translate factors related to growth projections, which may not always be related to catastrophe metrics into a planned portfolio, which represents the portfolio congruent with the plan. For instance, underwriters may wish to increase the premium for a specific line of business in a specific territory e.g. 10% increase in New York for commercial lines. This business projection will need to be translated into a catastrophe projection factor, usually to the mid-point of the planning period considered. This may need to take (re)insurance rate changes into account to translate the premium change into an exposure change and ultimately into a PML. Similarly, for global business, exchange rates will need to be considered.

This planned portfolio is likely to be used for determining reinsurance purchase (see Section 2.12) and to examine the company's overall solvency position (see Section 2.10).

### 2.8.2  What Is Portfolio Optimization?

Portfolio optimization is the process through which a portfolio or book of business is built so that certain characteristics or performance metrics are maximized or minimized. For instance, a portfolio optimization process may be constructed to maximize premium while keeping within certain risk accumulation constraints. Alternatively, it may be used to seek a portfolio that maximizes coverage (e.g. sum insured or limit) while keeping to a minimum revenue constraint. A diverse array of problems that can be classed as portfolio optimization.

Optimization problems are hard because there are always opposing forces at play. Increasing a desired trait in the portfolio, like premium, also results in the increase of undesired effects such as the accumulation of risk, for instance. The skill in solving the optimization problem lies in finding the optimal balance between these positive and negative effects. Often even the definition of this optimal balance is hard to find.

The first step in the design of a portfolio optimization process is the definition of appropriate metrics that represent the performance of the portfolio. For example, the definition of precise risk and return metrics is usually crucial. Metrics to quantify the risk may typically be some measure of the tail distribution of loss such as TVaR or VaR PML at a particular return period. Return may well be represented by the premium, profitability, or number of policies. Often the framework will also need to consider additional metrics such as volatility or some other projection of the uncertainty involved in the process.

The balance between the positive and negative forces at play is often represented by the set of solutions that are not evidently superior to one another. In other words, this set may include solutions that have high premium and high risk and solutions that have low premium but also low risk. Because from an 'objective' (i.e. mathematical) standpoint one cannot identify a solution that is universally better than the other within this set, we typically refer to these solutions as the **efficient frontier**. Any solution picked from this set is not quantitatively inferior to any other solution within this set.

Since the optimization metrics can usually be represented in a two-dimensional space (such as premium and risk), a curve in this space represents this efficient frontier set of solutions. The curve in Figure 2.17 represents this optimal relationship of risk and return, and the set of all optimal portfolios (those not inferior to any other) can be found along this curve. Obtaining this efficient frontier is not a trivial mathematical exercise and may involve convoluted and lengthy operations. This, in itself constitutes an extensive space in the field of operations research.

If we consider the portfolio represented by the blue dot in Figure 2.17, see that an upwards vertical movement in the premium-risk space would increase the return for the same risk. Similarly, a horizontal move to the left would decrease risk while obtaining the same return (green dots). Within this simple context, an optimal portfolio maximizes return at a defined, acceptable level of risk and/or minimizes risk at a defined rate of return.

Even though the concept is well accepted, the complex decisions involved in the selection and computation of the appropriate metrics make portfolio optimization difficult to apply in practice. As mentioned above, all portfolio managers and underwriters are, in fact, conducting a subjective optimization in their activities. Their decisions to write or not to write are guided by their expected profitability and loss and constrained by overall company objectives that measure overarching performance through Risk-Adjusted Return on Capital (RAROC) or others (see Section 2.6.2.3).

### 2.8.3  Using Catastrophe Models in Optimization

Catastrophe modelling plays a critical role in the portfolio optimization processes in (re)insurance operations. However, a portfolio optimization process that is completely dependent on catastrophe model output is likely to be short-sighted unless the uncertainty in these models is considered. As the mathematical portfolio optimization process seeks an extreme of performance (very high return and very low risk, see Figure 2.17), it may lead to underwriting

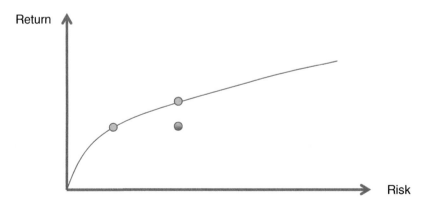

**Figure 2.17** Example of risk-return plot, where a metric for return is used on the x-axis and a metric for risk on the x-axis.

strategies that emphasize certain traits in the portfolio that may be highly susceptible to factors which are uncertain in the catastrophe models (e.g. the presence of category 5 hurricanes in the NE United States).

Experience has shown that it is difficult, but not impossible, to capture all the 'hard' metrics important in decision-making; these hard metrics are those that are mathematically defined through precise equations that take into account the capital requirements and usage as well as all the risk metrics derived from catastrophe modelling and from other sources. However, 'soft' issues also matter, and these are more difficult to quantify in a mathematical framework. For instance, in conducting portfolio optimization exercises with insurance companies, it is important to consider the legacy of a policy or contract that has been underwritten over many years, or any socially-responsible promise to the policyholders to provide coverage. An algorithm driven only by quantitative metrics only will not account for the value in certain business strategies that experienced underwriters would be able to implicitly identify. In order for the mathematical optimization process to be sensitive to these other factors, they need to be quantified and embedded into the process, which is often extremely difficult to do.

Companies find it harder to quantify these soft issues; these are often handled through human intuition and so there can be resistance to the idea of relinquishing control of the underwriting process to an algorithm programmed to simply maximize return metrics and minimize risk metrics. However, the appropriate approach is to develop algorithms that can *guide and inform* the underwriting process to provide solutions which are *influenced and constrained* by requirements aligned with the company's history and processes. Thus, pragmatically, portfolio optimization is often best introduced in small, limited steps. By restricting the scope of the mathematical portfolio optimization process, companies can make use of these techniques step by step, without imposing a sudden overhaul of their current underwriting practices.

In conclusion, in portfolio optimization based on catastrophe modelling, the optimization process must seek solutions that are adaptable to hard-to-quantify factors. It must also provide solutions whose performance is relatively impervious to deviations in catastrophe modelling results that may arise due to model changes or model error.

Given these challenges, what are some reasonable advances that a company can make in order to derive more sophisticated insight into their portfolio?

### 2.8.4 Optimization Methods

Lixin Zeng at Validus (Zeng, 2001, 2010) proposed use of the steepest ascent/descent methodology to solve a problem of portfolio optimization. By adding one policy/contract at a time to the total portfolio ensuring that the performance in each step was maximized, Zeng managed to obtain an overall portfolio that performed better than most portfolios that were not constructed with a structured process. Since the computational demands of such a method are also limited, the technique results in a speedy solution that can guide the analyst to a reasonably well-performing solution within the domain of the possible. This process simulates a **marginal impact analysis** in which one policy is added to the portfolio at a time, ensuring that certain overall metrics are maximized or minimized.

These successive marginal impact analyses are limited by the fact that in every step most of the portfolio remains unaltered and does not utilize the performance of the new policy or contract to search for new combinations that may perform betters and depend on the order. Adding the new policies in an established order and not allowing for the removal of previously added policies restricts the search space in the optimization process enormously.

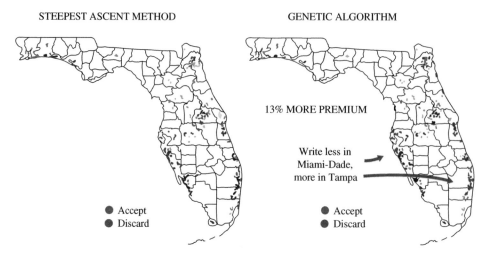

Figure 2.18 Comparison of two methods of portfolio optimization. Steepest ascent method (a) is a deterministic approach whereas the genetic algorithm (b) is a heuristic approach. *Source*: Reproduced with permission of AIR Worldwide Corporation.

Truly tackling the portfolio optimization problem with a more flexible search methodology requires more complex heuristics: examples of these include genetic algorithms, evolution strategies, simulated annealing and other methods that use an iterative approach to seek points of high performance in a highly complex and large domain of solutions (e.g. Franco, 2011a, 2011b; Ha, 2013).

Figure 2.18 from Franco (2011a, 2011b) compares portfolios of insurance policies optimized according to two different procedures. The first is a deterministic procedure (i.e. the steepest ascent), and the second is a heuristic one (i.e. a genetic algorithm). The aim was to achieve a maximum premium-to-loss ratio while maintaining constraints on losses so that maximum PML thresholds at certain return periods were not exceeded. The solutions obtained with the two different optimization approaches are significantly different. The deterministic algorithm suggests discarding policies in Tampa and central Florida, while the heuristic solution achieves a greater yield by discarding policies in the Miami-Dade area on the eastern seafront. This has important potential repercussions on the underwriting strategy of a company in terms of risk selection and shows the sensitivity of the process to the algorithms used; this sensitivity is on top of uncertainty in catastrophe modelling results.

Optimizing the same portfolio of properties in Florida using the heuristic approach results in a 13% increase in premium over the solution obtained with a deterministic method. The choice of one strategy over the other is an important financial decision for the company. Of course, there may be practical constraints to apply either proposal: both algorithms assume that a company can rebuild the portfolio with ease, which is often not realistic.

Portfolio optimization must always be undertaken in context; the constraints under which the algorithms will operate must be considered so that the solutions are viable and of practical use to the company. For instance, operational constraints could be integrated into the algorithm by defining a 'static portfolio', a part of the business that cannot be altered by the optimization process, showing where the company is not able to change underwriting decisions. Another constraint is that the solution is 'robust', meaning that small changes to the underwriting guidelines dictated by the optimization process should result also in small changes in portfolio performance. In other words, if the performance of a portfolio hinges on whether one particular contract or policy is included or excluded, the solution is not robust since failure to execute this decision may result in severe underperformance of the portfolio as a whole.

## 2.9 Event Response and Integration with Claims Team

*Kirsten Mitchell-Wallace*

A company's claims team is responsible for producing loss estimates for all claims. There should be a strong connection between the catastrophe risk management team and the claims team. This is especially important when a catastrophe event occurs or is imminent. This section presents the role of the catastrophe risk management team in the early estimation of claims, then presents claims stresses and inflations that catastrophe risk analysts should be aware of, before mentioning the role of lessons learnt analysis.

### 2.9.1 Early Estimation of Claims

Early estimates of event losses are important in (re)insurance companies. The estimates are required for internal business planning purposes, for regulators and ratings agencies, and for markets and analysts in the case of stock insurance companies. These may be confidential when first produced as an estimate can affect share price. There are also different methods of communication depending on the size of the loss, the reinsurance bought by the company and their philosophy, e.g. if the company has a low retention and loss is fairly certain to fall within the programme, the loss estimate will simply be the retention and can be reported quite quickly. However, other companies may wait more than a month before releasing loss estimates externally. The first loss estimate may also be used to book losses in the quarterly accounts of a (re)insurance company and ensure that appropriate money is set aside to pay the losses (**reserving**). The loss estimates and the estimated number of claims for an event are also used for post-event planning in insurance companies; this includes the mobilization of loss adjusters, claims handling staff or other contract staff to cope with the influx of claims and ensure their adequate processing. This early estimation process can be part of **event response**. Often a range of potential impact is provided here (e.g. 10–15 million dollars) rather than a single figure.

Most organizations will have an event response policy where the actions required by different parties within specific timescales will be defined. Event response differs depending on the peril; for meteorological perils monitoring will start when an event of interest is identified. Websites such as NOAA's National Hurricane Centre (http://www.nhc.noaa.gov) will be useful, but the company may also subscribe to alerts services such as Tropical Storm Risk (TSR) and Euro-tempest or WindJeannie from PERILS. Modelling companies will also offer bulletins. Since seismological events are not predictable, updates are only issued after events, although sometimes very quickly but the overall process is much the same.

A range of initial loss estimates will usually be required when there is still little information available: either before landfall or in the immediate aftermath of the event. The approaches used will therefore be pragmatic and a better estimate will emerge as more technical information becomes available over time, for instance as more detailed footprints are provided by model vendors or better data quality available is accessible. Methods used to estimate losses tend to follow a pattern of increasing complexity and robustness as more information emerges. For example, the progression of methods may be:

- apply damage factor on aggregate exposed value (Section 2.9.1.1);
- analyse likely industry market share (Section 2.9.1.2);
- use modelled footprints on portfolio, or apply the company's own footprint to the portfolio. These should be adjusted for non-modelled sub-perils, lines of business etc. (Section 2.9.1.3).

In all cases, losses must also include non-modelled loss components. These, however, are likely to be implicitly included in the first two methods to some degree.

Note that these steps may happen in parallel, or in a different order. For example, the catastrophe analyst may immediately try to identify an event similar to the one that has occurred from the catastrophe models' event set. This is known as an **analogue event**. Such identification could be done by comparing the event characteristics with those available in the event set (e.g. location, magnitude, depth for an earthquake) instead of applying a damage ratio on the aggregate exposure. The event response process will also differ between an insurer and a reinsurer given the different granularity of data, although the processes are likely to converge in the future as better systems are commonly available. In both cases, the impact of retrocession will be critical in determining the final number communicated externally; this said, the gross number must be estimated first.

### 2.9.1.1  Damage Factor on Exposure

The first task of the catastrophe analyst when any catastrophe occurs is to identify the area at risk: depending on the peril, this can be identified from the forecast, the touch-down data (Storm Prediction Center in the United States), news sources or a service such as USGS pager (www. earthquake.usgs.gov/earthquakes/pager). The location of the event is required to identify the total exposure in the area at risk in its footprint; this provides the upper bound for the loss estimate. Of course, while 100% damage is possible for an individual risk or a small area after a very severe event, this is not at all likely within an entire event footprint (see Chapter 1.8.1), except in extreme cases such as tsunami loss (see Chapter 4.3.3.5). In some cases, detailed exposure data may be available for loss estimate. However, since an estimate using a simple damage factor is very approximate, the catastrophe analyst may also choose to aggregate the exposure (e.g. at county, CRESTA, postcode).

This method is often used for perils for which there is no model (see Chapter 5.4.7.3.1). The aggregation level may be determined by the storage of aggregate exposures in an internal system. The level will also depend on the amount known about the extent of the event when the estimate is being made, given that the granularity of an estimate is always constrained by the weakest assumption.

A range of simple damage ratios should be applied to the aggregate exposures to create a range of outcomes. These damage ratios will be derived from experience, analysis of modelled damage rations and a large dose of judgement. The thought process in the specific case of a hail event can be illustrated by the catastrophe risk analyst asking the following questions: What is a credible average damage ratio across a county for a hail event? What damage ratios did we have in previous events? Is this event worse in terms of the area affected or its intensity? If so, how much worse? Once the damage ratios have been applied to the exposure, a process which is essentially a very simplistic event footprint, financial conditions must be applied to derive the total losses for the portfolio.

The resulting loss must then be sense-checked: Does it seem credible given the types of exposure? How does it compare to the probabilistic output? What is the return period of this loss compared to the portfolio VaR OEP curve? Of course, sometimes a particular event has a loss return period that is completely out of alignment with the underlying distribution because the event illustrates that something was wrong with the assumptions underpinning the loss distribution of the original model. At this point, it is common for the model development organization to consider a reformulation of the model in light of new information (see Chapter 4, and 'Key Past Events' in Chapter 3.X.6).

### 2.9.1.2  Market Share Analysis

One of the first external inputs will come when a range of agencies give their view on industry losses; usually these come from modelling companies, brokers and reinsurers and local insurance organizations. Knowing the likely market shares for different loss levels for different event types and geographies allows quick calculation of a range of losses using these industry loss estimates. This market share analysis allows cross-checking with more detailed approaches, such as footprint losses.

To develop an early estimate of loss, a study of company event losses as a proportion of the industry losses can provide a view of the company's share of market loss for events 'similar' to the one experienced. This study can be informed by historical events or probabilistic events. In the latter case, modelled portfolio loss is compared to the modelled industry loss for the whole probabilistic event set, segmented by appropriate geographies so that the area studied is representative of the event considered.

Similarity will be judged on the available event characteristics; for this task, as well as exposure analysis, an industry exposure database may need to be constructed for the region in question since many smaller countries do not have licensable industry exposure databases. The catastrophe risk analyst should also classify the modelled events into smaller geographical regions (sub-regions) of interest to ensure that the range of loss outcomes considered is representative of the particular region where the loss occurred. For instance, when considering hurricane, the catastrophe analyst should consider cutting the coastlines into segments (known as applying gates) so that the company loss and market share are appropriate for the subset of events that cross that particular gate. The location of the current event can then be compared with the sub-segmentation, in this case, the gate. The ratio of the company losses to the industry losses will be very variable, and particularly so for the smaller losses, so judgement and familiarity with the portfolio are essential in determination of an appropriate market share. Once a range of market shares have been applied to a range of losses, a range of outputs should be derived.

### 2.9.1.3 Applying Event Footprints to the Portfolio

Applying modelled footprints to the portfolio provides the most detailed insight into the potential loss (also see Chapter 5.4.3.1). The catastrophe management team will use a modelled footprint on the current portfolio to derive a loss estimate, examining the policies affected.

Selection of appropriate modelled footprints can be done in-house at the time of the event based on filtering criteria from the existing event set (analogue event), pre-selected (as in the gate analysis described above) or provided by modelling companies. Modelling companies may release new tailored event footprints to represent the events or identify event matches from their existing event sets that share similar footprints and have a similar loss (their own analogue events). Several footprints using different approaches are often released during the lifetime of a catastrophe event response process as more information is known about the event. Recently, modelling companies also have provided a range of outputs associated with the event so that an exceedance probability curve of potential portfolio loss is available (see Figure 2.19(a)). This is designed to help communicate the uncertainty around the loss estimate, but in reality the range is often larger than the stakeholders' tolerance for uncertainty so the catastrophe analysts will be asked to give their best estimate of a narrower range around the mean.

Sophisticated insurance companies will supplement this use of commercially available footprints with in-house metrics, for instance, they will develop their own footprint from scientifically available data, or use a footprint from another source (e.g. TSR; NOAA), and overlay with exposures and vulnerabilities. They may also interrogate the event sets to identify alternative analogue events from the catastrophe model event set comparing the characteristics of the event (see Figure 2.19(b)). Hazard footprints derived from satellite measurements can be useful, particularly when it is difficult to access areas; examples of this are after the 2011 New Zealand earthquake in Christchurch, and for tsunami events. Satellite footprints were very

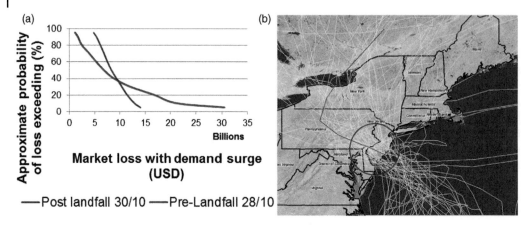

**Figure 2.19** (a) Event response output; (b) selection of an appropriate event for Hurricane Sandy from a major reinsurer. *Source*: (SCOR, 2013b).

successfully used to estimate loss for the Tianjin explosion in 2015 (Guy Carpenter, 2015), albeit this was not a natural catastrophe event.

### 2.9.2 Claims Stresses and Inflation

There are further factors that will influence an insurer's losses. After an especially large event, claims experience may be different from expectations, especially where those expectations are based on attritional claims levels (i.e. those in the absence of a catastrophe). It is neither practical nor cost-effective to staff claims departments for an extreme event. There are three major pressure points on an insurer after a catastrophe: resource pressures, extra expenses, and external pressures that may be regulatory or legal.

In some cases, the insurance company, its claims professionals and specialist loss adjusters may be unable to deal with the large number of claims effectively. This pressure on resources often leads to **claims inflation**. That is, claims values increase because normal claims handling techniques cannot be enforced with their usual rigour. This will usually be the case for lower value claims. Here insurers may suspend their basic rule of verifying cause, quantum and coverage to focus available resources on higher value and complex claims. For instance, in 1999, after cyclones Lothar and Martin in France, and Anatol in Denmark, the number of claims was so high, that many insurers elected to pay claims as presented without adjustment or validation below a threshold; a tenth of the Danish population was affected in the case of Anatol (Tätge, 2009), and 10% of residential policies were claimed on in France (RMS, 2000) Although this change in procedures allowed the insurers to focus on the more material claims, it is believed to have resulted in an increased industry loss (e.g. Piserra, 2000). Insurers may also incur additional professional costs for third party professionals such as adjusters and engineers, i.e. additional loss adjustment expenses may increase.

Economic demand surge can also cause **claims inflation**. This is the tendency for claims to become more costly because of a shortage of labour and/or materials after an event; if demand outstrips supply for the resource required to repair damage, the cost of settling the same claim is increased. This may be mitigated by the insurer having contracts in place with local suppliers at fixed costs.

There may also be political or regulatory pressure to apply a more liberal view to the interpretation of policy language or exclusions, known as **coverage expansion**. The courts may also be willing to support this, particularly in the case of personal lines policies where a consumers' rights issue may be cited. For example, in the United States, there are specific deductibles that apply depending on the perils. The deductible for hurricanes is much larger than for other wind perils for standard homeowners' policies (see Insurance Information Institute, 2015). After the very active hurricane season in 2004, the Office of Insurance Regulator (Kevin McCarty) decreed that the hurricane deductible could apply only once, even if a property experienced multiple losses (FLOIR, 2005). After Hurricane Sandy, hurricane deductibles were waived across multiple states for personal lines policies following the fact that Hurricane Sandy was no longer a named storm at landfall in New York, a condition for hurricane deductibles to apply there, e.g. Pinkser Gladstone (2012). Some believe this was a case of the regulators pre-empting the inevitable, whereas others believe it was politically motivated. In contrast, there have been a number of commercial cases where coverage expansion has been declined by the courts and regulators.

A further example of claims escalating due to environmental factors is related to the behaviour of contractors in Texas where the particular legal environment is especially punitive in case of claims handling failures (Badger *et al.*, 2014). The claims handling requirements of certain jurisdictions will drive insurers into relaxing their claims handling because the professional resources required to properly handle the claims in compliance with the applicable law will simply not be available.

Reinsurers generally agree to **follow the fortunes** of their clients; this is a doctrine where the reinsurer must accept the insurer's reasonable claims decisions provided there is coverage under the terms of the original insurance contract and unless the reinsurer has also agreed to cover **ex gratia** claim payments i.e. claim payments falling outside the terms and conditions of the contract or policy. The reinsurer will therefore pay the insurer's claims unless it can be shown that the insurer did not act in good faith or did not act reasonably. Thus, in practice, claims pressure on insurers following a catastrophe will likely also be experienced by the reinsurer. A further source of claims inflation is behavioural risk; this is where a cedant may present a claim to a reinsurer that is aggregated in a different way from that originally intended. An example of this is aggregation of winterstorm losses across a 'season', rather than by event or specific storm as experienced by some reinsurers in respect of the Q1 2015 north-eastern US winterstorms. Sometimes the reinsurer will agree to pay losses it believes to have been improperly aggregated as a commercial accommodation. This may be reflective of the insurer/reinsurer partnership arrangement or driven by soft-market conditions.

Coverage expansion (such as the Lothar/Martin case) is modelled in addition to economic demand surge as post-event loss amplification in some models, but is not applicable for all territories or perils nor necessarily comprehensive where it is applied.

Catastrophe claims are **short tail**. This means that the loss does not take a long time to develop, that is, that the final claim size will be known and paid in a reasonably short time after the event (tens of months to years). This contrasts with long-tail claims (e.g. third party liability), where the true extent of the final loss may not be known for decades. Although catastrophe claims are short tail, different claims development patterns are observed for perils and sub-perils. A particular example of this is for hail losses where the claims from roof damage may not be evident immediately. There may be differences by line of business with large commercial and industrial claims taking longer to settle.

### 2.9.3   Lessons Learnt Analysis

Close liaison with the claims team is also required to link policy exposure information with claims information, to inform lessons learnt analysis feeding into the development of each company's *own view of risk* (the subject of Chapter 5). For example, the modellers may wish to overlay estimates of loss derived from hazard footprints of an event with policy losses to understand the variability of claims by different policy attributes (e.g. cause of loss by windspeed band, loss adjustment expenses by loss type, losses by construction, year built, etc.). Claims counts and vulnerability functions can also be derived in this way. This analysis is extremely valuable in not only refining each organization's view of how models should be properly used, but also in beginning to understand the kinds of challenges faced by model developers when creating vulnerability functions (see Chapter 4.5).

There are practical challenges in post-event analysis, not least the association of the policy and location data, which is often much more onerous than expected.

This type of analysis will reveal that small claims are subject to larger uncertainty and that the number of claims at higher hazard intensities in the development of vulnerability functions is comparably small. The claims will be compared with the assumptions and the modelled output to understand:

- What's different?
- How did the model perform?
- If not well, can we understand where not (coverage, LoB)?
- Is it something that needs to be incorporated in future or just a quirk? How representative do we think the portfolio we interrogated is of the wider portfolio? How unusual are the circumstances of the event?

Since this is closely related to the development of one's own view of risk, examples are provided in Chapter 5.4.3.1.

## 2.10   Capital Modelling, Management and Dynamic Financial Analysis

*Junaid Seria and Kirsten Mitchell-Wallace*

### 2.10.1   Risk Appetite and Risk Tolerance

As discussed in Chapter 1.2, the economic benefit of risk transfer comes from risk carriers' ability to pool different risks and benefit from diversification effects. Risk transfer also ensures financial stability (see Section 2.3) by reducing financial shocks and allows businesses to continue operating in adverse circumstances.

(Re)insurance companies assume risk as the core of their business. To balance the amount of risk accepted from policyholders, the board of directors of a (re)insurance company defines a **risk appetite statement**. This risk appetite statement helps to define the level of downside risk that the company is comfortable accepting relative to the level of return desired. It helps to determine where companies wish to position themselves on the risk-return spectrum, between extremely risk averse (i.e. low risk-low return) and extremely risk prone (i.e. high risk-high return) (e.g. SCOR, 2015a). The risk appetite framework translates the firm's desired level of risk appetite into an operational framework that can be easily implemented and monitored (e.g. GIRO, 2011).

The risk appetite statement is translated into **risk tolerances.** Risk tolerances are a series of quantifiable constraints that limit the amount of risk that the company is willing to assume. There are many ways to define risk tolerances, but for catastrophe risk management these are typically based on measures of capital and/or profitability.

Catastrophe risk limits are typically defined by a risk metric drawn from the *tail* of a catastrophe loss distribution. The risk tolerance can be expressed as a limit, or maximum threshold, based on the maximum tolerable impact of modelled losses (at a given percentile) on available capital or regulatory capital. For example, Company ABC may state in their risk appetite framework: 'aggregate annual net losses (post-tax) at the 1-in-200 for UK Floods should not exceed 10% of ABC's available capital'.

For global insurers and reinsurers, risk tolerances are often defined for peak peril-regions based on the extent of exposure and hazard potential. These specific risk tolerances are likely to include:

- hurricanes (i.e. tropical cyclones) in the North Atlantic;
- earthquakes affecting North America;
- European windstorms (i.e. extra-tropical cyclones);
- Japanese earthquakes;
- Japanese typhoons (i.e. tropical cyclones).

These risk tolerances, or limits, are not always based on probabilistic measures. If no probabilistic models are available, a deterministic approach based on an extreme historical or hypothetical scenario may be used instead. Examples of deterministic approaches include a, a repeat of a severe historical event or particular stresses e.g. a 25% reduction in premium rates. The resulting impact of these individual (or composite) scenarios on the company's profitability and capital is then used to help set a firm's risk tolerance. Although deterministic approaches help to visualize a poor claims year, the downside ofusing them to define risk limits is that limit derived depends heavily on the appropriateness and relevance of the scenario.[1]

### 2.10.2 Why Capital Models?

Section 2.10.1 described risk limits as an effective tool to mitigate the chance of assuming too much of a particular risk. Financial (e.g. capital) models help to quantify these risk limits to allow effective risk management. Capital models are a subset of financial models that quantify the capital required using statistical and mathematical techniques based on company's risk profile (i.e. the number of type and type of risks underwritten).

As well as quantifying capital requirements and setting risk limits, capital models have other strategic uses, including:

- portfolio optimization, where firms evaluate various risk retention or reinsurance strategies in order to maximize the return on risk-adjusted capital;
- capital allocation to measure the return on risk-adjusted capital achieved by the company's business units (see Section 2.6.2.5);
- determining the cost of capital for pricing risks (see Section 2.6.2.5).

As an introduction to capital models, we will briefly revisit why (re)insurers need to hold capital (also discussed in Section 2.6.2.5). On average, a (re)insurer with a large and suitably diversified pool of risk should make a profit from its underwriting. This assumes good underwriting and claims management so that premiums exceed claims and expenses. However, natural catastrophes and other large events can, infrequently, result in years with extreme losses where claims are many multiples of the premium received in that year. Given these uncertain liabilities, (re)insurance companies must hold capital for these poor claims years. Capital models quantify the amount of capital that must be held under particular stress scenarios.

The limitations of using experience in a capital model are similar to the limitations of experience in catastrophe pricing (see Section 2.6). Where probabilistic approaches are used to model risk, historical data is required to fit a distribution to estimate potential loss outcomes. As historical observations are limited (as is the case for many types of low frequency, high severity risks such as catastrophe risk) the fitted or hypothesized loss distribution may be unstable and change in shape (particularly the tail thickness) as new observations emerge. This is because the tail of the fitted loss distribution represents observed historic volatility or 'known' volatility. Future outcomes, however, could be significantly more extreme than those implied by tail of the fitted loss distribution.

However, exposure-based models capture unobserved possible outcomes, thereby ensuring that the modelled loss distribution is suitably representative of extreme outcomes that may erode capital. This is one important reason for the use of vendor catastrophe models in capital models.

### 2.10.3   What Is a Capital Model?

A capital model is a financial model that helps to assess a company's financial condition. The company's liabilities based on the risks assumed are modelled using statistical and mathematical techniques. These liabilities are then compared to a profile of the firm's assets to derive a *balance sheet* view of the range of possible *economic outcomes*, i.e. capital models estimate the range of adverse outcomes arising from asset and liability stresses. This is called **Asset-Liability Modelling** (**ALM**). Stresses can include factors such as equity market decline and interest rate decline. Where companies have internal data available to model key variables, the capital model is referred to as an 'internal capital model'.

However, specialist third-party vendors provide 'external' modelling platforms to help improve the modelling of certain types of risks. The most commonly used external models are either Economic Scenario Generators (ESGs), such as those provided by vendors like Barrie and Hibbert (http://www.moodysanalytics.com/About-Us/History/BH-History) and GEMS (https://www.conning.com/risk-and-capital-management/software/gems.html), as well as natural catastrophe models. The value of these external models, including natural catastrophe models, is arguably to provide access to best practice scientific, mathematical and statistical techniques that are reportedly applied to extensive market asset or insurance exposure and loss data. This helps where companies have limited reliable internal data and are therefore at risk of spurious results. These external models provide input to the internal capital model.

However, external vendor models may be unaffordable or the development of an internal capital model too resource-intensive. In these circumstances, companies may use a simpler approach to quantify the risk-based capital required, for instance, a deterministic scenario where a factor is applied to an exposure measure (like premium or reserves). These factors provide an extreme loss (typically at 1-in-200 VaR AEP) on the exposure measures. These are typically calibrated at an industry level. This factor-based approach is one basis for calculating regulatory capital in many regulatory regimes and is known as a **standard formula** approach (see also Section 2.11.3.1).

### 2.10.4   The Structure of Capital Models

Capital models are well described in a number of academic papers and industry publications (e.g. Zurbuchen *et al.*, 2010; Morin, 2011). Two free sources of other publications include the Casualty Actuarial Society (www.casact.org) and the Institute and Faculty of Actuaries (www.actuaries.org.uk) websites.

A very brief overview of capital model construction is provided below. There are seven aspects core to the construction of a capital model:

1) *Risk factor definition:* In the first instance, capital models are based on key risk factors that need explicit definition. These are the risk factors considered to drive the performance of the insurance portfolio. Capital models will likely include one or more of the following risks: insurance or underwriting risk (including catastrophe risk), market risk (i.e. investment risk), counterparty default risk and operational risk.

2) *Risk segmentation:* Depending on the available data inputs, underlying behaviour of the risk factors and required model outputs, the modeller needs to define homogenous risk segments that can be easily calibrated.

3) *Method of risk factor quantification:* With risk factors and segments identified, the method by which data will be used to estimate loss outcomes must be selected so that the tail risk metric can be referenced. The following choices are typically considered:

   1) Whether to use a frequency-severity (e.g. for catastrophe-type losses) or an aggregate loss approach (e.g. attritional-type losses). In some cases (e.g. lack of data), loss calibration is inappropriate. In this case a discrete stress test approach may be used, which is where an economic variable (e.g. interest rates) is stressed upward or downward and the resulting economic loss is computed. The choice will need to consider the materiality of the risk factor and the availability of data in order to estimate the distribution parameters.

   2) Which distribution to assume and its parameterization. This choice will be based on the behaviour of the risk factor and the fit of the chosen distribution, particularly at the tail of the loss distribution. The selection is typically based on a goodness-of-fit testing process that compares observed and modelled losses. Parsimony is also important in parameter selection; that is, a distribution that provides an appropriate representation of risk with fewer parameters is likely to be preferred.

   3) Whether to use a stochastic or deterministic approach to generate loss outcomes. Stochastic Monte Carlo simulation techniques are favoured due to computational ease and convenience as these are standard in capital modelling software.

4) *Dependency and aggregation:* Once risks are calibrated, a basis for combining results is needed to compute the overall capital results. Simply adding loss results does not account for diversification of insurance portfolios. For instance, it would be overly conservative to assume that all bad events happen at the same time. Therefore, mathematical statistical approaches are employed to represent the diversification benefits of being exposed to multiple risk factors, some of which are independent, partially correlated or perhaps negatively correlated.

5) *Valuation basis:* The basis on which balance sheet assets and liabilities are valued is not always consistent. Broadly, an economic basis stipulates that assets and liabilities are valued using a market-consistent framework. Assets can typically be valued on a market-consistent basis, with adjustments for tax or 'marked to market' (i.e., where the fair value of assets are valued based on current market value). Insurance liabilities, however, need to be marked to model. In this case, the mean or best estimate reserves are based on the present value (see Section 2.6.2) of future cash-flows discounted using a **risk-free rate** plus a market-based risk margin that compensates a would-be buyer for the cost of capital required to accept these risks. This is the difference between the economic balance sheet used in in (re)insurance companies and the standard balance sheet subject to accounting principles only.

6) *Risk measure selection:* A key objective of capital modelling is to determine the level of capital to maintain to ensure the risk of insolvency is kept to some prescribed minimum. The risk measure selected is therefore typically a measure of the tail risk faced by the company. Two commonly used tail risk measures are the Value at Risk (VaR) or Tail Value at Risk (TVaR) (see Chapter 1.10 for definitions). The risk measure ideally should be coherent allowing comparison of different risks (see Chapter 1.11.3).

7) *Time horizon:* Assets and liabilities to which the company is currently exposed can either be projected into the future until all liabilities are settled (that is, their ultimate position) or they can be projected over a shorter one-year period in order to construct an economic balance sheet (i.e. determine the value of assets and liabilities) one year into the future. These two approaches will yield more disparate results where the firm holds longer-tail liabilities. The

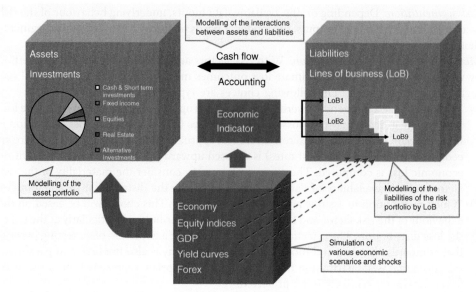

⟹SCR = VaR 99,5%, i.e. amount of capital needed to absorb losses by the end of the year with a 99,5% probability

Figure 2.20 Representation of a capital model. *Source*: (SCOR, 2015a).

ultimate and one-year position is likely to be aligned for a company with a portfolio of very short-tailed risks (Morin, 2011) such as a catastrophe risk (see Section 2.9.2 for a definition).

Figure 2.20 illustrates the link between assets and liabilities within a capital model. The left grey box represents the asset profile, for a particular mix of assets. The right grey box represents the liability profile. The Solvency Capital Requirement (or SCR) is computed by modelling the downside impact of key economic variables (such as interest rate and equity market indices (e.g. FTSE, Dow Jones) on the asset profile as well as the extreme loss potential arising from insurance liabilities (such as the impact of natural catastrophes).

### 2.10.5   Capital Models and Catastrophe Models

Capital models help to translate the company's risk appetite into an operational framework that can be easily implemented and monitored. Catastrophe risk is an important loss driver that can cause the (re)insurer insolvency if not managed appropriately. Therefore, catastrophe models are used in this process to define specific limits for particular geographical regions and insured perils. In addition, catastrophe models are used to quantify the catastrophe risk component of the capital model. For companies with an appetite for catastrophe risk with a low level of diversification and risk mitigation, catastrophe risk is likely to be the key risk driving the capital required for solvency purposes.

### 2.10.6   What is Dynamic Financial Analysis (DFA)?

In **Dynamic Financial Analysis (DFA)**, a (re)insurance company estimates their range of economic outcomes by using internal and external data to calibrate statistical distributions for key economic variables and applying Monte Carlo techniques. The term DFA can be used to mean capital modelling, although DFA tools can be used for accumulation (see Section 2.7.3) or reinsurance pricing (see Section 2.6.4.2) particularly where structures are complex. Another term for the same process is Asset Liability Modelling (ALM).

DFA (or ALM) tools are provided by reinsurance brokers such as:

- Willis Towers Watson (Igloo, https://www.towerswatson.com/en/Services/Tools/igloo);
- Aon Benfield (Remetrica, http://www.aon.com/reinsurance/analytics/remetrica.jsp);
- Guy Carpenter (MetaRisk, http://www.guycarp.com/content/guycarp/en/home/managing-risk/analytics/metarisk.html).

Stochastic simulation overcomes the limitation of using only historical losses, noted in the previous Section, by sampling a random variable from a pre-defined distribution to create any number of realizations of an output variable of interest. The process generates a large volume of 'synthetic' or 'pseudo' observations that yield more stable model outputs assuming a sufficient number of simulations are run. Since the outputs are generated from a distribution of inputs, the range of potential outcomes is likely to be diverse and more representative of the potential extremes than a deterministic approach (e.g. Kaufmann *et al.*, 2001).

Catastrophe model output (typically in the form of an ELT or YLT (see Chapter 1.10.4) is an input to the DFA models. For an ELT, the DFA's stochastic generator simulates catastrophe losses by sampling from the loss frequency and severity distributions based on the ELT. Loss distributions for different risk types are then combined or aggregated in order to generate an output distribution of economic surplus values over a defined time horizon. There are three main benefits of using DFA models:

- they are based on robust, accepted and recognized mathematical, statistical and economic principles;
- they provide a rich range of economic outcomes;
- they are accepted measures of assessing the risk of insolvency.

## 2.11 Regulation and Best Practice in Catastrophe Modelling

*Junaid Seria, Claire Souch, and Paul Nunn*

This section explores the drivers of the continued professionalization of catastrophe modelling. It starts with a summary of significant developments related to events then provides a brief summary of sources of information on current best practice. This is followed by a look at two important external drivers of changing modelling practice: the role of rating agencies and regulators.

### 2.11.1 The Evolution of Catastrophe Modelling as a Profession and Best Practice

#### 2.11.1.1 Introduction

Catastrophe models have existed for over 25 years. The discipline of catastrophe modelling is now widespread, and essential, in the (re)insurance industry. However, there is no professional body of catastrophe modellers with associated standards and qualifications in the same way as, for example, exists for the actuarial profession. Yet the practice of catastrophe modelling within the (re)insurance industry has evolved substantially, particularly since the mid-2000s. Many companies, particularly those with high catastrophe exposures, have realized the benefits of improving their knowledge, i.e. using catastrophe models to better manage capital, reduce surprises, and seek competitive advantages in underwriting and portfolio management.

Catastrophe modelling has, therefore, been undergoing a process of professionalization, even if still not officially a profession in the full sense of the word.

Today, many companies with significant catastrophe exposure employ their own scientists in-house. This allows them to explore the models more deeply, understand the data and assumptions going into the models, and create their own view of the risk (see Chapter 5) or build their own models. In the industry as a whole, there is arguably more expertise and a far deeper understanding of the models than there was even 10 years ago. However, in the absence of a professional governing body, catastrophe modelling 'best practice' still varies considerably. In general, however, standards appear to be increasing over time as a result of competition, recognized benefits and by financial regulation in some parts of the world. The best-known of these regulatory regimes is Solvency II in Europe (discussed in Section 2.11.4), the principles of which are spreading to other regulatory authorities across the world. There have, however, been other drivers of this evolution in catastrophe modelling practice in the past decade; these are outlined below.

### 2.11.1.2    Significant Developments

Over the years, catastrophe events have highlighted sources of loss not covered by models, the uncertainties associated with the understanding of both hazard and vulnerability and, at times, have shown systematic biases in models requiring correction. Some of the developments associated with these events are described briefly below.

*Lothar and Martin 1999*

Lothar and Martin were two strong windstorms which swept across Europe within 36 hours of each other on December 26–28, 1999. Of the two, Lothar caused the most significant damage (see Chapter 3.3.6); its track passed over Paris where it also reached its maximum intensity, arguably representing an almost worst-case scenario. These storms struck after a period of relative calm for European windstorms since 1990, and were a sharp reminder to the industry of the risk from such windstorms. The storms highlighted the lack of consideration of windstorm clustering in models at the time, i.e. the occurrence of events close together in both time and space. Clustering is driven by larger-scale atmospheric conditions such as the jet stream, which drive the development and trajectory of storms in a similar manner (see Chapter 3.3.1). This phenomenon challenges the assumption of event independence (i.e. Poisson) on which many models are built. Together with the 1990 cluster of four storms (i.e. Daria, Herta, Wiebke, and Vivian) Lothar and Martin prompted academic research and several initiatives to understand and model clustering; this remains ongoing more than 15 years later.

*The 2004 and 2005 Hurricane Seasons*

The 2004 and 2005 hurricane seasons produced a total of seven catastrophic hurricanes and around $100 bn of insured losses (Swiss Re, 2005; Swiss Re, 2006). Additionally, the 2005 season produced both the most intense hurricane ever recorded in the Atlantic basin at that time (Wilma), and the highest number of storms in one season (28 tropical and sub-tropical storms). The events highlighted multiple issues with the models across wind hazard, vulnerability and storm-surge-driven coastal flooding including Hurricane Katrina. Later, some of these components proved inadequate when Hurricane Ike struck Texas in 2008, again causing a high level of damage beyond the model predictions. These events also opened debate on whether hurricanes were going through a phase of, or even permanent, elevation in frequency and intensity, and if this should be considered by the industry (see Chapter 3.2.1).

These hurricanes also underscored the issue of data quality in the industry; the accuracy and completeness of data on the risks themselves, such as information on construction, line of business, and location was found often to be missing or inaccurate. One of the most famous examples was that of 'floating casinos'. At the time, laws in Mississippi confined all casinos to mobile marine vessels. Many casinos were built on barges on the waterfront and so were extremely vulnerable to storm surge-driven flooding. However, the information supplied about their location and building usage as 'casino' resulted in them being represented as more structurally sound properties on land in the models, and the damage to these expensive properties was substantially underestimated. This, and similar examples, prompted heavy emphasis on data quality across the industry.

### 2011 Model Changes

In 2011, one of the predominant providers of catastrophe models to the insurance industry, Risk Management Solutions (RMS), rebuilt both their North Atlantic hurricane model and Europe windstorm model, integrating new research, data and assumptions. In both cases the loss output from the models increased substantially. For some specific books of business, the new model results were more than 80% higher than the previous model at the 1-in-100-years level. Such dramatic shifts highlighted the degree of uncertainty that is inherent in catastrophe models (see Section 2.16.1). These uncertainties remain largely unquantified.

In recognition of this uncertainty and the validity of having multiple views of risk, many in the industry moved to using more than one model, while also conducting research to derive their own view of the risk. Both steps were taken in order to be less exposed to unvalidated changes in external models. Appreciation of the value of multiple views opened the door for new modelling companies. This also led to initiatives to build open modelling platforms from which many different models can be accessed, and open models that allow the user access to and ability to change components of the models (see Chapter 6.2.3).

### The 2011 Catastrophe Events

A series of catastrophe events in 2011 caused a combined insured loss of an estimated US $116 billion (Swiss Re, 2012), the most costly year for the insurance industry in history at the time. The majority of this insured loss was from perils not, or only partially, covered by the major catastrophe models in use at the time. Damage from the Thailand flooding, the tsunami from the Japan earthquake, and extensive ground failures and liquefaction from the Christchurch earthquake were all unexpected or non-modelled sources of loss. Combined with some large model changes in 2011, this proved to be a landmark year in the evolution of catastrophe modelling. This record year drove the recognition of the need to develop an 'own view of risk' to include non-modelled sources and to stabilize -impacts of external model change (Chapter 5).

### 2012 Hurricane Sandy

Hurricane Sandy (see also Chapter 3.2.6) had weakened to be a post-tropical storm by the time it made landfall over New York, according to the definition of National Hurricane Center. Yet, it caused an estimated US$20–25 billion of insured loss (Swiss Re, 2013). A very large storm in terms of its physical size, Sandy produced significant coastal flooding, which caused over 50% of the total insured loss (see Chapter 3.2.5). The storm reinforced the need for more advanced models of coastal flooding, as well as the capture and modelling of high-value equipment and contents in basements.

### 2.11.1.3 Industry Initiatives, Professional Bodies and Accreditations

Although becoming more important, catastrophe modelling remains a relatively niche specialism in global terms. The number of catastrophe modellers is not known, but the total is likely to be in the low thousands. In contrast, globally there are in excess of 40,000 actuaries, according to statistics from the International Actuarial Association (IAA). To put this further into context, it is estimated that for every actuary there are over 100 accountants and 200 doctors in the world.

However, there is a drive to professionalize this relatively niche specialism. Here we present a number of industry initiatives towards accreditations and practice standards that aim to help improve the way catastrophe risk is assessed by (re)insurance firms. Some examples are:

- The International Society of Catastrophe Managers (ISCM) is a professional association launched in 2006 to promote catastrophe management professionalism within the (re)insurance industry. The society had 450 members in 2015. The ISCM co-organizes the annual catastrophe risk management conference in America in association with the Reinsurance Association of America, with typically around 450-550 attendees. The society currently has a predominantly US focus and membership, so has more limited traction in international, and emerging markets in particular.
- Within the actuarial community, two sets of standards are worth noting. First, the Actuarial Standards Board in US published a practice standard (called ASOP 38, Actuarial Standards Board, 2013) entitled 'Using Models Outside the Actuary's Area of Expertise (Property and Casualty)'. This includes in its scope the use of catastrophe models and has led to the publication by both RMS and AIR of specific ASOP 38 reports on US catastrophe models aimed at actuaries that need to rely on catastrophe models. Second, the Financial Reporting Council (FRC) issue and maintain Technical Actuarial Standards (TAS). This covers a wide range of areas of actuarial work, including data (TAS D) and modelling (TAS M) (Financial Reporting Council, 2009, 2010). A number of actuarial working parties have focused on catastrophe modelling. Their publications are accessible from the publications Section of the Institute and Faculty of Actuaries website (https://www.actuaries.org.uk/) as well as the Casualty Actuarial Society website (http://www.casact.org/).
- The two predominant catastrophe model vendor companies, RMS and AIR Worldwide have created their own training programmes with associated certification. The RMS Certified Catastrophe Risk Analyst (CCRA®) Training Program leads to the CCRA designation. Holders of this designation are entitled to credits at the Advanced Diploma in Insurance level from the Chartered Insurance Institute (CII), and for continuing professional development (CPD) credits with the Chartered Property Casualty Underwriters Society. The AIR Institute Catastrophe Modeling Certification Program is a similar programme run by AIR Worldwide. Those who pass earn recognition as an AIR Institute Certified Catastrophe Modeler (CCM™), and gain eligibility for Continuing Professional Development credits from the American Institute for Chartered Property Casualty Underwriters (CPCU) and the American Academy of Actuaries (AAA). These training programs explore many aspects of model building and use, but also reflect the views of the commercial organizations.
- The Association of British Insurers (ABI) published its *Industry Good Practice for Catastrophe Modelling* (ABI, 2011). It serves as a guide to the incorporation of catastrophe models within a firm's economic capital model.
- The Lloyd's Market Association (LMA) produced in 2012 (LMA, 2012) an illustrative validation document for US Windstorm, aimed at providing practical guidance on how to validate an external catastrophe model in a way that complies with European regulation.
- Reinsurance brokers and (re)insurers publish technical thought leadership articles that serve as a useful resource to improve understanding of catastrophe models and the underlying hazards.

In this section various industry initiatives, professional standards and accreditations have been outlined. In Section 2.11.2 we consider the role of the rating agencies before focusing on regulation.

### 2.11.2  Rating Agencies

As the influence of catastrophe models on (re)insurance companies' management and their assessment of business solvency requirements has increased, so has scrutiny from external stakeholders such as rating agencies and regulators. The role of the regulator as an external stakeholder is discussed in Section 2.11.3. Here we describe the rating agency approach in brief and consider how catastrophe risk is incorporated in their assessment.

Rating agencies (such as A.M. Best Company, Standard & Poor's, Moody's Investors Service and Fitch) issue credit ratings for corporations, including (re)insurers, but also for governments and certain financial instruments (e.g. Standard & Poors, 2015).

A credit rating reflects an opinion about the ability of (re)insurers to meet their financial obligations. It is a forward-looking opinion on credit risk that considers a number of factors in the assessment of credit quality and so is not an exact measure of the likelihood of default.

Risk-based capital (RBC) adequacy models are used in the assessment of capital adequacy (and hence likelihood of default) for life, property/casualty (P & C) and health insurance companies, as well as reinsurance companies worldwide. Models are factor-based with the factors calibrated on industry data (e.g., US Schedule P data, which is an annual statement of losses and reserves, in the case of Standard & Poor's RBC.)

Qualitative factors that influence the risk of insolvency such as the proportion and nature of capital are included. These include: reinsurance versus direct property; that sufficient reserves are held to cover the final cost of losses; how easy it is to access assets and the risk concentrations.

While the model is factor-based, it aims to capture the present value of expected economic losses (change in shareholder equity/policyholder surplus i.e. assets – liabilities or net worth) experienced over a year, to a degree of certainty that is commensurate with the rating. The Standard & Poor's approach is briefly described in Box 2.5.

---

**Box 2.5  Case Study: Standard & Poor's (S&P) Rating**

---

Taking S&P as an illustration, the confidence levels establishing the degree of certainty for each individual risk are: 97.2% for 'BBB', 99.4% for 'A', 99.7% for 'AA', and 99.9% for 'AAA'.

In the case of S&P, there is an exposure-driven property catastrophe charge. They note that the charge is based on:

> the tax-adjusted aggregate one-in-250-year property-line-only probable maximum loss, calculated net of reinsurance and other forms of mitigation such as catastrophe bonds. This probable maximum loss must include demand surge, fire following (attached to earthquake and fire policies), sprinkler leakage, storm surge, and secondary uncertainty losses. The capital charge covers global catastrophe exposures: hurricanes (wind), flood, earthquake, tornadoes, and hail. The charge should capture the impact of investments in catastrophe bonds, as well as those issued by the insurer.

Despite containing an exposure-based charge, the RBC is calibrated on external industry data and so may not be a fully appropriate representation of the underlying risk profile of the company under review.

Rating agencies are increasingly considering, where available, the firm's own (economic) capital model and comparing these results to the RBC results. In certain cases, the RBC-based assessment of catastrophe risk may then be altered to reflect factors unique to the firm as represented in their economic capital model.

### 2.11.3 Regulation and Catastrophe Modelling

In this section we explore the link between regulation and catastrophe modelling. One objective for regulators is policyholder protection. Although financial conduct is also an important aspect of regulation, in the context of catastrophe models we restrict our discussion to the major risks facing a (re)insurance company. These are:

- the risk of *insolvency* as a result of adverse catastrophe losses relative to what is expected or planned;
- the risk of *inadequate pricing* of insurance or reinsurance contracts that provide coverage against natural catastrophe risks. Additionally in the case of US homeowners, this includes the risk of excessive pricing.
- the risk of *failure of key processes or controls* related to the management of catastrophe risk.

While regulatory regimes around the world differ in the application of insurance financial regulatory principles and their reporting requirements, there are typically two common elements:

1) quantitative regulatory capital thresholds required to maintain a pre-defined level of solvency;
2) qualitative regulatory requirements related to the firm's risk management framework.

Typically, the quantitative capital requirements address the risk of insolvency and related risk of inadequate pricing, while the qualitative risk management requirements address the risks related to failure of a process, a system or a function.

This segmentation is, however, not clear-cut and these risks are inter-related. For example, the risk of inadequate pricing may well manifest due to a failure in the firm's risk management framework, e.g. peer reviews not completed for material pricings. In the following sections, we describe briefly how regulation attempts to address these three risks.

### 2.11.3.1 Risk of Insolvency

The risk of insolvency is the risk that liabilities valued at a particular point in time exceed the available assets. More conservatively, this is when liabilities exceed a percentage of the eligible assets since regulators discount assets that are very illiquid. Insurance regulation in the country where the company is domiciled will specify a capital amount that the firm needs to hold in order to mitigate the risk of insolvency. This is regulatory capital, also known as prudential or solvency capital. The requirement imposed is typically linked to the liabilities to which the company is exposed, while the insurance regulation enacted in the country will specify how the requirement is to be calculated.

For instance, in the case of Solvency II, companies have two options in the calculation of their regulatory capital. The first is to use a standard formula, where the capital requirement is based on pre-defined industry-calibrated factors that are applied to the company's exposure. Alternatively, firms can use an internal model to calculate a distribution of economic capital outcomes based on the firm's own profile of risks (see Section 2.10.4). The solvency capital requirement is set at the 1-in-200 return period economic loss distribution outcome as illustrated in Figure 2.21.

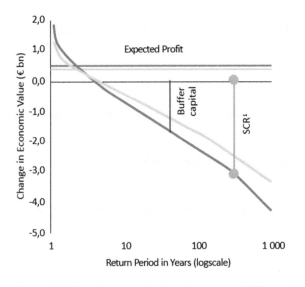

Figure 2.21 The solvency capital requirement and different risk appetites. *Source*: Reproduced with permission of SCOR.

For firms with catastrophe exposures that are material and complex, catastrophe model outputs are typically a major input in the calculation of the firm's capital requirement (see Section 2.10). For more sophisticated risk-based regulatory frameworks, such as Solvency II, the regulation will outline the technical specification for the calculation of the solvency capital requirement.

Here, the construction of the catastrophe model helps firms comply with these technical specifications by providing a framework to understand the range of extreme outcomes that may arise from the insurance coverage provided. As discussed in Chapter 1.4, in many cases these simulated yet plausible extreme events will not have been observed.

For the majority of insurance firms, however, a simpler factor-based approach as described in Section 2.10.3 is typically used to estimate a firm's solvency capital requirement; this is used because of cost and lack of data. This is referred to as the standard formula approach in Solvency II. These factor-based approaches are applied to the firm's expected exposures, for example, premiums, reserves and possibly asset values, depending on the regulatory regime.

### 2.11.3.2 Risk of Inadequate Pricing

Catastrophe models are also used to help estimate a company's catastrophe expected loss before the application of profit and expense loadings (Section 2.6.2). As a company's internal loss and claims experience is limited, catastrophe models provide insight into the range of extreme events that could emerge over the contract period for the perils insured.

Regulation does not typically specify that catastrophe models should be used in the pricing process, but there are notable exceptions (Section 2.11.5). As a general principle, however, insurance regulation notes that the valuation methods need to be proportionate to the nature, scale and complexity of the risks being assessed. Given the complexity of natural perils, catastrophe models arguably offer a proportionate approach to modelling these natural phenomena. A large catastrophe risk writer without access to models may well, therefore, be challenged by their regulator (see Table 5.1).

Further, inadequate pricing will result in a worse, and more volatile, underwriting result. This in turn will result in higher capital requirements as the higher volatility flows through to the company's economic loss distribution. The impact may take years to manifest, especially given that a company's catastrophe risk portfolio may fortuitously enjoy a number of good years with lower than expected losses, until a really bad year emerges. In the long run, however, continued under-pricing will likely result in incremental increases in capital requirements as the deviation of actual from expected losses widens.

### 2.11.3.3  Risk Management System Failures

In terms of the risk management system, regulation is concerned with the failure of the risk management framework to the extent that it threatens the viability of a firm to continue operating as a going concern (that is, with adequate resources to operate indefinitely). Typically, this manifests as breakdown in the company's internal processes, systems or due to misconduct of people overseeing these processes or systems. Risk management failures may in their extreme lead to wider business failure. Examples include the mis-estimation of liabilities (e.g. due to corrupt or missing data), severe loss of reputation or a loss of an entire catastrophe risk management team.

Insurance regulation will, therefore, typically set out qualitative governance requirements targeting the risk management framework with the aim of mitigating the risk of business failures.

These qualitative requirements typically cover:

- the organizational structure and in particular, the critical control functions within the company, such as risk management, internal audit, compliance and the actuarial function;
- how capital metrics should support decision-making to encourage the integration of models in the risk management process;
- reporting of relevant information;
- fit and proper requirements for individuals carrying out governance roles;
- outsourcing, including the use of external vendor models, which includes catastrophe models.

Catastrophe analysts make judgements on which switches and options to use in a catastrophe model or how to adjust the model based on an assessment of limitations in data and quantification techniques. These judgements are at risk of bias from catastrophe model vendors, brokers and internal experts. The qualitative requirements and governance structures outlined above serve to mitigate these biases, for instance, the requirement to have policies and procedures when using the model (this is discussed in more detail in Chapter 5.3), and via activities such as peer review or formal sign-off by a group of experts.

### 2.11.4  Case Study: Catastrophe Models and Solvency Regulation, Solvency II

While there are a number of different regulatory regimes, the two common themes linking them are the risk of insolvency and risk management failure. Given this commonality, we take Solvency II as a case study (EU LEX, 2015). Solvency II is a European legislative programme implemented by 1 January 2016 in 28 Member States, including the United Kingdom. It introduces a new, harmonized European Union-wide insurance regulatory regime. Solvency II is a framework based on principles rather than proscribed rules. There are two core aims of Solvency II.

1) Solvency II prescribes a deeper evaluation of the company's risk profile and risk management framework compared to earlier regimes where percentages of premiums and claims were used to assess capital requirements.
2) Solvency II aims to harmonize the approach of supervisory regimes through its application across the 28 member states. In addition, a Solvency II equivalence regime for 'third countries' (or non-EU countries) allows EU firms to use local regulation to report on their operations and enables non-EU firms to comply with local regulation in respect of their activities in EU countries.

The framework has three pillars:

- Pillar I – Quantitative requirements: two capital requirements are specified, a *Minimum Capital Requirement (MCR)* and a *Solvency Capital Requirement (SCR)*. Depending on the nature, scale and complexity of the company's risk profile, the SCR can be calculated using a standard formula or an internal model. In the case of the standard formula, the capital requirement is based on pre-defined industry-calibrated factors that are applied to the company's exposure. Alternatively, these risk factors can be derived from an aggregation of explicit loss distributions in respect of key economic variables modelled. The SCR is set at the 1-in-200 return period economic loss distribution outcome. The MCR is a function of the SCR: it cannot fall below 25% of the firm's SCR and cannot exceed 45% of the firm's SCR. The MCR is therefore lower than the SCR, representing the regulatory capital floor below which firms cannot operate.

- Pillar II – Risk Management (Qualitative) requirements: these focus on the company's system of governance and address the risk that a failure in the internal systems, processes or way that individuals carry out their functions, threatens the viability of the company. The regulation comprises five areas, namely: elements of the system of governance, function within the system of governance, fit and proper requirements, outsourcing, and remuneration policy. Firms also need to carry out a forward-looking assessment of the firm's own risks and overall solvency needs, to be disclosed in an Own-Risk and Solvency Report (ORSA) (see EIPOA, 2015 and PRA, 2015 for further details).

- Pillar III – Reporting and disclosure requirements: As part of providing evidence of on-going compliance, companies need to report to their supervisors on a regular basis. The regulation sets out four key reporting requirements: the Solvency and Financial Condition Report (SFCR), the Regular Supervisory Report (RSR), Own-Risk and Solvency Report (ORSA) and the Quantitative Reporting Templates (QRT) provided on a quarterly or annual basis. The scope of these reports includes qualitative and quantitative information on the company's business and performance system of governance, risk profile, valuation of assets and liabilities and capital management (EIOPA, 2015).

### 2.11.4.1 Selected Solvency II Pillar I Requirements in a Catastrophe Modelling Context

While firms need to demonstrate compliance with all regulatory requirements, Pillar I of the Solvency II requirements is arguably the most important in the context of catastrophe modelling. For firms using catastrophe models in their internal model, there are six articles that are of particularly importance in the context of catastrophe risk management. These six articles belong to the Solvency II Framework Directive and is sometimes referred to as the Level I text. The Framework Directive was proposed by the European Commission and approved by the European Council and Parliament in 2009. The Level II text refers to the Delegated Act. On 10 October 2014, the Commission adopted the Delegated Act containing implementing rules for Solvency II. Following approval of the European Parliament and Council, this was published and enforced in 2015, as Commission Delegated Regulation 2015/35 (EU-LEX, 2015).

The six level I Articles that are particularly important for catastrophe management are:

1) Use Test: assesses how catastrophe models are used in the decision-making process.
2) Statistical Quality Standards: assess the quantitative techniques used in catastrophe models.
3) Calibration Standards: assess the modelled time horizon, risk measures and approximations used in catastrophe models.
4) Profit & Loss Attribution: assesses whether financial results can be explained by the catastrophe model.
5) Validation Standards: assess how catastrophe models are validated.
6) Documentation Standards: assess how firms evidence aspects of the catastrophe model.

**Table 2.7** Level I: Article 120: Use Test.

| Level II Art. Ref | Article title | Summary requirements |
|---|---|---|
| 223 | Use of Catastrophe Models | Evidence use and consistency between different uses of models<br>If applicable, justify their non-use with regard to material risks |
| 224 | Fit to the Business | Level of complexity of the modelling needs to be proportionate to the nature, scale and complexity of the risks modelled<br>Evidence consistency between model outputs and reporting – both internally and externally<br>Ensure model outputs are suitably granular, e.g. US quake results to align to a US business entity<br>Ensure the model reflects changes in the underlying risk profile |
| 225 | Model Understanding | Demonstrate understanding of hazard, vulnerability and financial modules, scope/domain, purpose, modelled and unmodelled risks, quantitative methods, fit to business, integration with Enterprise Risk Management, limitations and diversification effects. This is related to model validation and evaluation covered in Chapters 4 and 5. |
| 226 | Support for Decision-making and Integration with ERM | Evidence how catastrophe models support relevant decision-making (e.g. risk mitigation, setting risk tolerance limits, business strategy)<br>Evidence engagement on catastrophe model (e.g. its limitations)<br>Demonstrate key risks are modelled<br>Show how model results are used in risk management and drive management action (e.g. reinsurance purchase decision)<br>Ensure validation can trigger changes to the model<br>Have a model change policy in place (see also Chapter 5.3) |

For each of these articles, the summary requirements are highlighted below in simple terms, with examples of what these requirements mean in the context of catastrophe modelling.

*Level I: Article 120 Use Test*
Table 2.7 shows the summary requirements for a Level I, Article 120 Use Test.

*Article 121: Statistical Quality Standards*
Table 2.8 shows Article 121 Statistical Quality Standards.

*Article 122: Calibration Standards*
Table 2.9 shows the Article 122 Calibration Standards.

*Article 123 Profit & Loss Attribution*
This requirement links the financial statements to the modelled results and assesses the degree to which the financial profits and losses may be explained by the model. Typically, compliance with this article is evidenced at the Internal Model level rather than the catastrophe model level. However, these requirements may still be embedded within the catastrophe modelling framework in order to support compliance across the internal model.

In addition, there are always lessons that may be extracted from new losses. In these cases, losses could help to identify new sources of risk that changes the way losses are modelled. In this way, actual experience becomes a feedback loop into loss calibration.

Table 2.8   Article 121 Statistical Quality Standards.

| Level II Art. | Article title | Summary requirements |
|---|---|---|
| 228 | Probability Distribution Forecasted (in this case, the projected, adjusted EP curve) | Ensure the model provides a representative distribution of loss outcomes that captures physical extremes |
| 229 | Adequate, Applicable and Relevant Actuarial Techniques | Use market-consistent actuarial techniques and timely information <br> Evidence understanding of the quantitative methods used in the model (e.g. contrast time-dependent and Poisson approaches to modelling earthquake event frequencies, see also Chapter 3.7.1) <br> Catastrophe models should reflect risk profile changes (for instance, in exposure, or business mix) <br> Unexplained change should be minimal <br> The model should represent the key risk drivers (e.g., fluvial and pluvial flood risk) <br> Techniques should fit data (e.g., use of a Poisson distribution where event rates are dispersive, potentially invalidating a pure Poisson approach) <br> Adjust the model for errors in sampling, or where modelled results do not converge on the vendor's ELT <br> Ensure transparent data, quantification methods and results |
| 230 | Information & Assumptions Used | Demonstrate information is realistic (of particular importance, when firms use default/unknown selections when specifying the catastrophe model) <br> Show how information used to generate the EP curve is credible, that is, show how it is consistent, reliably sourced, objective and generated in a transparent manner <br> Demonstrate that the assumptions used in the catastrophe modelling process are realistic, can be justified considering in the context of catastrophe modelling, their materiality, sensitivity and the alternatives considered |
| 231 | Data used in the Catastrophe Model | Evidence how data are complete, accurate and appropriate. Data here comprise site and source hazard data, vulnerability data, building and location data, values at risk and the insurance structure data. In addition, where applicable, internal claims data or external benchmark data also need to be included in the scope. |
| 232 | Ability to Rank Risk | Through the use of catastrophe models deployed across peril-regions, firms can rank region-perils using a consistent risk measure (for instance, 1% AEP TVaR). For an individual catastrophe model, the components of risk may be ranked, for instance, wind risk versus storm surge. <br> Demonstrate consistency of ranking with risk segmentation, across the business, over various time periods and with capital allocation process |
| 233 | Coverage of all material risks | At least on a quarterly basis, assess the extent to which the catastrophe model (including adjustments) covers all material risks <br> This assessment should consider qualitative indicators such as how risks not modelled are treated in the reinsurance programme, the ORSA risk register, or ERM framework. <br> Quantitative indicators of non-modelled risks should also be considered in the assessment: these include stress testing results, validation testing, financial losses unexplained by the model and allocated capital |
| 234 | Diversification effects | In evidencing that the methods of representing diversification effects are considered adequate, firms need to demonstrate that they have identified key dependency drivers, considered non-linear dependencies and characteristics of the risk measure used (e.g., 99.5th percentile AEP VaR or TVaR) |

Table 2.9   Article 122 Calibration Standards.

| Level II Art. Ref | Article title | Summary requirements |
|---|---|---|
| 238 | Choice of Risk Measure And Time Period Used | Catastrophe models produce EP curves from which one can extract the 99.5th percentile Value at Risk over a one year period. In these cases, an alternative risk measure or time period is not required. |
| | | Where companies use an alternative approach to calibrating the catastrophe loss distribution, the risk measure and time period should be consistent |
| | | Where companies opt to use a different risk measure or time period, additional requirements apply |
| 238 | Use of Approximations/ Simplifications | Approximations used in the process of generating the SCR should not introduce material error in the SCR |
| | | Similarly, it should not provide policyholders with any less protection than if the SCR was based on the probability distribution forecast derived from the Internal Model |
| | | Generally, for catastrophe models, the SCR is based on the Value at Risk measure derived from the output catastrophe risk loss distribution, rather than calculated by approximation |
| | | However, where a catastrophe model output is adjusted, for instance, for non-modelled risk or an alternative view of the underlying hazard using an approximate approach, firms would need to justify that the approximation of the adjustment does not materially misstate the resulting SCR and that policyholders are not adversely affected by the adjustment |

Therefore, we outline below four principles that can be considered in the catastrophe modelling context to support the firm's Profit and Loss attribution exercise:

1) *Granularity*: catastrophe losses should be able to be generated at the business unit and region-peril level of granularity.
2) *Categorization of risks*: there should be a clear distinction between the risks covered by the catastrophe model and those that are not covered by the catastrophe model (see Chapter 5.4.7).
3) *Consistency*: consistency between modelled losses and reported losses to enable meaningful comparison.
4) *Relevance for ERM and decision-making discussed*: by ensuring the granularity of modelled risk segments is relevant to the business, model results can support decision-making and risk management (see Article 120).

*Article 124 Validation Standards*
Table 2.10 shows the validation standards.

*Article 125 Documentation Standards*
Table 2.11 shows the documentation standards.

### 2.11.4.2   Conclusion
The selected Solvency II Pillar I requirements described here provide a high-level perspective on the increased level of scrutiny faced by companies using catastrophe models as an input to their capital models that subject to Solvency II.

Table 2.10  Article 124 Validation standards.

| L2 Art. Ref | Article title | Summary requirements |
|---|---|---|
| 241 | Validation Process Scope | The scope of the validation should cover all parts of the catastrophe model, including adjustments for data and model limitations |
| 241 (2,4) | Independent Validation | Validator should be 'free from influence' from those responsible for model development and operation<br>Independence may be assessed by considering:<br>• the responsibilities and reporting structures of those involved in validation<br>• the remuneration structure of the persons involved in the validation process. Independence means that the remuneration of the validator does not depend on the result of the validation. |
| 241 (3) | Validation Plan | Specify the validation processes and methods employed and the purpose of the validation<br>Specify the validation frequency and out-of-cycle validation triggers<br>Name persons responsible<br>Outline validation test fail procedure (escalation and resolution) |
| 242 (1) | Validation Tools | Test results against experience/other appropriate data (e.g., benchmarks) at least annually – at stand-alone region-peril level and aggregate or portfolio level<br>Justify deviation between assumptions and data and between observed and modelled results. For example, there have been some very large historical losses that when simply inflation-adjusted, could deviate significantly from the company's modelled losses. The justification here may reference:<br>• Changes in exposure over time, but after allowing for this, there may still be changes in the insurance or reinsurance covers provided that still result in material deviation between observed and modelled losses. For instance, firms implemented a number of contractual changes, such as reduced event limits, post the 2011 Thai floods(e.g. Aon Benfield, 2012). The pure inflation-adjusted loss may not even be a possible outcome in the current modelled loss distribution of flood losses from this region. In this case, one would need to adjust the loss to reflect the coverage changes.<br>• Changes in the underlying loss potential from a repeat of a historical event could also explain the deviation. For example, flood defence upgrades post Hurricane Katrina results in a lower 'as if' loss than if one simply inflated the historical loss (assuming the flood defence system holds). For certain catastrophe models for certain regions, firms are able to extract the deterministic as-if historical loss reflecting these updates. |
| 242 (3) | Statistical Testing in the Validation Process | The statistical process for validating the model should be based on:<br>• Current information including, where relevant and appropriate, developments in actuarial techniques and generally accepted market practice. This may include for example, modelled versus observed comparative plots (QQ plots), goodness of fit testing, etc.<br>• A detailed understanding of assumptions underlying methods used to produce the EP curve. For example, the independence assumption for Poisson distributions used in catastrophe models to model event frequencies. |
| 242 (4) | Key Assumptions | Explain why certain assumptions are sensitive (an example of a sensitive catastrophe model assumption is the assumed frequency of US hurricane landfalls)<br>Explain how sensitivity is considered in decision-making |
| 242 (5,6) | Stability and Appropriateness of Outputs | Test the stability of results by recomputing results based on the same data. Catastrophe models in general will produce the same results if the model is run on the same data. However, the results will change with a changed number of simulation runs. |

*(continued)*

Table 2.10 (*Continued*)

| L2 Art. Ref | Article title | Summary requirements |
|---|---|---|
| | | Test the appropriateness of results, and in particular the tail risk metrics, by identifying the probable stress scenarios that could threaten the viability of the firm. When compared to the modelled loss distribution, one would expect the stress scenario loss to correspond to a remote point on the modelled loss distribution, as opposed to beyond the range of possible loss outcomes. |

Table 2.11 Documentation standards.

| Level 2 Art. Ref | Article description | Summary requirements |
|---|---|---|
| 243 | General Documentation Requirements | Design and operational details of the model should be sufficient such that it can be understood by an '*independent knowledgeable third party*' <br> The documentation should allow for sound judgement on SII compliance <br> The documentation should be appropriately structured, detailed, complete and up-to-date <br> Model outputs should be capable of being reproduced using inputs to the model and documentation |
| 244 | Minimum Documentation Requirements | The following documented evidence is required: <br> • an inventory of documents <br> • a model change policy <br> • a description of processes, including risks and controls and staff responsibilities <br> • IT systems and contingencies <br> • a description of relevant assumptions, justification for these assumptions, method for setting assumptions, data used, limitations relating to these assumptions and validation criteria. <br> • A data directory that includes the data source, characteristics and usage <br> • Data flow, including collection, processing and application of data and treatment of inconsistencies and wider data deficiencies. <br> • Indicators used to evaluate model coverage <br> • Details of the risk mitigation <br> • Validation process and results <br> • Role of catastrophe models, justification for using a vendor model over an internally developed catastrophe model and the evaluation of alternatives. |
| 245 | Considerations when Assessing Model Effectiveness | • Consider model limitations: including non-modelled risks, limitations of risk modelling, IT, data and limitations arising from uncertainty in model results <br> • Consider sensitivity of results to key assumptions |
| 246 | Model Changes | Record all changes, including descriptions and rationale for changes and implications of change for the model design <br> Analyse material changes in model results, but also changes in the quantification methods, data and assumptions |
| 247 | External Models and Data | Monitor potential limitations of using catastrophe models and external data to ensure on-going compliance with the requirements set out above |

Further, the requirements show that catastrophe risk analysts supporting the internal model calibration cannot focus only on model results. They must also consider the process of producing these results. In particular, Solvency II compels practitioners to apply proportionate interrogation to the data inputs, assumptions, expert judgements, quantification methods, results and governance.

As outlined in Section 2.11.1.3, there is no professional body for catastrophe modellers with associated standards of practice. In addition to the various catastrophe modelling industry initiatives and catastrophe model vendor accreditations, Solvency II (and in particular the Pillar I Articles outlined above) provides another framework for raising, and potentially normalizing, professional standards among catastrophe risk analysts. Arguably, given the level of regulatory scrutiny faced by companies, Solvency II may be the strongest catalyst currently for change in behaviours among catastrophe risk professionals – only time will tell.

The outcome should be greater comfort in the approach to projecting catastrophe losses, even if there is no improvement in the skill of models. Detailed evaluation of models by firms should also improve communication between scientists, underwriters, actuaries and other risk professionals.

### 2.11.5 Case Study: Regulation of Catastrophe Models for Ratemaking

This section considers regulating catastrophe models themselves, In particular, the situation where a formal approval process determines if a model can be used as the basis for setting insurance premiums. In many states of the United States, homeowners' insurance premiums (known as rates) must be filed and approved with the Department of Insurance before any insurance company can write business. In the state of Florida, the Florida Commission on Hurricane Loss Projection Methodology (FCHLPM) reviews and approves catastrophe models for use in the rate-setting process. In 2014, South Carolina introduced the requirement that all property insurers must use one or more of seven state-sanctioned models when making rate filings.

The FCHLPM was created in the aftermath of Hurricane Andrew (1992); this event led to a number of insurance company insolvencies. It also served as a catalyst for the adoption of the nascent catastrophe modelling techniques (see also Chapter 3.2.6).

The FCHLPM is an independent body of experts created by Legislature in 1995 to develop standards and review hurricane loss models used in the development of residential property insurance rates and the calculation of probable maximum loss levels.

In the FCHLPM process, approved models must be based on the long-term historical average of hurricane activity. Additionally, models cover wind losses only; flood loss has typically been covered by the government-backed National Flood Insurance Program (National Flood Insurance Program, 2015). However, the FCHLPM (FCHPLM, 2015) is in the process of developing flood standards to be introduced in 2017. This is in recognition of the increasing private market coverage for flood, as well as the significance of coastal flood-driven losses in recent catastrophe events in the United States such as Hurricanes Katrina (2005), Ike (2008) and Sandy (2012).

### 2.11.5.1 Membership of the FCHLPM
The commission consists of twelve members with the following qualifications and expertise in the State of Florida:

- the Insurance Consumer Advocate;
- the senior employee of the State Board of Administration responsible for operations of the Florida Hurricane Catastrophe Fund;
- the Executive Director of the Citizens Property Insurance Corporation;

- the Director of the Division of Emergency Management;
- the actuary member of the Florida Hurricane Catastrophe Fund Advisory Council;
- an actuarial employee of the Florida regulator (Office of Insurance Regulation) who is responsible for property insurance rate filings and appointed by the Director of the OIR;
- five members appointed by the Chief Financial Officer, as follows:
  - an actuary who is employed full-time by a property and casualty insurer which was responsible for at least 1% of direct written premium for homeowner's insurance in Florida;
  - an expert in insurance finance who is a full-time member of the faculty of the State University System and who has a background in actuarial science;
  - an expert in statistics who is a full-time member of the faculty of the State University System and who has a background in insurance;
  - an expert in computing who is a full-time member of the faculty of the State University System;
  - an expert in meteorology who is a full-time member of the faculty of the State University System and who specializes in hurricanes;
- a licensed professional structural engineer who is a full-time faculty member in the State University System and who has expertise in wind mitigation techniques. This appointment shall be made by the Governor.

This is a broad and deep set of expertise as required to approve the models.

### 2.11.5.2 FHCLPM Model Standards

An important part of the work of the FCHLPM is to create and maintain the detailed standards by which catastrophe models will be assessed and these are reviewed and up on a two-year cycle.
The detailed standards fall into the following categories:

- *General Standards*: relate to model scope, qualification of modelling personnel, geo-resolution aspects, independence of model components.
- *Meteorological*: relate to the methodology for modelling the wind hazard.
- *Statistical*: relate to sensitivity/uncertainty analysis and how well the model results replicate historical and known event losses.
- *Vulnerability*: on what basis have the damage functions been derived and how contents and time element coverages are derived relative to direct damage to buildings; also a focus on whether and how mitigation measures (e.g. installation of storm shutters to protect windows) are reflected in the model.
- *Actuarial*: relate to the translation of calculated damage into financial losses including the capability of the model to reflect terms and conditions of insurance policies.
- *Computer*: these standards set out requirements on the design and governance around the computing architecture, maintenance, updates and testing, as well as security considerations.

### 2.11.5.3 Approval Process

The model development company must submit a comprehensive submission of evidence in the form of an extensive report for meeting the Model Standards. This report and its contents are public under Florida's open government approach (known as the 'sunshine laws'). However, given the need to protect each company's intellectual property, some control over trade secrets is allowed. The commission employs a Professional Team of individuals with expertise and professional credentials in actuarial science, statistics, meteorology, computer science and engineering to review aspects of the model submission defined as 'trade secret'. They do this during a multi-day interactive on-site visit to the modellers' premises. During this visit the

Professional Team can request to review additional documentation and software code. The Professional Team provides recommendations and information back to the Commission on their findings during a closed session (i.e. not open to the public). The commission's final approval is in the form of a vote during a public hearing; a majority vote is required for approval. The standards are reviewed every two years. Thus, any approved model must be re-submitted for approval against the new standards every two years.

## 2.12  Case Study: Catastrophe Modelling for Reinsurance and Retrocession Purchase

*Juan England*

### 2.12.1  Introduction

This section touches on many of the issues already discussed and pulls them together into a practical study of how catastrophe modelling is used at a reinsurance broker to inform reinsurance and retrocession purchase.

Reinsurance strategies are defined through risk appetite statements linked to capital management, business objectives, and regulatory or rating agency requirements (see Sections 2.10 and 2.11). Catastrophe risk is usually an important factor in defining a company's reinsurance and risk transfer strategy. Establishing how much catastrophe risk to transfer and how much to retain is a complex and important decision for any (re)insurance company. An inappropriate strategy could result in the company not buying enough reinsurance and becoming insolvent after a large catastrophe event, or not achieving its annual earnings targets due to a series of smaller but more frequent retained losses aggregated throughout the year.

There are three commonly used methods for quantifying catastrophe risk in this process, which have already been discussed in this chapter:

1) *Statistical analysis of loss history*: this comprises burning cost analysis, which extrapolates losses to extreme values using experience-based methods (Section 2.6.2.1).
2) *Standard formula*: a fixed percentage of main zone exposure accumulation defined by a regulator or internal practices (see Section 2.11.4).
3) Catastrophe modelling of a portfolio of exposures:
   a) deterministic: 'as-if' historic events (see Section 2.7.1.2 or Chapter 5.4.3.1)
   b) probabilistic/stochastic: EP curves – losses at a return period, average and standard deviation annual loss (see Chapter 1.10).

Which method is used varies according to the size of each company, local regulatory requirements, a company's sophistication, and availability of data. There is no well-defined best practice, but in many cases, it is not the modelling alone but the combination of all three methods that provides most insight about the risk and therefore leads to a more robust decision.

This said, the use of catastrophe modelling in this process has been growing in importance over the past 15 years as the models have become more established. This is primarily attributed to the improvements in data quality and quantity. Notable enhancements include those in the underlying exposure information (e.g. from CRESTA zone to latitude-longitude coordinate and improvements in other primary risk characteristics) and new information about the hazard (e.g. science or observation, new secondary perils). The entry of new hazard data and model providers has accelerated this process.

Although catastrophe modelling has brought awareness and discipline to the industry in terms of understanding and quantifying catastrophe risk, a strong technical knowledge of the data and models is required when defining a reinsurance strategy. First and foremost it is imperative to understand their strengths and limitations. A number of issues need to be taken into consideration before making any reinsurance decision and these factors are considered in the own view of risk process described in Chapter 5. These issues include model miss (difference between actual and modelled loss because e.g. lack of completeness in model, for example secondary perils), model error, or model bias (e.g. preference for a particular model since it is the most commonly used in that market). As discussed in previous sections, catastrophe modelling should also be supplemented with loss history analysis and main zone accumulations in order to derive a more complete view.

The widest current use of catastrophe modelling in the reinsurance purchase process is to review changes in the underlying exposures. It is also commonly used to determine absolute loss potential, particularly for new portfolios. By using a catastrophe model to analyse the portfolio of risks at two different points in time (yearly, quarterly, etc.), it is possible to determine how the catastrophe risk has changed in that period. The comparison of EP curves at two exposure dates takes into account geographic movements and changes in underlying policy conditions relative to the hazards. Any changes in the model(s) and in data quality are also considered. Such analysis provides early insight into what could be required from the reinsurance coverage.

Different kinds of reinsurance treaty can be combined to meet different aims (Section 2.4.2.3), but here we consider only catastrophe excess of loss treaties. For these, by considering the outputs of the analysis it is possible to draw some conclusions: first, if the OEP losses at a high return period increase, then more reinsurance in the form of a higher catastrophe excess of loss limit might be required, or, second, if the AEP losses at low return period increases then more reinstatements might be an appropriate solution.

We consider the role of catastrophe models assisting the broker in the design of a CAT XL programme in four steps:

- Determining the total limit required.
- Deciding the layering.
- Setting the price.
- Cost allocation.

These are treated in turn below.

### 2.12.2  Determining the Total Limit Required

Once the exposure changes are understood, the first steps for defining the key parameters of a property catastrophe reinsurance programme can be taken. Determining the total limit required in a property catastrophe excess of loss programme varies globally depending on local and specific country laws. In some cases, the regulatory requirements are completely prescriptive and a specific standard formula or a specific type of model at a given return period needs to be used. In other cases, a framework which allows for any model to be used as long as it can be technically justified and is aligned to an internal capital model and risk appetite/tolerance can be used (i.e. Solvency II in Europe, APRA in Australia). The list below provides examples of some of these methods at the time of writing:

- *Regulators where using a 'standard formula' is applicable* (e.g. Section 2.10.3):
    - The minimum catastrophe excess of loss reinsurance required is defined by a fixed percentage of the retained exposures in the geographic zone with largest accumulation.

- For example, in Colombia, it is 15% of the retained earthquake CRESTA Zone 1 exposure, or in Chile it is 10% of Buildings and Contents and 15% of Business Interruption of the retained CRESTA 3 exposures.
- *Regulators with selected approved models*:
  - The minimum catastrophe excess of loss reinsurance required is defined by the OEP loss at a specified return period using a model that has been pre-approved by the local regulator.
  - For example, in Mexico, the countrywide 1-in-1500-year return period loss of the retained hurricane and earthquake exposures using the regulator's reference model.
  - The regulator selects the models that can be used through a thorough review of their methodology and defines a return period in line with their solvency requirements.
- *Regulators and rating agencies with a model framework approach*:
  - The minimum total catastrophe excess of loss requirement or event limit is defined by the modelled OEP or AEP loss at a specific return period using any market recognized available model.
  - For example, the 1-in-250-year return period OEP loss of the countrywide retained exposures using Model A for earthquake and the 1-in-100-year return period AEP loss using Model B for windstorm. Model A and Model B were chosen by the reinsurance decision-makers as the models that represent the earthquake and windstorm risk of their portfolio.
  - In Puerto Rico, as an example, it is required to use the average of two recognized vendors' catastrophe models.
  - In addition, there might be further requirements relating to higher frequency but less severe perils such as the 1-in-10 year return period AEP loss from flood or hail using a different model. APRA in Australia has set some of these requirements.
  - The regulator establishes the return period for each peril on the basis of historic industry losses or minimum solvency requirements.
- *Insurance companies with defined risk tolerances* (see Section 2.10.1):
  - The company has defined specific return period metrics based on their internal capital model and risk appetite.
  - For example, the total limit purchased is determined as the 1-in-100-year return period TVAR OEP loss of Model C for windstorm and earthquake.
  - Multinational companies with centralized reinsurance strategies maximize their catastrophe reinsurance purchase through the consolidation of multiple geographical portfolios to obtain greater diversification and efficiency of scale. For example, diversification will mean that they will often only need a limit equivalent to requirements of their peak exposure and not the sum of requirements for all portfolios, although factors such as frequency in different regions will also affect the design of the programme.

Defining the retention level in a catastrophe excess of loss programme is less prescriptive. This is closely related to a company's risk appetite and how much it is willing to lose in one event. There is no prescribed best practice to define this metric. Some companies define it as a fixed percentage of shareholders' funds, others at a low modelled return period, such as 1-in -5 or 1-in -10 years. Loss history, if available, is also used to define the retention through statistical analyses of the frequency of loss at different retention levels.

Catastrophe modelling is also used to benchmark reinsurance buying behaviour across industry peers. This analysis compares catastrophe modelling outputs from different companies and it has multiple benefits. At a high level, it provides insights into different companies' risk appetites and approaches to defining reinsurance strategies. On a more practical level, it can further support the definition of reinsurance programme's limits and retentions or act as a second check to what has already been defined.

The most common metrics used in a benchmarking analysis are the entry and exit return periods of the programmes which are being compared; this is also used as a check in reinsurance pricing (Section 2.6.4.2.2). These are the points on the OEP curve that correspond to losses at the programme attachment point and occurrence limit. The benchmarking of entry points shows how different companies define their retention levels, so this provides an insight into their risk appetite (see Section 2.10.1). For example, if the entry point of one company is higher than that of its peers, that particular organization is willing to take on more risk. A high entry point reduces the overall spend by assuming a higher retention on the programme; this is at the cost of potentially having a greater loss before getting a reinsurance recovery. On the other hand, the lower the return period of the entry point, the more risk averse the company is, i.e. it wants to buy more reinsurance than its peers. Similar conclusions can be drawn from the benchmarking of programme exit points, but in the opposite direction; if the exit point corresponds to a higher return period, the organization prefers greater reinsurance coverage and a lower appetite to take risk, and vice versa.

### 2.12.3 Layering of a CAT XL Programme

Once the overall limit and retention of a catastrophe excess of loss programme are set, then the number of layers must be chosen. Layering is usually necessary from a marketing perspective. This is because, depending on the size of the programme, it might not be possible to place it all as a single layer due to individual reinsurers' line size and capacity restrictions, that is, the share and limit they can take on an individual client. The programme is therefore divided into layers to attract a larger number of reinsurers and be able to optimize the price of the programme by transferring the risk to a larger panel of carriers.

The choice of layering for catastrophe excess of loss programmes varies by company and market conditions, but catastrophe modelling can provide insights. Bottom layers tend to be between 1-in-5 and 1-in-30-year return periods, middle layers between 1-in-30 and 1-in-60-year return periods, and top layers above 1-in-60-year return periods. Since bottom layers are exposed to more frequent perils they might have further requirements such as additional reinstatements or recover losses on an aggregated basis. Top layers provide coverage for the extreme events and due to their low probability do not normally need additional reinstatement provisions.

Defining the layers based on return periods can also help to find the optimal reinsurer panel. Reinsurers sometimes participate in shares across a whole programme; this is called 'writing across the board' (see Section 2.8.1). However, depending on the type of peril and region, they might have a stronger appetite for lower layers than for higher layers. The opposite is also possible. Therefore, layering the programme also allows reinsurers to better deploy their capital by allowing them to match the layers they underwrite to their risk appetite preferences and achieving further price efficiencies.

### 2.12.4 Price

Once the programme layers are agreed, the next step is defining the price. This is heavily influenced by the market conditions. However, catastrophe modelling can be used to determine the technical price.

Section 2.6.2 describes how the expected loss must include other factors to obtain the technical price, including expenses, profit and capital loading. In some cases, these loadings are simply achieved by loading the standard deviation (see Equation (2.1)), although this is increasingly uncommon where companies employ more complex capital methods. Reinsurance pricing is described in Section 2.6.4.2.

However, considering Equation (2.1), and remembering the difference between the market RoL (actual premium/limit) and the LoL (expected loss/limit) including any expected reinstatement premium (labelled as $LOL_t$), will represent the actual profit, i.e. the target profit overlaid with market factors, different pricing can be compared by considering the loading factor.

$$loading\ factor = (market\ RoL - LoL_t)/\ \sigma \tag{2.17}$$

The loading factor varies depending on the type of model, peril and region, and requires adjustment every year as the market conditions change. This comparison provides insights into the price efficiencies of different programmes.

Another price comparison analysis is to determine the risk-adjusted price change to the programme which is the price change accounting for differences in underlying exposure, sometimes measured by change in expected loss, which is also useful for benchmarking reasons.

### 2.12.5   Cost Allocation

Once the reinsurance structure has been decided and the price decided, the cost of that reinsurance is often distributed through the organization who has bought it. Usually more of the cost will be borne by the segments of business units that benefit from the reinsurance.

Determining which units benefit is not always a simple process. This is because the different entities and business units that are benefiting from the reinsurance protection have different needs and expectations. Usually a metric that indicates the benefit of the programme will be chosen to distribute the cost. This may be the expected loss to the layers caused by each entity or business unit, although TVaR is also commonly used for this purpose. As a result, it is possible to determine how much of the expected loss of the programme corresponds to each business unit. This division provides a sensible rationale for a percentage split that can be used to determine which units pay what for the reinsurance.

### 2.12.6   Conclusion

Designing a robust catastrophe reinsurance strategy is fundamental to successfully achieving insurance or reinsurance company financial objectives. This is a complex decision-making process involving the definition of a number of key parameters that need to meet regulatory requirements but also need to fit with a company's risk appetite. There are a number of decision-making methods for evaluating and defining these parameters and catastrophe modelling is one of them. It is important to have a deep understanding of the model(s) that are being used as model error, bias or mismatch could result in defining an inappropriate reinsurance strategy with potentially catastrophic results. However, aside of the potential model limitations, there are a significant number of usages to a catastrophe model that can still inform the decision-making process of defining a reinsurance strategy.

## 2.13 Government Schemes and Insurance

*Matthew Eagle*

### 2.13.1 Introduction

The roles of government and the private sector in providing and supporting insurance, risk mitigation and disaster recovery are heavily debated. A useful comparative analysis of government and market insurance is provided by Priest (2003). Arrow (1963, 1978) makes a strong economic justification for government insurance. Although his arguments relate to the provision of medical care, many of the points raised are general ones which also be apply to natural catastrophe risk. Arrow highlights some reasons why private market insurance is sometimes unavailable. Key factors are **adverse risk selection** and an inability to control **moral hazard**. Adverse risk selection is the tendency for more demand for insurance from high-risk compared to low-risk individuals, and moral hazard is a potential for change to imprudent behaviour when the policyholder no longer needs to face consequences of a risk. Adverse selection can be overcome by compulsory insurance; Arrow argues that this could be of overall benefit to society as the benefits to the high-risk individuals outweigh the disadvantages for the low-risk ones. Von Peter *et al.* (2012) analysed the extent to which risk transfer to the insurance markets can facilitate economic recovery. They conclude that while major natural catastrophes can have large negative impacts on economic activity it is mainly the *uninsured* losses that drive subsequent macroeconomic cost.

The private insurance sector generally provides an efficient framework for individuals and businesses to pool risk (Chapter 1.2). It has built scale and expertise to manage the sale, administration and accounting of policies and claims. Although private sector shareholders also require an adequate profit margin to generate a favourable return on their investment, the rationale is that the benefits of these competitive efficiencies are passed on to end-customers, i.e. creating an effective insurance market. The system must provide a net benefit to all so that even the benefits for the lower-risk individuals outweigh their subsidy of higher risk policyholders and the insurers' expense and profit loadings (see Section 2.6.2).

However, in some cases, particularly in the developing world, the insurance industry has not been able to distribute policies effectively or offer cover at affordable levels. Although many people might like to have some form of insurance, it is not a financial priority for all. Incentivizing insurance purchase by individuals can be difficult when governments provide financial aid post-loss. For example, illustrating a case of moral hazard, what happens when the government steps in to help where some people have paid for insurance coverage and others have not? An assumption that the government, or aid agencies, will indeed intervene following a catastrophe may make the benefits of catastrophe risk insurance less tangible to low-income householders, particularly with regard to infrastructure. However, in many cases, governments may not be best-placed at the time to find and use what is essentially taxpayer revenue to indemnify those who have suffered a loss.

Low insurance penetration is, however, generally a residential homeowners' issue since most businesses require insurance to operate, particularly those that rely on financial loans. Even so, low penetration has led to wide discrepancies between overall economic losses from natural disasters and the losses recoverable through insurance cover. This was observed in the devastating earthquakes in Nepal in 2015.

Affordability is also related to the effect of risk mitigation measures. Typical insurance policies include some standard policy wordings that require the insured to exercise due care to avoid loss, e.g. installing suitable locks, fire detectors and alarms. As well as this, if buildings are constructed and maintained to a sufficiently high standard this usually mitigates the risk from natural catastrophes. However, in many areas, building standards and land use planning rules are neither well developed nor fully enforced so properties may be of lower quality in higher risk areas. Insurers may therefore require higher and consequently unaffordable premiums to cover the risk in these areas. Further recent research on this topic includes Orie and Stahel (2013).

Over the years many governments, particularly in catastrophe-prone areas of the world, have sought to establish schemes that provide all households with a minimum level of protection. In some instances, these schemes have been encouraged and supported by organizations such as the World Bank, the Organization for Economic Cooperation and Development (OECD), the Asia-Pacific Economic Cooperation (APEC) and the Association of Southeast Asian Nations (ASEAN) as they seek to create a financially stable and resilient environment. Instead of raising taxes and holding a fund within government budgets, a mechanism can be created with an insurance premium or levy paid by all households and pooled to provide the cover to all who require it.

In most schemes, it takes time for funds to be built up to provide financial resilience against major disasters. In some cases explicit financial guarantees from government are required, but reinsurance plays an important role, particularly in the early years.

Catastrophe modelling is used in many aspects of creating and maintaining a government-backed scheme. These are similar to the applications of models at a private (re)insurer, and include:

- *Setting a fair tariff.* The models can be used to estimate annual economic, direct and insured losses together with the potential cost of reinsurance to protect the scheme from large accumulated losses from natural catastrophes, based on specific proposed deductibles and limits.
- *Understanding the impact of risk mitigation features.* This includes modelling how the loss may vary with different construction types, building age, building height and other features.
- *Managing the fund/capital of the scheme and organizations running the scheme.* Reinsurance is a core component of any risk management strategy. Models provide information on PMLs and reinsurance decisions with regard to attachment points, limits and layering (see also Section 2.12).

Each scheme is different and structured to suit specific country requirements and constraints. There are differences between the schemes, which can be explored by asking the following questions:

- Is the insurance compulsory or optional? Compulsory schemes are often encouraged to avoid anti-selection and to broaden the premium fund and diversification of risk as much as possible. However, in some cases it may be difficult to convince homeowners to purchase the policy and to enforce any mandatory laws.
- How is the policy sold or distributed? Is the policy sold by insurance companies (who may receive a fee/commission for managing the policy), funded through an insurance levy (part of the premium) or possibly through a tax (e.g. utility)?
- Who manages and pays the claims? Is this handled by a central company established to run the scheme or by the private insurers who have the direct policyholder contact?
- What perils does the policy cover?

- Do the policies include some form of deductible or coinsurance?
- Do the policies have a limited pay-out?

It is not possible to examine all current schemes in detail within this book but this section examines a range of schemes. It also attempts to group them into types, although this is difficult given some of the differences. A common theme is that the funds have been set up in response to natural catastrophe events, e.g. the 1999 Izmit earthquake for TCIP and the 1999 Magnitude 7.3 earthquake in Taiwan.

### 2.13.2 Government Schemes with Standalone Products Managed by a Central Organization

This sub-section details government schemes with standalone product that are managed by a central organization. These include the Turkish Catastrophe Insurance Pool (TCIP), Pool-ul de Asigurare Impotriva Dezastrelor Naturale (PAID) – Romania and the New Zealand Earthquake Commission (EQC) (Table 2.12) These schemes were set up with standalone products sold by private companies and agents but managed centrally. All are mandatory schemes, although this does not ensure full take-up by the public.

### 2.13.3 Government Schemes Where Catastrophe Cover Is Provided as an Add-on to a Fire Policy

This sub-section details government schemes where catastrophe cover is provided as an add-on to Fire. These schemes include the Taiwan Residential Earthquake Insurance Fund (TREIF), Japanese Earthquake Reinsurance (JER) and the California Earthquake Authority (CEA) (Table 2.13).

### 2.13.4 Government-Backed Reinsurance/Pooling Schemes

In some countries mechanisms and organizations have been created to provide reinsurance and financial capacity to governments and/or the insurance industry. These include Fondo de Desastres Naturales (FONDEN) in Mexico, the Caribbean Catastrophe Risk Insurance Facility (CCRIF), Casse Centrale de Reassurance (CCR), in France, the Florida Hurricane Commission Fund (FHCF), in the United States, the CONSORCIO de Compensación de Seguros (CCS) in Spain (Table 2.14).

### 2.13.5 Private Insurance Company Pools Supported by Government Legislation

In a number of countries private insurance companies have created pools where the premiums and losses for natural catastrophe insurance are pooled and managed centrally. These are typically supported by government legislation to provide the framework for the policies to be priced, sold and managed. Three examples include the Norwegian Natural Perils Pool (NNPP), the Swiss Natural Perils (Elemental) Pool (SVV) and the Interkantonaler Rückversicherungsverband (IRV) (Table 2.15). Flood Re in the United Kingdom is covered in detail in Section 2.13.6.

   This section has not covered all current government schemes and risk pooling arrangements. Others to note include the Algerian Catastrophe Insurance Programme (ACIP) and TARSIM in Turkey which supports crop and livestock exposures.

**Table 2.12** Government schemes with standalone products.

| | Established and management | What is covered? | Terms of coverage | Distribution and take up | Capacity |
|---|---|---|---|---|---|
| TCIP, Turkey http://www .tcip.gov.tr/ | 2000 with support from World Bank. Managed by a private company (currently Eureko Sigorta following regular tender process) on behalf of TCIP Board. | Residential buildings only, against earthquakes and fire, tsunami and landslides caused by the earthquake. | Limited (TL 150,000).[1] Exact cover is based on re-building calculation, depending on type and size of property. Risk tariff applied to sum insured based on earthquake risk zone (5 defined zones), and type of property. | Sold through insurance companies and their network of agents. Although compulsory take-up is <<100%, but growing significantly to almost 7 m policies by end of 2014. Enforced through new laws making policy a prerequisite for utility subscription. | USD 4.3 billion of coverage per event up to a modelled return period in excess of 1 in 250 years (does vary according to model used). |
| PAID, Romania https:// paidromania .ro/en | 2009 with support from World Bank. PAID is made up of existing insurers in the local market and operates as a company in its own right. | Residential buildings against flood, earthquake, landslide. | Sum insured of either €10,000 for a €10 premium or €20,000 for a €20 premium, depending on the type and quality of construction of the building. | Although compulsory take-up not 100%. Just over 1.5 m policies by first quarter 2015 (90% of which have €20 K cover) and expected to grow further. | Over €400 m capacity, which is about 1 in 250 EQ, and it is expected that they may need to purchase up to €1.5bn capacity in the future. |
| EQC, New Zealand http://www .eqc.govt .nz/ | Natural Disaster Fund was set up via Earthquake and War Damage Commission in 1945. Earthquake Commission Act in 1993. Policy is government-guaranteed – so any shortfall is met by government. | EQ Cover policy covers residential property against earthquake, natural landslip, volcanic eruption, hydrothermal activity and tsunami. | Maximum cover of NZ$100,000 for buildings, NZ$20,000 for contents, and cover for insured residential land for 15c/$, max NZ$180 premium (rates were increased after 2011). | Automatically included in any household fire policy so take-up is close to 100%. | From mid-2014, EQC had NZ$4.5bn of reinsurance protection, with reinstatement of protection available should a second event occur within the policy year. The Natural Disaster Fund was also used to cover claims from Canterbury earthquakes in 2010 and 2011. |

Note:
1) As of 1 January 2011.

**Table 2.13** Government schemes where catastrophe cover is provided as an add-on to Fire.

| | Established and management | What is covered? | Mechanism | Terms of coverage | Capacity |
|---|---|---|---|---|---|
| TREIF www.treif .org.tw/treif/content/ insurance/TREIP .htm | Government review following 1999 EQ and cover included in residential fire policies since 2002. Cover ceded by private insurers to TREIF who retain or transfer the risk. | Residential property against fire, explosion, landslide, land subsidence, land movement, land fissure, land rupture, or tsunami, sea surge and flood caused by an earthquake. | If total loss (building demolished, requiring demolition, inhabitable or where the rebuilding costs >50% of replacement costs), underwriting insurer will pay insured amount and contingent living expense. | Flat premium (since 2009 this is NT$1,350) with a maximum sum insured of NT$ 1.5 million and the maximum contingent living expense of NT $200,000. | NT$70bn; the co-insurance pool covering the first NT$3.0 billion, and TRIEF (Tier 2) the next NT$67.0 billion, some of which is transferred to reinsurance and capital markets. |
| JER http://www .adbi.org/files/ 2014.03.14.cpp .sess6.6.harada .earthquake .insurance.japan .pdf | 1966 (after 1964 Niigata EQ). Operates as a pool for insurers enrolled in the scheme. Cooperative mutual insurers do not participate. | Residential Buildings and/ or Household Property, against fire, destruction, burial or flood following an earthquake, volcanic eruptions or tidal wave (tsunami). | Earthquake insurance from an insurance company available as optional rider to fire policy (estimated that <50% of fire policies include earthquake cover). | Usually covers between 30% and 50% of the fire sum insured to a maximum of JPY 50 m for buildings and JPY 10 m for contents. Rate varies depending on prefecture location and wooden or non-wood construction. Claims are shared between insurers and government – structured so government takes larger share when annual | Maximum JER pay-out in a single year is JPY 5.5 trillion (~$40bn). If claims exceed this amount then claims are pro-rated among all claimants. |

aggregate claims are higher.

| CEA http://www.earthquakeauthority.com/Pages/default.aspx | 1996. Cover withdrawn or limited in private market after 1994. Northridge earthquake. So new law created 'mini-policy' which all insurers had to offer when writing or renewing a residential fire policy. CEA is non-profit and privately funded but is publicly managed. | Residential property against earthquake CEA currently write 860,000 policies (over 75% of all EQ policies sold) but estimated that less than 12% of homeowners have EQ insurance. | Individuals must hold a residential insurance policy through a participating insurer before they can purchase a CEA policy. CEA policies are sold and serviced (including the managing of all claims) by >20 participating insurance companies who (or their agents) receive a fee and commission on each policy. | Two policies offered. The 'Homeowners Choice' policy includes cover for house, building code upgrades and emergency repairs together with optional contents cover (under a separate lower deductible) and living expenses (with no deductible). The second 'Standard Homeowners' policy bundles all the above coverages into one package but all (except living expenses) subject to the single large dwellings deductible. | CEA annual premium over $500 m with a claims paying capability exceeding $11bn and an A- rating from AM Best. If damage exceeds CEA claims capacity then policyholders' claims may be pro-rated (or paid in instalments). |

Table 2.14 Government-backed reinsurance/pooling schemes.

| | Established and management | What is covered? | Mechanism | Capacity |
|---|---|---|---|---|
| Consorcio (CCS), Spain http://www .consorseguros .es/web/ | Established in 1945 to cover extraordinary risks. In 1991 it ceased to be autonomous and became a Public Corporate Entity, and lost exclusivity for cover of extra-ordinary risks | Provides compensation for damage caused to people and property (including loss of profits) as a result of certain natural phenomena such as flood, earth (sea)quake, volcanic eruption and atypical cyclonic storms (>120 km/hr). It also covers losses from terrorism, riots, rebellion, sedition and uprising. Exclusions include rainfall, hail and windstorms below 120 km/hr. | Cover is obligatory (based on principles of compensation and solidarity). Cover is either included in insurance policy or where explicitly excluded then with the Consorcio directly. The Consorcio surcharge is based on a tariff calculated against the sum insured. The surcharge is collected by insurance companies as part of premiums and deposit monthly with the Consorcio after retention of a 5% commission. The current tariff has been in force since 2008. | In addition to the property cat function the CCS is reinsurer of the Agroseguros Pool, it reinsures some credit businesses, acts as liquidator of insurance companies and provides MTPL (Motor Third Party Liability) guarantee. While it is authorized to buy reinsurance CSS does not currently do so. Over the last 10 years CCS has paid an average of just over €350 m per annum in claims. |
| FONDEN, Mexico www .proteccioncivil .gob.mx/.../ Almacen/ libro_fonden.pdf | Established in late 1990s to support the rapid rehabilitation of federal and state infrastructure affected by adverse natural events | The FONDEN Program for Reconstruction is designed to provide financial support to rehabilitate and reconstruct public assets | Combination of Parametric cover for quick emergency disaster relief, plus traditional Indemnity based insurance cover | Parametric cover of ~$315 m. Indemnity based cover of ~$440 m (modelled to provide cover beyond the 1-in-200-year level) for reconstruction of infrastructure and government property. |
| CCRIF http://www.ccrif .org | Formed in 2007 as the first multi-country pool, under technical leadership of the World Bank, with mission to assist Caribbean governments and communities in understanding and reducing the socio-economic and environmental impact of natural catastrophes. CCRIF segregated portfolio company (SPC) is registered in Cayman Islands and | Earthquake, Tropical Cyclone and excess rainfall for Caribbean governments. Soon to also offer loan portfolio coverage to financial institutions in the region. | Fund is capitalized through contributions to a multi-donor Trust Fund (Govt. of Canada, EU, World Bank, governments of UK and France, Caribbean Development Bank and the governments of Ireland and Bermuda) as well as through membership fees from participating governments (currently 16). Parametric | Payouts to date of $35.6 m from 12 events to eight member governments |

| | Description | Coverage | Operation | Additional information |
|---|---|---|---|---|
| | operates as a virtual organization, supported by a broad network of service providers. | | insurance mechanism allows fund to provide rapid payouts for initial disaster response. | |
| CCR http://www.ccr.fr/index.do?langue=gb | Created in 1946. A public limited company owned by the French state, tasked with designing, implementing and managing efficient reinsurance cover for exceptional perils. | Provides cover against all natural disasters except for windstorm in France | CCR acts as a global reinsurer but approximately 67% of turnover is for the French market. With guarantee from the state they can provide unlimited cover for specific classes of business. Insurers in France are free to decide whether to participate in CCR cover and most do, paying a premium based on an agreed formula. | Now one of top 25 reinsurance companies in the world with capital of approximately €60 m and 2014 turnover of €1.323bn. Rated AA (stable outlook)[1] by the Standard and Poor's and A+ (stable outlook) by AM Best.[2] |
| FHCF http://www.sbafla.com/fhcf/ | Created in 1993 following Hurricane Andrew. Purpose is to protect and advance the state's interest in maintaining insurance capacity in Florida. | Provides cover for damage from hurricanes | Insurers pay a part of the policy premium (identified within their rate filings) to FHCF in return for aggregate catastrophe reinsurance cover. Deductible (attachment point) and limit is determined by the size of the ceded FHCF premium. | In 2015, provided approximately $17bn of hurricane capacity. In the event of a shortfall FHCF would be able to raise revenue bonds. |

1) As of 21/10/2016
2) As of 21/07/2016

Table 2.15  Private insurance company pools supported by government legislation.

| | Established and management | What is covered? | Mechanism | Terms of coverage | Capacity |
|---|---|---|---|---|---|
| NNPP http://www .naturskade .no/en/ | 1961 with updates to the law in 1979 and 1989. Organized as a distribution pool to equalize losses for insurance companies (since 1995 the market has been open to foreign companies who are obliged to participate in the pool) | All natural perils. | Insurance companies have all the contact with policyholders. The sums insured, premiums and claims/losses are reported on a monthly basis. Every quarter claims are distributed according to market share and each insurer either has to pay more or get money back. | An agreed premium for natural perils is separately identified. Premiums sit in a fund which covers administrative expenses and reinsurance premium. | NNPP purchases reinsurance cover of over NOK 12.5bn (as of 2015) on behalf of the member companies (members can choose to act a reinsurer to the fund with participation limited by their market share). |
| SVV http://www.svv .ch/en/consumer-info/non-life-insurance/swiss-natural-perils-pool | 1939. Pooling of private insurance companies to equalize the risk. Currently 15 companies participate covering approximately 95% of the market. | Covers flood, storm, hail, avalanche, snow pressure, landslide, rockfall and earthslip (but not currently earthquake) | Natural perils cover is include in the fire insurance for buildings and moveables | Buildings cover only supplied for Cantons of Geneva, Uri, Schwyz, Ticino, Appenzell Innerrhoden, Valais and Obwalden and in Liechtenstein. Contents are covered in all cantons except Vaud and Nidwalden. | The fund is protected with reinsurance purchased on behalf of the participating members. |

| IRV http://www.irv.ch/(German/French) | 1911. The 19 cantonal Public Insurance Companies for Buildings (PIB's) are combined in the Intercantonal Reinsurance (IRV) through which they jointly cover their reinsurance requirements in terms of fire and natural hazards. | Natural hazard losses that are caused by storms, hail, flooding, landslips/ collapses, snow pressure and avalanches are insured, but not losses caused by earthquakes. | As mentioned above all building owners in the other 19 cantons are obliged to insure with the PIB's, and in Glarus, Nidwalden and Vaud the PIB also insure moveable (contents). The PIBs are obliged to provide insurance cover for all buildings in their territory irrespective of their exposure to risks and hazards. In this way, adverse selection and risk-selection are prevented effectively (no 'fighting' over 'good risks' and mutual solidarity is at a maximum) with the greatest possible risk collective to ensure that premiums remain affordable for all policyholders. | Premiums and reinsurance cover between each PIB and the IRV are negotiated individually and the IRV itself therefore bears part of the bundled risks. The IRV places parts of the risks assumed on the national and international reinsurance market – all the losses of a particular year are totalled up and reinsured together. | The insurance cover of the PIBs is unlimited. A state guarantee for this only exists in one case, namely the canton of Nidwalden. |
| --- | --- | --- | --- | --- | --- |

### 2.13.6 Case Study: UK Flood Re

The United Kingdom is one of the few countries where flood cover is included as a standard part of every fire insurance policy issued for homeowners and commercial properties. However, there has been pressure on the continued provision of flood cover at affordable prices for the highest-risk properties. In response, flood underwriting and pricing has become increasingly sophisticated as insurers look to control their exposure, especially following recent flood losses (e.g. see Chapter 3.5.6). There has also been considerable debate surrounding the public investment required to create and maintain flood defences in the face of these financial pressures.

#### 2.13.6.1 Background

In 2000, UK insurers, through the Association of British Insurers (ABI), agreed to follow the 'Flood Insurance Statement of Principles', designed to ensure continued cover for all house-holders while flood prevention measures were evaluated and implemented. However, a decade on, the ABI believed that these had become unsustainable and a new approach to help flood-risk households obtain affordable flood insurance was required. The Statement was only ever intended to be a temporary measure and restricted customer choice; insurers only had commitments to their existing customers, and new insurers could decide to whom they offered flood insurance. In 2011, the industry formally recommended Flood Re as a long-term flood insurance solution in the United Kingdom.

#### 2.13.6.2 Legal Establishment

The ABI received government support for their position and after lengthy negotiations a Memorandum of Understanding (MoU) between the Government and the insurance industry was reached in June 2013. This described how to develop a not-for-profit scheme to allow flood insurance to remain widely affordable and available, while allowing a sustainable transition to risk-reflective pricing over 25 years. The Water Bill, legislating the powers to set up Flood Re, passed through Parliament and became law in May 2014. Further legislation enabled Flood Re to become the scheme administrator, with the Flood Re scheme launched in the spring of 2016. While Flood Re was developed, ABI members voluntarily continued to meet their commitments to existing customers under the old Flood Insurance Statement of Principles agreement.

#### 2.13.6.3 Structure and Mechanisms

Flood Re is a not-for-profit flood reinsurance fund, owned and managed by the insurance industry. Insurers will sell policies to their customers in the usual way, but have the option to selectively pass the risk carried by those policies to Flood Re for a set fee per policy. Those risks are then pooled into a fund which pays out to the insurer if claims are made. Insurers will, therefore, retain the incentive to compete for the business of customers with high flood risk because they know they can pass the flood risk associated with the policy into Flood Re. If a claim is made to a policy that is passed (ceded) to Flood Re, the contractual responsibility for paying the customers is still with the original insurer who can claim from the reinsurance pool, so allowing the risk to be better spread.

#### 2.13.6.4 Costs and Premiums

The insurance industry is paying the £10 million set-up costs for Flood Re. The Flood Re pool itself has two sources of income. The first is the flood portion of the original insurance premium for the policies passed into it. Flood Re premium rates are set by tariff according to the property's local Council Tax band (based on estimated property value). There is no 'risk reflection' in these premiums, but the Council Tax band is intended to act as a proxy for household financial strength and affordability. Insurers will have the option to pass policies which they feel should attract a flood premium higher than the premium needed to cede the risk to Flood Re into Flood Re, retaining the non-flood risks such as fire, theft and subsidence. Insurers are expected to use this for the highest-risk homes that would have struggled to find any affordable cover in a normal market; this is the top 1%–2%, an estimated 300,000–500,000 homes. Customers will deal with their insurer in the usual way to get their claim paid and Flood Re will reimburse the insurer behind the scenes. As mentioned in Section 2.3, the overall premium charged by insurers will also take into account various other risks including fire, storm, burst pipes, subsidence and (for contents) theft, the overall claims experience and other factors in a competitive market.

The second element is a levy on the industry, intended to be equivalent to the informal cross-subsidy that exists today in a 'typical' insurers' portfolio. This will be a proportion of the insurer's income, around 2.2% of an insurers' household gross written premium (GWP), about £10.50 per customer. If Flood Re requires additional capital, its management can call a second levy (Levy II) on the market to increase its capital resources.

### 2.13.6.5 Reinsurance

Flood Re will provide reinsurance to the market and purchase its own reinsurance and hold reserves and capital so that it can fully cover all claims in at least 99.5% of years (in line with the current UK and European regulation, see Section 2.11.4). In the event of a loss above Flood Re's limit, the remaining exposure is unprotected and falls back on the individual insurance companies to fund claims. If this happens, as well as the option to call on a second levy, the government has committed to work with the insurance industry and Flood Re to distribute any available resources to customers. Flood Re will be reviewed every 5 years to ensure that it is performing to requirements and make any necessary amendments. The UK government has also committed to a specified level of flood defence spending for 2015/16 following the government's spending round in June 2013. They will also ensure adherence to planning guidelines to avoid developments in areas at high risk of flooding.

### 2.13.6.6 Which Properties Are Covered by the Fund?

Flood Re is designed to provide support to those homeowners who are most likely to face problems in obtaining affordable flood insurance without it. Homes built after 1 January 2009 will not be covered, as in the old Flood Insurance Statement of Principles, to avoid incentivizing unwise building in known high flood risk areas. This is an example of avoiding moral hazard.

Commercial properties, including commercial leasehold properties, will not be included. The individual nature and assessment of business and commercial property risks mean that available and affordable flood insurance is less of an issue than for homes. While some individual firms may experience problems in accessing flood insurance, these can usually be resolved by using, as most do, an insurance broker. Flood Re will establish clear rules for borderline cases such as 'Bed and Breakfast' properties.

Those in commercial leasehold property who take out contents insurance (buildings cover is usually through the landlord) can have the flood risk element covered by Flood Re. Leasehold blocks with three residential units or fewer are eligible, providing the freeholder responsible for purchasing the buildings insurance lives in the block and the building meets the other required eligibility criteria.

Within the above eligibility criteria, an insurer is free to choose which properties it cedes to Flood Re.

### 2.13.6.7 Flood Re's Operating Lifetime

Flood Re is designed to have a 25-year operating existence and aid the transition to an open market over this period.

## 2.14  Catastrophe Models and Applications in the Public Sector

*Rashmin Gunasekera*

### 2.14.1  Introduction

In recent years, increased recognition of the impacts of natural catastrophes has reinforced the need to quantify catastrophe risk. Disaster (or catastrophe) risk assessment within the public sector progressed significantly during the Hyogo Framework for Action (HFA) period from 2005 to 2015. The transition from *qualitative* approaches to disaster risk assessment using risk indices, e.g. the World Bank Hot Spots project in 2006, to *quantitative, probabilistic* approaches, such as those used in Global Assessment Report 2015 (UNISDR, 2015) has been a major focus. There is now a demand for openly available catastrophe models and assessments in the public sector. The current practice of public sector catastrophe risk assessment together with advances made over the last decade are detailed in 'Understanding Risk in an Evolving World: Emerging Best Practices in Natural Disaster Risk Assessment' (GFDRR, 2014a). This study was developed to inform post-HFA discussions and the 2015 Global Assessment Report on Disaster Risk Reduction (GAR).

This section outlines the scope, availability and challenges associated with catastrophe models and tools used to quantify probabilistic catastrophe risk in the public sector. The discussion then focuses on the applications of these catastrophe models in the public sector. It highlights the different public sectors that use these models and their applications. A short discussion of the future direction of development in public sector catastrophe models can be found in Chapter 6.2.4.

### 2.14.2  Public Sector Catastrophe Models

Public sector quantitative risk models extend traditional financial loss estimates to cover some economic, social and infrastructure losses. Many software packages are available for these risk assessments. The quality of these software packages (or models) has improved considerably in the last few years as shown by the assessment of over 80 open source or freely available models on a variety of criteria (GFDRR, 2014b). For example, RiskSCAPE (https://riskscape.niwa.co.nz/) and CAPRA (http://ecapra.org/) allow for probabilistic, modular multi-hazard risk analysis. As a deterministic risk assessment, communication and geo-visualization tool, InaSAFE (http://inasafe.org/en/) ranked number one due to its simplicity and intuitiveness in the World Bank/Global Facility for Disaster Reduction and Recovery (GFDRR, 2014b) study of earthquakes and tsunami, as the user could simply install the plug-in and undertake a risk calculation. As part of this study, for single hazard public sector models, GEM's OpenQuake (http://www.globalquakemodel.org/), Hec (suite) (http://www.hec.usace.army.mil/software/) and TCRM (https://github.com/GeoscienceAustralia/tcrm/wiki/Tropical-Cyclone-Risk-Model) scored highly for earthquake, flood and windstorm respectively (GFDRR, 2014b).

Initiatives such as the USGS PAGER initiative, Global Earthquake Model (GEM) and Global Assessment Report (GAR15) have provided not only open and freely available datasets for catastrophe models but also loss estimates and different loss metrics, such as exceedance probability curves. ELTs and YLTs are not output since the primary purpose of these models has been risk identification and communication. However, since most of these models are at the global scale, they have inherent weaknesses relating to scale, fitness for purpose and calibration of results. They are also limited to assessing direct damage. These aspects need to be considered when considering the choice and the application of these models.

### 2.14.3 Applications of Public Sector Catastrophe Models

There are many similarities in the application and construction of the probabilistic catastrophe models used within the public and private sectors, particularly (re)insurance catastrophe models. The (re)insurance industry catastrophe model applications include (re)insurance pricing, solvency capital requirements and portfolio optimization (Sections 2.6–2.12). Within the public sector, probabilistic catastrophe models have also been largely used for financial protection measures. However, applications also extend to casualty estimation, urban planning, structural risk reduction measures and developing strategies for resilient reconstruction among others, albeit it at a much smaller scale.

Governments and sub-national public entities (including municipalities) are increasingly aware that natural disasters pose a significant risk (or liability) to public finance management in numerous developing countries, particularly Small Island Developing States. Ministries of Finance in certain countries allocate some form of economic loss calculation for emergency preparedness and response during the budgeting process. However, in most developing countries the quantification of contingent liabilities has not assessed the probabilistic risk using quantitative models. This risk could significantly impact and destabilize the fiscal balance of country's economy and so impede development (Yamin *et al.*, 2013).

Institutions such as the World Bank are increasingly using probabilistic catastrophe risk models to assess this sovereign risk from natural disasters, offer the government innovative tools to examine their disaster risk financing options and develop a framework for financial and fiscal protection strategy. For example, a Country Disaster Risk Profile (CDRP) used by the World Bank is a coarse level probabilistic catastrophe risk analysis that presents an estimate of risk at the national level. This type of product could be the starting point for a dialogue on disaster risk management (DRM) policy within a country, to assess fiscal responsibility, raise public awareness of disaster risks, and provide momentum to undertake more resource-intensive and detailed risk assessments for specific financial decision–making or disaster risk reduction (DRR) measures.

These could, in turn, be used for measures such as cost benefit analysis of development projects. Another application is in the provision of **building codes**. Numerous developing countries use the Universal Building Code (UBC) as a building standard. However, this code is not always applicable, particularly in Small Island Developing States (SIDS) due to the heterogeneity in the building stock in a small geographical area. There is a need to update building codes and probabilistic modelling provides great insights into the financial, and other, benefits of adapting a country-specific UBC or similar code. In Turkey, probabilistic risk modelling has been used to inform safety standards and socio-economic loss potential of public buildings from earthquakes (World Bank, 2012). In the public land-use and urban planning sectors, robust analysis provided from probabilistic risk analysis assists in site selection, developing early warning systems and investment in flood protection (e.g. the Mekong River flood project, Viet Nam). An added benefit of datasets from these catastrophe models is scenario impact analysis in post-disaster situations. For example, data and results from probabilistic modelling were useful in disaster response and resource mobilization after Typhoon Haiyan in the Philippines in late 2013. However, in terms of risk communication and enforcement of particular disaster risk reduction measures within the public sector, deterministic or scenario risk assessment approaches are more suitable. Applications include population impact scenarios, sectoral planning capacity to address population impact and simulations and training of disaster first responders. They could also relate to Climate Change Adaptation (CCA) measures (GFDRR, 2014a).

### 2.14.4 Case Study: Country Disaster Risk Profiles (CDRP)

One of the missions of the World Bank's Global Practice for Social, Urban, Rural Resilience Practice Group is to support national and sub-national governments to identify, assess and analyse the potential economic and fiscal impact of disaster and climate risks so they can be effectively managed. Since 2008, the World Bank has promoted the Comprehensive Approach Probabilistic Risk Assessment Initiative (CAPRA). This aims to raise awareness among in-country stakeholders by providing them with a set of tools to enable them to better understand the risk of adverse natural events, with a particular focus on Latin America and Caribbean. One way it does this is by providing a free and open source software suite to analyse the magnitude, distribution and probability of potential losses from disasters. Without such evaluations, governments encounter major obstacles when trying to identify, design and prioritize vulnerability reduction measures.

Under the CAPRA program in the Central-American region, the World Bank is also developing a series of Country Disaster Risk Profiles (CDRP). CDRPs assist in informing government and other stakeholders by quantifying the impact of disaster through both structural (e.g. building stock) and non-structural measures (e.g., flow of goods e.g. Gross Domestic Product (GDP)). CDRPs also provide different metrics, including Annual Average Loss (AAL), Probable Maximum Loss (PML) (see Chapter 1.10), mainly for earthquake and windstorm hazards.

In addition to the country-level risk profile, higher resolution urban or sector disaster risk profiles are also becoming increasingly important for public-sector decision-makers. These are important not only for making risk sensitive decisions incorporating potential impact of future losses, but also for viability and cost benefit analysis of urban and/or infrastructure projects. Since the CDRPs offer flexibility in their method the resulting risk profiles themselves vary in the level of detail, scale and end-users. Some output datasets also have multi-purpose applicability. This can include downscaled exposure datasets or methodological adaptations to assess the risk at city level or at sector level, such as for transport or education.

To derive these outputs, hazard, exposure, vulnerability and risk components were assessed for each country. This included developing a new open exposure modelling methodology for the CDRP project (Gunasekera *et al.*, 2015). The exposure model outputs consists of: (1) a building inventory model; (2) an infrastructure density and value model that also corresponds to the fiscal infrastructure portion of a country; and (3) a spatially and sector disaggregated GDP model (see Figure 2.22). The underlying datasets are based on population, country-specific building type distribution, global engineering construction reference guides and economic indicators such as World Bank indices that are suitable for natural catastrophe risk modelling purposes. The output is a GIS grid (raster) at approximately 1 km spatial resolution which identifies the degree of urbanized areas (population, building typology distribution and occupancy) for each cell at sub-national level and associated economic (stock) or asset value for each cell. These exposure models could be used in conjunction with hazard and vulnerability components to create views of risk for multiple hazards that include earthquake, flooding and windstorms for national level disaster risk profiling for the public sector in a probabilistic and historical scenario basis. An example of the output of this project is shown in Figure 2.23.

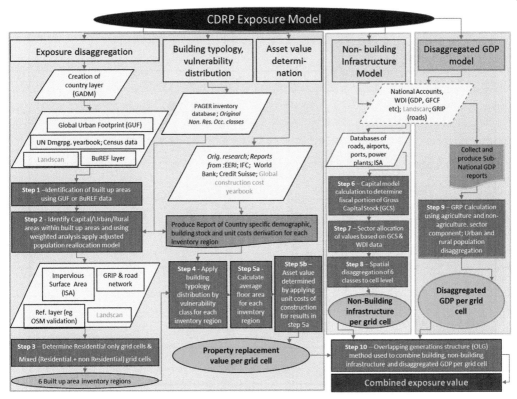

**Figure 2.22** Flow chart outlining the components and steps in creating the composite exposure model. Required input datasets are in yellow parallelograms, model processing steps are in blue boxes, and intermediate and final outputs are shown in red and green ellipses respectively. Demographic, building typology, floor area and unit costs of construction information and sub national GDP datasets compiled and developed for each country are shown in orange coloured folder. *Source*: Gunasekera *et al.* (2015). Reproduced with permission of Elsevier.

In terms of earthquake vulnerability, the CDRP project also aims to rate different seismic vulnerability functions. This work expands on Rossetto *et al.* (2014) to assess the statistical significance, variability and implications for economic loss evaluation of choice of seismic vulnerability functions for the Central American region (Stone *et al.*, 2015). Rossetto *et al.* (2014) compiled databases of existing analytical and empirical fragility functions worldwide and developed a method for rating these functions in terms of relevance and quality as part of Analytical and Empirical Physical Vulnerability component of Global Earthquake Model (GEM) project. This work could assist model developers to select the most reliable functions given certain conditions.

While the CDRP project incorporates authoritative existing models such as Resis II which is a probabilistic seismic hazard model for Central America (Resis II, 2010), the CDRP project has also produced a state-of-the-art probabilistic windstorm model for Central America (Pita *et al.*, 2015). A key aspect of these models is that the methodologies are open, therefore allowing the end user to understand and potentially improve the model. This also strengthens capacity for assessing, understanding and communicating disaster risk to improve the integration of vulnerability reduction policies into the countries' development.

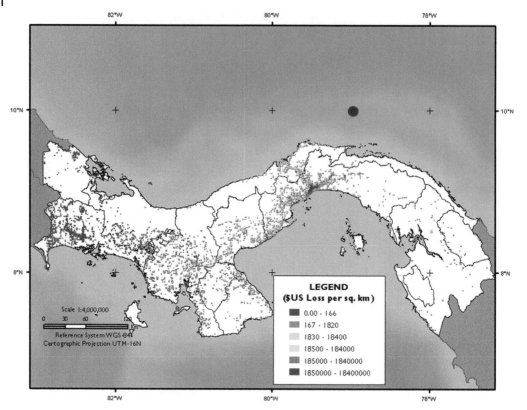

Figure 2.23 Distribution of potential direct economic losses in US$ from a repeat of the 1882 Mw 7.8 offshore Panama earthquake (epicentre shown as a red dot) in present time. *Source*: Gunasekera *et al.* (2015). Reproduced with permission of Elsevier.

## 2.15 Insurance Linked Securities

*Arnab Chakrabati*

### 2.15.1 What Are Insurance Linked Securities?

An **Insurance Linked Security (ILS)** is a financial instrument whose price is determined by insurance losses. It is a type of **derivative**, which is a security that derives its price from an underlying quantity, the underlying quantity in this instance being insurance losses. ILS instruments are issued by an entity seeking insurance protection; as with reinsurance, this entity is called the cedant because it cedes risk. The buyer of an ILS instrument is, therefore, a provider of capital, and gets paid a premium in excess of the risk-free rate by the cedant. Here, the **risk-free rate** is what a capital provider can earn without taking on any type of risk. The return on US Treasury Bills is sometimes used as the risk-free rate. The premium is much like the coupon paid by a bond that carries a credit risk but with the difference that loss of the principal sum results from natural catastrophes instead of financial default. The most common type of ILS instrument is the catastrophe bond, commonly referred to as a 'cat bond' (e.g. Yoon and Scism, 2014). ILS is currently a relatively small subset of the broader catastrophe risk transfer asset class (see Section 2.3).

The first insurance linked security was issued in 1996 by St Paul Re and the first fund, Nephila, was set up in 1997. Since the 1990s, there has been rapid growth in this area (see Section 2.15.7).

### 2.15.2 From Insurance to Reinsurance to ILS

As discussed in Section 2.4.1, to insure something is to hold capital against its value to pay for losses when they happen. The provider of capital is compensated for the opportunity cost of capital in the form of a premium. The premium must compensate the capital provider by paying the risk-free-rate, plus the expected loss, costs and a risk premium, as for (re)insurance.

Capital for catastrophe protection can be provided in a number of different ways. Shareholders of reinsurance companies act as providers of equity capital. Reinsurance companies can also issue debt. ILS instruments provide yet another way for capital to find its way to cedants – through securitization. **Securitization** is the process of taking an illiquid asset and using financial engineering (i.e. mathematical methods applied to financial problems) to transform it into an instrument that can be traded. Securitization divides the risk of catastrophes into layers with different loss probabilities and correspondingly different risk premiums. One of the key benefits of ILS is that these instruments facilitate more direct participation of capital markets in catastrophe risk transfer, thereby removing some barriers between cedants and capital providers. Since catastrophic events happen independently of financial markets, ILS losses are largely uncorrelated with other financial assets, and this makes ILS attractive for investors (Bryjolfsson and Dorsten, 2007; GIRO, 2008).

Strictly speaking, ILS instruments are securities. Cat bonds are securities that trade in secondary markets. Basically, this means that after a cat bond is issued, its owner can sell it to someone else, and therefore does not have to hold the bond until its maturity. Sometimes instruments that are not strictly securities but which enable capital markets to invest in catastrophe risk are broadly labelled as ILS. Collateralized reinsurance, which we define in Section 2.15.3, is often not securitized, but is sometimes labelled as a class of ILS. The broader definition of ILS is sometimes termed **alternative capital** to differentiate it from the narrower definition of a security. Brief descriptions of ILS instruments follow in Section 2.15.3.

The underlying catastrophic events (e.g. hurricanes and earthquakes), the hazards they create (e.g. strong winds and ground shaking), and the damage they cause to life and property are modelled the same way for traditional reinsurance and ILS. The development of catastrophe models and their adoption by market participants have been key reasons behind the success of ILS. These models promote a more transparent view of risk than was possible before.

A cat bond's **Offering Circular** (OC) describes the underlying risk of loss with statistics derived from catastrophe models. Although OC formats differ, some statistics, such as the expected loss probability, can be found in every offering circular. Even though loss statistics and other descriptive information are included in the OC, the raw exposure data itself is not included. A reinsurance contract, by contrast, includes the cedant's exposure data, which the reinsurer then processes to evaluate the cedant's risk.

Having access to the raw data affords increased visibility into a cedant's book. A prudent reinsurer should have the expertise to form a well-informed opinion of the risk based on the exposure data. In the case of cat bonds, the issuing entity and modelling firm hired for the purpose do much of this work. Cat bonds are often assigned a credit rating by an entity such as S&P or Moody's, and these securities are subject to rules such as the Securities and Exchange Commission's Rule 144A. Having said that, expert investors are able to form opinions on the modelling of cat bonds too based on their understanding of differences in models used by modelling firms and other factors.

On the other hand, cat bonds can be traded unlike reinsurance contracts, which opens up opportunities to profit from buying and selling cat bonds in secondary markets. The ability to evaluate cat bonds based on their prevailing bid and offer prices and residual risk of loss can enable a cat bond investor to generate profits through secondary trading.

### 2.15.3 Common ILS Instruments

This section describes the most common ILS instruments:

- **Catastrophe bonds:** A catastrophe bond is like a regular bond in that the issuer (the cedant) pays premium coupons (interest payments), usually in excess of a floating benchmark yield (such as 3-month LIBOR or EURIBOR). The principal (notional amount) of the bond serves as collateral to pay for potential losses from catastrophes.
- **Industry Loss Warranties (ILW):** These contracts pay when the loss to the entire industry crosses specified thresholds. For example, an ILW could be triggered if the US industry loss resulting from a catastrophic event exceeds US$40 billion.
- **County-Weighted Index Loss (CWIL) deals:** CWIL contracts pay when industry losses multiplied by weights assigned to each county crosses specified thresholds. The county weights are chosen to reflect the cedant's exposure. These are mainly US-only products.
- **Collateralized reinsurance:** Reinsurance companies use the same dollar (or capital) to sell protection to multiple cedants. Collateralized reinsurance, by contrast, provides that each dollar of collateral will only be used once. The collateral is usually set aside in a trust for the term of the contract while the cedant continues to pay a premium to the reinsurer.
- **Sidecar:** A sidecar is a collateralized vehicle that allows alternative capital to invest in an existing book of business, in effect, sharing the risk and (some of) the return of the existing book. These may be thought of as a collateralized twist on 'quota-share' or 'proportional reinsurance' where two reinsuring entities share the premium and the risk of a portfolio in proportion to their capital.

Catastrophe bonds, ILWs (Industry Loss Warranties) and CWIL (County Weighted Insurance Loss) deals are generally considered ILS because they combine the elements of collateralization and securitization. If we expand the definition of ILS to include more traditional means of risk transfer that are now backed by collateralized capital, then we may include collateralized reinsurance and sidecars into the ILS fold as well. These are not strict distinctions; some sidecars and collateralized reinsurance contracts are securitized. There are introductory chapters on several of these ILS instruments in Lane (2012).

### 2.15.4 Preliminaries of ILS Instrument: Measurement and Layering of Losses

#### 2.15.4.1 How Losses Are Measured

ILS instruments are annual or multi-year contracts. Within the period of the contract there can be zero, one or more catastrophic events impacting the instruments. As elsewhere, the loss from a single event is termed the occurrence loss. The loss from all events occurring in a year is termed aggregate loss for the year. Contracts can be triggered either on an occurrence basis or on an aggregate basis, but the former is more common.

There are a few different ways to measure loss for catastrophe bonds:

1) **Ultimate Net Loss (UNL,** or **indemnity-based) contracts** are structured to indemnify the cedant based on the loss it has actually suffered, as in reinsurance. The cedant prefers this type of contract, because it is the most reliable way for it to hedge (or limit) catastrophic risk. The (re)insurer, on the other hand, is exposed to additional uncertainty stemming from the quality of the cedant's book and is generally at an informational disadvantage regarding the cedant's underlying exposure and loss settlement practices. As a result, contracts that measure losses in this way tend to carry higher premium than other types of ILS instrument.
2) In **parametric** contracts, losses are measured indirectly using a parameter such as wind speed. These contracts are less attractive to the cedant because the parameter chosen is

unlikely to correspond exactly to the cedant's actual loss. Parametric contracts therefore tend to carry lower premiums. The term **basis risk** is used to describe the uncertainty that remains when there is no exact correspondence between the insured quantity and the true exposure. It is worth noting that basis risk affects both the cedant and the reinsurer, as both are vulnerable to underestimating their exposure.

3) Another common way to measure losses uses loss indexes such as the Property Claim Services (PCS) index and the PERILS index. PCS is an independent organization that tracks losses in the United States, and publishes an index update of aggregate insured property losses after each event. Index losses often serve as the basis for Industry Loss Warranties (ILW). PERILS is a European organization with a similar mandate. Contracts based on index losses also carry basis risk because such indexes measure industry losses, which can differ significantly from the cedant's losses in an event.

4) A product that aims to reduce the cedant's basis risk without increasing the reinsurer's risk is a **County-Weighted Insurance Loss (CWIL)** contract. The loss used to trigger such a contract is a weighted sum of county-wise industry losses, where the weights are tailored to the cedant's exposure by county. CWIL was introduced in 2009 by Guy Carpenter and Nephila.

### 2.15.4.2 Layering of Losses

Now we examine the basics of how losses are divided into layers for securitization. Many risk transfer contracts are excess of loss (XL) structures, in the same way as non-proportional reinsurance (see Section 2.4.2.2), i.e. the cedant's losses must exceed a pre-specified attachment amount to cause a payout and the contract will suffer a full loss should the cedant's losses exceed an exhaustion amount. As with XL reinsurance, the difference between the attachment and exhaustion amounts is called the limit. Figure 2.24 shows the attachment, exhaustion and limit for a hypothetical catastrophe risk-transfer layer. These are juxtaposed with a loss probability density function. All the analysis that culminates into producing this type of loss probability density (or equivalently, the cumulative loss exceedance curve) falls under the umbrella of modelling. On the other hand, analysis that accepts this type of probabilistic loss characterization and uses it to calculate premiums falls under pricing as discussed in Section 2.6. ILS shares its modelling framework with traditional reinsurance, but is priced differently. The concept of a layer is central to the securitization and pricing of ILS, as we shall see in Section 2.15.5.

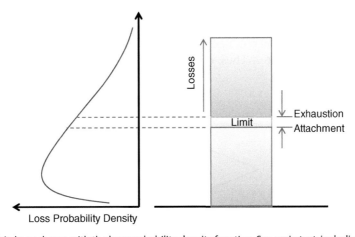

**Figure 2.24** A thin layer shown with the loss probability density function. See main text, including Section 2.15.5.

### 2.15.5 Pricing an ILS Contract

Pricing of ILS is a ripe area for research. Early pioneers in this area include Litzenberger *et al.* (1996), Canabarro *et al.* (1998), and Froot *et al.* (1997), who provide a broad introduction to ILS while emphasizing different aspects. Lane *et al.* (2008) perform an empirical study of cat bonds that introduces the reader to useful techniques; however, since its publication the cat bond market has become more efficient so that the pricing conclusions presented there may not represent today's realities. By improved efficiency, we mean that as investors have become more comfortable with cat bonds, more cat bonds have been issued and the premium that a cedant has to pay to receive protection for a given amount of risk in cat bond form has gone down. There is also a chapter on the basics of cat bonds and ILS pricing in Lane (2012). Recent work by Jaeger *et al.* (2010) presents a good overview of ILS investment strategies.

Below, we explain premium calculation for a 'thin' layer of catastrophe loss. By pricing a layer in this way, one arrives at a formula for premium that is supported by empirical data (e.g. in cat bond prices) and has also been studied in the literature (see Jaeger *et al.*, 2010).

#### 2.15.5.1 The Concept of a Thin Layer

We define an infinitesimally thin layer as one where the exhaustion is just barely more than the attachment. It follows that the probability of attachment is almost the same as the probability of exhaustion, so that if an event triggers a thin layer loss, then the layer is used up completely. The probability of loss is constant throughout the thickness of the thin layer – from the attachment all the way to exhaustion. Consequently, the probabilities of attachment and exhaustion are both equal to the expected loss of a thin layer.

#### 2.15.5.2 Pricing a Thin Layer

We label the probability that a thin layer will pay out as $p$, and the discount margin (premium or Rate on Line[2] in excess of the benchmark rate such as LIBOR) associated with the thin layer as $R$. The return from insuring a thin layer is captured by the binary model in Figure 2.25.

For every dollar of collateral we hold, we either receive $(1 + R)$ if there is no event, or just $R$ if there is an event. The expected loss, $E[L]$ (or simply $EL$), of the above model is

$$E[L] = 1.p + (1 - p).0 = p \tag{2.18}$$

and the standard deviation of loss, $\sigma$ is

$$\sigma = \sqrt{p(1-p)^2 + (1-p)(-p)^2} = \sqrt{p(1-p)} \tag{2.19}$$

The basic premise in pricing catastrophe risk is that investors seek to be compensated for the expected loss plus an added premium that is in proportion to the standard deviation of the loss, as also discussed in Section 2.6.2. Thus, the postulated model is:

$$R = p + \gamma \sqrt{p(1-p)} \tag{2.20}$$

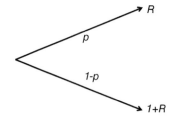

**Figure 2.25** Binary loss model for a thin layer, where R is the premium or Rate on Line and p is the probability.

where $\gamma$ is the risk premium coefficient for the risk being undertaken. Since the expected loss $E[L]$ is the most commonly used (and arguably most fundamental) basis for pricing catastrophe risk and p is equivalent to $E[L]$ as shown above and since the premium is the difference between the Rate on Line and Cost of Capital

$$R = RoL - CoC \tag{2.21}$$

$$RoL = CoC + E[L] + \gamma\sqrt{E[L](1 - E[L])} \tag{2.22}$$

### 2.15.6 Pricing Cat Bonds with the Thin Layer Model

Since most cat bonds provide protection for thin layers of losses (Lane, 2012), it is reasonable to price them with the above model. The green line in Figure 2.26 shows the variation of premium with expected loss for a single-**metarisk**[3] thin layer bond. Pricing can be extended to bonds that cover multiple metarisks by summing the risk premiums across metarisks. It is empirically observed that different metarisks pay different risk premiums for the same amount of risk; hence we have a separate $\gamma$ for each metarisk. For example, the premium for protecting against a hurricane in Florida is higher than the premium for protecting against earthquakes in Japan. In summary, the premium (Rate on Line or $RoL$) for a multi-metarisk cat bond is given by:

$$RoL_{cat-bond} = CoC + \sum_{i=1}^{metarisks} E[L]_i + \gamma_i\sqrt{E[L]_i(1 - E[L]_i)} \tag{2.23}$$

The $CoC$ as well as $\gamma_i$ are estimated by regressing the observed market premiums against the expected losses and standard deviations for all bonds trading in the market. $CoC$ is the $RoL$ intercept.

Various assumptions are implicit in pricing a cat bond using the simple model above. Here are a few that a pricing analyst should be aware of:

- A real layer is not infinitesimally thin, and the difference can be significant for the pricing of wider layers.
- The **cost of capital** is defined as the return that the market expects from a cat bond with zero expected loss. Although called cost of capital, this term actually includes other premiums including basis risk and extreme tail risk and its value is often significantly higher than the risk-free rate. Its value also changes with time.

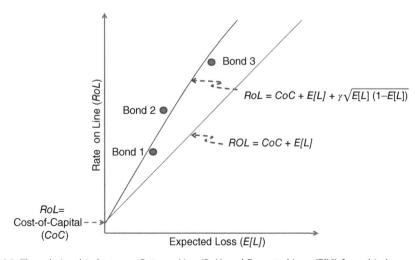

**Figure 2.26** The relationship between Rate on Line (RoL) and Expected Loss (E[L]) for a thin layer cat bond.

- The risk premium coefficient $\gamma$ has historically been different for different metarisks, so different metarisks are modelled differently. Over time, the risk premium coefficients for different metarisks have also been changing, providing a reason to track their evolution over time.
- The regression sometimes produces negative values of $\gamma_i$ for some metarisks, usually 'non-peak' metarisks, mostly outside the United States. This does not mean that the market is willing to pay (as opposed to being paid) to hold those risks. It simply means that the risk premium included in the cost of capital allows the market to hold the metarisk with a negative $\gamma_i$.

### 2.15.7   Growth of the Market for ILS

Figure 2.27 shows the relative size of the different segments of the property catastrophe reinsurance market in recent years. Two things stand out. First, the market share of traditional reinsurance (UNL) is still much larger (~80%) than the share of the market served by alternative capital (~20%). Second, the market share of alternative capital has more than tripled since 2002 at the cost of traditional UNL's market share. It is worth mentioning here that in absolute dollar terms, even the traditional reinsurance market has grown in size; it is just that the growth of ILS has been much more rapid.

Of the various ILS securities, cat bonds are the most liquid, and the market for bonds is also relatively transparent, which makes them useful for tracking the growth of ILS. Cat bonds were first issued in the 1990s. Issuance has grown multi-fold over the years as Figure 2.28 shows. As a case in point, Citizens, Florida's state-run insurer, completed a cat bond transaction for $1.5 billion in 2014, the highest to date at that point.

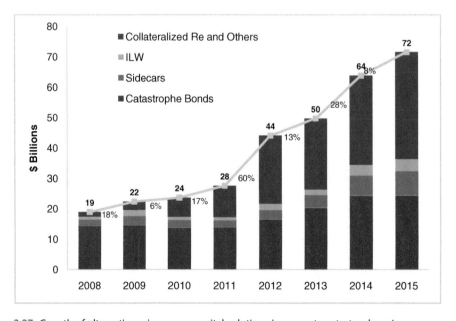

**Figure 2.27** Growth of alternative reinsurance capital solutions in property catastrophe reinsurance market, 2008–2015. *Source*: Reproduced with permission of Aon Benfield.

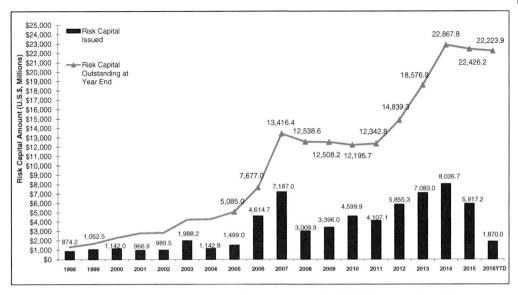

**Figure 2.28** Cat bond issuance and capital outstanding, e.g. outstanding if $10 billion in cat bonds is issued each year, and each bond has a life of 3 years, then one would have $30 billion on any given date. *Source*: Reproduced with permission of Guy Carpenter, LLC © 2016.

Cat bond yields, on the other hand, have been moving lower steadily in recent years (e.g. Bantwal *et al.*, 1999). This is illustrated by the US cat bond market multiple shown in Figure 2.29. The **multiple**, which we define as dollars of premium for each dollar of expected loss, helps to normalize the market premium against changes in the average expected loss as the market's composition may change with time. The inflow of alternative capital and the simultaneous lowering of premiums are two observable phenomena in a cyclically reinforcing process that will ultimately lead to a more efficient market for catastrophe risk transfer.

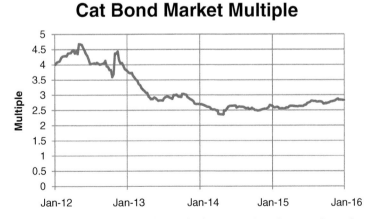

**Figure 2.29** Notional-weighted cat bond market multiple with a clear downward trend in recent years. Courtesy of Nephilia.

---

**Box 2.6  Case Study: Everglades Re**

A series of Everglades Re cat bonds have been issued by Florida Citizens Property Insurance Corporation ('Citizens' in short). Citizens, the cedant for the purpose of these bonds, was created by the Florida legislature in 2002 as an insurer of last resort. It is a not-for-profit, tax-exempt government entity. Its purpose is to provide insurance to Florida policyholders who are unable to find property insurance coverage in the private market. The need for a Citizens-like entity arose because the value of property in Florida that is exposed to damage through hurricanes is large, and therefore the demand for insurance is high. Citizens, in turn, needs protection against losses from hurricanes that can damage many of the properties it insures.

The first Everglades Re bond was issued in 2012, and there has been an Everglades issue each year since then. In the first year, Citizens intended to obtain $200 million protection in cat bond form, but the bond grew in size to $750 million making it the largest single cat bond at the time. Part of the reason for its success was its attractive high yield. Everglades 2012 had an expected loss probability of around 2.5% (according to the modeling in its offering circular), but the bond paid a coupon of 17.75%, leaving investors with an expected return (the difference of the coupon over the expected loss) of over 15%. Subsequent Everglades issues were successful too. Everglades 2013 had an issue size of $250 million. Everglades 2014, with an issue size of $1.5 billion remains the largest single cat bond issue ever. Everglades 2015 was issued to provide $300 million of protection to Citizens.

As the high yield and non-correlation to other asset classes brought more capital into the cat bond space, the cat bond market became a more efficient space for cedants to seek protection, meaning that the price for seeking protection fell. In contrast to Everglades 2012, which had an expected return of 15+% (the bond has since expired), Everglades 2015 has an expected return of only around 4%. The lower return can partly be attributed to the fact that Everglades 2015 has a smaller chance of losing its principal (its expected loss probability is 1.31% whereas the figure for Everglades 2012 was 2.5%), but even after adjusting for the difference in risk, it is obvious that yields have fallen significantly.

Citizens is currently undergoing 'depopulation', which is an initiative to reduce the number of policies covered by Citizens. Private insurers are encouraged to take over policies that were previously covered by Citizens. As Citizens' exposure reduces, it is likely that the amount of reinsurance protection Citizens will need will decrease, and the protection private insurers participating in depopulation will need will increase.

---

### 2.15.8  Conclusion

ILS instruments are securities that facilitate catastrophe risk transfer. The most common type of ILS instrument is the cat bond. Sometimes, instruments that are not strictly securities but which enable alternative capital to invest in catastrophe risk are broadly labelled as ILS. An example would be collateralized reinsurance. Catastrophe risk is finding growing acceptance as an asset class that is uncorrelated with stocks and bonds. Moreover, securitization of catastrophe risk transfer is gaining momentum at the cost of more traditional risk transfer methods. The latter trend is partly because securitization is more 'efficient' than the conventional use of equity capital in the reinsurance industry. The aforementioned factors are collectively responsible for the growing popularity of ILS instruments in recent years.

While ILS has succeeded in attracting investors with its uncorrelated return, it is important for investors to invest in what they understand. Doing so enables investors to manage their basis

---

**Box 2.7 Case Study: Amtrak Storm Surge Cat Bond**

Cat bonds are increasingly being used to cover potential infrastructure losses. Two recent examples are the Metrocat deal for the New York Metropolitan Transit Authority, which was the first storm surge deal issued in the wake of Hurricane Sandy, and the Amtrak cat bond. The Amtrak cat bond is planned to have a three year per occurrence trigger and to cover earthquake, storm surge and wind from named storms (Okocha, 2015).

---

risk effectively. Investing through trusted intermediaries who combine deep expertise with efficiency and transparency is a good solution.

## 2.16 Effective use of Catastrophe Models

### 2.16.1 Treatment of Uncertainty in Catastrophe Models

*Ian Cook, Matthew Jones, Adam Podlaha, and Kirsten Mitchell-Wallace*

It can be difficult to grasp model uncertainty when trying to understand and use the output of catastrophe models. There are many reasons for this, including:

- ambiguous or misunderstood terminology relating to uncertainty;
- the complexity of uncertainty;
- the lack of widespread uncertainty quantification from the model vendors, possibly fuelled by the lack of potential application or concerns in compromising the usability of the output;
- a variety of approaches used to incorporate and propagate uncertainty within the catastrophe models;
- requirements for a definitive, certain, answer for many end-users, e.g. CFOs, accountants, reinsurance and retrocession buyers who want a nominal value and not a range to determine profitability, efficacy, etc.;
- no commonly accepted framework for the communication of model uncertainty in the catastrophe modelling community.

To explore the issue of uncertainty in catastrophe models, we need to step back and consider uncertainty more widely.

#### 2.16.1.1 Philosophical Classification of Uncertainty

Some philosophers of science consider that risk or uncertainty can be categorized as one of two fundamental types: **aleatory uncertainty** and **epistemic uncertainty**. There is also, in fact, a third classification, **ontological uncertainty**, which refers to unknown unknowns and as such defies quantification and, therefore will not concern us further beyond observing that its existence implies that even if all known uncertainties are measured, there is still additional uncertainty that cannot be quantified.

Aleatory uncertainty is defined as the inherent uncertainty in a random process or phenomenon. This uncertainty is therefore always present and cannot be reduced. However, it can be quantified using various modelling methods.

Epistemic uncertainty, on the other hand, is defined as the uncertainty that comes from our lack of knowledge about the underlying process or phenomenon. This lack of knowledge can encompass many things, including lack of understanding of the physics behind the process or sub-processes, imprecise data and sampling uncertainty. Since this uncertainty is due to our lack of knowledge it can, at least in theory, be reduced, given more information, and so can be quantified and taken into account.[4]

In practice though, while our understanding of the physics of the underlying process being modelled can improve and more available computing power can increase model resolution, the amount of useful information will continue to be limited by the availability of historical hazard, insured exposure or claims data.

### 2.16.1.2 Actuarial Classification of Uncertainty

When it comes to understanding the uncertainty in a model, actuarial science, where the practice of building models of different risks is common, is a good place to start. There, uncertainty in a model of losses from a phenomenon is considered to come from three sources:

1) Process Error is caused by actual future losses being random (e.g. that the losses come from natural variability);
2) Model Error is defined as the error arising from when the methodology used does not accurately reflect the underlying loss/physical processes (e.g. the statistical distributions chosen might not behave as the real world does);
3) Parameter Error arises when the model parameters are calibrated from a finite amount of data (i.e. a sample) and do not represent the real-world 'parameters' (i.e. whole population), for instance, the value of lambda estimated for the mean of a Poisson distribution is only an estimate of the real world mean.

Some actuarial texts (e.g. Parodi, 2015) also include assumption/data errors and approximation errors as additional categories of uncertainty for clarity. Alternatively, these effects can also be thought of as increasing the parameter and model errors respectively.

Loosely speaking, process error is equivalent to aleatory uncertainty and model and parameter errors (together with data and approximation errors) correspond to epistemic uncertainty since they can be decreased with more knowledge.

A good discussion of uncertainty and error analysis can be found in Taylor (1996).

Some sources of uncertainty underlying an individual model can be easily overlooked. This includes the consideration that much of the historical 'observational' data underlying catastrophe models is not actually data in the strict sense of the word. Instead, it is the output of a model applied to a much smaller set of historical data. For example, many models of extra-tropical cyclones in Europe are based in part on a dataset called 'ERA Interim' (see Chapter 4.3.3). This dataset is in itself a product, derived from the more limited observations; the re-analysis is used to 'fill in' (geographically and temporally) the more limited observations. Even historical windspeed gust 'observations' for a single site might themselves be adjusted based on a model to allow for the exact location of the measuring instruments being changed (e.g. increased in height by 10 m due to new building).

### 2.16.1.3 Source Classification of Uncertainty

Before we turn to the treatment of uncertainty in practice it may be helpful to further classify uncertainty by its origin in the modelling process. This is closely linked with its practical use. Here three main sources can be defined:

### 1) *Model, i.e. the Model Developer*

The source of most of the uncertainty discussed is imperfect models, in other words, it comes from the process in which a model developer is tasked to create a model with limited resources whether time, people, costs or knowledge.

As an example, consider the assumption taken by the Dutch flood model developer when estimating the probability of flooding of Dyke-ring No. 1: the only information available is the design specification of failure with a probability of 1-in-10,000 years (see Figure 2.30). The modeller, could, for example, ignore the inherent uncertainty in this parameter and develop a model solely based on this best estimate assumption. However, this approach will not quantify the uncertainty present in this best estimate.

The other option is to make alternative assumptions about the probabilities, either based on expert local opinion or estimated by the model developer carefully judging the possible realistic

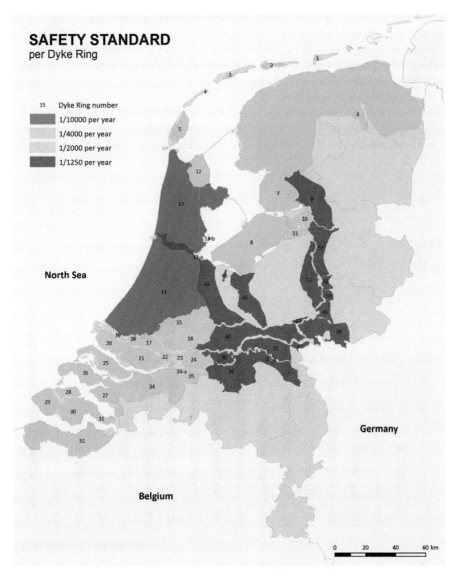

Figure 2.30 Dike rings in the Netherlands. Courtesy of Aon Benfield Impact Forecasting.

options. This approach can, for instance, lead to three different models based on different failure scenarios: best estimate, optimistic and pessimistic, which significantly improves the understanding of this particular uncertainty for both the model developer and the model user (see also Figure 2.33 for a practical example).

### 2) *Operational Input Data*

The source of this uncertainty is portfolio data or any input to the model used on an operational basis, i.e. data that are not used in building the model. Various assumptions and generalizations are taken when, for example, the input data are reported by the insureds, when they are entered into the various underwriting systems, and when they are extracted from the system for modelling.

Again, the model user may not have control of the data received, so the best way to understand the potential uncertainty is to test the impact of various options in the model itself. Take, for example, the insured value for a given mass-market household insurance product. Say that the typical insured value assumed for a house in a postal code is assumed to be £300,000. The ideal solution would be a reassessment of the insurance company's assumption about the average value of the property for this postal code. However, in the absence of that, the modeller can assume different average values and create a distribution based on these sensitivity tests to evaluate the effect of this assumption.

An example of how the sum insured values assumed by different insurers can vary for very similar buildings is given in Figure 2.31. Given most models assume a linear relationship between sum insured and loss, the uncertainty from this source alone could be significant.

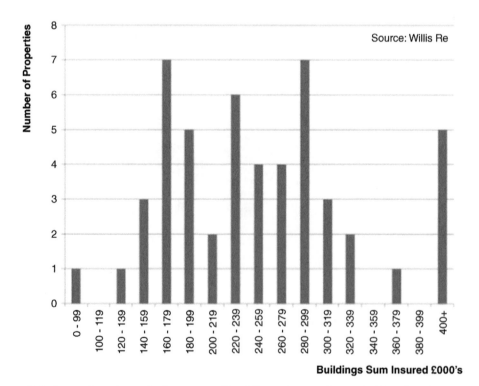

Figure 2.31 Range of sum insured values assumed by different insurers. *Source*: Smith and Cook, 2013.

3) *Model User*

Processes controlled and followed by the model user include the model input data preparation process and the selection of model options. Aside from possible errors during the many processes, assumptions taken by the model user can have a large influence on how certain the analysis results are.

Examples of these assumptions include the mapping of occupancies, constructions and other modifiers in the original data to the classes supported by the given model (see also Section 2.5.4) or the choice of model options, such as analysis sampling methods or selection of event set.

To conclude, in all three cases the model developer or the model user can take measures to evaluate the effect of many model assumptions. Although the uncertainty does not necessarily decrease, it is then certainly better quantified.

### 2.16.1.4   Catastrophe Model Treatment of Uncertainty in Practice

The discussion of uncertainty in catastrophe models is also confused by the different terminology adopted by model vendors in discussing uncertainty. For example, uncertainty in catastrophe models is often defined as follows:

- primary uncertainty: the uncertainty associated with the event generation process, e.g. does that event happen or not?
- secondary uncertainty: the uncertainty associated with the estimation of loss, given that an event has happened.

The primary uncertainty explicitly represented within the model represents part of the aleatory (process) uncertainty. The secondary uncertainty explicitly represented within the model represents part of the aleatory and part of the epistemic (parameter and model) uncertainty. Explicit representation of primary epistemic uncertainty (i.e. uncertainty in the event generation) is usually missing. Some models contain alternative events sets (e.g. to represent a warm sea surface temperature conditioned view of hurricane risk) which help to better understand this uncertainty, but it is rarely, if ever, quantified in a systematic way. The scale of this uncertainty is likely larger than most of catastrophe model output realize or can implement realistically into practical calculations. For example, the two standard error range for a 100-year return period for a National US Hurricane portfolio is roughly half to double the 100-year loss (Guy Carpenter, 2011) (see Figure 2.32).

Another example can be taken from the Impact Forecasting Dutch flood model which shows the difference between flood loss estimates based on three different assumptions about the probability of dyke failure. Figure 2.33 shows the impact of the assumptions which varies by return period, i.e. the largest differences between losses are observed at smaller return periods, whereas the results are more similar at longer return periods.

The most important question for a user being asked to choose a model is not, in fact, which model to use but what the difference is between the choices available (e.g. what impact does it have if the user chooses the optimistic, pessimistic or realistic assumption in respect of dyke ring failure?). This translates into knowledge of whether a particular assumption is important and if so, how this can, or should, be included in the representation of the risk and downstream decisions. However, this understanding can only be achieved if (1) the user knows about this assumption and (2) if the possibility of stress-testing it exists.

Table 2.16 shows the sources of uncertainty within a catastrophe model, together with an indication of whether these sources are usually represented within a model, and a rough classification into actuarial and scientific terms. The sources are also classified by model component, means of expression and model agent to link back to the text. Table 2.16

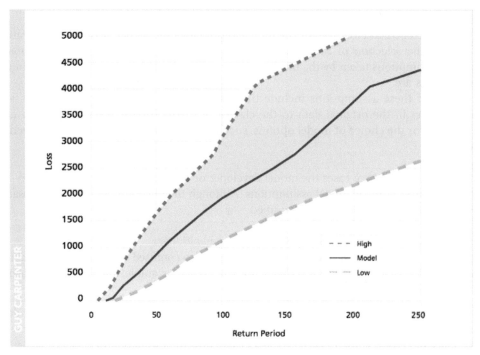

**Figure 2.32** Range around catastrophe model output. *Source*: Reproduced with permission of Guy Carpenter, LLC © 2016.

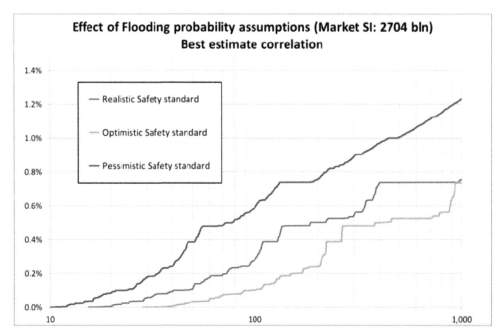

**Figure 2.33** Alternative views of Dutch flood risk resulting from different assumptions about dyke failure. Courtesy of Aon Benfield Impact Forecasting.

Table 2.16 Classification of uncertainty sources in catastrophe models. *Source:* Adapted from Sarka and Jones (2015).

| Component | Uncertainty source | Explicitly represented | Means of expression | Comment | Actuarial uncertainty terminology | Scientific uncertainty terminology | Model agent |
|---|---|---|---|---|---|---|---|
| **HAZARD** | The natural process itself | **Yes** | NA | The purpose of the model (and the reason for insurance)! | Process | Aleatory | N/A |
| | Footprints modelling | **Not normally** | Alternative sub-events or event sets | Uncertainty arises from input data (DTM, gauge data), choice of model (distribution type, GMPE, hydraulic model) and selected parameters or their estimation | Parameter & Model | Epistemic | Model developer |
| | Time dependence | **Sometimes** | Alternative event sets | Some peril regions have alternative event sets (e.g. near term, warm SST, time dependent earthquake, windstorm clustering) | Parameter & Model | Epistemic | Model developer |
| | Local geophysical conditions (e.g. surface roughness, soil type) | **Partly** | Alternative sub-events or event sets | Some uncertainty represented through the secondary uncertainty | Parameter & Model | Epistemic | Model developer |
| **EXPOSURE** | Poor data quality and/or geocoding precision | **Sometimes** | Alternative assumptions in input data | In some models secondary uncertainty increases with unknown data or poor geocoding (location uncertainty) | Parameter | Epistemic | Model user or operational input data |
| | Variation in the hazard and location of property within the modelled area (grid cell, postal code, CRESTA etc.) | **Partly** | Alternative assumptions in values distribution (via event sets etc.) | In some models explicitly represented, in some through the secondary uncertainty | Parameter & Model | Epistemic | Model developer |
| **VULNERABILITY** | Damage curve definition/selection | **Yes** | Alternative damage functions | Secondary uncertainty | Process, Parameter & Model | Aleatory & Epistemic | Model developer |
| | Location/coverage uncertainty correlation | **Sometimes** | Alternative treatments of correlation | Models often assume 100% coverage correlation and 0% location correlation – but this varies | Parameter & Model | Epistemic | Model developer |

*(continued)*

Table 2.16 (*Continued*)

| Component | Uncertainty source | Explicitly represented | Means of expression | Comment | Actuarial uncertainty terminology | Scientific uncertainty terminology | Model agent |
|---|---|---|---|---|---|---|---|
| **PLATFORM** | Implementation/ discretization | **No** | Alternative implementation where appropriate | Should be reduced by proper testing during model development | Model | Epistemic | Model developer |
| | Sampling | **Sometimes/ Not normally** | Alternative runs with different assumptions (# of samples etc.) | Can be reduced, but only with increased run-times | Model or Parameter | Epistemic | Model/ platform developer or model user depending on if can be changed |
| **USER** | Insufficient understanding of the model | **No** | Alternative runs based on different assumptions | Portfolio data preparation, parameters mapping, model settings, results interpretation | Model or Parameter | Epistemic | Model user |

summarizes the discussions above and provides a tool in which the different classifications of uncertainty can be related to the practical elements of the model.

As well as its different definitions and treatment, the challenges in discussion of uncertainty are compounded since questions on uncertainty asked in respect of catastrophe models are often vague and do not specify what type(s) of uncertainty is/are meant or even what metric they are asking for the uncertainty in.

A good example of this can be seen in the responses of model vendors to the FHCLPM (see also Section 2.11.5) who asked 'What is the uncertainty in this 1-in-100 year occurrence loss?' for hurricanes affecting Florida: Three major vendors provided the quite different uncertainty ranges ±5%, ±30%, ±50% in their answers (Major, 1999a, 1999b).

Unfortunately at present, catastrophe models do not fully represent all the sources of uncertainty. Given the significant range in uncertainty of catastrophe model outputs, and the fact that much of this uncertainty is not quantified within the model, there is a danger that decisions have been, and are being, made based on over-confidence in model results, without appreciation of the inherent uncertainty. In the face of the uncertainty discussed in this section, one may ask why catastrophe models have been developed and are used at all, especially given their often-significant cost. The following list recalls some reasons why catastrophe models are used in practice:

- they still provide insight: For regions with no historical exposure, and perils with no claims experience the only approach available is to build a model, which, even if uncertain, provides insight;
- there is pressure from rating agencies and regulators: Even if a (re)insurer thinks the uncertainty is too significant to warrant using such models, the rating agencies and regulators will likely not share this view and therefore a pressure exists to use them: an uncertain method is better than no method;
- they are useful for comparative analytics: When performing comparative analyses (e.g. PML of a portfolio with and without a specific account), some of the uncertainties may cancel out meaning that an informed decision (e.g. on whether an account diversifies or concentrates a portfolio) can still be made. The catastrophe model provides a framework for decision-making;
- use of these models helps drive improved data capture: This is useful for exposure management purposes and better understanding a portfolio even if the models are uncertain.

However, modelling companies are now starting to provide more insights into uncertainty and some practical options are considered below.

### 2.16.1.5 Practical Approaches to Deal with Uncertainty

The uncertainty coming from the model itself based on choices from the model developer can be quantified in more detail. One possible solution is based on the following three principles:

1) Define different uncertainties and link them to model components.
2) Be transparent in how uncertainties are handled and define multiple options.
3) Document the uncertainties and allow the catastrophe risk analyst to interact with them.

The three principles will be illustrated using the development of the Impact Forecasting flood model for the Netherlands.

#### *Define Different Uncertainties and Link Them to Model Components*

Different model developers may make different choices and assumptions for a given model and therefore it is helpful to assign these to the model components consistent with the classic components of a catastrophe model:

a) *Hazard* – this classification could also be called 'the model developers' choices' and can be the most numerous depending on the focus of the model developers. The idea is that every time the model developers make a particular choice in developing a solution, the uncertainty in that choice is quantified. For the Dutch flood model example, the following aspects were tested: probability of flooding, spatial correlation between dyke rings, hazard map.

b) *Exposure (Location)* – what is the uncertainty in locating the insured risk and what approach has been selected?

c) *Vulnerability* – could be limited in this case to how the loss is calculated and the vulnerability curve selection. In the case of the Dutch flood model, two parts were used:

    1) damage function and

    2) chance of loss.

As already alluded to above, ensuring the full range of choices is recorded is more important than its assignment to model components, which is relatively arbitrary.

### Be Transparent in How Uncertainties Are Handled and Define Options

Although defining the uncertainties associated with specific choices is important, defining how those uncertainties are treated and propagated through the model is at least equally so. There are different options for treatment of these uncertainties and those used in Impact Forecasting's model are shown in Table 2.17.

Taking the example of the Dutch flood model again, the major uncertainties identified were treated in the following way:

- probability of flooding: incorporated with multiple options;
- spatial correlation among dike rings: incorporated with multiple options;
- vehicle evacuation: incorporated with multiple options;
- hazard map: identified and stress-tested.

As shown in Table 2.17, if the given uncertainty was 'incorporated with multiple options' it means that the user can execute different options in the catastrophe modelling platform.

### Document the Uncertainties and Allow the Modeller to Interact with Them

When the uncertainties, their handling and the options are defined, this information should be revealed to the catastrophe analyst in a simple, universal and user-friendly way, for instance, through the software interface. The results of the options can be summarized to illustrate the uncertainties and quantify their effect on a key metric (in this case the AAL, see Figure 2.34). This framework allows the user to see the impact of potential choices on results and provides a basis for communication around the uncertainties.

Table 2.17 Classification of uncertainty handling in the Impact Forecasting models. Courtesy of Aon Benfield Impact Forecasting.

| Value | Explanation |
| --- | --- |
| Identified and documented | Identified and documented, no in-depth quantification was performed |
| Identified and stress-tested | Identified, quantified in depth and documented. The default model version follows official guidelines |
| Incorporated implicitly | Identified, quantified in depth and documented. One (default) version expresses our expert opinion |
| Incorporated with multiple options | Identified, quantified in depth and documented. Multiple model versions available |

| Uncertainty | Component | Quantified? | Size | Implemented? |
|---|---|---|---|---|
| **Three flood defence failure possibilities**<br>1. Realistic<br>2. Pessimistic<br>3. Optimistic | **Hazard** | YES | Pessimistic: losses **increase by 150%** Optimistic: losses **decrease by 50%** | YES |
| **Three dyke rings correlations**<br>1. Normal<br>2. Weak<br>3. Strong | **Hazard** | YES | Strong: **increase by 10%** Weak: **decrease by 15%** | NO |
| Portfolio data available on 4-digit postal code level only | **Location** | YES | Between +17% and -11% compared to average | YES |
| 2 metres of inundation depth can cause different losses | **Vulnerability** | YES | Between +5% and -4% compared to the average | YES |

Figure 2.34 Example of practical uncertainty quantification matrix for the Impact Forecasting flood model for the Netherlands. Impact on losses is quantified in terms of AAL. Courtesy of Aon Benfield Impact Forecasting.

One of the challenges of using multiple options is that multiple runs are required to investigate this, which requires considerable resource, both computationally and operationally. It may also be impractical to propagate the options to the level required to quantify the impact on the metric of interest. For instance, a reinsurer might be interested in the impact of such choices on the portfolio level net perspective at the 1 in 200 AEP VaR to consider its impact on regulatory solvency. This would require multiple versions of loss accumulation and would not be practical unless very sophisticated systems beyond the scope of anything available today were able to automate multiple option choices through into an accumulation framework. Instead these uncertainties would need to be inferred from uncertainties in other metrics or at other perspectives.

Quantification of the uncertainty around the EP curve can be achieved using a different method. If sampling is part of the financial module component, the impact of uncertainty can be represented using a so-called 'spaghetti' plot (e.g. Figure 2.35). This shows alternative EP curves

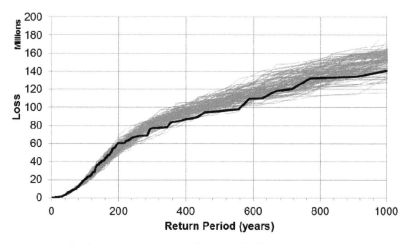

Figure 2.35 'Spaghetti' plot for EP curve. Courtesy of Aon Benfield Impact Forecasting.

which have been derived from sampling the uncertainty in the damage and the location. In this example, the observed range of 1-in-500-year PML from 90 million to 120 million is just from uncertainty in location and damage *alone*.

A method which can be used to communicate uncertainty is the tornado diagram (Porter, 2002; 2016) which compares the sensitivity of a result to different variables. The outcome is a horizontal bar chart with the values sorted in descending order: the gradation of bar length resembling a tornado. This is a technique taken from decision analysis (Howard, 1988) that is used to determine how much the uncertainty associated with different parameters affects the result, by choosing a base, high and low values for the variable and testing the impact of one variable while the others are held constant.

As described above, model vendors are providing increased ability to test alternative (built-in or user-defined) parameterizations. Examples include event rates sets such as long-term vs near-term/warm SST, or alternative vulnerability curves based on the amount of loss data available to calibrate them, or alternative attenuation functions, or time-dependent vs time independent rate sets for earthquake within the models. The Netherlands example provided earlier is a further example of this. While this kind of approach makes the uncertainty more transparent, at present it only addresses a limited part of the full epistemic uncertainty.

One method advocated to understand model uncertainty (Cook, 2011) is to compare the outputs of multiple catastrophe models for the same peril. Although the differences between models at the higher return periods of interest can be surprisingly large, they are still likely to underestimate the uncertainty because the models rely, to some extent, on the same data sources. In time the differences between models tend to reduce. This may be partly due to convergence in methodology and, perhaps partly subconsciously on the part of the model vendors in not wanting to be outliers. However, there have been notable occasions when a model update or rebuild by one vendor has increased the gap again. Blending the outputs of multiple models should reduce uncertainty, assuming there is some independence between the models and the blending is done with care. See Chapter 5.5.4 and Calder *et al.* (2012) for more information on model blending techniques and their appropriateness in the context of catastrophe models.

It is part of the role of the catastrophe manager to educate the users of model output on the uncertainties associated with catastrophe modelling. The catastrophe manager should be aware of the drivers of uncertainty and these should be explored when deriving a view of risk. However, as more information is available, communication will become increasingly complex. Training in the psychology and communication of uncertainty may well become necessary.

In conclusion, on uncertainty, as we have shown, there are no easy answers. The approaches above can be used to understand elements of the uncertainty and can help to communicate certain aspects of it. However, we recommend that users of catastrophe models and their output remain aware that the true uncertainty is unquantifiable and likely much larger than they think.

### 2.16.1.6 Future of Uncertainty Reporting

Uncertainty reporting has two main constituents: the model developer and the end user. Every model developer needs to make many assumptions during the model development process. There are always multiple solutions to a problem. If one solution is chosen without any exploration of alternatives, then the opportunity to better understand this problem is limited.

In the future, it is very likely that model developers will be more thorough in identifying, quantifying and documenting various uncertainties. In time, methodologies will be standardized and visualization methods will enable the effect of different assumptions to be more clearly understood, increasing transparency to the end user.

In the meantime the model users can help to initiate such developments by asking the model developers questions, such as:

- Which uncertainties have you identified?
- Which ones have you quantified?
- Why have you not quantified all identified uncertainties?
- Which ones have you implemented in the model?
- Why have you not implemented all quantified uncertainties?
- What is the effect of all the quantified uncertainties?
- What is the technical solution for implementing different uncertainties?
- Can I quantify these myself for my portfolio? Does the platform allow it?

As more different models, and open models come to the market, multi-modelling will provide more insight, but also become more of an overhead in terms of costs and resources.

With more tools available to understand the sources of uncertainty and their impacts, the choices and model options will surely need to be more formally linked to risk appetite and tolerance, the view of risk and regulatory context. This process is in its infancy.

However, there may be something for catastrophe modellers to learn from meteorology and climate science, where uncertainty is an important part of communication. For instance, the fifth IPCC report uses calibrated language to communicate the degree of certainty in key findings where confidence is expressed qualitatively, and measures of uncertainty in a finding are expressed probabilistically (IPCC, 2015). The sensitivity testing and the spaghetti plot shown in Figure 2.35 are reminiscent of ensemble approaches in weather forecasting, where there is uncertainty in model initial conditions and in the propagation of these through the models. Better methods of communicating the uncertainty associated with models and its impact on the downstream business decisions dependent on them will be required.

### 2.16.2 Importance of Framework: A Tool, Not an Answer

This section draws on much of the discussion earlier in the chapter to provide some thoughts on the use of catastrophe models. Effective decision-making using catastrophe models depends on understanding the nuances of the modelling approach, the data, and the uncertainties associated with the inputs and outputs of catastrophe models. This is underpinned by effective model validation which is discussed in Chapter 5.4.

However, one of the most important benefits of models is that they provide a *framework* within which to treat the risk. This brings consistency, a way to measure and compare different risks and discipline. The ability to quantify the risk in a framework with further downstream applications is arguably the most important aspect of modelling and the usefulness of this framework approach has more than compensated for some of the limitations of any individual model, allowing their widespread adoption.

A fundamental tenet of risk management is that the act of managing the risk instils discipline in the stakeholders. The premise is that if stakeholders are asked to think about, and engage with, the risk, this in itself modifies their behaviour. It is certainly true that since catastrophe modelling has developed, the process has forced more detailed capture of exposure data, including ever-more differentiation on policy forms to capture the physical risk characteristics which influence catastrophe modelling results (e.g. roof equipment for wind perils). The portfolios themselves and the assets insured are better understood and recorded than ever before.

Modelling has increased transparency and challenged existing rules of thumb. Although initially regarded as a 'necessary evil' by many, modelling is now completely embedded in the (re)insurance world and becoming ever more important in the disaster reduction and resilience communities. Although the science behind the models themselves will change, the software chosen and used will alter, catastrophe models are now seemingly irrevocably intertwined with the risk transfer process. The persistence and dominance of catastrophe modelling can be illustrated when it is considered that every event that could be considered to demonstrate a 'fatal flaw' in existing modelling approaches has, in fact, simply underscored the need for more, and better, modelling. Some practical examples are listed in Section 2.11.1.2 and 'Key Past Events' in Chapter 3.X.6.

As discussed above, it is important that model results are contextualized and that sufficient sensitivity testing is done to supplement the model's use in both underwriting and risk management. The models have known limitations, as well as unknown limitations, and the organization's sensitivity to these should inform the models' use within an organization. As discussed in Chapter 5.4.3, evaluation of the model for the specific portfolio written is a critical aspect in establishing its fitness for purpose. Initial over-confidence and a belief that models provided 'the answer' to hitherto intractable problems caused significant distress when organizations had based their business plan or underwriting guidelines on the output of one model which then changed due to new, or better understanding of the physical concepts underlying that model. Examples include changes in the science on e.g. inland filling or transitioning for hurricane (see Chapter 3.2.1) or the discovery of a new fault source for earthquake (see Chapter 3.7). These views will evolve as science progresses and catastrophe events continue to provide sometimes unexpected insights.

Detailed models provide the ability to understand the portfolio behaviour in a very granular way and encourage collection of very detailed physical risk attributes, since these can be entered into the model and will often significantly influence the results. The models allow fine differentiation of the risk and are particularly applicable for perils with a high hazard gradient, i.e. where the hazard intensity changes a large amount over a small distance. In these cases, the exact location of the exposure is especially important. For instance, for flood, the precise location and elevation of the property relative to the flood plain will significantly influence loss outcomes or for hurricanes which fill after landfall, distance to coast is an indicator of the risk. Aggregate models are designed to be used with less detailed exposure data, at a larger geographical area (e.g. CRESTA) and where less detail is known about the individual risks (just line of business, i.e. residential/commercial/industrial with no information about the physical risk characteristics). Detailed models are therefore more suitable for insurance purposes, whereas aggregate models lend themselves more to treaty reinsurance applications where a large number of similar risks are considered. Complex policy terms (such as layered business) cannot be represented in aggregate models, since these models tend to depend on standardized assumptions about the relationship between insured limit and the sum insured, which break down in these cases. However, an aggregate model may be sufficient, particularly in emerging markets at a reinsurance level for standard exposure types.

Understanding the strengths and weaknesses of the individual modelling approaches for your portfolio and deriving one's own view of the risk are critical in model use (see Chapter 5). To do this, one must understand the physical characteristics of the perils (see Chapter 3) and the model development process (Chapter 4).

As this chapter on applications has shown, catastrophe modelling is an input into many areas of risk management, but most notably insurance and reinsurance at present. As the following chapters describe in detail, much complex and detailed knowledge has gone in to building these models. However, the modelled output remains a tool in order to meet specific objectives in terms of risk quantification, whether from a risk selection, risk pricing or capital approach. All of these objectives exist to support the business and paradoxically the use of catastrophe models increases the need for judgement in their use and application.

## Notes

1. According to Mandelbrot and Taleb (2006):

   Traditional 'stress testing' is usually done by selecting an arbitrary number of 'worst case scenarios' from past data. It assumes that whenever one has seen in the past a large move of, say, 10 per cent, one can conclude that a fluctuation of this magnitude would be the worst one can expect for the future. This method forgets that crashes happen without antecedents. Before the crash of 1987, stress testing would not have allowed for a 22 per cent move.

2. Cat bonds pay coupons, and the coupon rate is the *RoL* for a cat bond at issuance. However, post issuance, the price of a cat bond can fluctuate so that the effective *RoL* may change with time.
3. Metarisk simply means 'type of risk'. Florida Hurricane and California Earthquake are two examples of metarisks.
4. There is one school of thought that says that there is no difference between aleatory and epistemic uncertainty because, with more understanding, everything becomes in the limit predictable. Beyond commenting that if we ever reach this philosophical 'utopia' the world will be very different and the insurance industry will cease to exist, this discussion is beyond the scope of this book.

## References

Association of British Insurers (2014) *Non-Modelled Risks: A Guide to More Complete Catastrophe Risk Assessment for (Re)Insurers.* Association of British Insurers, London. https://www.abi.org.uk/~/media/Files/Documents/Publications/Public/2014/prudential%20regulation/Nonmodelled%20risks%20a%20guide%20to%20more%20complete%20catastrophe%20risk%20assessment%20for%20reinsurers.ashx (accessed 10 Jan. 2016).

Actuarial Standards Board (2004) *Actuarial Standard of Practice No. 23* Data Quality. http://www.actuarialstandardsboard.org/wp-content/uploads/2014/07/asop023_097.pdf (accessed 10 Jan 2016).

Actuarial Standards Board (2013) *Standard of Practice No. 38 Using Models Outside the Actuary's Area of Expertise (Property and Casualty).* http://www.actuarialstandardsboard.org/asops/catastrophe-modeling-practice-areas/(accessed 4 Dec. 2015).

Aon Benfield (2012) *2011 Thailand Floods Event Recap Report.* http://thoughtleadership.aonbenfield.com/Documents/20120314_impact_forecasting_thailand_flood_event_recap.pdf (accessed 4 Dec. 2015).

Arrow, K.J. (1963) *Uncertainty and the welfare economics of medical care.* American Economic Review, **53**, 941–973.

Arrow, K.J. (1978) *Risk allocation and information: some recent theoretical developments.* The Geneva Papers on Risk and Insurance Theory, **8**, 5–19.

Association of British Insurers (2011) *Industry Good Practice for Catastrophe Modelling: A Guide to Managing Catastrophe models as Part of an Internal Model under Solvency II.* Association of British Insurers, London. https://www.abi.org.uk/~/media/Files/Documents/Publications/Public/Migrated/Solvency%20II/Industry%20good%20practice%20for%20catastrophe%20modelling.ashx (accessed 10 Jan. 2016).

Badger, S., Zelle Hofmann, V. and Mason L.L.P. (2014) *The Emerging Hail Risk: What the Hail Is Going On?* http://www.claimsjournal.com/news/national/2014/05/02/248354.htm (accessed 22 Sept. 2015).

Bantwal, V. J. and Kunreuther, H.C. (1999) *A Cat Bond Premium Puzzle?* Wharton Financial Institutions Center. http://fic.wharton.upenn.edu/fic/papers/99/9926.pdf.

Bellerose, R.P. and Paine C. (2003) *Reinsurance for the Beginner,* 5th edn. Witherby, Edinburgh.

Bodoff, N.M. (2007) *Capital allocation by percentile layer.* https://www.casact.org/pubs/forum/08wforum/Bodoff_Capital.pdf (accessed 10 Jan. 2016).

Brynjolfsson, J. and Dorsten, M. (2007) *Do natural disasters affect the stock market?* Pimco, July.

Calder, A., Couper, A. and Lo, J. (2012). *Catastrophe model blending: techniques and governance.* Paper presented at General Insurance Convention, The Actuarial Profession, Brussels.

Canabarro, E., Finkemeier, M., Anderson, R.R., and Bendimerad, F. (2000) *Analyzing insurance-linked securities.* The Journal of Risk Finance, **1**, (2), 4973.

Coles, S. (2001) *An Introduction to Statistical Modelling of Extreme Values.* Springer, London.

Cook, I. (2011) *Using Multiple Catastrophe Models.* http://www.actuaries.org.uk/documents/using-multiple-catastrophe-models, (accessed 29 Nov. 2015).

Cordier, T. and Lacoss, D. (2011) *Capital and Excess of Loss Reinsurance Pricing.* GIRO, UK Actuarial Profession, London.

Diacon, S.R. and Carter, R.L. (2005) *Success in Insurance,* 3rd edn. John Murray, London.

EIOPA (2015) *Guidelines on Own Risk and Solvency Assessment (ORSA).* https://eiopa.europa.eu/publications/eiopa-guidelines/guidelines-on-own-risk-solvency-assessment-(orsa) (accessed 4 Dec. 2015).

EU LEX (2015) *Commission Delegated Regulation (EU) 2015/35 of 10 October 2014 Supplementing Directive 2009/138/EC of the European Parliament and of the Council on the Taking-up and Pursuit of the Business of Insurance and Reinsurance (Solvency II).* http://eur-lex.europa.eu/legal-content/EN/TXT/?uri=OJ%3AL%3A2015%3A012%3ATOC (accessed 4 Dec. 2015).

FCHLPM (2015) *Florida Commission on Hurricane Loss Projection Methodology.* http://www.sbafla.com/methodology/Home (accessed 4 Dec. 2015).

Financial Reporting Council (2009) Board for Actuarial Standards. *Technical Actuarial Standard D: Data* https://frc.org.uk/getattachment/1d08e3b5-00bc-4793-b457-284162f002af/TAS-D-Data-version-1-Nov-09.pdf last accessed 4/12/2015 (accessed 4 Dec. 2015).

Financial Reporting Council (2010) Board for Actuarial Standards. *Technical Actuarial Standard Modelling: M* http://frc.org.uk/getattachment/3a8825b8-4560-4750-955f-30f740960c7f/TAS-M-Modelling-version-1-Apr-10.pdf (accessed 4 Dec. 2015).

FLOIR (2005) *Consent Order.* www.floir.com/siteDocuments/Nationwide_FL_77987-04-CO.pdf (accessed 17 Oct. 2015).

Franco, G. (2011a) *Portfolio optimization for insurance companies.* AIR Worldwide. http://www.air-worldwide.com/Publications/AIR-Currents/2010/Portfolio-Optimization-for-Insurance-Companies/(accessed 1 Nov. 2015).

Franco, G. (2011b) *Optimización de carteras y análisis de decisiones.* AIR Worldwide. https://piensaama.files.wordpress.com/2011/10/optimizacion-de-carteras-expuestas-a-riesgos-catastrficos.pdf (accessed 27 Nov. 2015).

Froot, K., O'Connell, P.G.J. (1997) *On the Pricing of Intermediate Risks: Theory and Application to Catastrophe Reinsurance.* Wharton Financial Institutions Center. http://fic.wharton.upenn.edu/fic/papers/97/froot.pdf (accessed 4 Dec. 2015).

Froot, K.A. and Stein, J.C. (1998) *Risk management, capital budgeting and capital structure policy for financial institutions – an integrated approach.* Journal of Financial Economics, **47**, 55–82. http://scholar.harvard.edu/files/stein/files/financial-risk-management-jfe-jan-98_0.pdf (accessed 17 Jan. 2016).

GIRO (2008) *Securitization of Non-life Insurance Working Party, Zero Beta Sub-group.* https://www.actuaries.org.uk/documents/3-zero-beta-assumption-sub-group (accessed 17 Jan. 2016).

GIRO (2011), *Risk Appetite Working Party: Risk Appetite for a General Insurance Undertaking.* http://www.actuaries.org.uk/sites/all/files/documents/pdf/c5-paper-risk-appetite.pdf (accessed 17 Jan. 2016).

Global Facility for Disaster Reduction and Recovery (GFDRR) (2014a) *Understanding Risk: The Evolution of Disaster Risk Assessment.* World Bank, Washington, DC. http://www.wcdrr.org/wcdrr-data/uploads/856/Understanding%20Risk_GFDRR.pdf (accessed 17 Jan. 2016).

Global Facility for Disaster Reduction and Recovery (GFDRR). (2014b) *Understanding Risk: Review of Open Source and Open Access Software Packages Available to Quantify Risk From Natural Hazards.* World Bank, Washington, DC.

Goldfarb, R. (2010) *Risk-Adjusted Performance Measurement for P&C Insurers.* http://www.casact.org/library/studynotes/goldfarb8.2.pdf (accessed 17 Jan. 2016).

Gorge, G. (2013) *Insurance Risk Management and Reinsurance.* Bibliothèque Nationale, Paris.

Gunasekera, R., Ishizawa, O., Aubrecht, C., Blankespoor, B., Murray, S., Pomonis, A., and Daniell, J. (2015) *Developing an adaptive global exposure model to support the generation of country disaster risk profiles.* Earth-Science Reviews, **150**, 594–608.

Guy Carpenter (2011) *Managing Catastrophe model Uncertainty: Issues and Challenges.* Guy Carpenter & Company, London. http://gcportal.guycarp.com/portal/extranet/getDoc?vid=1&docId=10068 (accessed 17 March 2016).

Guy Carpenter (2014) *Currents Trend in the Global Reinsurance Market.* March, 2014, Guy Carpenter, London.

Guy Carpenter (2015) *Port of Tianjin Explosions: CAT-VIEW Event Briefing.* Guy Carpenter, London.

Ha, S. (2013) Optimal insurance risk allocation with steepest ascent and genetic algorithms. *The Journal of Risk Finance*, **14** (2), 129–39.

Harrison, C.M. (2010) *Reinsurance Principles and Practices.* AICPCU, Malvern, Pennsylvania.

Howard, R.A. (1988) *Decision analysis: practice and promise.* Management Science **34** (6), 679–695. http://goo.gl/0LVJaA (accessed 14 March 2016).

Insurance Information Institute (2015) *Hurricane and Windstorm Deductibles.* http://www.iii.org/issue-update/hurricane-and-windstorm-deductibles, (accessed 19 Nov. 2015).

Intergovernmental Panel Climate Change (2015) *Fifth Assessment Report, AR5.* https://www.ipcc.ch/report/ar5/index.shtml (accessed 29 Nov. 2015).

Jaeger, L., Müller, S., and Scherling, S. (2010) *Insurance-linked securities: what drives their returns?* The Journal of Alternative Investments, *Fall.*

Kaufmann, R., Gadmer, A. and Klett, R. (2001) *Introduction to dynamic financial analysis.* ASTIN Bulletin, **31** (1), 213–249. https://www.casact.org/library/astin/vol31no1/213.pdf (accessed 19/11 2015).

KCC (2012) *KCC introduces characteristic events.* http://www.karenclarkandco.com/press-releases/2012/01/31/KCC-Introduces-Characteristic-Events.html (accessed 17 Feb. 2016).

Kiln, R. and Kiln, S. (1996) *Reinsurance Underwriting*, 2nd edn. LLP, London.

Kiln, R. and Kiln, S. (2001) *Reinsurance in Practice*, 4th edn. Witherby, London.

Klugman, S.A., Panjer, H.H., Willmot, G.E. (1998). *Loss Models: From Data to Ddecisions*, 4th edn. Wiley, Hoboken, NJ.

Kreps, R. (2005) *Riskiness Leverage Models.* CAS Proceedings 2005. https://www.casact.org/library/05pcas/kreps.pdf

Lam, J. (2014) *Enterprise Risk Management: From Incentives to Controls.* Wiley, Hoboken, NJ.

Lane, M. (ed.) (2012) *Alternative (Re)insurance Strategies*, 2nd edn. Risk Books, New York.

Lane, M. and Mahul, O. (2008) *Catastrophe risk pricing: an empirical analysis.* World Bank Policy Research Working Paper No. 4765. November.

Litzenberger, R.H., Beaglehole, D.R. and Reynolds, C.E. (1996) *Assessing catastrophe reinsurance-linked securities as a new asset class.* Journal of Portfolio Management. December.

Lloyd's (2015a) *Lloyd's Minimum Standards.* https://www.lloyds.com/~/media/files/the%20market/operating%20at%20lloyds/minimum%20standards/minimuds.pdf (accessed 21 Nov. 2015).

Lloyd's (2015b) *RDS scenario specification 2015.* https://www.lloyds.com/the-market/tools-and-resources/research/exposure-management/realistic-disaster-scenarios/rds-scenario-specification-2015 (accessed 19 Nov. 2015).

Lloyd's Market Association LMA (2012) *External Catastrophe Model Validation a Illustrative Document No. 1, US Windstorm.* https://www.Lloyds.com/~/media/files/the%20market/operating %20at%20Lloyd's/solvency%20ii/2012%20guidance/validation%20of%20external%20catmodels% 20%20illustrative%20example%20for%20u%20s%20windstorm.pdf (accessed 21 Nov. 2015).

Major, J.A. (1999a) *Uncertainty in catastrophe models, Part I: What is it and where does it come from? Financing Risk and Insurance* (February). International Risk Management Institute.

Major, J.A. (1999a) *Uncertainty in catastrophe models, Part II: How bad is it?* Financing Risk and Insurance (March). International Risk Management Institute.

Mandelbrot, B. and Taleb, N. (2006) *A focus on the exceptions that proves the rule. Financial Times,* March 23.

Mango, D. (2003) *Capital Consumption: An Alternative Methodology for Pricing Reinsurance.* CAS Forum, Winter 2003, 351–379. www.casact.org/pubs/forum/03wforum/03wf351.pdf

Mango, D.F. (1997) *An Application of Game Theory: Property Catastrophe Risk Load.* http://www .casact.org/pubs/forum/97spforum/97spf031.pdf (accessed 27/11 2015).

Morel, D. (2013) *Pricing catastrophe excess of loss reinsurance using market curves,* Casualty Actuarial Society E-Forum, Spring 2013,Volume 2.

Morin, F. (2011) *Introduction to Economic Capital Modelling.* https://www.casact.org/education/ clrs/2011/handouts/ERM2-Morin.pdf (accessed 19 Nov. 2015).

Munich Re (2004) *A Basic Guide to Facultative and Treaty Reinsurance.* https://www.munichre. com/site/mram/get/documents_E96160999/mram/assetpool.mr_america/PDFs/3_Publications/ reinsurance_basic_guide.pdf (accessed 27 Nov, 2015).

Myers, S.C. and Read, J.A. Jr. (2001) *Capital Allocation for Insurance Companies.* https://math. illinoisstate.edu/krzysio/MAT483/myersreadnotes.pdf (accessed 27 Nov. 2015).

National Flood Insurance Programme (2015) *The National Flood Insurance Programme.* https:// www.fema.gov/national-flood-insurance-program (accessed 27 Nov. 2015).

Okocha, W. (2015) *Amtrak launches $200 mn cat bond.* Insurance Insider, 17 September.

Orie, M. and Stahel, W.R. (2013) *Insurers' contributions to disaster reduction—a series of case studies.* The Geneva Reports, Risk and Insurance Research, No. 7, May.

Parodi, P. (2015) *Pricing in General Insurance.* CRC Press, Baton Rouge.

Pinkser Gladstone, B. (2012) *Homeowners dodge Sandy deductibles, face other costs.* http://www .claimsjournal.com/news/national/2012/11/02/216681.htm (accessed 22 Sept. 2015).

Piserra, M. (2000) *European Storms in December 1999.* http://www.mapfre.com/ccm/content/ documentos/mapfrere/fichero/en/trebol-en-num15-2.pdf (accessed 10 Oct. 2015).

Pita, G.L., Gunasekera, R., and Ishizawa, O.A. (2015) *Windstorm hazard model for disaster risk assessment in Central America,.* Paper presented at 14th International Conference on Wind Engineering (ICWE), Porto-Alegro, Brazil. June 21–26.

Porter, K. (2016) *A Beginner's Guide to Fragility, Vulnerability, and Risk.* University of Colorado, Boulder. http://spot.colorado.edu/~porterka/Porter-beginners-guide.pdf (accessed 14 March 2016).

Porter, K.A., Beck, J.L. and Shaikhutdinov, R.V. (2002) Sensitivity of building loss estimates to major uncertain variables. *Earthquake Spectra,* **18** (4), 719–743. www.sparisk.com/pubs/ Porter2002-Sensitivity.pdf (accessed 14 March 2016).

PRA (2015) *Solvency II: applying EIOPA Set 2, System of Governance and ORSA Guidelines – CP30/15.* http://www.bankofengland.co.uk/pra/Documents/publications/cp/2015/cp3015.pdf (accessed 4 Dec. 2015).

Priest, G.L. (2003) *Government insurance versus market insurance.* The Geneva Papers on Risk and Insurance, **28** (1), 71–80.

Qiu, J., Li, M., Wang, Q. and Wang, B. (2012) *Catastrophe Reinsurance Pricing – Science, Art or Both?* https://www.towerswatson.com/en/Insights/Newsletters/Global/emphasis/2012/ Catastrophe-Reinsurance-Pricing-Science-Art-or-Both (accessed 17 Oct. 2015).

Randall, E. (1994) *Introduction to Underwriting.* Insurance Institute of America, Malvern, PA.

Resis, II. (2010) *The 2010 RESIS-II Seismic Hazard Model for Central America.* https://hazardwiki. openquake.org/resisii2010_intro (accessed 17 Oct. 2015).

Riley, K. (2013) *Reinsurance: The Nuts and Bolts*. Witherby, Edinburgh.

RMS (2000) *Windstorm Lothar and Martin*. https://ipcc-wg2.gov/njlite_download.php?id=6144 (accessed 17 Oct. 2015).

Robbin, I. (2013) *Catastrophe pricing: making sense of the alternatives*. Paper presented at Casualty Actuarial Society E-Forum, Spring 2013. https://www.casact.org/pubs/forum/13spforum/ Robbin.pdf (accessed 10 Jan. 2016).

Rossetto, T., D'Ayala, D., Ioannou, I. and Meslem, A. (2014) *Evaluation of existing fragility functions*. In *SYNER-G: Typology Definition and Fragility Functions for Physical Elements at Seismic Risk*, Springer Science & Business Media, Dordrecht. pp. 47–93.

Ruhm, D. and Mango, D. (2003) *A risk charge based on conditional probability*. Paper presented at the 2003 Bowles Symposium. www.casact.org/coneduc/specsem/sp2003/papers/.

Ruhm, D., Mango, D. and Kreps, R. (2003) *A method of implementing Myers-Read capital allocation in simulation*. www.casact.org/education/spring/2003/handouts/ruhm2.doc (accessed 14 Nov. 2016).

Sarka, P. and Jones, M. (2015) *Cat Models: Useful Tools or Budget Sinkholes?* GIRO. UK Actuarial Profession, London.

Schmutz, M. and Doerr, R. (1998) *The Pareto Model in Property Reinsurance: Formulas and Applications*. Swiss Re, Zurich.

SCOR (2013a) *Cat modeling and cat underwriting: the non-identical twins?* Cat Risk Management. SCOR, London.

SCOR (2013b) *Lessons learned from Hurricane Sandy*. Cat Risk Management, London.

SCOR (2015a) *Gaining a strategic edge through capital management: key issues faced by the P&C (re)insurance industry*. http://scor.com/images/stories/pdf/library/focus/FocusPC_ CapitalManagement.pdf (accessed 19 Nov. 2015).

SCOR (2015b) *US pricing for non-quake and hurricane risks*. Cat Risk Management, SCOR, London.

Smith, A. and Cook, I. (2013) *Is your cat model a dog?* GIRO. https://www.actuaries.org.uk/sites/ default/files/documents/pdf/e5-presentation.pdf, (accessed 14 Feb. 2016).

Standard & Poor's (2015) *About Credit Ratings*. https://www.standardandpoors.com/ aboutcreditratings/RatingsManual_PrintGuide.html (accessed 5 Dec. 2015).

Stone, H., D'Ayala, D., Gunasekera, R. and Ishizawa, O.I. (2015) *A review of seismic vulnerability assessments in Central America*, paper presented at SECED 2015 Conference: Earthquake Risk and Engineering towards a Resilient World, 9–10 July 2015, Cambridge.

Swiss Re (1997) *Proportional and Non-Proportional Reinsurance*. Swiss Re, Zurich. http://media. swissre.com/documents/pub_proportional_non_proportional_en.pdf (accessed 22 Sept. 2015).

Swiss Re (1999) *Designing Property Reinsurance Programmes: The Pragmatic Approach*. Swiss Re, Zurich. http://media.swissre.com/documents/pub_designing_property_reinsurance_ programmes_en.pdf (accessed 22 Sept. 2015).

Swiss Re (1999b) *From Risk to Capital: An Insurance Perspective*. Swiss Re, Zurich. http://media. swissre.com/documents/pub_risk_to_capital_1999_en.pdf (accessed 22 Sept. 2015).

Swiss Re (2002) *An Introduction to Reinsurance*. Swiss Re, Zurich. http://media.cgd.swissre.com/ documents/pub_intro_reinsurance_en.pdf (accessed 22 Sept. 2015).

Swiss Re (2003) *Natural Catastrophes and Reinsurance*. Swiss Re. Zurich available from http:// media.swissre.com/documents/Nat_Cat_reins_en.pdf (accessed 22 Sept. 2015).

Swiss Re (2004) *Exposure Rating*. Swiss Re, Zurich. http://media.cgd.swissre.com/documents/ pub_exposure_rating_en.pdf, (accessed 4 Dec. 2015).

Swiss Re (2005) *Natural Catastrophes and Man-Made Disasters in 2004: More Than 300 000 Fatalities, Record Insured Losses*. Swiss Re, Zurich.

Swiss Re (2006) *Natural catastrophes and man-made disasters in 2005: high earthquake casualties, new dimension in windstorm losses*. *Sigma*. http://media.swissre.com/documents/ pr_20062402_sigma_cat05_en.pdf (accessed 22 Sept. 2015).

Swiss Re (2007) *Facultative Non-Proportional Reinsurance and Obligatory Treaties*. http://www .swissre.com/publications/94922554.html (accessed 22 Sept. 2015).

Swiss Re (2012) *Natural catastrophes and man-made disasters in 2011: historic losses surface from record earthquakes and floods. Sigma.* http://www.swissre.com/sigma/?year=2012#anchor0 (accessed 22 Sept. 2015).

Swiss Re (2013) *Natural catastrophes and man-made disasters in 2012: A year of extreme weather events in the US. Sigma.* http://www.biztositasiszemle.hu/files/201303/sigma2_2013_en.pdf (accessed 22 Sept. 2015).

Swiss Re (2014) *The Essential Guide to Reinsurance.* Swiss Re, Zurich. http://www.swissre.com/library (accessed 22 Sept. 2015).

Tasche, D. (2000). *Capital Allocation to Business Units and Sub-portfolios: The Euler Principle.* http://arxiv.org/pdf/0708.2542 (accessed 22 Sept. 2015).

Tätge, Y. (2009) *Looking back, looking forward: Anatol, Lothar and Martin ten years later.* AIR Worldwide. http://www.air-worldwide.com/Publications/AIR-Currents/Looking-Back,-Looking-Forward–Anatol,-Lothar-and-Martin-Ten-Years-Later/(accessed 22 Sept. 2015).

Taylor, C. (1996) *An Introduction to Error Analysis: The Studies of Uncertainties on Physical Measurements.* University Science Books, Sausolito, CA.

Thoyts, R. (2010) *Insurance Theory and Practice.* Routledge, London.

UNISDR (2015) Global Assessment Report (GAR) on Disaster Risk Reduction 2013. *Global Risk Assessment: Data, Sources and Methods.* Geneva: UNISDR.

Upton, G. and Cook, I. (1996) *Understanding Statistics.* Oxford University Press, Oxford.

Vaughn, T.R. (2007) *Comparison of Risk Allocation Methods: Bohra Weist DFAIC Distributions.* CAS Forum, Winter 2007 pp. 329–337.

Venter, G. (2010) *Non-tail measures and allocation of risk measures.* www.casact.org/library/studynotes/Venter_Non-Tail.pdf (accessed 22 Sept. 2015).

Venter, G., Major, J.A. and Kreps, R.E. (2006) *Marginal decomposition of risk measures.* ASTIN Bulletin, **36** (02), 375–413.

Von Dahlen, S. and von Peter, G. (2012) *Natural Catastrophes and Reinsurance: Exploring the Linkages.* Bank for International Settlements. http://www.bis.org/publ/qtrpdf/r_qt1212e.pdf (accessed 22 Sept. 2015).

Von Peter, G., von Dahlen, S. and Saxena, S. (2012) *Unmitigated disasters new evidence on the macroeconomic cost of natural catastrophes.* Bank for International Settlements, Working Paper 394. http://www.bis.org/publ/work394.pdf (accessed 22 Sept. 2015).

Walters, M.A. (2007) *Catastrophe Ratemaking.* Casualty Actuarial Society Study Notes. http://www.casact.org/library/studynotes/5_WaltersJune2007.pdf (accessed 22 Sept. 2015).

World Bank (2012) *Consultancy for Prioritization of High Seismic Risk Provinces and Public Buildings in Turkey, by Proto Engineering.* World Bank, Washington, DC.

Yamin, L.E., Ghesquiere, F., Cardona, O.D. and Ordaz, M.G. (2013) *Modelación probabilista para la gestión del riesgo de desastre.* World Bank, Washington, DC http://documents.worldbank.org/curated/en/2013/07/18100020/colombia-probabilistic-modeling-disaster-risk-management-modelacion-probabilista-para-la-gestion-del-riesgo-de-desastre.

Yoon, A. and Scism, L. (2014) *Investors embrace 'catastrophe bonds'.* The Wall Street Journal, April 23, 2014 http://www.wsj.com/articles/SB10001424052702304049904579517710350913016 (accessed 22 Sept. 2015).

Zeng, L. (2001) *Using cat models for optimal risk allocation of P&C liability portfolios.* Journal of Risk Finance, **2** (2), 29–35.

Zeng, L. (2010) *Catastrophe reinsurance risk – a unique asset class,* Financial Engineering seminar. Colombia University, New York.

Zurbuchen, B., Radoff, R. and White, S. (2010) What makes a good economic capital model? http://www.casact.org/education/reinsure/2010/handouts/cs24-zurbuchen.pdf (accessed 19 Nov. 2015).

# 3

# The Perils in Brief

*John Hillier*

## 3.1 Overview

3.X   Structure of the Sections
METEOROLOGICAL PERILS (i.e. 'Wind Driven')
3.2   Tropical Cyclones (TC) (a.k.a. typhoon, hurricane)
3.3   Extra-Tropical Cyclones (ETC) (inc. US winter storms)
3.4   Severe Convective Storms (SCS)
HYDROLOGICAL PERILS (i.e. 'Rain Driven')
3.5   Inland Flooding (FL)
3.6   Shrink-Swell Subsidence (SS)
GEOLOGICAL PERILS (i.e. 'Solid Earth Driven')
3.7   Earthquakes (EQ)
3.8   Mass Movement (MM) (a.k.a. landslide)
3.9   Tsunami (TS)
3.10 Volcanoes (V)
References

The hazards, also known as **perils,** are grouped here according to the nature of the dominant process that drives them. This classification is by necessity somewhat arbitrary; for example, despite currently only being modelled with respect to earthquakes, mass movements are arguably mainly driven by rainfall. However, any typology used to classify natural hazards is ambiguous. There is no unique, correct, fundamental or natural typology of natural hazards, or indeed subtypes of them (e.g. flooding, Section 3.5). Dependencies between individual hazards contribute substantially to this ambiguity, and different companies use different categorizations.

### 3.1.1   What Is Included

Not all perils are included in this chapter, which is not just an abridged hazards textbook. The content was specifically selected according its relevance to catastrophe modelling. The space devoted to each peril is primarily allocated to reflect its currently perceived significance or likely significance in driving future modelling developments. Significance can be considered a number of ways. Table 3.1 lists the top 10 recent insured losses, dominated by tropical cyclones

*Natural Catastrophe Risk Management and Modelling: A Practitioner's Guide*, First Edition.
Edited by Kirsten Mitchell-Wallace, Matthew Jones, John Hillier and Matthew Foote.
© 2017 John Wiley & Sons Ltd. Published 2017 by John Wiley & Sons Ltd.

Table 3.1   Ten costliest recent events (1980–2014) in terms of insured losses.

| Peril Type | Event name/ description | Year | Area(s) Impacted | Insured Losses US$ billion (Original Values) | Overall Losses US$ billion (Original Values) | Fatalities |
|---|---|---|---|---|---|---|
| *Tropical Cyclone* | Hurricane Katrina and storm surge | 2005 | USA | 62.2 | 125.0 | 1322 |
| *Earthquake* | Tohoku or GEJET | 2011 | Japan | 40.0 | 210.0 | 15880 |
| *Tropical Cyclone* | Hurricane Sandy | 2012 | USA and region | 29.5 | 68.5 | 210 |
| *Tropical Cyclone* | Hurricane like | 2008 | USA and region | 18.5 | 38.0 | 170 |
| *Tropical Cyclone* | Hurricane Anderew | 1992 | USA and region | 17.0 | 26.5 | 62 |
| *Earthquake* | Christchurch | 2011 | New Zeland | 16.5 | 24.0 | 185 |
| *Flood* | Thailand | 2011 | Thailand | 16.0 | 43.0 | 813 |
| *Earthquake* | Northridge | 1994 | USA | 15.3 | 44.0 | 61 |
| *Tropical Cyclone* | Hurricane Ivan and strom surge | 2004 | USA and region | 13.8 | 23.0 | 120 |
| *Tropical Cyclone* | Hurricane Wilma | 2005 | USA and region | 12.5 | 22.0 | 44 |

Note: Areas impacted simplified to 'USA and region' where the USA is the driver of the financial loss. *Source:* Figures are from Munich Re (2015).

impacting the United States, with occasional earthquakes in 'high-income' (i.e. rich, insured) countries such as New Zealand and Japan (Munich Re, 2015); so, tropical cyclone and its modelling are found in Section 3.2. The earthquake peril is included as it causes some of the largest overall economic losses (Munich Re, 2015). The stark contrast to the types and locations of event that cause most fatalities (Table 3.2) reflects no lack of compassion on the part of the

Table 3.2   Ten costliest recent events (1980–2014) in terms of fatalities.

| Peril Type | Area(s) Impacted | Year | Insured Losses US$ billion (Original Values) | Overall Losses US$ billion (Original Values) | Fatalities |
|---|---|---|---|---|---|
| *Earthquake* | Haiti | 2010 | 0.2 | 8.0 | 222,570 |
| *Earthquake* | Indonesia and region | 2004 | 1.0 | 10.0 | 222,000 |
| *Tropical Cyclone* | Myanmar | 2008 | - | 4.0 | 140,000 |
| *Tropical Cyclone* | Bangladesh | 1991 | 0.1 | 3.0 | 139,000 |
| *Earthquake* | Pakistan and India | 2005 | 0.005 | 5.2 | 88,000 |
| *Earthquake* | China | 2008 | 0.3 | 85.0 | 84,000 |
| *Heat Wave* | Europe | 2003 | 1.1 | 13.8 | 70,000 |
| *Heat Wave* | Russia | 2010 | - | 0.4 | 56,000 |
| *Earthquake* | Iran | 1990 | 0.1 | 7.1 | 40,000 |
| *Earthquake* | Iran | 2003 | 0.02 | 0.5 | 26,200 |

*Source:* Figures are from Munich Re (2015).

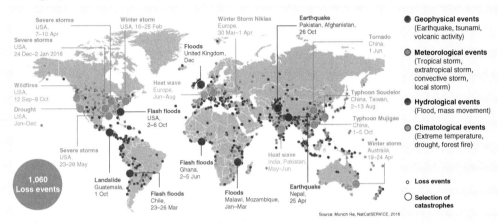

Figure 3.1 Geographical overview of loss events worldwide in 2015. Larger circles are losses >US$1.5 billion. In this classification 'geophysical' events are red, storms are green, 'hydrological' events (flood and landslide) in blue, and 'climatological' (extreme heat, drought, wildfire) in orange. *Source:* Reproduced with permission of Munich Re, NatCatSERVICE (2016).

modelling community, simply the priorities of those providing resources and their ability to provide those resources (e.g. insurers, government, non-governmental organizations). The attributes of the perils and lessons about modelling them described in this chapter will assist risk assessment whatever the required definition of 'cost' (e.g. lives, money, travel time lost, deaths of an endangered species, happiness).

In any multi-hazard environment, it is essential to understand which perils govern likely losses. This will vary by region (Figure 3.1), the definition of loss, the concerns of each stakeholder, and the underlying distribution of assets at risk (see Chapter 1).

Examination of total global insured losses in the past 10 years (Table 3.3) highlights the importance of 'Severe Weather' and flooding; although these may not be largest losses in individual events, they occur relatively frequently and so cause substantial average annual losses. Table 3.4 again directly compares the perils but provides more insight into the distributions of potential losses. For instance, although earthquakes cause major insured losses relatively rarely, they dominate annual losses in years when they do happen. On this basis, extra-tropical cyclones, both in the United States and Europe, and flooding are commonly modelled perils and included here.

Implicit in the use of distinct models for individual **peril-regions** (see Chapter 4.3) is that the key hazards in the regions that are seen as independent, unrelated to, and not triggered by other hazards. These are referred to as **primary hazards** (e.g. Gill and Malamud, 2014) and include some of the largest threats (e.g. US hurricane); thus, these are considered independently of each other even in state-of-the-art multi-hazard risk analysis, for example, for insurance or resilience planning (Marzocchi *et al.*, 2007; Kappes *et al.*, 2012; Swiss Re, 2014; The Royal Society, 2014). Hazards and the processes driving them (e.g. in the atmosphere) are well studied in isolation, including work to understand their impact (e.g. Munich Re, 2008; Corti *et al.*, 2009; Donat *et al.*, 2011; Swiss Re, 2014). This is the mainstay of our understanding and modelling as detailed in

Table 3.3   Yearly aggregate insured losses from 2005 to 2015 from natural perils.

| Peril Type | 2005 | 2006 | 2007 | 2008 | 2009 | 2010 | 2011 | 2012 | 2013 | 2014 | 10-yr average (2005-2014) |
|---|---|---|---|---|---|---|---|---|---|---|---|
| *Tropical Cyclone* | 105.48 | 2.58 | 2.58 | 20.03 | 1.48 | 1.55 | 10.29 | 32.80 | 4.09 | 3.62 | 18.45 |
| *Several Weather* | 5.19 | 9.20 | 4.83 | 15.06 | 13.40 | 16.29 | 29.98 | 15.61 | 18.07 | 18.08 | 14.57 |
| *Flooding* | 6.40 | 1.46 | 9.96 | 7.94 | 4.04 | 9.20 | 26.55 | 3.94 | 14.66 | 6.17 | 9.03 |
| *Earthquake* | 0.29 | 0.10 | 0.53 | 0.46 | 0.71 | 13.15 | 53.54 | 1.96 | 0.43 | 0.56 | 7.17 |
| *Drought* | 0.99 | 2.75 | 1.72 | 6.00 | 2.13 | 1.14 | 5.22 | 18.65 | 5.11 | 3.07 | 4.68 |
| *Winter Weather* | 1.46 | 1.95 | 3.31 | 3.05 | 1.13 | 4.37 | 2.97 | 0.93 | 2.29 | 7.00 | 2.85 |
| *EU Windstorm* | 3.28 | 0.00 | 7.08 | 2.03 | 4.17 | 3.95 | 0.73 | 0.79 | 3.12 | 0.94 | 2.61 |
| *Wildfire* | 0.04 | 0.19 | 1.80 | 0.55 | 1.35 | 0.37 | 1.54 | 0.61 | 0.71 | 0.14 | 0.73 |
| *Other* | 0.14 | 0.47 | 0.23 | 0.46 | 0.07 | 0.19 | 0.37 | 1.18 | 0.12 | 0.02 | 0.33 |
| **Total** | **123.27** | **18.70** | **32.04** | **55.58** | **28.48** | **50.21** | **131.19** | **76.47** | **48.60** | **39.60** | **60.41** |

Note: Figures in billion US$. 'Severe Weather' refers to the sum of tornado, hail, and thunderstorm events. Highlighted cells represent 20% or more of losses within that year. Courtesy of Aon Benfield (2015).

Table 3.4   The range of losses for the 10 costliest events for each peril in the time period, 1980–2014.

| Peril Type | Insured Losses US$ billion (Original Values) | Overall Losses US$ billion (Original Values) |
|---|---|---|
| *Hurricane* | 62 - 6 | 125 - 12 |
| *Earthquake* | 40 - 1 | 210 - 14 |
| *Flooding* | 16 - 1 | 43 - 10 |
| *Typhoons* | 6 - 1 | 11 - 5 |
| *Europe winter storms* | 6 - 2 | 12 - 3 |
| *US storms and winter damage* | 2 - 1 | 5 - 1 |

*Source:* Figures from Munich Re (2015).

Chapters 3.2–3.10. Spatial and temporal patterns within the primary perils are also increasingly being included in models (e.g. clustering) (see, e.g. Vitolo *et al.*, 2009; Wang and Lee, 2009). Such work is detailed for each peril where appropriate. However, dependencies between perils exist.

Dependencies can, and do, take the form of multiple hazards affecting the same location (e.g. simultaneous wind and flooding; Trapero *et al.*, 2013; Jansa *et al.*, 2014). Although the existence of multiple perils within events is likely to become increasingly more explicitly recognized and modelled (e.g. rain in US hurricane models), the most common way to incorporate this so far has been through defining the related contemporaneous peril as a 'secondary peril' (e.g. coastal surges produced by storms). This is currently the status of storm surge, tsunami, and mass movements (i.e. landslides); the material impact of these can be large, and latter two of these have short sub-sections.

A further aspect of multi-hazard interactions are interdependencies in impacted systems, and a number are recognized (cascades or hazard chains, e.g. INTERAct, 2007; Gill and Malamud, 2014). Currently, secondary perils that impact with some time delay once triggered by a primary event are rarely considered in commercial models; an example of such a time lag is the ability of large triggers (e.g. earthquakes) to enhance the landslide susceptibility of a region for years (e.g. Hovius *et al.*, 2011). Furthermore, even primary perils currently treated as distinct (e.g. flooding, windstorm) could interact (e.g. Gill and Malamud, 2014), modifying aggregated losses (i.e. AEP) from that estimated under the assumption of independence; an example of this is the weather-driven hazards in the United Kingdom (i.e. flooding, windstorm, and shrink-swell subsidence; Hillier *et al.*, 2015). Assessment of these in models largely remains a challenge for the future.

Although business interruption linked to direct damage is included in models, contingent business interruption (i.e. related to a supplier or customer being affected) is not; this is recognized to be difficult to do well, as illustrated by the large and unexpected losses to flooding in Thailand in 2011 (see Table 3.1; Munich Re, 2012). Hazard dependencies mediated by indirect (e.g. supply chain) mechanisms are not currently considered in catastrophe models.

Future catastrophe models will likely cover new perils and regions. Some perils may have only been tackled in a limited way because of the computational expense required to make modelling reliable, or simply the lack of a large recent loss to prompt the development. A few seen as most likely to drive risk assessment capability, although perhaps more for humanitarian purposes than insurance ones, have been give sub-sections: shrink-swell subsidence, volcanic risk, and severe convective storms.

### 3.1.2 What Is Not Included

For reasons of materiality or the current lack of commercial catastrophe models, there are a number of perils driven by natural processes that have not been included in this chapter; these include snow avalanche, drought, wildfire, temperature extremes, and solar flares. Storm surge has no explicit section, but it is included with cyclones (tropical and extra-tropical). Further, we do not consider perils with man-made origins (e.g. fires, explosions, road traffic and rail disasters, aviation and space catastrophes, mining accidents, shipping accidents, collapse of buildings/bridges, terrorism, pandemic, industrial accidents). Although these were estimated to cost $9 billion in 2014, this is a small fraction of the total losses of $110 billion (Swiss Re, 2015).

### 3.1.3 Why Read This Chapter?

Catastrophe models blend statistical approaches with science-based modelling to compensate for the fact that our experience (personal or corporate) is insufficient to successfully judge risk from the least frequent and most severe events. That is, they attempt to compensate for a lack of data through assumptions and our, as yet incomplete, knowledge of our physical environment. Thus, there are several key points to note.

- Data (e.g. past losses, wind speed measurements) are always sparse and need interpolation.
- Model developer time is always limited.
- Computational power and incomplete scientific understanding always require that approximations and compromises are made.
- Models are only a simplification; hopefully simplified using the right approximations.

The model developers will have done their best to circumvent or overcome these difficulties, but any user must take a view on how well this has been done (Chapter 5); for this, it is critical to understand each peril and its anatomy. Models should not be trusted blindly and there is significant uncertainty in their use and application (see Chapter 2).

Insight into the state of modelling is provided by contrasting the state of scientific knowledge, in the first part of each sub-section with the description of the modelling in the second half. A contrast is particularly evident in some sub-sections (e.g. severe convective storm, mass movement). The open question is: how much complexity or sophistication is necessary? The answer may be different for those seeking a complete understanding of the physical system (e.g. the scientist), and those seeking to effectively and efficiently quantify material losses to a pragmatically useful level (e.g. a catastrophe risk modelling practitioner). A related question is: how much of the complexity is justified by the observational evidence available?

## 3.X Structure of the Sections

A common structure is used for each peril that follows. An $X$ is used to denote applicability to all perils; for instance, 3.X could denote flooding in 3.5 or tropical cyclones in 3.2. Some headings are omitted for less significant perils. The first three sections (i.e. 3.X.1–3.X.3) are the science; they describe and explain the peril, but entirely without consideration of implementation in a catastrophe model. The later sections (i.e. 3.X.4–3.X.7) detail the numerical and computational translation of this knowledge into a tool for risk estimation; these also include what is not yet covered (non-modelled), how key past events have driven the development of insurance industry models, and a sample of questions to use to probe a model provider when assessing a model (e.g. for Solvency II purposes, see Chapter 2.11.4).

### 3.X.1 What Is the Peril?

- **DESCRIPTION:** The kinematics of the physical process; what it is that makes a hazardous event for this peril is described, rather than explaining why. Key metrics are, for instance, its size (e.g. magnitude), location, spatial scale, speed of propagation, and frequency of occurrence. Basic descriptive terms are also defined (e.g. the 'eye' of a hurricane).
- **PHYSICAL DRIVERS:** The mechanics, explaining why the hazard acts as it does; a basic background to the physics is presented, restricted to what is scientifically well known. The phenomenon's life cycle, if it has one (e.g. genesis, transmission, persistence, transitioning, death), is also usually described here because its description is commonly tied to explanations of the underlying processes.
- **KEY BEHAVIOURAL SUBTLETIES:** An expansion of key elements of the physical behaviour that really matter to its impact, and in particular those where the science is more recent, less certain, or ongoing.

### 3.X.2 Damage Caused by the Peril

The main ways in which the peril causes damage, affecting either property or life. This includes summary loss statistics of insured loss, economic loss and deaths as they are distributed geographically. The focus is on the size and materiality of damage and what drives it.

### 3.X.3 Forecasting Ability and Mitigation

This ties together the last two sections, taking knowledge of the peril and its effects to relate them to two questions. First, is there any predictability, and if so, over what timescale in advance of the event causing loss? Second, can this predictability lead to mitigation? A brief evaluation of advance planning measures that may be taken to mitigate the impacts of an event (e.g. building regulations) may be included here. Timescales (e.g. 5 days, seasonal) will depend on the peril, as will the spatial scale (e.g. regional or global). Note that 'prediction' and 'forecasting' are distinct and different, for example, see Section 3.7.3.

### 3.X.4   Representation in Industry Catastrophe Models

This section describes how the peril, in both its physical process (hazard) and impact (risk) via vulnerability, is simplified and encapsulated in event sets and exposure-based portfolio risk management tools called 'catastrophe models'. General details, strategies, structures, and processes used to create such catastrophe models are detailed elsewhere, primarily Chapter 4. Here, the focus is on any peril-specific quirks and variations that are required for its representation (e.g. spatial resolution, temporal dependence). The focus is on current practice, highlighting best practice. Various levels of sophistication exist, so this is sometimes extended to include imminent developments or the need for catastrophe modelling to incorporate better practice from elsewhere. Three self-explanatory headings are used.

- **HAZARD DESCRIPTION AND EVENT SETS**
- **DAMAGE METRIC AND VULNERABILITY**
- **ADDITIONAL LOSS DRIVERS INCLUDED**

### 3.X.5   Secondary Perils and Non-Modelled Items

A brief summary of which other physical hazards are modelled as triggered by, or associated with, the primary one detailed for a peril; this includes how these compounding effects are approximated in current catastrophe models, which can vary quite substantially. Also included here are statements about what is not currently modelled. Both of these are important in forming a view of risk (Chapter 5).

### 3.X.6   Key Past Events

Often, a significant part of the explanation for why a catastrophe model is designed or structured as it is lies in the history or legacy of its development, and a dominant impetus in model development is commonly major events (see also Chapters 1 and 2). Some events are key because they caused surprises, highlighting limitations of existing models. Key events are summarized including, most importantly, what knowledge was gained from them in terms of lessons learnt for model design.

### 3.X.7   Open Questions/Current Hot Topics/Questions to Ask Your Vendor

Selected questions are listed concerning hot topics and pressing issues for each peril. Where possible, reference is made to current research areas that are changing our view of the peril or its impacts.

### 3.X.8   Non-Proprietary Data Sources

Details are given of important and accessible databases for each peril. Many data sets are, however, simply proprietary.

# METEOROLOGICAL PERILS (I.E. 'WIND-DRIVEN')

## 3.2 Tropical Cyclones

*James Done and Brian Owens*

*Reviewer: Greg Holland*

### 3.2.1 What Is the Peril?

**DESCRIPTION:** Systems of cyclonic air circulation that initiate within the tropical region (i.e. approx. ±30 degrees) and generate severe winds (i.e. > 62 km/h) are defined as **tropical cyclones (TCs)**. TCs with sustained winds exceeding 118 km/h are known as **hurricanes** over the North Atlantic and Central and East Pacific oceans and **typhoons** over the western North Pacific Ocean. Over the Indian Ocean and South Pacific, TCs are referred to as cyclones. TCs mainly occur in a hemisphere's summer, although rarely over the South Atlantic (Figure 3.2 (a)). Hurricane severity is described by categories 1–5 on the Saffir-Simpson scale (Figure 3.2 (b)).

Mature TCs are approximately symmetrical about a central **eye**, which is a calm area delimited by a wall of strong winds and rain known as the **eye wall** (Figure 3.3). Wind speeds peak at a radius of 5–100 km from the centre and decrease slowly thereafter over a few hundred km (Holland, Bellanger, and Fritz, 2010). Surrounding the eye wall are outer bands of thunderstorms that spiral into the core of the storm. Typically, TCs move westward and

(a)

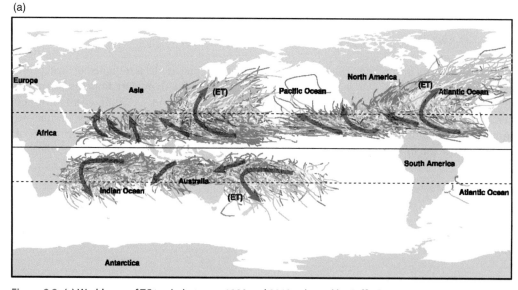

**Figure 3.2** (a) World map of TC tracks between 1980 and 2013 coloured by Saffir-Simpson intensity category in (b). IBTrACS data (Knapp *et al.*, 2010). Arrows indicate general TC tracks and '(ET)' indicates the major regions where TCs can undergo extra-tropical transition. TCs are most prevalent from June to November in the Northern Hemisphere peaking in mid-September, and from November to April in the Southern Hemisphere peaking in mid-February. Courtesy of Ming Ge, NCAR.

(b) **Saffir-Simpson Hurricane Scale**

| Category | Wind speed | |
|---|---|---|
| | mph | km/h |
| 5 | > 156 | > 250 |
| 4 | 131 - 155 | 210 - 249 |
| 3 | 111 - 130 | 178 - 209 |
| 2 | 96 - 110 | 154 - 177 |
| 1 | 74 - 95 | 119 - 153 |

**Non-Hurricane Classifications**

| | | |
|---|---|---|
| Tropical storm | 39 - 73 | 63 - 118 |
| Tropical depression | 0 - 38 | 0 - 62 |

Figure 3.2  (*Continued*)

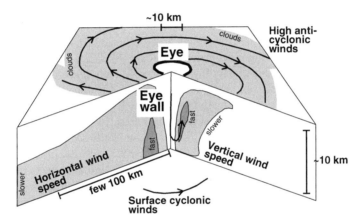

Figure 3.3  Basic structure and terminology associated with a TC. Horizontal winds spiral cyclonically inwards and are fastest just above the surface and in the eye wall where the vertical winds loft air high into the upper troposphere and lower stratosphere fanning out anti-cyclonically (Emanuel, 2003). In the Northern Hemisphere, cyclonic is anticlockwise and anticyclonic is clockwise, as shown, while the reverse is the case in the Southern Hemisphere.

poleward in the deep tropics at 10–60 km/h, sometimes erratically, and arc poleward and eastwards at higher latitudes. This translation increases the surface wind speeds on the right side of the TC in the Northern Hemisphere and on the left side in the Southern Hemisphere (Figure 3.4) and may cause damage affecting areas several hundred km wide (Czajkowski and Done, 2014).

**PHYSICAL DRIVERS:** TCs are driven by a transfer of energy from the surface, and typically form over oceans where sea surface temperatures (SSTs) exceed 26°C (Palmen, 1948) in a process called **cyclogenesis**. A number of other factors must also align for a TC to form (Gray, 1979). **Atmospheric instability**, specifically warm moist air with the tendency to rise, is essential (Emanuel, 1986, 1988; Holland, 1997) to transport the high-energy air aloft through deep clouds and establish the storm's structure. Humid conditions at mid-levels are needed to limit the amount of storm-retarding dry air brought down to the surface (Rappin, Nolan, and Emanuel, 2010). Strong **wind shear**, a difference between winds at low and high altitudes, can

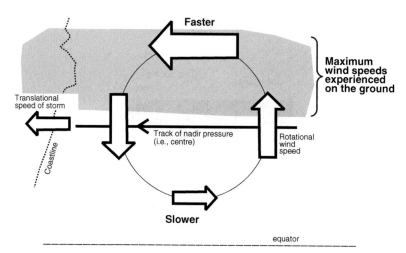

**Figure 3.4** Plan view of a Northern Hemispheric tropical cyclone showing how the translational motion of the cyclone increases the winds experienced at the ground on the right side of the track and decreases the winds on the left side.

tear apart a developing TC (DeMaria, 1996; Frank and Ritchie, 2001; Wu and Braun, 2004; Tang and Emanuel, 2010). Finally, a **trigger** is needed to initiate the TC. Triggers vary in importance by basin and can take many forms, including trailing cold fronts, monsoonal circulations, or pulses of energy in the atmosphere known as tropical waves (Frank and Roundy, 2006).

Similar to a stick floating down a stream, the route or **track** of a TC is largely controlled by the environmental winds, with prevailing direction from East to West (Trade Winds) at low latitudes (< 30°) and the reverse (West to East) at higher latitudes (see Figure 3.2 (a)). The poleward drift results from interaction of the TC circulation with a latitudinal change in the effect of Earth's rotation (Holland, 1983). Thus, where TCs originate affects their potential to cause damage. For instance, the tracks of TCs that form far from land have a greater chance of being directed away from land by passing weather systems than those forming closer to a coastline, but they also have more time to intensify (Dailey *et al.*, 2009). Variability in tracks is caused by interactions between a TC and the daily weather patterns in its locality (Galarneau and Davis, 2013), leading to erratic motion when the large-scale guiding winds are weak (Holland, 1983; Fiorino and Elsberry, 1989).

**KEY BEHAVIOURAL SUBTLETIES:** TC activity is influenced by both environmental conditions and processes internal to the TC itself across a range of timescales. For intense TCs, the eye wall can decay and be replaced by a new outer wall in an **eye-wall replacement cycle**. This process is difficult to predict but can occur several times in the TC's lifetime and is generally accompanied by fluctuations in maximum wind speed intensity (e.g. Hurricane Andrew; Willoughby and Black, 1996) with implications for potential losses. TCs can undergo **extra-tropical** (or post-tropical) **transition** (see Figure 3.2 (a)), namely developing into a new, but potentially also damaging extra-tropical system (see Chapter 3.3). During this process the primary energy source for the cyclone switches to energy arising from interacting air masses, and typically results in an expanded, asymmetrical footprint. Given favourable conditions, TCs can also undergo **rapid intensification**, which is defined as an increase in maximum sustained winds of at least 55 km/h in 24 hours. Typhoon Chebi, for example, intensified explosively with a 60 hPa pressure drop in just 6 hours from before making landfall in the Philippines in 2006. Once a TC makes landfall and is cut off from its energy source, it will start to lose vigour or **decay** as it moves inland.

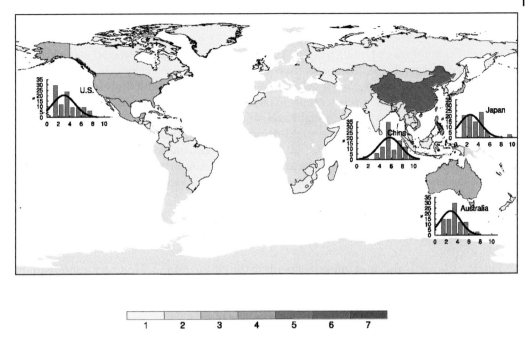

**Figure 3.5** Annual average numbers of landfalling tropical cyclones of tropical storm intensity or greater per country over the period 1980–2013 (IBTrACS data; Knapp *et al.*, 2010). Inserts show histograms of the number of landfalling tropical cyclones per year (grey bars) for the United States, China, Japan, and Australia. Poisson distributions (black lines) assume independence between events, and illustrate what is expected without clustering. Courtesy of Ming Ge, NCAR.

TCs can **cluster** in time and space with more TCs arriving in a given period and region than is expected by chance for independent events (Jagger and Elsner, 2012; Mumby, Vitolo, and Stephenson, 2011; Figure 3.5). Clustering raises the chance of a second landfall, so ignoring clustering could underestimate aggregate annual losses. Clustering can arise from westward-moving tropical waves that track pulses of energy (triggers) through favourable environments. Clustering has also been attributed to the Madden-Julian Oscillation (Maloney and Hartmann, 2000), a 30- to 60-day disturbance that propagates eastward in the tropical belt influencing TC activity (Kossin, Camargo, and Sitkowski, 2010; Hall, Matthews, and Karoly, 2001). This is notorious for producing twin TCs either side of the equator.

On longer seasonal to decadal timescales, TC activity has been linked to variations in the atmosphere and oceans known as **climate modes** (e.g. Jewson *et al.*, 2008), the most important of which is the El Niño Southern Oscillation (ENSO). Opposite phases of ENSO, **El Niño** and **La Niña**, occur irregularly every 3 to 8 years and influence the number, general locations, intensity, and landfall likelihood of TCs in many ocean basins. For example, the probability of two or more landfalling US hurricanes is 28%, 48%, and 66% during El Niño, neutral, and La Niña years, respectively (Bove *et al.*, 1998), strongly affecting economic losses (Pielke Jr. and Landsea, 1999; Jagger, Elsner and Burch, 2011). Similarly, the Indian Ocean Dipole (Saji *et al.*, 1999) affects TC activity over the North Indian Ocean and Australia, and typhoon activity over the western North Pacific may be influenced by the Pacific Decadal Oscillation (Chan, 2008; Goh and Chan, 2010). The **North Atlantic Oscillation** (Wang *et al.*, 2011; Elsner, Liu and Kocher, 2000), the **Atlantic Multi-decadal Oscillation** (Wang, Lee and Enfield, 2008), and the **Atlantic Meridional Mode** have all been suggested as affecting TC activity levels over the North Atlantic (Chang and Li, 1997; Vimont and Kossin, 2007).

Annual global TC numbers are steady at ~90 (Frank and Young, 2007), but individual ocean basins may exhibit *trends*. For instance, over the past 30 years, the North Atlantic has seen increasing TC numbers and an increase in TC intensity (Holland and Webster, 2007; Kossin, Olander, and Knapp, 2013). The proportion of the most intense hurricanes, however, has increased both globally and regionally (Webster *et al.*, 2005; Holland and Bruyère, 2013). There is consensus that TC intensity will continue to increase with warming sea surface temperatures (Knutson *et al.*, 2010; Intergovernmental Panel on Climate Change, 2012). Thus, intertwined with intensity trends are decadal and multi-decadal swings in the numbers of major TCs in many ocean basins (Goldenberg *et al.*, 2001) and the origin and impact of these swings are poorly constrained and remain an active area of research (Knudsen *et al.*, 2011). Overall, there is sufficient evidence for stakeholders to take the possibility of significant future changes into serious consideration (Emanuel, 2011).

### 3.2.2 Damage Caused by the Peril

US hurricane is one of the most important perils to the global insurance industry, representing 42% of all US insured catastrophe losses (Insurance Information Institute, 2013). Annualized direct economic losses are estimated at $10 billion in the United States (Pielke *et al.*, 2008), with single event losses of up $125 billion – 'as-if' modelling of 1926 Florida event, the Great Miami Hurricane (Karen Clark and Company, 2012). Damage is typically assumed (e.g. Schmidt, Kemfert, and Hoppe, 2010) to be primarily driven by the maximum speed of winds ($v$) since damage increases with its cube (i.e. $v^3$), representing the power dissipated at the ground by the winds and engineering relationships between winds and damage (Nordhaus, 2010). Clearly, with all else equal, storms causing damage over a larger area (e.g. Hurricanes Ike or Sandy) will cause greater losses (Zhai and Jiang, 2014). Slowly moving or larger storms can also arguably have increased damage impact (e.g. Czajkowski and Done, 2014) because each exposed structure experiences damaging winds for longer. Rainwater ingress into buildings through wind-driven damage to the roof, windows, or building envelope can cause additional damage to the wind effects alone and can contribute significantly to contents losses.

Damage associated with TCs can also be caused by secondary perils: tornadoes-Hurricane Ivan in 2004, for example, produced 117 tornadoes across a broad swath of the United States that generated $97 million in property damage (Watson *et al.*, 2005; Belanger, Curry, and Hoyos, 2009; National Climate Data Center, 2009); riverine flooding; flash flooding particularly in urban areas; landslides, for example, Typhoon Morakot in Taiwan in 2009 (Tsai *et al.*, 2010); and coastal flooding caused by storm surge, for example, Typhoon Haiyan in the Philippines in 2013 (Mori *et al.*, 2014). Some nonintuitive complexities, however, exist. For example, Hurricane Sandy in 2012 illustrated how even a storm with low wind-speeds at landfall can raise water levels by up to 4.28 m (Forbes *et al.*, 2014), and topography can enhance surges such as in the shallow and semi-enclosed Bay of Bengal (Dube *et al.*, 2004) during Cyclone Phailin in Odisha, India in 2013. For inland flooding, rainfall relates to ascending moisture (Langousis and Veneziano, 2009), enhanced by slow or meandering progress (e.g. Tropical Storm Allison in 2001; Evans, Gudes and Kelly, 2001), interaction with the jet stream (e.g. Floyd in 1999; Daley, Baker and Kelly, 2000), or passage over mountainous terrain, typhoons impacting Japan and Taiwan (Tan, Lim and Abdullah, 2012; Rogers, Marks and Marchok, 2009).

### 3.2.3 Forecasting Ability and Mitigation

Short range (3- to 5-day) forecasts of TC tracks now have significant predictive power, or **skill** (National Hurricane Center, 2014). However, properties like the maximum wind speed or area of damaging winds, which are important for preparations to protect human life and mitigate economic and insured losses (e.g. Demuth *et al.*, 2012), are less well forecast (Zhang, 2011). Loss is also influenced by behavioural factors including moving property out of harm's way, property protection such as storm shutters, and on a longer timescale, adhering to building codes (Dehring and Halek, 2013).

Seasonal forecasts of total basin and season numbers of TCs and have been shown to be skilful on average over many years (Owens and Landsea, 2003; Elsner and Jagger, 2006; Jagger, Elsner and Saunders, 2008). Individual years, however, are poorly forecast because of incomplete understanding and the inherent uncertainty of the climate system (Done *et al.*, 2014). For example, one of the most reliable seasonal predictors of hurricane activity used to be West African precipitation, but this deteriorated 15 to 20 years ago (Fink, Schrage and Kotthaus, 2010) and illustrated that linkages are not well understood. Furthermore, since a large number of TCs do not reach land, the total number of TCs generated, or **basin activity**, cannot be directly translated into landfall numbers (Chang and Guo, 2007; Holland, 2007; Dailey *et al.*, 2009). Current seasonal forecasts therefore show substantially reduced skill for loss-related metrics such as landfall numbers or damage potential.

### 3.2.4    Representation in Industry Catastrophe Models

**HAZARD DESCRIPTION and EVENT SETS:** The occurrence of a TC is known as an **event**, and catastrophe models calculate losses from ensembles (e.g. >10,000) or **event sets** (see Chapter 4.3.1) generated using stochastic combinations of cyclogenesis, track, size and intensity. Typically, this does not explicitly consider the physical mechanisms of cyclogenesis, rather model developers resample climatological distributions of SSTs, SST gradient, and wind shear to seed random TCs. Then, the number of the modelled genesis events and their location are calibrated against the historical record (e.g. Hall and Jewson, 2007). Tracks are typically calculated using a stepwise algorithm where the track length and direction in each step are derived after superimposing a mean component, based on historical data, and a random deviation around it. Both the mean and random displacements depend on the previous positions of the storm (e.g. Hall and Jewson, 2007) as well as related meteorological variables; the key variables include central pressure, radius to maximum winds, maximum wind speed, and post-landfall **inland filling rate**, that is, the rate at which the central pressure 'fills' and the storm decays (Vickery, 2005). Intensity is modelled in a similar stepwise fashion to track, again perturbed according to meteorological variables. Finally, a spatial wind field for each event is modelled by applying a characteristic wind field footprint along and around the track approximated from inner core and outer storm wind observations and pressure drop to the storm centre (e.g. Holland, Belanger and Fritz, 2010). See Chapter 4.3.6.4 for an illustrative example of the modelling. Note that events are not assigned a fixed duration, which may have potential reinsurance implications for enduring events (e.g. >72 h; see Chapter 2.4.2). An alternative approach (Emanuel, Sundararajan and Williams, 2008) seeds the environment with potential TCs, generating a wind profile using a simple numerical model guided by environmental conditions.

**DAMAGE METRIC AND VULNERABILITY:** TC damage is typically modelled as a **damage ratio (DR)** that is, a function of wind speed. Damage ratios are estimated repair cost (i.e. modelled loss) over the replacement cost of the property. The wind speeds used in the vulnerability portions of the model vary from model to model with 3-second gusts and 1-minute sustained wind speeds being the most common. Many models use the peak gust, as research has shown that that metric is most highly correlated to the observed damage (Vickery, Masters, Powell and Wadhera, 2009), and that is the same metric used in most modern building codes today (Florida Building Code, 2014). Loss modification as a result of duration of the wind is included in some models (e.g. AIR; Jain, 2010). See Pita *et al.* (2015) for a comprehensive review.

**BEHAVIOURAL SUBTLETIES INCLUDED:** TC activity deviates from its long-term average in multi-decadal cycles of higher and lower activity (e.g. Holland and Bruyère, 2013; Caron, Boudreault and Bruyère, 2014; Zhao, Wu and Wang, 2014), and this is often captured in commercial models in modified event sets presenting this as an alternative view of risk. This is most common for the North Atlantic and usually done by producing stochastic events sets that are conditioned on periods in the historical record that exhibit above- or below-average activity. Climate model ensemble forecasts of TC activity based on basin conditions such as sea surface temperatures have also been used.

Implementation of clustering of TCs in space and time varies; methods include combining multiple footprints into a single low-probability event and assigning events a time signature to effectively create a time series. Changes to wind fields during extra-tropical transitions of TCs are sometimes accommodated by modified event wind footprints. Eye-wall replacement cycles are not currently included in any model. Event sets, however, typically contain a selection of rapidly intensifying storms.

### 3.2.5   Secondary Perils and Non-Modelled Items

Coastal flooding associated with storm surge and inland flooding resulting from heavy rainfall are the two most important secondary perils associated with TCs, representing 65–70% of the $35 billion losses in Sandy (Swiss Re, 2013); these are increasingly explicitly included in catastrophe models (e.g. RMS, 2013). Storm surge can drive losses for some storms and is included in most US models and increasingly in the Asia Pacific region. The most sophisticated models use full life cycle hydrodynamic modelling to estimate surge heights, and these are included as a flood footprint linked to the windspeed, one for each TC, evaluated with the model's flooding vulnerability curves. Modelling storm surge based on a hurricane's entire life cyle is important as large surges can be developed while the hurricane is out at sea many days prior to landfall, leading to surges that are disproportionately large for the hurricane's landfall intensity (e.g. Hurricane Sandy; Forbes *et al.*, 2014). For this reason, storm surge was removed from the Saffir-Simpson scale in 2009.

Tornadoes, hail, and lightning are not explicitly modelled as part of TCs; separate models may exist for these (sub-)perils. Damage from falling trees is not yet generally explicitly modelled. Neither is downwind debris from failed buildings; however, these are implicitly included through their impact on the basic damage metric and its damage curves are calibrated to real losses which are not granular enough to isolate and remove these complications. Debris impact can also be captured through model secondary modifiers such as impact on building cladding and openings. Contamination and pollution are considered to be a nonmaterial fraction of losses. In terms of insurance policy coverages or lines of business, loss adjustment expenses, interruption to trade or **contingent business interruption (CBI)** and marine cargo are generally not currently modelled.

### 3.2.6   Key Past Events

Historical events are used by model developers and users to justify the need for, and to validate models, cross-check between them, and identify factors that should be included. Differences in modelled industry loss estimates, such as $10 billion to $25 billion for Hurricane Sandy (AIR, 2012; EQE, 2012; RMS, 2013) result from differing assumptions, methodologies, and non-modelled sources of loss.

Looking at the Americas and Caribbean, in 1992, the huge losses ($25 billion, $15.5 billion of which was insured) from Hurricane Andrew contributed to the insolvency of some insurers

(Lecomte and Gahagan, 1998), demonstrating the need for catastrophe modelling. Six US landfalls in the 2004 season (see also Chapter 2.11.1.2) illustrated clustering, with the four hurricanes affecting Florida causing increased prices due to a shortage of supply known as **demand surge** (Olsen and Porter, 2011). In 2004, Hurricane Ivan showed that the strongest hurricanes could form closer to the equator than previously thought, whereas Hurricane Juan making landfall in Canada as a category 2 in 2003 showed that Canada was not immune to strong hurricanes. Hurricane Katrina's storm surge and associated flooding re-emphasized this as a major contributor to loss of human life and property (Jonkman *et al.*, 2009; Rappaport, 2014) and the role of coastal defence vulnerability (Brunkard, Namulanda and Ratard, 2008). In 2008, Hurricane Ike re-intensified as an extra-tropical system as about 46% of North Atlantic tropical storms do (Hart and Evans, 2010), emphasizing that large related losses can occur far beyond coastal areas and for a long time after landfall. In 2012, Hurricane Sandy caused significant business interruption at least in the short term (Henry *et al.*, 2013) and unusually high insured damage to movable property (e.g. $2.7 billion for cars estimated by the Property Claim Services [PCS]).

In the Asia and Pacific regions, Cyclone Tracy in 1974 devastated Darwin and illustrated the mitigating impact on insured loss of measures such as building code updates and enforcement, and thus the importance of good quality exposure data as a model input. More recently, Typhoon Bart in 1998 provided the impetus for the first models of Japan. Extra-tropical transitioning cyclones, particularly series of cyclones, cause significant flooding. For example, Cyclone Wanda in 1974 flooded Brisbane, Australia; floods from Typhoon Fitow in 2013 devastated Wenzhou, China, and a succession of cyclones contributed to the catastrophic 2011 Thai floods (Takahashi *et al.*, 2014).

### 3.2.7 Open Questions/Current Hot Topics/Questions to Ask Your Vendor

Some open or academic models (e.g. Vickery, 2000; Powell *et al.*, 2005) are documented so as to make the work reproducible, but others may not be at this level of detail. Key current questions are:

- Using all historical data to model next year's risk may mask recent changes (e.g. in number, intensity). How are decadal fluctuations in TC hazard, that is, non-stationarity, accounted for (e.g. Holland and Webster, 2007; Bonazzi *et al.*, 2014)?
- What is the relationship used, if any, between basin and landfall rates (Lonfat, Boissonnade, and Muir-Wood, 2007; Hall and Jewson, 2008; Villarini, Vecchi and Smith, 2012)?
- Is clustering important for this basin? If so, how is it included (e.g. Mumby, Vitolo and Stephenson, 2011)?
- How is wind speed decay with progress inland parameterized (Kaplan and DeMaria, 1995; Vickery, 2005; Leith and Nolan, 2010), and are extra-tropical transitions considered (Loridan *et al.*, 2014)?
- How are wind footprints calibrated? What are the major historical events used for model calibration or validation and are sufficient windspeed, loss, and exposure data available for these (Holland, Belanger and Fritz, 2010)?
- How is the spatial and intensity distribution of TCs by region calibrated?
- How are topography and land-surface roughness accounted for (e.g. Zhu, 2008)?
- Are wind loss drivers other than the maximum speed considered (e.g. Jain, 2010; Czajkowski and Done, 2014) for example, is precipitation with associated riverine or flash flooding considered (Langousis and Veneziano, 2009; Grieser and Jewson, 2011)?
- How are storm surge effects quantified? What is the grid size used, the propagation methodology, and linkage to the wind field over time (Dinesh, 2012; Mori *et al.*, 2014)?
- How is the event set 'boiled down' from the initial simulation to manageable numbers?

Many testing questions are asked by Florida's Commission on Hurricane Loss Projection Methodology, with many technical details publicly available in 'Modeler Submissions' on their Web site (more detail in Chapter 2.11.5).

### 3.2.8  Nonproprietary Data Sources

Authorative hazard data sources of intensity and location for TCs include: HURDAT2 (Atlantic, Eastern, and Central Pacific Oceans; Jarvinen, Neumann, and Davis, 1984) extended to include TC size variables by Demuth, DeMaria, and Knaff (2006) and Knaff, Longmore, and Molenar (2014), and NOAA's International Best Track Archive for Climate Stewardship (IBTrACS; Knapp *et al.*, 2010) covering all basins. IBTrACS attempts to synthesize and quality control datasets from the Japanese Meteorological Agency, the Joint Typhoon Warning Centre, the Korean Meteorological Administration and others. Older historical records may contain artificial trends (Landsea *et al.*, 2006; Vecchi and Knutson, 2011), and change as events are modified or removed from a data set.

> Demuth: http://rammb.cira.colostate.edu/research/tropical_cyclones/tc_extended_best_track_dataset/
> HURDAT: www.nhc.noaa.gov/data/#hurdat/
> IBTrACS: https://www.ncdc.noaa.gov/ibtracs/

Detailed data for calculating vulnerability, that is, contemporaneous loss and exposure data, are normally proprietary and not publicly available. Typically, vulnerability curves based on engineering principles, research studies, or available empirical data that vary by construction type are used (e.g. Unanwa *et al.*, 2000; Pinelli *et al.*, 2004; Li and Ellingwood, 2006). Historical loss/claims and exposure data can be difficult to obtain, but US insured losses are collated and archived by the PCS at the US state and event level. PCS: https://www5.iso.com/pcs/app/start.do.

### Acknowledgements

This section was developed from a publication by SCOR by Iakavos Barmpadimos (SCOR, 2013).

---

**Further Reading**  Aguado and Burt [2012] give a clear introductory explanation of the types of atmospheric hazard and the physical processes driving them, Holton and Hakim [2012] give a more technical guide, while Emanual [2005] gives a scientific and historical perspective on hurricanes.

---

## 3.3  Extra-Tropical Cyclones

*Len Shaffrey and Richard Dixon*

*Reviewer: Mark Dixon*

### 3.3.1  What Is the Peril?

**DESCRIPTION:** Mid-latitude cyclonic weather systems (i.e. 30–80°) are known as **extra-tropical cyclones** (ETCs). Although there is no widely accepted definition, ETCs that generate

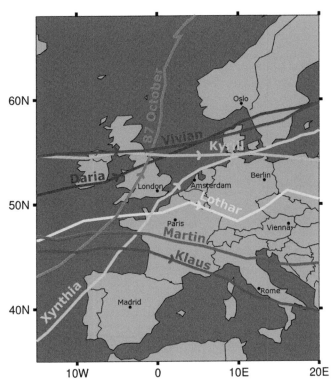

Figure 3.6 Paths of selected major European windstorms (1979–2013) from XWS windstorm database. Lines are tracks that are the trace of the location of the lowest pressure in the ETC through time. *Source:* Roberts *et al.* (2014).

strong surface winds (potentially up to 100 mph inland, up to 130 mph at coasts) are known as **windstorms** (WSs).

ETCs typically have radii of 1000–2500 km. ETCs develop over the western side of major oceanic basins, strengthening as they cross the ocean. They may or may not decay on making landfall on the eastern side of the basin. In particular, many of the WS that hit north-west Europe are still in the most damaging phase of their lifecycle as they pass over the major population centres 300–1,000 km inland (Figure 3.6). WSs are the major insurance peril for Europe (Barredo, 2010), but WSs also cause damage in other areas of high population density such as the northeast and Pacific Coastline of the United States and Japan (Figure 3.7). Storms are more prevalent in and near winter (e.g. Oct.–Mar. for the United Kingdom; Lamb, 1991). This said, intense WSs can also occur in other months, for example, the windstorm in August 1979 which struck during the Fastnet yacht race around the United Kingdom (Lamb, 1991).

ETCs and WS are generally more asymmetric in structure than TCs (see Chapter 3.2, Figure 3.4), containing several regions where damaging winds may develop. At the surface, winds rotate cyclonically around a central low pressure (see Figure 3.8). The flow of warm, moist air ahead of the cyclone is known as the **warm conveyor belt (WCB)**, while the flow of cold, dry air into and around the back of the central low pressure is the known as the **cold conveyor belt (CCB)**. Strong winds arise from the intensification of both belts. The **propagation** (or **translation**) **speed** with which the ETC travels enhances winds on the equatorward side of the ETC and diminishes winds on the poleward side, particularly where the CCB wraps around the central low pressure. Smaller-scale embedded features such as Sting Jets and embedded convection can also lead to strong winds. Sting Jets tend to form in air descending from above the CCB (Browning, 2004).

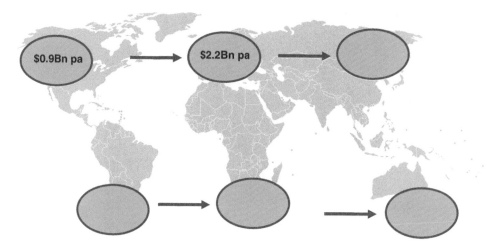

Figure 3.7 Regions of substantial windstorm hazard. Numbers refer to one estimate of annual expected losses in the North America and Europe, which are the regions driving insurance losses.

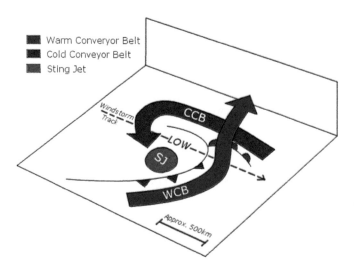

Figure 3.8 Illustration of where strong winds develop in a windstorm, namely the warm and cold conveyor belts (WCB and CCB) and the Sting Jet (SJ) region. In the Northern Hemisphere, cyclonic air flow is anticlockwise, as shown, while it is clockwise in the Southern Hemisphere. *Source:* Adapted from Browning (1997).

**PHYSICAL DRIVERS:** ETCs derive their energy from two primary sources. The first source is the potential energy associated with the equator-to-pole temperature gradient, and the second is from the release of latent energy through moisture condensing (e.g. leading to rainfall) within the storm itself. Importantly, this means they do not necessarily decay and, often even intensify, after landfall (e.g. Lothar in 1999; WillisRe, 2015). ETCs tend to form, a process known as **cyclogenesis**, where an equator-to-pole atmospheric temperature gradient is strong and moisture is available. Cyclogenesis therefore tends to occur on the western side of the North Atlantic and North Pacific oceans where ocean currents (the Gulf Stream and Kurishio) lead to strong atmospheric temperature gradients (Hoskins and Valdes, 1990). ETC growth can be enhanced by **seeding**, that is the presence of existing meteorological features that can extract energy from the environment and develop into fully-fledged ETCs. These features can be (i) features in the upper atmosphere from previous ETCs (Pettersen and Smebye, 1971); (ii) disturbances along trailing cold fronts (Renfrew, Thorpe and Bishop, 1997), or (iii) the extra-tropical transition of TCs that have moved from lower latitudes and become entrained in the mid-latitude jet stream (Jones *et al.*, 2003).

**KEY BEHAVIOURAL SUBTLETIES:** The frequency and location of ETCs and WSs can be modulated by large-scale modes of climate variability (e.g. Hoerling and Ting, 1994; Rogers, 1997). In the North Atlantic, the primary mode is the North Atlantic Oscillation (NAO), which is characterized by changes in the surface pressure gradient between lower and higher latitudes over the North Atlantic (Hurrell, 1995). When the pressure gradient is stronger than usual, the North Atlantic jet stream (Holton, 2004) steers more ETCs into Western Europe. ETCs and WSs also tend to exhibit **clustering** in space in space and time (aka 'seriality'), which may increase with their intensity (Mailier *et al.*, 2006; Vitolo *et al.*, 2009). For example, the windstorms Anatol, Martin, and Lothar all struck Europe during December 1999 (Munich Re, 2002). Clustered WSs have greater socioeconomic impact through the failure of already weakened or damaged infrastructure and processes such as increased prices because of a shortage of supply known as **demand surge** (see Chapters 2.9.2 and 5.4.2). Clustering of ETCs may arise from processes such as steering by atmospheric modes (e.g. the NAO) (Vitolo *et al.*, 2009) or from Rossby Waves breaking on both sides of the jet stream (Gómara *et al.*, 2014).

One key question is whether there are long-term trends in ETCs and WSs. Studies of long-term observations from weather stations (Alexandersson *et al.*, 1998; Wang *et al.*, 2011) found large decadal variability in European storminess, with a maximum in activity in the late nineteenth century, a comparative lull during the 1960s, and an increase in activity in the 1990s after which it has been relatively quiet again. Climate model projections suggest that the number of ETCs will decrease across the North Hemisphere by the end of the twenty-first century (Lambert and Fyfe, 2006). Regionally there might be increases, for example, Zappa *et al.*, (2013) found climate models project a 5–10% increase in the number of wintertime ETCs over North Western Europe by the end of the twenty-first century.

### 3.3.2 Damage Caused by the Peril

European WSs on average lead to economic losses of about $4 billion per year at 2008 values (1970–2008; Barredo, 2010), with individual events costing greater than €10 billion (i.e. Daria) and yearly totals reaching €16 billion (i.e. four events in 1990; Munich Re, 2002). So, while windspeeds are relatively low, with the strongest WS only equating to category 2 on the Saffir-Simpson Hurricane scale, damage is significant. ETCs' ability to maintain their strength over land allows them to affect large areas of high-population density (e.g. in NW Europe) creating the potential for significant insurance losses, typically through many relatively small claims (e.g. <€5,000; Munich Re, 2002). Typical WS damage is to roofs, such as the loss of tiles, although trees falling on houses and cars is also common, damage being akin to Force 10 or 11 on the Beaufort Scale (Beaufort, 1805). In extremes, mobile homes can be severely damaged and weaker roofs can be torn off. As with TCs, damage is typically assumed (e.g. Schmidt, Kemfert and Hoppe, 2010) to be primarily driven by the maximum winds ($v$) and to increase with its cube (i.e. $v^3$), representing the power dissipated at the ground by the winds and engineering relationships between winds and damage (Nordhaus, 2010). Some studies also point to duration of wind having an impact on resultant damage (Jain, 2010).

Again like TCs, WS can also cause coastal flooding through **storm surge**, where the combined force of the wind and the low pressure acts to lift and push the water onshore (Lowe and Gregory, 2005). These are severe when concurrent with high tides. The 1953 North Sea floods are regarded as a benchmark, with likely losses for a repeat in the present day estimated at £5.5 billion insured (RMS, 2003) and £3.3 billion economic (AIR, 2013). However, as demonstrated in the North Sea storm surge event of December, 2013 (Sibley, Cox and Titley, 2015) improved sea defences can protect effectively. WSs may also generate heavy precipitation (e.g. up to 100 mm in single events) and flash flooding or flooding in river catchments with short response timescales (e.g. Lavers, 2011). Typically, WSs are dry compared to the average 340 mm dropped by a TC (Roth, 2015), and move fast over any given site, so flooding is typically neither great nor widespread. However, multiple WSs over the course of a season (e.g. the winters of 2000–2001 and 2013–2014) can lead to more widespread flooding in larger river catchments (e.g. Hillier, Macdonald, Leckebusch and Stavrinides, 2015; Kendon and McCarthy, 2015) due to saturated catchments and high water tables.

### 3.3.3 Forecasting Ability and Mitigation

Weather forecasts for ETCs have skill on timescales of a few hours to a few days (Inness and Dorling, 2013) with the large-scale circulation being more predictable than smaller embedded features. Improvements in observations, data assimilation, and models have all led to substantial increases in forecast skill over the past few decades. For example, the path and development of the Great October Storm of 1987 was poorly forecast, which resulted in little to no preparation from civil response agencies. In contrast, the path and winds from windstorm Christian (the St. Jude's Day storm) in October 2013 was well forecast a few days ahead. Early warnings of high winds are regularly used in Europe to mitigate risks from WSs. Examples include the closing of bridges to high-sided vehicles, emergency response preparation, and preparation for power outages.

There is very little skill in forecasting ETCs on seasonal timescales. Recently, however, there has been some progress in forecasting on seasonal timescales the large-scale modes of climate variability that modulate ETCs. For example, some seasonal forecast centres have reported modest but significant skill for forecasts of the wintertime NAO for lead times of three months ahead (Scaife *et al.*, 2014).

### 3.3.4 Representation in Industry Catastrophe Models

**HAZARD DESCRIPTION and EVENT SETS:** The occurrence of a WS is known as an **event**, and catastrophe models calculate losses from **event sets** intended to simulate thousands of years of events (see Chapter 4.3.1). However, how these event sets are produced can vary considerably. Of the two most contemporary methods, the first (also see Chapter 4.3.6.5) uses several thousands of years of WSs simulated in a Global Circulation Model (GCM), which are then statistically 'down-scaled' (e.g. Della-Marta *et al.*, 2010; Haas and Pinto, 2012). From this, wind data are extracted and applied to high-resolution surface roughness data to generate wind footprints for each event, accounting for the known biases in GCMs. In the second approach (e.g. Keller, Dailey and Fischer, 2004), historical events are identified within a re-analysis product (e.g. NCAR-NCEP; Kistler *et al.*, 2001), which coherently synthesizes diverse weather observations for 40 or 50 years. Then, as in weather forecasting, initial conditions for each historical event are used as a seed in a regional numerical weather prediction model. Randomness (e.g. in how surface pressure changes) is used to perturb the events as they grow, creating a large set of events that could have happened. This 'ensemble' methodology (e.g. Svillo, Ahlquist and Toth 1997; Berliner, 2001) is limited by the assumption that the initial conditions for historical events well sample those of all scenarios. Both these approaches are argued (e.g. Keller, Dailey and Fischer, 2004) to improve on the time-stepping windfield models used for TCs that link wind fields to parameters such as the central pressure of the storm (see Chapter 3.2), a method less suitable for the more complex wind fields of ETCs. Tree damage in windstorm Gundrun in 2005 illustrates that correctly introducing smaller-scale spatial variations into hazard footprints (e.g. using roughness, land cover, wind profile, topography) is non-trivial (Chapter 4.3.3.3). Return periods and statistical uncertainty should be considered robustly (e.g., Della-Marta and Pinto, 2009; Della-Marta *et al.*, 2008).

**DAMAGE METRIC AND VULNERABILITY:** ETC damage is modelled as a **damage ratio** (DR), the ratio of the insured loss to the rebuild cost of the property. Damage ratios are typically a function of wind speeds, with 3-second gust speed the most frequently used hazard metric; these **vulnerability curves** are underpinned by engineering studies (e.g. Cook, 1985), but loss data have been used, largely since the storms of 1999, to help tune the curves even if loss data for winds > 40 m/s are rare. At the relatively low wind speeds that are common in WSs, there is typically a less strong correlation between wind speed and damage ratio. This is because damage tends to be fairly similar (e.g. minor roof damage) across this wind speed range, which is a problem for vulnerability curves. So, the chance of a claim being filed, or **claims ratio**, is increasingly being used to better describe this damage. Some models also use damage relationships that include duration of the storm. For the secondary peril of storm surge, the depth of water is linked to damage to the property. Given the relatively lower windspeeds compared to TCs, and the drier nature of ETCs, water damage from rainfall ingress is not so much of an issue.

**BEHAVIOURAL SUBTLETIES INCLUDED:** Since the mid-2000s, modelling companies have sought to simulate the clustering that causes multiple-event years such as 1990 and 1999 (Karreman *et al.*, 2014). Companies take different approaches to simulating clustering. One approach includes using a statistical 'block bootstrapping' approach that captures the intra-seasonal variability including the clustering behaviour (e.g. AIR, 2010). Other methods take the clustering from a GCM, or vary its strength according to a typology of storms. Insurers with loss data that date back 10–20 years have found that their reported losses typically have not matched the level of loss suggested by WS models. These models have been based on broad-scale WS activity since, typically, the mid-twentieth century. The suggestion is that WS activity, which is modulated by the NAO (see previous discussion), has waned since the mid-to-late 1990s. Thus, shorter-term perspectives conditioned upon a subset of data to represent the current state of this oscillation are included in some models.

### 3.3.5 Secondary Perils and Non-Modelled Items

Storm surge is usually modelled, but only for part of the length of European coastlines. Most often coverage is down the North Sea and coastlines where the potential (and historical precedent) are greatest. Recent events, however, such as Xynthia's flooding in the Bay of Biscay in 2010 (Guy Carpenter, 2010; Kolen, Slomp and Jonkman, 2012) have highlighted the peril is not limited to the North Sea. Insurance coverages for flood vary significantly by country, and thus while the peril may exist, it may not be of concern from an insured loss perspective.

The suspected correlation between stormy winters, wind damage and flooding (e.g. Hillier *et al.*, 2015), illustrated by NW Europe in 2013–2014 (e.g. Kendon and McCarthy, 2015; Matthews *et al.*, 2015; Huntingford *et al.*, 2015), is currently not in models. Modellers, however, are working towards this as part of a broader solution where multiple perils are simulated concurrently. In areas where WSs co-exist with cold air (e.g. US interior, Japan), heavy snowfall and freezing rain compound the threat, leading to collapsed roofs and power outages, respectively, and these effects are modelled.

### 3.3.6 Key Past Events

Historical events are used by model developers and users to justify the need for and to validate models. Differences in modelled industry loss estimates result from differing assumptions, methodologies, and non-modelled sources of loss. Estimates are €7.2–7.8 billion for Lothar 'as if' re-run in 2014 for Germany and France (Willis Re, 2015), and €7–11 billion 'as if' in 2015 (Waisman, 2015).

Although Daria on 25 January 1990 (e.g. McCallum, 1990; Munich Re, 2001) pre-dates Europe WS catastrophe models, it is a benchmark for European WSs. It is the largest WS loss ($8.5 billion in 2014 values; Swiss Re, 2015), of four >€1 billion events storms that made the winter of 1990–1991 the most expensive ever for WSs (Munich Re, 2001). Strong winds of >80 mph were felt across a wide area including Ireland, the United Kingdom, the Netherlands, France, and Germany. Its passage across the United Kingdom was during daylight hours and as a result there were close to 100 fatalities.

Furthermore, re-visiting this winter has led to catastrophe models considering the impact of multiple damaging events in one season (see Chapter 2.11.1.2). Lothar on 26 December 1999 (e.g. Wernli, 2002; Munich Re, 2001), with 'as-if' 2014 losses of ~$8.2 billion (Swiss Re, 2015), reinforced this message by being one of three significantly damaging storms in December 1999. Its strongest winds directly hit the Paris region, with some winds as strong as 90 mph even in central Paris (RMS, 2000). Martin followed 36 hours later (€5 billion damage), highlighting issues with a standard reinsurance clause that treats losses occurring within a time-span of 72 hours as a single event (AIR, 2009; see Chapter 2.4.2).

Other events are notable locally. Anatol in December 1999 is thought to have been a 1-in-100-year event for Denmark (Ulbrich *et al.*, 2001). Similarly, the October Storm (sometimes called *87J*) that crossed SE England was damaging for this highly populous area with some suggestions it was a 1-in-250-year event for this region of the United Kingdom (Burt and Mansfield, 1988). The 'Nyttarsdag' storm in 1992 was similarly a 200-year event for Norway (Meteorologisk Institutt, 2008). In the United States, the 'Superstorm' of 1993 (NOAA, 2015) highlighted the potential for multi-peril (wind, snowfall, ice) damage when it impacted much of the Eastern Seaboard of the Unites States. It was responsible for 318 deaths, and based on industry exposure, a repeat of such an event in 2008 would have given around $6.5 billion insured losses (RMS, 2008).

### 3.3.7  Open Questions/Current Hot Topics/Questions to Ask Your Vendor

Here are some questions worth asking of vendors either on existing hot topics or questions that will help inform risk-based decisions on how complete the model is:

- What are the historical datasets that inform the hazard event set?
- Does the model contain a shorter-term view on the current low levels of windstorm activity in the twenty-first century? If not, are there reasons or beliefs for it not being included?
- How does the model incorporate the clustering of multiple events within seasons (Khare *et al.*, 2015)?
- Is there a relationship between intensity with clustering in your model (Vitolo *et al.*, 2009; Hunter, 2014)?
- Which events in which territories have you used to calibrate the model? And, what loss data do you use? Also, across what range of the wind speeds, and for territories, do you have loss data to constrain the model?
- How does loss data availability vary between territories and across the range of plausible wind speeds?
- How is land-surface roughness accounted for?
- At low wind speeds where damage is typically scattered, does your model consider the importance of claims frequency as well as loss ratios?

### 3.3.8 Nonproprietary Data Sources

Data sources of historical windstorms and ETCs include:

- Lamb (1991) is a classic book describing major historical ETCs over North Western Europe since ~1570.
- The XWS extreme windstorm catalogue (Roberts *et al.*, 2014) is a freely available catalogue of European WSs between 1979 and 2013. The XWS catalogue contains cyclones tracks from the ERA-Interim reanalysis and wind gust footprints from the UK Met Office (www .europeanwindstorms.org).
- The Extratropical STORM Atlas (Dacre *et al.*, 2012) is a freely available database of 200 cyclone tracks of intense North Atlantic storms from the ERA-Interim reanalysis (www.met .reading.ac.uk/~storms/).

Data sets of detailed (e.g. house-by-house) insurance losses are not publicly available. Aggregated data are available from many national insurance associations (e.g. Association of British Insurers), but formats vary.

> **Further Reading** Lamb (1991) gives a comprehensive overview of the major historical WSs that have affected Europe since 1570. McIlveen (2010) provides a clear description of the structure and dynamics of extratropical cyclones. Innes and Dorling (2013) cover the basics of weather forecasting for mid-latitude storms.

## 3.4 Severe Convective Storms

*Michael Kunz and Peter Geissbuehler*

*Reviewer: Juergen Grieser*

### 3.4.1 What Is the Peril?

**DESCRIPTION:** Of the 16 million thunderstorms estimated to occur each year around the globe those with winds stronger than 90 km/h, or hail larger than 2.5 cm in diameter, or tornadoes, are usually termed **severe convective storms (SCS)** (NSSL, 2015). Basically, SCSs can occur wherever thunderstorms are observed (Figure 3.9). They have been reported on all continents except Antarctica (Fujita, 1973; Rauber, Walsh and Charlevoix, 2014) and mainly occur in each hemisphere's spring (e.g. May in the United States) and summer (e.g. June/July in Central Europe). Only in the Inter-Tropical Convergence Zone (ITCZ), which roughly coincides with the broad Trade Wind current near the equator, are they year-round events. In the United States, convective events occur most frequently east of the Rockies. In Europe, SCSs occur most often south of 50° N and substantially decrease to the east (Punge *et al.*, 2014). Hot spots are identified over north-eastern Spain, parts of France, southern Germany, Switzerland, Austria, and northern Italy (Dotzek *et al.*, 2003; Bissolli *et al.*, 2007; Hand and Cappelluti, 2010). In Asia, SCS occur most frequently in east China, South-east Asia, India, and Bangladesh. The majority of Southern Hemisphere tornadoes and hail events are reported in Australia (mainly New South Wales; Schuster, Blong and Speer, 2005; Schuster *et al.*, 2005), South Africa, and Argentina/

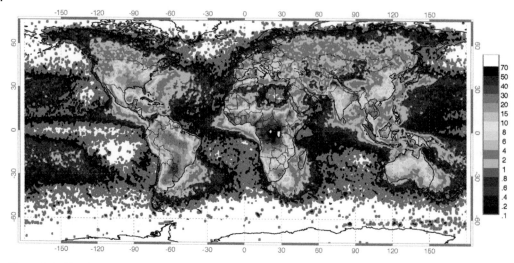

**Figure 3.9** Thunderstorm locations, as indicated by lightning flash rates (strikes per km$^2$ and year) observed by the spaceborne Optical Transient Detector (OTD) and Lightning Imaging Sensor (LIS) instruments, combined to cover 1995–2010 (cf. Cecil, Buechler and Blakeslee 2014). *Source:* Image obtained from http://thunder.nsstc .nasa.gov/data.

southern Brazil (Munich Re, 2011). Note, however, that records of SCS and hazardous weather events are fragmentary in many areas, making estimates of the global frequency incomplete.

SCSs cause various local-scale (i.e. <10 km) hazardous weather events (Markowski and Richardson, 2010). Rain can be intense (>100 mm/h), and large hailstones with diameters in excess of 5 cm may reach terminal fall velocities of 100 km/h or more (Knight and Heymsfield, 1983). Each **hailstreak**, the area affected by a hail event, covers only a few 10 s of km$^2$. In the United States, for instance, 80% of observed hailstreaks affected areas of less than 40 km$^2$ (Changnon, 1970, 1977). The size distribution for hailstones is fitted best by an exponential distribution (Sánchez *et al.*, 2009; Pruppacher and Klett, 2010).

Violently rotating columns of air in contact with the ground called **tornadoes** may form, often (but not always) visible as a funnel cloud (AMS, 2013). Tornadoes typically have widths between 50 and 1000 m (averaging 100 m), have lifetimes of only a few minutes, and have track lengths of several 100 meters up to a few 10 s kilometres. In very rare cases tornadoes can last up to 1 hour while propagating more than hundred kilometres (Rauber, Walsh and Charlevoix, 2014). Because direct wind speed measurements are usually not available, tornadoes are classified by damage patterns caused to various structures using the six-tier **Fujita scale** (F0 >64 km/h to F5 >419 km/h; Fujita, 1971). In the United States and in other countries, the Enhanced Fujita Scale (EF0 >105 km h$^{-1}$; EF5 >323 km h$^{-1}$) has been in use since 2007 (Potter, 2007). Tornado observations approximately follow a Weibull distribution with intensity so that in the United States around 85% are weak (EF0 + EF1) and only 1% are violent (EF4 + EF5; NOAA, 2015a).

Other severe convectively-induced ground-level winds are called **straight-line winds** to distinguish them from tornadoes. Straight-line winds are usually local phenomena (Wakimoto, 2001). In rare cases, however, they may extend over several hundred kilometres related to larger convective systems (e.g. Przybylinski, 1995; Atkins *et al.*, 2005). Straight-line winds in excess of 94 km/h that propagate over distances of more than 400 km are termed **derechos** (Johns and Hirt, 1987).

SCS form either as isolated cells or as larger complexes (Houze and Hobbs, 1982; Markowski and Richardson, 2010). The organizational form has important implications for the threat they

Table 3.5  Overview of thunderstorm forms, their characteristics, and damage potentials.

| Organizational form | Temporal scale | Spatial scale | Environmental conditions | Track direction | Hazardous weather events |
|---|---|---|---|---|---|
| Single-cell | 30 min–1 h | 1–10 km | Atmosphere slightly unstable, weak wind, weak wind shear ($<10\,\mathrm{m\,s^{-1}}$) | ~ With mean wind | Low risk |
| Multi-cell | >1 h | <50 km | Atmosphere highly unstable, wind shear, especially speed shear ($10–20\,\mathrm{m\,s^{-1}}$) | Combination wind/cell formation vectors | Heavy rain, small to medium hail, straight-line winds |
| MCS/MCC/Squall Line | 6 h–1 day | >100 km | Atmosphere highly unstable, strong wind shear, large-scale lifting | ~ Large-scale wind field | Flood-triggering rainfall, small to medium, rarely large hail, straight-line winds, weak tornadoes |
| Supercell | 1–8 h | ~50 km | Atmosphere highly unstable, directional wind shear ($>20\,\mathrm{m\,s^{-1}}$), large-scale lifting | Deviation to mean wind; difference between cyclonical/anticyclonical rotating mesocyclone | Heavy rain, large hail, severe straight-line winds, severe tornadoes |

pose (see Table 3.5). **Single-cell thunderstorms** are most common but not a major hazard. **Multi-cell thunderstorms** are clusters of single cells at different stages of their development that are dynamically interconnected, repeatedly spawning new cells (Chisholm and Renick, 1972). **Mesoscale convective systems (MCS)** are ensembles of thunderstorms with contiguous precipitation of 100 km in diameter in at least one direction (Houze *et al.*, 1989; Houze, 2004). Special types of an MCS are **squall lines (SL)**, which are narrow bands of convective cells extended in one direction, and **mesoscale convective complexes (MCC)**, horizontally extended over a large area (Maddox, 1980). **Supercell thunderstorms** (Figure 3.10) are the most dangerous convective storms, capable of producing the largest hailstones and the most violent tornadoes (EF3 and higher; Chisholm and Renick, 1972; Beatty, Rasmussen and Straka, 2008).

**PHYSICAL DRIVERS:** Latent heat released by condensation of rising air drives SCS, a process referred to as **deep moist convection** (Doswell, 2001; Bluestein, 2013). Warmed air is less dense than its surroundings, reinforcing the updraft (see Figure 3.10). Thus, prevailing instability and high moisture content at low levels are prerequisites for thunderstorm development (Groenemeijer and van Delden, 2007). To initiate convection, an additional **trigger** (e.g. low-level flow convergence, large-scale lifting by a low pressure system) is required to lift air parcels to a level where condensation can occur (Kottmeier *et al.*, 2008). **Wind shear**, how horizontal wind speed and direction vary with height (usually between the surface and 6 km), dictates the organizational form of the convective systems (Weisman and Klemp, 1982; see Table 3.5). If a thunderstorm is cut off from the inflow of warm and moist air, it begins to dissipate (e.g. Houze, 2014). The prerequisite of both latent heat and wind shear explains the

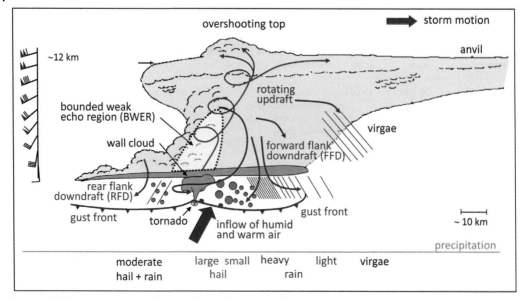

**Figure 3.10** Schematic vertical section through a supercell thunderstorm and characteristic wind profile. The bounded weak echo region (BWER), where hydrometeors remain small due to the high ascent speed, causes the typical hook echo form of supercells. The two downdrafts (FFD and RFD), caused by evaporation and sublimation of hydrometeors, produce the gust fronts near the surface, where wind speed is highest. More details can be found, for example, in Markowski and Richardson (2010) and Houze (2014).

highest probability of SCS on the continents in the mid-latitudes (equatorward of 60° N and S; Cecil and Blankenship, 2012).

**KEY BEHAVIOURAL SUBTLETIES:** Condensation in the updraft merges into small liquid water droplets, some of which freeze upon reaching the 0° C level if they have an insoluble aerosol nucleus (i.e. fine particle) included. Since ice-forming nuclei occur relatively sparsely, and water cannot freeze homogeneously (i.e. with no nucleus) until −38° C, a significant proportion of cloud droplets remain **supercooled** as a subzero liquid (Pruppacher and Klett, 2010). The few ice particles present in the convective cloud can grow rapidly by accretion as the supercooled droplets freeze onto them, forming first **graupel** particles and, from these, hail. Large hail (>2 cm) requires persistent (>30 min) conditions with a strong updraft and a high concentration of supercooled droplets.

Straight-line winds are driven by **downdrafts** of cold air that spread out horizontally upon reaching the ground, forming a **gust front** of high wind speeds bordering the SCS (Wakimoto, 1982). Downdrafts are induced by cooling associated with the evaporation of rain and sublimation of ice particles, that is, latent heat absorption (Wakimoto, Kessinger and Kingsmill, 1994). Severe downdrafts are referred to as **downbursts**, and small-sized events (< 4 km) as **microbursts** (Fujita, 1981). A bow-shaped line of convective cells, termed a **bow echo** (Fujita, 1981; Weisman and Rotunno, 2004), indicates the most severe straight-line gusts of an MCS and can span 20–200 km and last for 3–6 hours.

The most severe tornadoes are produced by supercells in environments with strong vertical wind shear, in particular directional wind shear (Lemon and Doswell, 1979; Davies-Jones *et al.*, 1990, 2001). Supercells have a highly organized structure (see Figure 3.10) with a single rotating updraft called a **mesocyclone** flanked by two downdrafts (forward [FFD] and rear flank downdraft [RFD]). Non-supercell tornadoes (i.e. landspouts or waterspouts) with weaker intensities ($\leq$ EF2) can arise from convergence of small-scale pre-existing vortices (Straka *et al.*, 2007).

Whether convective extremes have increased (in number or severity) due to climate change is debated (IPCC, 2012, 2013). Over recent decades, atmospheric instability has been increased due to an increase in low-level moisture (e.g. Kunz, Sander and Kottmeier, 2009; Brooks, 2013; Mohr and Kunz, 2013). In contrast, a number of studies found an increase in the variability of SCSs (e.g. Brooks and Dotzek, 2008; Kunkel *et al.*, 2009) and related phenomena such as hail (e.g. Xie, Zhang and Wang, 2008; Berthet, Dessens and Sanchez, 2011; Eccel *et al.*, 2012) or tornadoes (Diffenbaugh, Trapp and Brooks, 2008), but no overall trend. For future decades, the atmosphere may change towards environments that favour SCS (Leslie, Leplastrier and Buckley, 2008; Kapsch *et al.*, 2012; Sanderson *et al.*, 2014; Mohr, Kunz and Keuler, 2015), but there is low confidence in climate model projections of SCSs (Brooks, 2013).

### 3.4.2 Damaged Caused by the Peril

Compared to tropical cyclones, single SCS events cause modest losses to buildings, crops, and vehicles (i.e. <$15 billion, see Table 3.6). However, since SCSs occur frequently, total annual losses may be very high. In the United States, for example, 1,400 tornado, 6,200 hail, and 13,100 straight-line wind events are reported each year on average (2005–2014; NOAA, 2015b). Corresponding annualized economic losses collectively are estimated at $7.9 billion (1993–2013), similar to those from hurricanes. **Lightning** causing local or large-scale fires occurs most frequently (see Figure 3.9), and intense rain (>50 mm/h) can trigger floods in small river basins.

Hail causes damage because of the high kinetic energy of the solid hailstones (Gessler and Petty, 2013), roughly increasing with the fourth power of a hailstone's diameter. Thus, the largest proportion of damage is caused by the few storms spawning the largest hailstones. Damage increases if high horizontal wind speeds are present when walls and vertical structures also become exposed to hail (Schuster, Blong and McAneney, 2006). In the United States, hail (including lightning) causes ~70% of insured losses related to SCSs, mainly because these happen more frequently and affect larger areas than tornadoes (RMS, 2015). In several European regions, for example, south-west Germany, most of the insured damage to buildings is related to hail (Kunz and Puskeiler, 2010). Damage to buildings is thought to be mainly to roofs (40% of total) and shutters (26%), with windows (1.5%) and solar panels (1%) contributing little (Kantonale Gebäudeversicherungen, 2012), although the minimal contribution of solar panels has been questioned (e.g. Swiss Re, 2014).

Straight-line winds reaching peak gusts in excess of 180 km/h (Lemon, 1998) may become dangerous for two reasons: they induce strong wind shear, which is extremely dangerous for air traffic (Fujita and Caracena, 1977), and can in rare cases affect large areas, particularly in the case of a derecho. In the United States, around 10% of the insured damage is related to straight-line winds (RMS, 2015).

Because of their enormous (rotational) wind speed of up to 500 km/h, tornadoes are the most destructive meteorological phenomena and can destroy all but the best-built structures. As with other wind perils, the damage potential is assumed to increase with the cube of maximum wind speed (i.e. $v^3$; Dotzek *et al.*, 2005). However, even though the United States experiences more tornadoes than any other region in the world (e.g. Central Europe; Dotzek, 2003), they account only for ~20% of all SCS-related insured losses due to their localized nature. The few intense tornadoes (i.e. EF > 4) cause 70% of all fatalities (NOAA, 2015a).

**Table 3.6** Key past SCS events.

| Country | Date | Major loss driver, number of events | Total loss | Insured loss | Fatalities |
|---|---|---|---|---|---|
| USA[1,2,3] | 22–28 April 2011 | Tornadoes (428 EF0-EF3; 16 EF4+5) | 15 | 7.3 | 354 |
| USA[1,2,3] | 20–27 May 2011 | Tornadoes (121 EF0-EF3; 4 EF4+5) | 14 | 6.9 | 178 |
| USA[2,3] | 02–11 May 2003 | Tornado, Hail (various) | | 3.2 | 45 |
| USA[5] | 10 April 2001 | Hail (single supercell) | >2 | 1.4 | |
| USA, Joplin, Missouri[7,8] | 22 May 2011 | Tornado (single EF5 multiple-vortex) | | 2.4 | 158 |
| USA, Tri-State (MO, IL, IN)[8] | 18 March 1925 | Tornado (single F5) | | | 695 |
| France, Belgium, Germany[1] | 9–10 June 2014 | Hail, gust (different supercells) | 3.5 | 2.8 | |
| Germany[1] | 27–28 July 2013 | Hail (two supercells) | 4.95 | 3.85 | – |
| Germany Munich[1] | 12 July 1984 | Hail (single supercell) | 0.95 | 0.45 | |
| Australia, Sydney[1,4] | 14 April 1999 | Hail (single supercell) | 1.48 | 1.1 | 1 |
| Australia, Brisbane[6] | 27 Nov 2014 | Hail (single supercell) | | >0.87 | 0 |

Note: All losses in billion US$ as at date incurred.
*Sources:*
[1]Munich Re NatCatSERVICE; [2]Swiss Re *Sigma* Reports; [3]Insurance Information Institute; [4]Schuster *et al.* (2005); [5]Glass and Britt (2002); [6]ABC (2015); [7]propertycausalty360.com; and [8]NOAA.

### 3.4.3 Forecasting Ability and Mitigation

Neither hail nor tornadoes can be reliably forecast by numerical weather prediction (NWP) models, mainly because of (i) large uncertainties in the initial conditions; (ii) low spatial (e.g. $2 \times 2 \, \text{km}^2$) resolution of regional NWP; and (iii) the need for parameterization of subgrid-scale processes (Markowski and Richardson, 2010). NWP models struggle to predict SCSs 1–3 days ahead, but can predict environmental conditions favouring SCS formation (Romero, Gaya and Doswell, 2007; Barthlott *et al.*, 2010; Johnson and Sugdon, 2014). Short-range prediction (1–12 hours) of larger convective systems by NWP models that assimilate radar or satellite data is now reasonably skilful (Brown *et al.*, 2012; Martinet *et al.*, 2014). 'Nowcasting' methods for short lead times (0–2 hours) have a high prediction skill, but only for cases where the convective systems have already developed (Lakshmanan *et al.*, 2007). In several countries, severe thunderstorm or tornado warnings are issued by the National Weather Services when predicted conditions favour SCS/tornado formation. Severe thunderstorm/tornado watches are issued by the local storm prediction centres when the weather events have been observed either by trained spotters or indicated by radar (Doswell, 2005).

Hail suppression programmes are conducted in several regions of the world, using aircraft or ground-based systems to seed cumulonimbus clouds (Browning and Foote, 1976; Smith, Johnson and Priegnitz, 1997; Khain *et al.*, 2011). In theory, a potential hail cell has to be seeded with an appropriate number of aerosols (e.g. silver-iodide) before hail is produced. However, without good predictive power to say that hail would have occurred, it is not possible to measure success or failure.

Property can be protected against damaging hail (e.g. use of hail-resistant construction materials for buildings, hail-protection nets to protect crops and fruits). Indeed, a number of official building codes in countries such as the United States require mitigation against this element of SCS (e.g. Crenshaw and Koontz, 2001). For tornadoes, protection is more difficult and more expensive. Damage to motor vehicles is theoretically readily avoidable because they are mobile; but in practice car owners cannot yet be warned sufficiently in advance.

### 3.4.4 Representation in Industry Catastrophe Models

**HAZARD DESCRIPTION AND EVENT SET:** SCS models cover only a few regions, but for the United States, Canada, and Australia they incorporate tornadoes, hailstorms, and straight-line winds, for which losses are calculated separately. In Europe, a limited number of SCS models currently exist (i.e. RMS, 2007; Punge *et al.*, 2014) and for hail alone. Flash floods and lightning are not considered explicitly, but may be included implicitly depending upon how the model is calibrated (e.g. loss data in vulnerability curves, total losses for past events). As an SCS event may occur spatially and temporally independently and rapidly, high spatial ($1 \times 1$ km$^2$) and temporal resolution (15 min) of the data is desirable.

The occurrence of an SCS system defines an **event**. Ensembles of hazard footprints (~10,000) or **event sets** (see Chapter 4.3.1) are generated stochastically from a database of historical occurrences. A caution is that the historical record of SCS incidents is typically very short and incomplete (i.e. ~ 10 years for Europe and < 30 years for the United States). Therefore, these observations are usually supplemented by remotely sensed proxies such as radar reflectivity (RMS, 2007), lightning (Guy Carpenter, 2014), or overshooting cloud tops from satellite (Punge *et al.*, 2014; see Figure 3.9). Some models use NWP outputs to overcome large gaps in observational records (e.g. RMS, 2015). A complementary approach is to identify meso- to synoptic-scale environmental variables (e.g. in reanalysis products) that favour SCSs and to use these parameters as proxies (e.g. Huntrieser *et al.*, 1997; Brooks, Lee and Craven, 2003; Kunz, 2007; Allen, Tippett and Sobel, 2015; Mohr, Kunz and Keuler, 2015).

Stochastic events' footprints are typically created by resampling the historical SCS occurrences and their atmospheric conditions. The mean values of key parameters (i.e. duration, genesis location, shape, orientation, width, direction and distance of travel, storm intensity) are perturbed randomly according to spatially localized probability distributions, including correlations where appropriate (cf. Deepen, 2006; Otto, 2009; Punge *et al.*, 2014). Some cut-off thresholds are used (e.g. Punge, 2014). Then, for each event, hazard values for hail and straight-line winds are placed in each grid cell as specified by an appropriate probability distribution. Similarly, tornado streaks are added with an intensity distribution matching historical data. Typically, each subperil is assigned independently, meaning that hail, wind, and tornado will not necessarily all occur in the same location as part of the same event. Overall, wind speeds are usually adjusted for the local effects of roughness estimated from satellite data, and the number of storms per year is set to historical observed values.

**DAMAGE METRIC AND VULNERABILITY:** With insufficient scientific studies into the mechanisms linking each subperil of an SCS (e.g. hail, wind) and property or motor loss, vulnerability functions are currently mainly constructed using past insurance claims (Hohl, 2001; RMS, 2008; Punge *et al.*, 2014). State building design codes (International Code Council), technical reports (Marshall and Herzog, 2006) and analytical building simulation (Reinhold, Reinolds and Morrison, 2014) are also considered where available. In the United States, numerous functions have been estimated. In Europe, sparse and simple data (e.g. no lines of business details, no location) allow only very generic functions to be created; for instance, one for motor and one for property. Hail damage is parameterized as a function of the integrated kinetic energy of the stones experienced. Damage increases due to high horizontal wind speeds are not factored in. Tornado and straight-line winds are parameterized by maximum 3-second gust wind speeds. See Chapter 4.5.8 for a case study on hail vulnerability.

**BEHAVIOURAL SUBTLETIES INCLUDED:** Subtleties, or parts of them, are likely implicitly included by the use of historical record. None, however, are included explicitly, and nor is climate change.

### 3.4.5 Secondary Perils and Non-modelled Items

When SCS travel slowly across the ground, or when subsequent storm cells pass over the same area, lots of rainfall can accumulate and trigger flash floods (e.g. Akaeda, Reisner and Parsons, 1995). This, however, is not considered in SCS catastrophe models. Furthermore, neither possible amplification of losses due to interactions between simultaneously occurring extremes (e.g. hail in combination with heavy rain causing the wetting of a building) nor cascading effects (e.g. fire in the aftermath of a tornado) are considered in the models. Due to the limited extent, SCS events demand surge is expected to be minimal.

### 3.4.6 Key Past Events

The deadliest tornado in US history was the so-called Tri-State F5 tornado on 18 March 1925, which caused 695 fatalities on a track of 470 km. The costliest tornado to date was the Joplin, Missouri, EF5 tornado on 22 April, 2011, with insured losses estimated at $7.6.billion and 160 fatalities (see Table 3.6). Overall, in 2011 a record 1690 tornado outbreaks caused economic damage estimated at $28 billion (Swiss Re, 2012). Furthermore, two hailstorms at the end of July 2013 caused Germany's largest ever insured loss to natural hazards, estimated at $3.85 billion (Munich Re, 2014). In Australia, the most damaging event was the Sydney hailstorm in 1999 with a loss amounting to $1.1 billion (Schuster *et al.*, 2005). These severe hailstorms provided additional impetus for the insurance industry to develop commercially operational SCS models (e.g. Swiss Re, 2015). The Sydney hailstorm showed that the dynamics of traffic flow for the hail component of SCS models is key; if the hailstorm had occurred ±1 hour either side of the rush-hour, the losses would have been substantially smaller.

### 3.4.7 Open Questions/Current Hot Topics/Questions to Ask Your Vendor

The key challenge is to increase the quantity and quality of tornado, hail, and straight-line wind observations around the globe, to better constrain theory and models. This will likely be aided by exploiting new data collection opportunities (e.g. by crowd sourcing or smartphone apps) and new observing systems such as dual-polarization radars, which are able to distinguish among several precipitation classes (Straka, Zrnic and Ryzhkov, 2000). Key current questions are:

- Which subperils (i.e. hail, tornado, flood, lightning, straight-line wind) are included explicitly, and which implicitly?
- Building realistic stochastic events likely requires a very high spatial ($1 \times 1 \text{ km}^2$) and temporal resolution (15 min.), so what is used and how is this achieved?

- Vulnerability is a key uncertainty for SCS models. To what extent are the loss data used to create the vulnerability curves sufficient to distinguish properly between different sub-perils and coverage types (e.g. residential, commercial)? Is any knowledge from engineering incorporated?
- Given the small spatial and temporal scale of hailstorms, traffic dynamics (e.g. lunch break, diurnal commuting cycle) can lead to significant mismatches between modelled and reported losses (e.g. Australia in 1999). So, how might this be integrated into event sets?
- Exposure data quality for motor and buildings is typically very poor, usually with only the number of policies available, but neither details about the building type or type of cars nor of the geographical distribution of the portfolio. How might this best be accounted for?
- To what extent is the occurrence probability of SCSs determined by temporal clustering? Is there any relation to teleconnections?

### 3.4.8 Nonproprietary Data Sources

The best source of data including intensity, location and spatial extent of convective events is available for the United States, either from SPC/NOAA or from NSSL. Each summer, the student-run NSSL/CIMMS Severe Hazards Analysis and Verification Experiment (SHAVE) collects hail, wind damage, and flash flooding reports through phone survey. Public precipitation reports are collected within the frame of the meteorological Phenomena Identification Near the Ground (mPING) project. For Europe, the European Severe Weather Database (ESWD), operated by the European Severe Storms Laboratory (ESSL), collects and provides detailed and quality-controlled information on SCS-related events. Because the database and most of the spotters are located in Germany, so are most of the reports, too. There are some other databases providing information on SCS, which, however, are restricted to single countries/perils (e.g. KERAUNOS in France).

ESWD: http://www.eswd.eu/
KERAUNOS: http://www.keraunos.org
mPING: http://mping.nssl.noaa.gov/
NOAA SPC: http://www.spc.noaa.gov/gis/svrgis/
NSSL: http://www.nssl.noaa.gov/
SHAVE: http://www.nssl.noaa.gov/projects/shave/

Detailed information on loss, vulnerability, or exposure is not publicly available. The same applies to different types of vulnerability curves. Aggregated loss data separated into country, type of event or line of business are available via publications from insurance and reinsurance companies such as Munich Re or Swiss Re.

Munich Re: http://www.munichre.com/en/reinsurance/business/non-life/natcatservice
Swiss Re: http://www.swissre.com/sigma

---

**Further Reading** Markowski and Richardson (2010) and Houze (2014) give a detailed overview of the dynamics of clouds and the physical processes of thunderstorms and associated weather phenomena. Rauber and colleagues (2014) provides a good overview of weather-related extremes. Bluestein (2013) and Doswell (2001) focus on severe convective storms and tornadoes.

# HYDROLOGICAL PERILS (I.E. 'RAIN-DRIVEN')

## 3.5  Inland Flooding

*Jane Toothill and Rob Lamb*

*Reviewer: Edmund Penning-Rowsell*

### 3.5.1  What Is the Peril?

**DESCRIPTION: Flooding (FL)**, the accumulation of water on land where it is not usually present, is often classified by the dominant source of water (Figure 3.11), or the pathways water takes through natural **hydrological systems** (e.g. soils, aquifers, hillslopes, rivers), engineered **drainage systems** (e.g. pipes, floodways), and the built environment. There is, however, no definitive or 'correct' typology of flooding, with various definitions adopted in insurance or governance (Australian Government, 1984; European Parliament, 2007; Federal Emergency Management Agency [FEMA], 2015) and the technical literature (e.g. Merz and Blöschl, 2003; Barredo, 2007; Pender and Faulkner, 2010). Water escaping from river channels, over natural riverbanks or man-made defence systems, is known as river or **fluvial flooding**. **Surface water flooding** is water moving or ponding outside established flood plains, or yet to enter a watercourse. It is known as **shallow flooding** in FEMA flood zones (either **ponding** on low ground or **sheet flow** on slopes) and is often linked directly to intense rainfall (**pluvial** flooding). **Flash flooding** denotes events where the onset of flooding is rapid. Flooding caused by rising water tables, such that flow appears close to or above the ground, is **groundwater flooding**.

Labels given to flood events can be ambiguous because real hydrological systems are interconnected, therefore sources and pathways of flooding are not easily separated (Cabinet Office, 2008). The types of flooding discussed in this section are fundamentally driven by

**Figure 3.11**  Sources of inland flooding: (a) lake outburst; (b) volcanic glacier melt; (c) fluvial; (d) surface water; and (e) groundwater.

**precipitation** (e.g. rain, hail, snow), so the following are *not* covered: Elevated sea levels (**surge**) or waves can cause **coastal flooding**, driven by wind and atmospheric pressure changes (e.g. Lamb, 1991; Sections 3.2 and 3.3). **Volcanic flooding** results from heat causing glaciers to melt (Section 3.10). Earthquakes or landslides cause **tsunamis (TS)** (Section 3.9). **Lake outburst floods** occur when natural geomorphological features that act as dams (e.g. ice barriers, landslide deposits) collapse, and can be sudden and destructive (Costa and Schuster, 1988; Korup, 2002; Worni, Huggel and Stoffel, 2013). The failure of man-made assets designed to hold back water such as canals, reservoirs, and flood defences can cause or exacerbate flooding, and the associated risks can be considered using reliability analysis methods (US Army Corps of Engineers [USACE], 1999; Vrijling, 2001; Buijs *et al.*, 2009) within the context of broader-scale investment and planning decisions (Van Heerden, 2007).

Because water is essential for life, most habitable locations are at some risk of flooding (Fleming, 2002; Ward *et al.*, 2013), even if flooding has not previously occurred. There is a disproportionate human impact of river flooding in developing countries with high urbanization rates, with 15 countries including India, Bangladesh, and China, accounting for 80% of the population exposed to flooding worldwide (Luo *et al.*, 2015).

The **magnitude** or severity of flooding is measured ultimately in terms of its impacts (Figure 3.12). Spatial extent, water depth, velocity, duration, volume, and speed of onset may all be used to describe the magnitude of flooding hydrologically, as well as rainfall depth, which is accumulated over specified durations, and intensity, which is the amount per unit time (Chow, Maidment and Mays, 1988; Shaw *et al.*, 2010). There is no standardized index of severity in common use comparable to, say, the Beaufort (wind) and Richter (earthquake) scales (although see proposals by Feng and Luo, 2010). The area impacted by a flood is known as its **footprint** (see Figure 3.12).

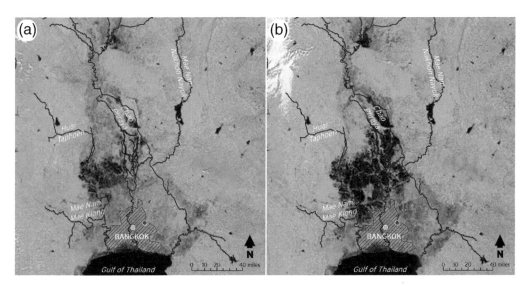

Figure 3.12 MODIS images illustrating the 2011 flood event impacting Bangkok, Thailand. (a) 13 November 2008, showing typical seasonal conditions and (b) on 8 November 2011, during flooding that lasted for approximately three weeks in the city. Visible and infrared light composite image with water shown in shades of electric to navy blue, vegetation in green, bare ground and urban areas in earth-tones, and clouds in pale blue-green. Yellow and red dots show locations of Rojana and Bang Chan, respectively. *Sources:* NASA images (Thailand_tmo_2008318_lrg.jpg, Thailand_tmo_2011312_lrg.jpg) by Jeff Schmaltz, LANCE/EOSDIS MODIS Rapid Response Team at NASA GSFC, courtesy of NASA's Earth Observatory. Images created using ArcGIS software by ESRI.

Flood **frequency analysis** refers to an assessment of the probability of a flood event, defined by a relevant measure of its magnitude such as peak flow rate (in $m^3s^{-1}$, also called **discharge**), and sometimes expressed as an equivalent long-run frequency of occurrence. An event that has a 1/T chance of occurring in any year is said to have a T-year **return period** (**RP**, also known as **recurrence interval**). Probabilities may be empirically determined from historically observed peak flows and analysed statistically using extreme value theory (e.g. Gumbel, 1941; Stedinger, Vogel and Foufoula-Georgiou, 1993; Coles, 2001). Other variables (e.g. water volume) can be assessed in this way. All vary spatially, potentially leading to an array of RP values for any event. Concurrent flooding at multiple locations, or from multiple sources, requires consideration of joint probabilities. This can be handled using computer modelling to simulate river flows for many realizations of possible weather events (Cameron *et al.*, 1999; Lamb, 2005; Sampson *et al.*, 2014). Alternatively, statistically, observables (e.g. rainfall or river flows) can be modelled using copula functions (De Waal, Van Gelder and Nel, 2007; Chen *et al.*, 2012) or conditional probability models (Lamb *et al.*, 2010). Probabilistic analysis of flooding is complicated by the different definitions of what constitutes a distinct inundation episode or flood **event**. Hydrological analysis can suggest definitions based on spatial-temporal dependence in river flows (Lamb *et al.*, 2010), cluster analysis (Eastoe and Tawn, 2012), or weather events (Stephenson, 2008). However, the criteria used to define events vary according to the application. In the insurance industry, an hours clause (Munich Re, 2005) that specifies a maximum time window (e.g. 168 hours) is used to identify discrete loss-generating events (see Chapter 2.4.2).

**PHYSICAL DRIVERS:** Inland flooding is driven by precipitation mediated by complex landscape processes (Hornberger, 1998; Beven, 2011). The overall fraction of precipitation that eventually drains into rivers at continental scales ranges between 0.16 and 0.43 (Pagano and Sorooshian, 2005). However, in a particular weather event, the flow in a river may be a much greater proportion of recent precipitation, and the relationship between flow and precipitation is often not linear because of factors such as

- how quickly the ground absorbs rainfall, and how much water it can hold (Ward and Robinson, 2000);
- snow accumulation and melting (DeWalle and Rango, 2008);
- water removed by transpiration (i.e. vegetation) and evaporation (i.e. temperature and wind), or **evapotranspiration** (Kabat *et al.*, 2004).

Thus, antecedent soil moisture conditions affect the potential for flooding; rain falling on saturated rather than dry ground typically produces more runoff.

Researchers are discovering how large-scale atmospheric circulation influences flooding (Allan and Soden, 2008; Woollings, 2010; Delgado, Merz and Apel, 2012). Warmer air can hold more moisture (Hartmann, 1994; Barry and Chorley, 2009), leading to robust findings of increases in water vapour of around 6–7% per Celsius degree of near surface global warming (Allan and Soden, 2008; Westra, Alexander and Zwiers, 2013). Many future projections therefore suggest changes in rainfall related to global warming, particularly increasing risks of shorter-duration, intense rainfall events (Alexander *et al.*, 2006; Trenberth *et al.*, 2007; Liu *et al.*, 2011; Jones *et al.*, 2013; Westra *et al.*, 2014).

**KEY BEHAVIOURAL SUBTLETIES:** The area draining into a river above any given point (i.e. the **catchment area**) is an important determinant of how precipitation translates into flood flows. Around the world, peak flows ($Q$) appear to scale with catchment area ($A$), according to a power law relationship $Q = cA^b$, with $b$ in the range −0.1 to −0.4 (Jothityangkoon and Sivapalan, 2001). This relationship appears to integrate and summarize information about complex rainfall→runoff→flood processes operating within different regions (Gupta and Dawdy, 1995; Robinson and Sivapalan, 1997; Gupta, 2004).

Runoff is also controlled by local variations in topography and soils, including artificial drainage systems (urban or agricultural). Watercourses draining urban areas tend to rise faster than in rural areas, especially for intense rain (e.g. Rose and Peters, 2001). However, despite knowledge that soil and vegetation processes affect runoff locally (e.g. Verstraeten and Poesen, 1999), relationships between land use and flooding at river basin scales remain complex and uncertain (Beven *et al.*, 2008; O'Connell *et al.*, 2007). Furthermore, the effects of engineered structures are difficult to predict, owing to the apparent randomness of phenomena such as blockages of bridges or culverts, and the erosion or collapse of flood defences.

### 3.5.2 Damage Caused by the Peril

Flooding has been estimated to account for around a third of economic losses caused by natural hazards globally, as well as being among the most frequent of natural disasters now, and potentially increasing in the future (United Nations International Strategy for Disaster Reduction [UNISDR], 2009; Munich Re, 2010; Winsemius, 2013; Contestabile, 2012).

Fatalities and injuries are common (Xia *et al.*, 2011). Jonkman (2005) reported more than 175,000 fatalities globally in 1816 inland floods from 1975 to 2001. Long-term effects on health are also well documented (e.g. Few and Matthies, 2006; Jonkman *et al.*, 2009; Di Mauro and De Bruijn, 2012).

Monetary damages occur through the inundation of properties, agricultural land, and infrastructure (power, transport, water supply systems); see Penning-Rowsell and colleagues (2013) and FEMA (2008) for damage appraisal methods. The primary hazard in flood impact assessments is usually water depth, but flow velocity and flood duration can also be important (Penning-Rowsell *et al.*, 2013). Contamination from sewerage or chemical discharges can cause significant harm (Veldhuis *et al.*, 2010; Arthur, Crow and Karikas, 2009). Sediments carried and deposited by flooding can also be damaging (Baker, Kochel and Patton, 1988).

Water-driven 'flood' damage within buildings can be caused by factors such as leaking windows and broken pipes or domestic appliances, which might not be distinctively labelled as such in insurance claims data records. Such damage is unlikely to be included in an industry flood model, and may or may not be covered by flood insurance. Thus, there is potential for a mismatch between the presence of damage, modelling, and the insurance cover in place (Figure 3.13).

### 3.5.3 Forecasting Ability and Mitigation

In the short term (days to weeks), it is often possible to **forecast** flooding (i.e. determine the location and timing of future inundation), but with decreasing accuracy (or **skill**, a formal measure) as the time frame increases. Peak flows can be predicted from (observed or modelled) weather data using **hydrological models** (Singh, 1995; Lamb, 2005; Boughton and Droop, 2003; Beven, 2011). These flows are often used as inputs to **hydraulic models** that calculate how water spreads out across a terrain in order to determine inundation extent or depth; see Environment Agency (2013) and Hunter and colleagues (2008) for reviews of the performance of hydraulic flood models.

**Ensemble forecasts**, defining the spectrum of what could happen, are probabilistic and based on hydrological modelling driven by multiple runs of a weather model, each perturbed slightly so as to capture uncertainty about the predictions (Schaake *et al.*, 2007). The skill of a forecasting system is typically assessed in terms of the balance between false alarms and the likelihood of making a correct prediction.

Longer-term (seasonal and beyond) uncertainties in weather, climate, hydrological systems, and infrastructure mean that prediction of flooding is typically framed as a risk analysis (Mascarenhas, 2005; Beven and Hall, 2014; Schumann, 2011), with possible future events

Figure 3.13 Types of damage after flooding: (a) Railway embankment eroded by river flooding in Tanzania, 1 Feb. 2012 (United States Army Africa, 2012); (b) Water contamination in Lat Karabang, Thailand, 2 Jan. 2012 (JBA Risk Management, 2012c); (c) Discarded flood-damaged furniture in Shoal Creek, TX, 25 May 2015 (Ploughmann, 2015); (d) Internal building damage: wrack marks inside Rojana industrial site, Thailand, 3 Jan. 2012 (JBA Risk Management, 2012b); (e) Delays and cancellations on Long Island Rail Road due to surface water flash floods, 13 August 2014 (Metropolitan Transportation Authority of the State of New York, 2014).

considered probabilistically or as scenarios of events that could happen over a given time scale. Insurance industry flood models are generally of this type.

Flood risk can be mitigated either by reducing the chance that flooding will occur, protecting individual assets if it does, or accepting it cannot be avoided and instead reducing the associated harm (Rijkswaterstaat, 2006; Michel-Kerjan, 2010; Pritchard *et al.*, 2014; Botzen, 2013). Broadly, this leads to four, non-exclusive, risk-mitigation strategies each with associated costs and benefits (e.g. Aerts *et al.*, 2013):

- flood defences (e.g. Mens, Klijn and Schielen, 2015);
- water flow management (e.g. deliberately releasing flood waters to protect towns downstream);

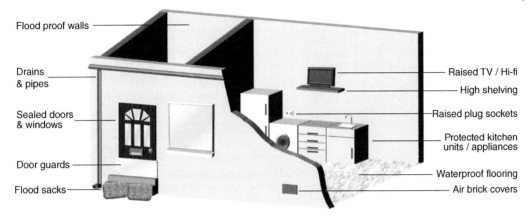

Flood proof walls

Drains & pipes

Sealed doors & windows

Door guards

Flood sacks

Raised TV / Hi-fi

High shelving

Raised plug sockets

Protected kitchen units / appliances

Waterproof flooring

Air brick covers

Figure 3.14 Property-level protection.

- property-level protection (e.g. Figure 3.14) or other local resilience measures;
- flood warnings (Alfieri *et al.*, 2013).

The impact of any residual risk may be spread through insurance where flood cover is available.

Generally, flood risk cannot economically be eliminated. In the United Kingdom, for example, the Environment Agency (2014) shows that continued investment in flood defences is beneficial, but even if all cost-effective investments are made, the residual risk is expected to increase eventually. Attention is turning also to smaller-scale protection for individual properties (e.g. see Figure 3.14), which appears to be economically justifiable even for relatively infrequent (e.g. 2% annual probability) flooding (Kreibich, Christenberger and Schwarze, 2011; Owusu, Wright and Arthur, 2015). Flood gates and temporary mobile flood barriers can offer more flexibility and may be easier to adapt to existing buildings and infrastructure, than larger capital schemes.

### 3.5.4 Representation in Industry Catastrophe Models

**HAZARD DESCRIPTION AND EVENT SETS:** Inland flood (i.e. river and surface water) catastrophe models provide stochastic **event sets** (see Chapter 4.3.1) that contain flood impact footprints for an ensemble of synthetic events, together intended to describe the full plausible range of future flood scenarios. Flood catastrophe models require a high level of precision; this is because flood intensity (e.g. depth) can vary dramatically in locations separated by just a few (e.g. < 5) metres in response to variations in the terrain. Pragmatically, therefore, computational limitations drive many of the decisions made during the production of flood models; see Chapter 4.3.6.4 for an illustrative case study. For instance, small watercourses that might ideally be included in a river component could be approximated as surface water flooding in a coarse-resolution model. As such, Digital Terrain Model (DTM) resolution and the minimum size of catchment modelled are useful measures of flood model quality. Computational limitations also drive a different approach to flood model development from that used for other hazards.

With unlimited computational resources, flood models might be developed by physically modelling each synthetic event individually as is illustrated with the grey arrow in Figure 3.15. That is, the runoff into rivers for each of a set of representative rainfall episodes would be modelled to estimate flow in the river, each quantified in terms of flow per point on the river network. Hydraulic modelling of each event would then enable flood severity at locations to be determined throughout the model domain for each event. This approach has been used in some precipitation-driven flood models associated with tropical cyclones (e.g. Lohmann and Yue,

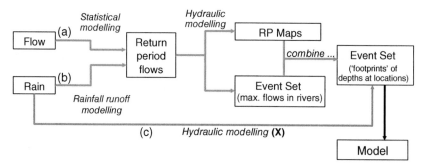

**Figure 3.15** A schematic of approaches to river flood model development. The first approach is based on either (a) river gauge data or (b) rainfall (orange pathways). Statistical or rainfall-runoff modelling is used to generate flows (cumecs) for all points in the river system corresponding to a specified set of RPs. Maps are created, using hydraulic modelling, by assuming flow of a fixed RP every point on the river network for each map. Separately, rain events that will lead to flooding are defined in terms of their RP flows in rivers. Then, these are combined into an event set of footprints of flood depth for each event (see text). The second approach is direct physical modelling (c, grey pathway), which remains highly computationally challenging.

2011). However, today's highest resolution models incorporate hydraulic modelling at ~5 m horizontal resolution; this requires significant computing power and is only just becoming possible in a realistic timeframe at national scale and for event set of tens or hundreds of thousands of events; this difficulty is marked **X** in Figure 3.15.

One way to circumvent this issue is to use hydrological modelling to estimate RP-flow relationships across the river network based upon either rainfall or flow gauge observations following the orange pathways (a) and (b) in Figure 3.15 (e.g. AIR Worldwide, 2013; Haseldine *et al.*, 2014). From these return period flows, two separate components are generated, which are later combined to form the event set of flood footprints. The first component is a series of floodplain maps that describe the extent of flooding associated with flows of a specific return period (e.g. 100 years) occurring at every point along the river network; up to ~10 maps may be generated using hydraulic modelling for RPs of between, 20 and 1000 years. The second component is a synthetic set of maximum river flows, corresponding to discrete rainfall episodes, defined in terms of return period, at points along the river network. In essence, for each event (rain→flow in river), these two components are then linked together to create the necessary map of flood severity per location as follows: first, for each location $(x, y)$ look up the river flow RP experienced nearby, perhaps at the closest location in the river; second, for the floodplain map of the appropriate RP, look up the flood severity (e.g. water depth). This is then repeated for all points $(x, y)$, interpolating between values at the point on two floodplain maps for intermediate RP flows. This approach allows the best data to be used to estimate RP flows, which may vary by territory. For modelling surface water flooding, an analogous method is used but using observed rainfall severity at a grid of locations across the modelled area in place of flow at points in the rivers.

Most event sets include the variability in the number of events to occur within a year. More advanced event sets will also consider the spatial clustering and seasonality of events, perhaps encoded by assigning a day of occurrence to each event.

To limit uncertainties, where good quality river gauge data exist, return period flows along a river's course can be estimated directly using gauge data (few locations only) and information on river catchment characteristics, such as area, geology, and annual rainfall. Indeed, this is considered good practice where suitable data exist (e.g. Castellarin *et al.*, 2012; Rosbjerg *et al.*, 2013; Smith *et al.*, 2015; Beck, de Roo and van Dijk, 2015; SMHI, 2015; Faulkner, Warren and

Burn, 2016), and there are officially sanctioned approaches for such flood frequency analysis (see Section 3.4.1) in the United Kingdom (Institute of Hydrology, 1999; Environment Agency, 2008), the United States (US Department of Agriculture, 1986), Australia (Engineers Australia, 2015), Canada (Alberta Transportation, 2014), and Ireland (Murphy *et al.*, 2014). Alternatively, rainfall data can be converted into river flows by physically modelling how water runs of the land in **rainfall-runoff models** (Beven, 2011). Because these complex models can potentially introduce large uncertainties, rain-derived flows are typically calibrated against river gauge data. A rainfall-based approach is likely to be preferred in areas where few or no river gauge data exist, or artificial influences are important (including urbanization).

Like river flow data, observed rainfall data records are often too short to capture very rare events. A number of approaches can assist in creating long 'records'. Statistical rainfall generators (e.g., Kilsby *et al.*, 2007) can make a reproducible rainfall time series for an arbitrarily long time period that should include plausible extremely severe events. Because these extrapolate outside the range of their observed calibration data, however, the caveat is that they will depend upon the robustness of the statistical assumptions in the model (Beven, 2011; Onof *et al.*, 2000). An alternative means of extrapolation is to incorporate the physical principles that govern the Earth's meteorological system, simplified to a series of equations in a **Global Climate Model** (**GCM**) (e.g. Arakawa and Lamb, 1977; McGuffie and Henderson-Sellers, 2005) or a **Numerical Weather Prediction model (NWP)** (Coiffier, 2011). GCMs can run globally, simulating many years but have a low resolution (e.g. hourly time resolution at approximately $67 \times 54$ km for HadGEM3 in the United Kingdom; Met Office, 2015). NWPs run at a finer scale to capture more spatial detail (e.g. approximately $14 \times 8$ km for ECMWF in the United Kingdom), but only for forecasts of up to several days (ECMWF, 2015), and both currently fail to capture flood-producing convective storm events, for example, the 2004 flooding in Boscastle, in the United Kingdom (see May *et al.*, 2004), and the 1999 derecho event in the United States (see Gallus *et al.*, 2005). NWPs tend to smooth out localized, high-intensity, short duration events that can cause flooding particularly in smaller or mountainous catchments, although **downscaling** (Wilby and Wigley, 1997) can help to recreate smaller, subgrid variability by incorporating more detailed information.

Disadvantages of GCMs and NWPs are the computation time required to run them, the uncertainty from the interaction of many poorly defined model parameters, and no transparent calibration against historical observations. In a hybrid approach, GCMs can be run over a specific historic period in order to assimilate available observations, producing a **reanalysis data** set (e.g. NCAR-40, see Saha *et al.*, 2010; ERA-40, see Uppala *et al.*, 2005). These 'observations' have the advantage of being regularly gridded but rely on the quality of the underlying model and the density of observation points. Some models use a perturbation-based statistical (e.g. copulas) resampling approach to simulate large event sets from reanalysis data.

Hydraulic modelling is used to calculate flood severities (e.g. depth) from the hydrologically estimated flow conditions. Traditionally, spatially one-dimensional (1-D) hydraulic models have been used as a basis for flood inundation mapping. These represent flow and water levels along a single direction (aligned with the river), discretized at cross-sections through the floodplain typically spaced hundreds or thousands of metres apart. Increasingly, it is common practice to use more sophisticated two-dimensional (2-D) models, which use the shallow water equations (SWEs) (Sleigh *et al.*, 1998; LeVeque, 2002; Environment Agency, 2013) and resolve the speed and depth of flow via fluxes in two spatial dimensions over a spatial grid (e.g. Lamb, Crossley and Waller, 2009; Environment Agency, 2013).

Simpler approaches have been developed based on either approximations of the SWEs or derived directly from terrain data using geomorphological regression equations, which use the DTM and catchments characteristics (but not flow) as a major inputs. These approaches are capable of providing reasonable results in areas of well-defined topography but can result in inaccuracies of flood extents in flat or featureless terrain (typically resulting in an overestimation of the floodplain).

Finally, the quality of **terrain data** used is extremely important for flood modelling. Terrain data quality may range from low-resolution Digital Elevation Models (DEMs); for example, the freely available global NASA SRTM data at 90 m (Jarvis *et al.*, 2008; Farr *et al.*, 2001), to LiDAR data at 1 or 2 m. Data with resolution of ~5 m or less (e.g. NextMAP) are sometimes available at a national scale, which is considered to adequately represent individual buildings because it is comparable with their scale. Note that DTM resolution is not necessarily the same as a flood model's spatial resolution, the spacing at which flow or flood depth are assessed; cell size in models is a separate trade-off between computational run time and precision.

DTMs used for flood modelling are typically filtered to remove the effects of trees and buildings that appear in raw sensor data (Sithole and Vosselman, 2004) but are not in reality significant obstructions to flow, with a significant effect on modelled flood data (Figure 3.16). Furthermore, much processing of the terrain data is required if the model is to be useful for flood modelling; for example, structures such as bridges must be removed from DTMs if flow paths under them are to be modelled correctly (Figure 3.16).

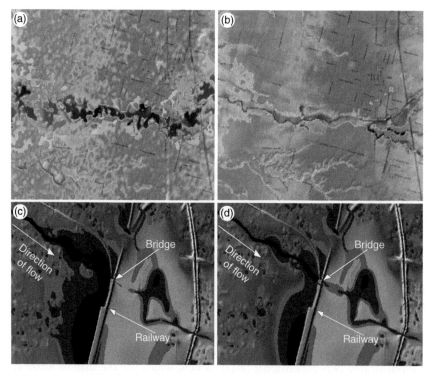

**Figure 3.16** Impact on modelled flooding (blue hues) of small-features such as vegetation and houses illustrated by contrasting (b), which uses a DTM cleaned of these features to an non-cleaned terrain in (a). In effect, false obstructions to flow are removed. (d) and (c) show the impact of a bridge under a railway line as shown in a DTM and after structure removal treatment, respectively. The volume of water modelled is identical and flows down a water course in a south-easterly direction from the north-west corner of the map (arrow). *Sources:* Terrain model courtesy of Intermap Technology; modelling by JBA Risk Management Limited; images created using ArcGIS software by ESRI.

Benchmark analysis (Enviorment Agency, 2013) shows that commercial codes based on the SWEs perform similarly well in controlled tests and are accurate enough for flood-risk management applications, given suitable input data. The same tests also show how some unavoidable subjective decisions in model configurations can significantly affect outputs. Assessing the many uncertainties and accuracy of a model requires careful analysis, dialogue with the model developer, and a high level of skill (Beven, 2009); in some countries formal guidance and standards apply in the development of flood models for use by government agencies (FEMA, 2003).

**DAMAGE METRIC AND VULNERABILITY:** Flood severity is typically quantified as a fraction of the asset's value or **damage ratio** (DR), that is a function of maximum water depth reached ($d$), perhaps modified by water velocity ($v$), or combined with it (e.g. $v \times d$). In contrast to $v$, which is difficult to measure during flooding, $d$ is readily recorded (e.g. in photographs) and often leaves visible 'tide-line' marks on buildings or trees (see Figure 3.13(d)). Thus, **vulnerability curves**, relating $d$ to damage can be created with some confidence (e.g. Davis, 1985; Davis and Skaggs, 1992; Kiefer and Willett, 1996; 2005; US Army Corps of Engineers, 2003; Penning-Rowsell *et al.*, 2005 and 2013) and verified against historical insurance claims data. Conversely, functions that incorporate $v$ may be poorly calibrated. Furthermore, many users find that combining $v$ and $d$ into a single measure of hazard intensity is not easy to understand. At best, modelled mean damage correlates to that observed only relatively weakly. Losses computed for single properties should therefore be viewed as highly uncertain, although this averages out for multi-location losses. As a consequence, some recent models simply use the distribution of past claims for a particular property type to make damage estimates, omitting vulnerability curves.

Flood losses are dependent on a number of building attributes, especially the presence or absence of a basement, the property height (e.g. contents above the first floor may not be damaged), and **occupancy** or the nature of the business in commercial or industrial properties. Including this differentiation is regarded as minimum best practice in today's models, although in many markets these data will not be available. Type of structure (e.g. bungalow, detached house, mobile home, etc.), building age, standard of maintenance, and building material (e.g. wood, brick, etc.) are also commonly included, requiring a range of vulnerability functions. Additional lines of business such as agricultural and marine may be covered by some flood models, and including the cost of being housed during repair work or **alternative living expenses** (ALE) is also usual (see Chapter 1.9.1). Because business interruption and ALE losses are governed by the duration of flooding, not $d$, a distinct vulnerability must be implemented for these coverages, even if these are best treated as rough estimates. Large or complex industrial sites are not considered to be suitable for generic catastrophe model analysis, and should properly be assessed for flood exposure on a site-by-site basis.

**ADDITIONAL LOSS DRIVERS INCLUDED:** Without flood defences, estimated losses are likely to be unrealistically high, whereas assuming defences that do not actually exist will underestimate losses. Unfortunately, information on defences is often, sparse, generic, and of variable quality. A generic statement of the standard of protection in place for a town or given area may be suitable to incorporate flood management schemes (e.g. withholding reservoirs, movable flood defences). If the precise location, structure and height of a fixed defence are known, its performance can be characterized by a fragility function, which expresses the probability of the defence failing conditional on it experiencing a given hydraulic loading. However, event data from failures may not be adequate to validate these curves statistically (Haven, 2013; JBA Trust, 2014) and maintenance information, critical to likelihood of failure, is rarely available. Thus, pragmatic assumptions are commonplace in models.

Modelling may be done on an 'undefended' basis, in which it is assumed that no defences are present (or all defences fail) or a 'defended' basis in which defence performance is taken into account. Defended models (e.g. RMS, JBA UK Flood Model) may assess the probability of failure in relation to the water level relative to the top of the defence, or may simply 'fail' the defence once the standard of protection (expressed as the return period of flooding against which the defence is designed to protect) considered to be offered by the defence is exceeded. Following failure, models may or may not calculate the volume of water able to pass across the failed defence. Failure scenarios for major defences (e.g. the Thames Barrier) may be included in models. Currently, however, even if information about flood defences were complete, detailed modelling is computationally infeasible; even ignoring the extent or depth of breaches, there are too many possible combinations of defences failures to consider explicitly, especially in combination with the many thousands of synthetic events in an event set. Storm sewers, about which very little information may be available, can be approximated in models by removing the amount of water they theoretically transport from the volume applied in the modelling.

### 3.5.5 Secondary Perils and Non-Modelled Items

Flood models typically do not include secondary perils. Climate change is not generally accounted for (Lloyd's, 2014), but recent changes in hydrology may be implicitly accounted for in models based on gauge records. Flooding caused by the failure of man-made structures (e.g. reservoir failures, canal breaches) is generally not considered, as each is unique and cataloguing case-by-case assessments (e.g. of RP) is not feasible at a national scale. However, these events are mapped in some regions and their impact on insurance portfolios may be considered using a scenario-based approach (e.g. Lloyd's, 2015; see Chapter 2.7.1.3).

### 3.5.6 Key Past Events

Flood catastrophe models have developed comparatively recently, in part due to the relatively high complexity and detail required. As with other perils, developments have frequently been prompted by high-impact events.

The development of the first inland flood catastrophe models was triggered by events in Continental Europe and the United Kingdom in the early 2000s. Early models were typically based on lower resolution digital terrain data and analysis cells than more recent models. Flooding in Poland and the Czech Republic in 1997 (economic/insured loss \$4.5 billion; Schanze, Zeman and Marsalek, 2006) and the United Kingdom in 1998 and 2000 (Bye and Horner, 1998; Environment Agency, 1998) and along the Elbe and Danube rivers in 2002 (economic/insured loss €15–20 billion/€3billion; Toothill, 2002) prompted the development of the first inland models for the United Kingdom in 2001 (Lohmann *et al.*, 2009) then subsequently Germany, Austria and Belgium (e.g. Toothill, 2002; *Insurance Journal*, 2005; *Business Wire*, 2011). Flood catastrophe model developments have focused on a limited number of countries with further models and model upgrades occurring in both the United Kingdom and Germany (e.g. AIR Worldwide, 2011; JBA Risk Management, 2015c). National probabilistic flood models have only recently become available in a wider range of territories, including Thailand (JBA Risk Management, 2012a), the United States (AIR Worldwide, 2013), and Malaysia (JBA Risk Management, 2015c). In some larger territories, models have focused only on the largest cities, for example, in India (e.g. RMSI, 2014; JBA Risk Management, 2015c). Most models include only river or a combination of river and surface water flood, whereas coastal flood is typically included as a secondary peril in windstorm models (e.g. Lohmann and Yue, 2011).

Flood differs from other perils in the ability to identify locations susceptible to flood damage at high resolution (i.e. down to property level). This has meant that insurers have sought to control flood exposures not only through catastrophe modelling but also through use of flood hazard maps, which are commonly used to identify, underwrite, and price flood risk at individual property level. The first flood maps available for insurers were developed in the Czech Republic in an initiative led by the Czech Insurance Association (Česká Asociace Pojišt'oven, 2003) and the United Kingdom, where the first maps were developed in 1999 by the Environment Agency and then in 2004 via a collaboration between Norwich Union and JBA Consulting (Aviva, 2004). Today, flood maps are widely used by insurers, but are not suitable for re-insurance as events in catastrophe models are required to capture spatial correlations and the accumulation of risk (see Chapter 4.3.1).

### 3.5.7 Open Questions/Current Hot Topics/Questions to Ask Your Vendor

When making decisions from the output of flooding catastrophe models, the following questions could be revealing:

- What flood types are included in the model, and is it probabilistic?
- From what data has the model been developed?
- How is an event defined? Specifically, how is the hours clause modelled (e.g. Munich Re, 2005; Oasis Loss Modelling Framework, 2014)?
- How are flood extents calibrated? What are the major historical events used for model calibration (e.g. American Academy of Actuaries, 2001)?
- What is not included in the model? Specifically, what regions, perils, lines of business, possible sources of flooding are excluded (e.g. Oasis Loss Modelling Framework, 2014)?
- Against what claims data has the model been calibrated (e.g. Oasis Loss Modelling Framework, 2014)?
- To what parameters is flood damage related? What evidence is there for a relationship between the chosen hazard intensity and claims values?
- What is the resolution of (i) the underlying digital terrain model; and (ii) the level at which analysis is carried out?
- What measure of hazard intensity is used? How are secondary factors that may impact the amount of damage (e.g. sedimentation, debris impacts) taken into account (e.g. Oasis Loss Modelling Framework, 2014)?
- What is the range of uncertainty associated with the level of damage resulting from a given hazard intensity (e.g. Oasis Loss Modelling Framework, 2014)?
- What quality assurance has been done on the Digital Terrain Model?
- What defence data have been used in the model, where are defence data not available, and what assumptions have been used to account for the impact of flood defences in areas where no information is available?
- How is defence failure modelled? What is the impact on the results of removing defences from the model?
- Is storm drainage accounted for in urban areas? How are buildings represented in a digital terrain model?

### 3.5.8 Nonproprietary Data Sources

Some historical and real-time flood gauge data are freely available. In the United Kingdom, for example, daily flow data are publicly available via the National River Flow Archive (http://nrfa.ceh.ac.uk/) and live data (a few hours old) from the Environment Agency for England (http://www.gaugemap.co.uk/) or the Scottish Environmental Protection Agency for Scotland (http://apps.sepa.org.uk/waterlevels/). However, access varies by country and region.

The NASA SRTM data can be used as a basis for flood model development. This provides a basic Digital Surface Model at a resolution of 90 m (Farr *et al.*, 2001).

Generic depth-damage flood vulnerability functions in the United States are provided by the US Army Corps of Engineers, for residential properties (Davis and Skaggs, 1992) and commercial properties (Davis, 1985; Kiefer and Willet, 1996). The residential functions were updated in 2003 (US Army Corps of Engineers, 2003). In the United Kingdom, similar information is available from Middlesex University in the Multi-Coloured Manual (Penning Rowsell *et al.*, 2005 and 2013).

The Centre for Research on the Epidemiology of Disasters (CRED) provides an International Disaster Database from which a comprehensive history of large flood events can be extracted. Tropical cyclone data are also available (see Section 3.2) for flooding of this origin.

The *UK Flood Estimation Handbook* (Institute of Hydrology, 1999) is available for purchase by commercial and academic organizations. This is a publication that provides a set of tools and data for flow estimation. The associated data sets are UK-specific, but it provides approaches that can be adapted for use elsewhere.

### Acknowledgements

We thank Cressida Ford, David Leedal, Simon Waller, Duncan Faulkner, Ye Liu, and JBA Risk Management.

---

**Further Reading** Shaw *et al.* (2010) is an authoritative text on the practical aspects of hydrology, while Mascarenhas (2005) is an up-to-date text on the numerical hydraulic simulation of floods in both rural and urban areas. Flood Risk Science and Management (Pender and Faulkner, 2010) provides a synthesis of current research in flood management, providing a multi-disciplinary reference text for the research community, flood management professionals, engineers, planners, and government officials.

---

## 3.6 Shrink-Swell Subsidence

*John Hillier*

*Reviewer: Thierry Corti*

### 3.6.1 What Is the Peril?

Ground movements induced by change in volume of soils rich in some silicate **clay minerals** are known as **shrink-swell subsidence (SS)** (e.g. Brady and Weil, 2002; Swiss Re, 2011; Pritchard, Hallett and Farewell, 2014). Water bonds to susceptible minerals (e.g. smectite and vermiculite), causing them to expand (i.e. swell; e.g. Brink, Partridge and Williams, 1982) when wet, and conversely the withdrawal of moisture leads to contraction (i.e. shrinkage). Soil moisture is broadly governed by evaporation and transpiration (evapotranspiration) and rainfall, leading to seasonal movement. This can be enhanced locally by changes in drainage or vegetation or, most

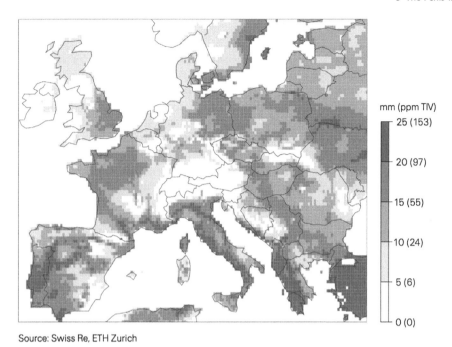

Source: Swiss Re, ETH Zurich

**Figure 3.17** Estimated loss potential from SS in Europe.
Notes: mm, movement in millimetres; ppm, parts per million; TIV, total insured value.
*Source:* Reproduced with permission of Swiss Re Management Limited.

importantly, regionally (100 s of km) by drought and subsequent re-wetting (e.g. BGS, 2012; Figure 3.17). Typically, SS can affect the top 1–1.5 m of soil but deeper during droughts (Pritchard, Hallett and Farewell, 2014), and its gradual onset (up to years) has led to it being described as 'a gradual catastrophe' (Cheetham, 2008).

### 3.6.2 Damage Caused Caused by the Peril

SS damage has been most prominent in the United Kingdom and France, causing >€8 billion of industry losses in the United Kingdom since 1976 (Cheetham, 2008), peaking at more than £500 million in 1991 (ISE, 1994; BGS, 2012), and an average €340 million per year in France (Corti *et al.*, 2009). Damage is typically assumed to be driven by an uneven distribution of shrinking or swelling beneath the foundations, also known as **heave** (e.g. Prichard, Hallett and Farewell, 2014), of domestic buildings with losses increasing non-linearly with the magnitude of ground motion (e.g. Corti *et al.*, 2009; Swiss Re, 2011; BGS, 2012). However, bridges, industrial sites, and other infrastructure (e.g. mains water pipes; Pritchard, Hallett and Farewell, 2014) are also at risk. A caveat is that damage could be cumulative (Cheetham, 2008; Swiss Re, 2011) or include a political or administrative component (Corti *et al.*, 2009).

### 3.6.3 Forecasting Ability and and Mitigation

Droughts are predictable two to four months in advance with partial skill using climate models (Hao *et al.*, 2014), but over decades (e.g. to 2050) the risk is expected to increase (e.g. by hundred million pounds) as the climate changes (Pugh, 2002; Jones, 2004; Swiss Re, 2011). Mitigation is by education (e.g. planting trees), engineering, and building standards (e.g. paved areas around buildings or **underpinning**; e.g. Page, 1998), and removing trees close to buildings is effective.

### 3.6.4 Representation in Industry Catastrophe Models

The occurrence of a year (or season) of enhanced SS is taken as an **event** (Swiss Re, 2011) and catastrophe models calculate losses from ensembles or sets of events generated using stochastic resampling of the meteorological variables related to historical droughts defined by how the hydrological balance affects the shallow subsurface (e.g. Swiss Re, 2011). Vegetation cover and proximity to large trees may also be included (e.g. see Lawson and O'Callaghan, 1995). The key driver of losses is considered to be **soil moisture deficit** (e.g. Cheetham, 2008) because this interacts with the spatial distribution of **bedrock** and **surficial deposits** (e.g. soil types; BRE, 1993; Corti *et al.*, 2009). Event footprints are in terms of soil movement, and SS damage is modelled as a damage ratio (DR) that is a function of soil movement (see Figure 3.17) (e.g. Swiss Re, 2011).

### 3.6.5 Key Past Events

In the United Kingdom, 1975–1976 was the first year to cause large claims (e.g. Lawson and O'Callaghan 1995), with 1991 causing the most loss (e.g. Cheetham, 2008). In 1989, tens of thousands of buildings in France were affected (Salagnac, 2007), after which subsidence was incorporated in the country's natural catastrophe insurance system (Cat-Nat). The 2003 European heatwave caused ~€1.1 billion losses as estimated by Caisse Centrale de Réassurance (Swiss Re, 2011).

## 3.7 Earthquakes

*Joanna Faure Walker and Guillaume Pousse*

*Reviewer: Rebecca Bell*

### 3.7.1 What Is the Peril?

**DESCRIPTION:** As the Earth's outer shell moves, elastic strain accumulates across friction-ally-locked **faults**, storing energy like a stretched elastic band. The sudden release of this energy in an **earthquake (EQ)**, when the fault **slips**, causes rocks in the Earth's crust to move violently. The shaking as this disturbance propagates outward in all directions from the source of the break, or **hypocentre** (red star on Figure 3.18 (a)), are known as *seismic waves*. The location on the Earth's surface above the hypocentre is the **epicentre**, and the break is known as the **rupture**.

The main types of seismic wave are primary or 'P' waves, secondary or 'S' waves, and surface waves (Rayleigh waves and Love waves; see Fowler, 1990). Surface waves have the largest **amplitudes** (see Figure 3.18 (b)) and therefore tend to cause the most damage from ground shaking. Amplitudes are bigger for larger earthquakes, and **attenuation functions** are used to describe the way amplitude decreases with distance away from the epicentre (Douglas, 2003). The vibrations typically occur at frequencies between 0.1 Hz and 10 Hz (Pousse, 2005), a range in which most civil engineering structures (e.g. houses) are sensitive to shaking (O'Connell and Ake, 2002), but larger earthquakes will typically generate lower frequency waves as well.

Earthquake **intensity** is the severity of shaking as experienced at a geographic location. It is commonly expressed as I to XII on the Modified Mercalli Intensity (MMI) scale, or quantified

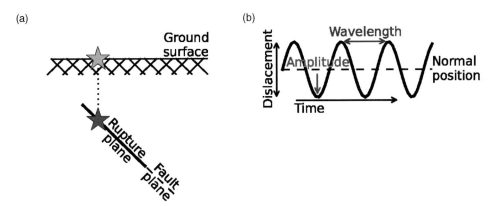

(a)                                                    (b)

Ground surface

Rupture plane — Fault plane

Wavelength

Amplitude

Normal position

Displacement

Time

Figure 3.18  (a) Illustration of a rupture on a fault plane. Usually, only part of a fault ruptures in a single earthquake, as shown. The red star denotes the hypocentre, and the orange star is the epicentre. (b) A simplified seismic wave (black), displacing rock from its rest position by a maximum amount defining the wave's amplitude. The distance between these peaks is the wavelength, and the time between them is the frequency in Hz (cycles per second).

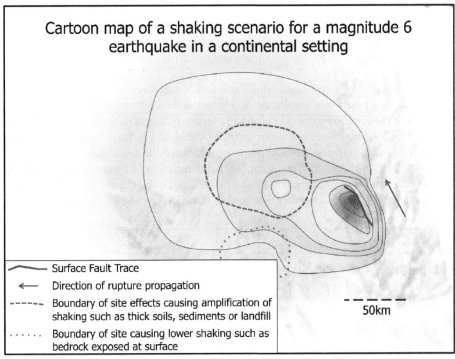

Cartoon map of a shaking scenario for a magnitude 6 earthquake in a continental setting

—— Surface Fault Trace

←— Direction of rupture propagation

------ Boundary of site effects causing amplification of shaking such as thick soils, sediments or landfill

...... Boundary of site causing lower shaking such as bedrock exposed at surface

- - - -  50km

| Perceived Shaking | Not felt | Weak | Light | Moderate | Strong | Very strong | Severe | Violent | Extreme |
|---|---|---|---|---|---|---|---|---|---|
| Potential Damage | None | None | None | Very Light | Light | Moderate | Moderate /Heavy | Heavy | Very Heavy |
| Peak Acceleration (%g) | <0.17 | 0.17-1.4 | 1.4-3.9 | 3.9-9.2 | 9.2-18 | 18-34 | 34-65 | 65-124 | >124 |
| Peak Velocity (cm/s) | <0.1 | 0.1-1.1 | 1.1-3.4 | 3.4-8.1 | 8.1-16 | 16-31 | 31-60 | 60-116 | >116 |
| Instrumental Intensity | I | II-III | IV | V | VI | VII | VIII | IX | X+ |

Figure 3.19  Spatial footprint of shaking due to an earthquake, coloured according to various common and equivalent measures of intensity. Red line represents surface rupture. Note that in general the most intense shaking occurs where there is the greatest deformation, and decreases away from the fault, but surface conditions affect the shaking, creating a non-uniform decrease in intensity away from the fault, such as due to thick soils west of the fault in this illustration.
Note: The scale is used in USGS ShakeMap.
*Source:* Adapted from Wald *et al.* (1999).

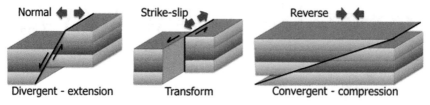

Figure 3.20 Types of faulting (*top*) and their tectonic settings (*bottom*)
*Source:* Reproduced with permission of Faure Walker.

as the peak velocity or acceleration experienced (Figure 3.19). Shaking intensity is affected by the distance from the epicentre, the hypocentre depth, the type of fault, the fault orientation, bedrock geology, the soil type, and thickness (Bonilla, 2000; Akkar and Bommer, 2007), rupture direction ('directivity'; Ameri *et al.*, 2009), and basement geometry (USGS, 1996). Reverse faults (Figure 3.20) and shallow buried ruptures generally produce the greatest ground shaking (Somerville and Pitarka, 2006). Thick sediments and loose soils slow down seismic waves increasing their amplitude in a **site amplification** effect. This phenomenon is well illustrated by the 19 September 1985 earthquake that caused US\$3–4 billion in damage in Mexico City, which was 400 km away from the epicentre, but built on a former lake with soft soils.

Earthquake magnitude quantifies its overall size and is measured on a logarithmic scale. The Richter Scale is based on wave amplitude at a given distance from the epicentre but is not accurate for large magnitude earthquakes ($M_L > 8$), and so moment magnitude ($M_w$) is now used instead in science because it does not suffer from this problem (see Kanamori, 1983; dePolo and Slemmonds, 1990). $M_w$ (Fowler, 1990, p. 88) is a measure of the work done by the earthquake (see Box 3.1), based on the seismic moment ($M_o$) and converted to a convenient number that approximates Richter Scale values, so that it can be reported more simply by the media, and it is comprehensible to the general public; that is, $M_w$ is what the media cite as a 'Richter Scale'

---

**Box 3.1  Mathematics of Earthquake Size**

$M_w$ is calculated from the seismic moment ($M_o$) as shown in Equation (3.1) (Fowler, 1990, p. 91), where $M_o$ is a measure of the work done: a product of the distance the fault moved, $u$, and the force needed to move the fault, $\mu A$ (Equation (3.2)). A is the area of the fault that moved; $A = Lw$, where $L$ is rupture length and $w$ is the rupture width. $\mu$ is the shear modulus (Pascals), which quantifies the resistance to movement.

$$M_w = \frac{2}{3}\log_{10}(M_o) - 6.0 \tag{3.1}$$

$$M_o = \mu Au \tag{3.2}$$

The width of the rupture area is constrained by the dip of the fault and the thickness of the seismogenic layer, the depth over which the tectonic plate remains brittle and can rupture in earthquakes. In smaller earthquakes, $A$ can increase its extent roughly equally in both depth and length, so $M_o \propto L^3$. For large earthquakes, $L$ is much greater than the thickness of the seismogenic layer so, geometrically, $M_o \propto L^2$ (see Wells and Coppersmith, 1994). Assuming that the intrinsic frictional properties of rock are constant, and hence that rocks experience a scale-invariant constant stress drop in earthquakes, $u \sim 10^{-5}L$ for interplate events and $u \sim 6 \times 10^{-5}L$ for intraplate events (Scholtz, 1982; Scholtz *et al.*, 1986).

earthquake size. Each integer increase (i.e., +1) in magnitude represents an earthquake 10 times the size in terms of seismic wave amplitude, releasing approximately 32 times the energy. As $M_o$ is proportional to the length of a fault that ruptures, the maximum earthquake magnitude that can occur on a fault is related to its length (e.g. Wells and Coppersmith, 1994). For a given size of earthquake, empirical relationships exist between the amount of slip, the displacement measured at the surface and the fault length (see Stirling *et al.*, 2013, for a review); these relationships vary depending on the type of fault (Wells and Coppersmith, 1994). In general, about 1 m of slip gives a $M_w$ 7 quake, while $M_w$ 8 to 9 quakes can slip 10 to 20 m (e.g. Delouis *et al.*, 2010; Loveless *et al.*, 2011).

Earthquake frequency, or the probability of occurrence of an earthquake in a region, is empirically a function of the magnitude and described statistically using a power law; namely, $N = 10^{a-bM}$, where $N$ is the number of earthquakes greater than magnitude $M$, and $a$ and $b$ are constants (Gutenberg and Richter, 1954; Kramer, 1996). This so-called 'Gutenberg-Richter' relationship applies for various tectonic settings, earthquake depths, and catalogue lengths, and $b$ is usually taken as ~1.0 (Marzocchi and Sandri, 2003), so that event frequency decreases tenfold for each unit increase in magnitude. There are estimated to be on average 134, 15, and 1 earthquakes per year worldwide with magnitudes 6.0–6.9, 7.0–7.9 and >8.0, respectively (USGS, 2012). Earthquakes less than magnitude 6 can cause local damage but are unlikely to cause large insurance losses (USGS, 2014).

**PHYSICAL DRIVERS:** Plate tectonic theory (see Chapter 2 of Fowler, 1990, for a review) describes the Earth's outermost shell as a small number of nearly rigid plates that move relative to each other; these motions and the physical properties of rocks dictate the size, frequency and location of earthquakes. The plates are driven by convective motions in the mantle, 'ridge push' from newly formed plates in the middle of the oceans, and 'slab-pull' from where old plates sink back into the Earth at subduction zones (Forsyth and Uyeda, 1975). Oceanic plates are young and strong, so deformation and earthquakes are generally confined to narrow zones along their boundaries such as along the circum-Pacific 'Ring of Fire' (Figure 3.21). However, the older and weaker nature of continental plates allows deformation and earthquakes to occur within as well as around the plates, across wide zones such as the Trans-Alpine belt (see Figure 3.21).

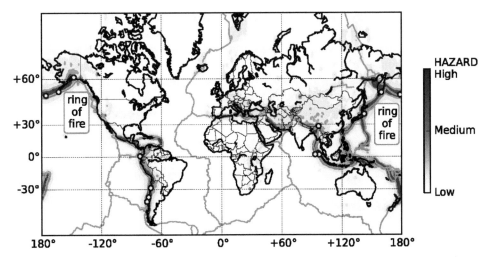

Figure 3.21 Illustration of the main tectonic plates delimited by the light grey lines and an indication of the level of seismic hazard in red shades. Yellow circles show the location of earthquakes with Mw > 8.5 since 1900. Almost all the earthquakes Mw > 8.5 have occurred on reverse faults at subduction zones. The 'Ring of Fire' has active volcanoes at its edge – circum-Pacific – and the hazardous region extending from Italy up to the north of Indonesia is called the 'Trans-Alpine belt'.
*Source:* After University of Texas at Austin; United Nations International Strategy for Disaster Reduction (UNISDR); and National Geophysical Data Center (NGDC).

Large earthquakes can only occur on long (e.g. 100–1000 km), pre-existing faults. A fault's size places an upper limit on the size of the earthquake possible, even if how exactly to determine this largest 'characteristic' size is debated partly because the whole fault need not rupture in an individual event (Wesnousky, 1994; Parsons, 2000; Jackson and Kagan, 2006). Fault length, which varies with fault type (see Figure 3.20), is, in turn, determined by tectonic setting. In **convergent** tectonic settings, motion is typically taken up on **reverse faults** also known as **thrust faults** if they dip at less than 30 degrees. These faults can be very large and, as rock is strongest in compression, produce the biggest earthquakes; the largest recorded earthquake was $M_w 9.5$ offshore Chile in 1960 (e.g. Kanamori, 1977; Scawthorn and Chen, 2002: 47; RMS, 2011). Nine of the largest 10 earthquakes since 1900, and most of those of $M_w > 8.5$ (see Figure 3.21), have been in subduction zones where denser ocean plates are forced towards and beneath continental plates. In **transform margins**, plates such as the North American and Pacific move past each other and are characterized by strike-slip faults such as the San Andreas Fault. Strike-slip earthquakes are typically up to $M_w 8$. Normal faults, typifying extensional settings (e.g. Italy, Greece, the Basin, and Range Province), have the smallest maximum $M_w$ typically up to about 7, usually along faults tens of kilometres long.

**KEY BEHAVIOURAL SUBTLETIES:** Faults have finite strength and in the context of slowly accumulating strain this leads to the concept of time-dependence and an **earthquake cycle** wherein the probability of an earthquake increases with concurrently increasing stress, strain, and time since the last earthquake on a given fault (e.g. Thatcher, 1984; Parsons, 2004). Assuming all the accumulated strain is released in large magnitude earthquakes, average recurrence intervals for a given earthquake magnitude along a fault can be calculated by dividing the slip during an earthquake by the rate of slip across the fault (e.g. Faure Walker *et al.*, 2010). Over a period of several seismic cycles, deformation rates across faults or fault zones are expected to reach the long-term tectonic rates (e.g. Bull *et al.*, 2006). This has the implication that if a part of a fault plane has slipped an unusually small amount in the past, a **seismic gap** is said to exist (e.g. Toksoz, 1979). This idea has been criticized (e.g. Kagan and Jackson, 1991), but some earthquakes have occurred in gaps identified after the event (e.g. Loma Prieta in 1989 by Housner and Thiel, 1990; Nicoya Costa Rica in 2012 by Protti *et al.*, 2014). The potential magnitude of the devastating Haiti earthquake of 2010 (Earthquake Engineering Research Institute, 2010) was identified in advance (Manaker *et al.*, 2008).

Recurrence intervals between large earthquakes on individual faults, however, are not regular (and hence currently not predictable) due to a number of other factors that are thought to affect the timing of individual earthquakes. Primarily, stress may be relieved on parts of a fault by smaller events (Kagan, Jackson and Geller, 2012). Furthermore, in some regions such as the Tokai region, Japan, a substantial proportion of plate motion is accommodated **aseismically** without earthquakes (e.g. by aseismic creep or in 'slow slip' events; Scholz, Wyss and Smith, 1969; Ozawa *et al.*, 2002). It is therefore unknown how much accumulated stress across a fault will be released during large damaging earthquakes. As the exact timing of future earthquakes cannot be predicted, earthquakes must be forecast probabilistically. Common practice is to use a time-independent model such as a Poisson distribution. This approach assumes that earthquakes occur entirely randomly in time (i.e., are **Poisson** events), implying that they are independent of past earthquakes even on the same fault and their probability of occurring (e.g. in the coming year) does not change with time. An alternative approach is to use a time-dependent model such as a Brownian Passage Time (BPT) distribution, but, in this approach, the time since the last event needs to be known.

Forecasting is further complicated because faults are not independent of their neighbours, that is, they **interact**. Fault interactions cause some proximal earthquakes to cluster in time, usually the time period between events is neither regular nor Poisson (i.e. entirely random) because an earthquake on one fault can alter the stress state on neighbouring faults. This is typically modelled using Coulomb stress changes (King, Stein and Lin, 1994; Stein, 1999; Toda and Stein, 2013) or through changes in pore fluid pressure (Sammonds, Meredith and Main, 1992; see Staecy, Gomberg and Cocco, 2005, for a review) and is thought to be capable of either bringing forward or delaying an event on these faults. The stress change brought about by nearby earthquakes could either be permanent or transient (e.g. Belardinelli, Bizzari and Cocco, 2003). Clustering has been observed off the coast of Sumatra after the 2004 Boxing Day Earthquake (e.g. Henstock *et al.*, 2010); along the section of the fault that later ruptured during the 2005 Nias-Simeulue Earthquake, scientists calculated an increased probability of a large magnitude earthquake because of Coulomb stress changes resulting from the Boxing Day Earthquake, and this was calculated before the 2005 earthquake occurred (McCloskey, Nalbant and Steacy, 2005). Sometimes, sequences can be systematic progressions such as the latest cycle of migration of large earthquakes westward along the North Anatolian Fault (NAF) in northern Turkey (Toksoz, 1979; Stein, 1997; Swiss Re, 2000). Based on long-term rates, about 20% of the NAF would have been expected to have ruptured since 1939, when in fact about 50% has; the most recent 1999 Izmit earthquake raised the risk of an $M_w > 7$ earthquake affecting Istanbul in the next 30 years to about 40% (Parsons, 2004; Armijo, 2005), with the potential of high losses (i.e. >$10 billion; Durukal, Erdik and Sesetyan, 2006) creating interest from stakeholders. Controversially, some recent research has also argued that seismic waves from very large earthquakes can trigger earthquakes around the globe (e.g. Pollitz *et al.*, 2012).

Given that the maximum possible rupture area (length and width) primarily dictates the maximum size of an earthquake in a particular region, determining this maximum is key to hazard and risk assessment. One approach is to use the Gutenberg-Richter frequency magnitude distribution which describes how the number of earthquakes with magnitude $\geq M$, N, changes with magnitude up to a maximum magnitude: $N = 10^{a-bM}$, where $a$ and $b$ are constants. However, an alternative approach that has been used is based on the idea that either whole faults rupture in individual earthquakes or for very large faults, the fault is divided into segments that rupture in isolation, leading to quasi-periodic earthquakes of a size 'characteristic' of the fault or **fault segment**. This **characteristic earthquake (CE) model** (McCann *et al.*, 1979; Schwartz and Coppersmith, 1984) was, for example, used to predict a 95% probability of an earthquake on the Parkfield Section of the San Andreas fault before 1993, based on previous moderately sized ($M_w\sim6$) earthquakes occurring every 20 years or so (i.e. 1857, 1881, 1901, 1922, 1934, and 1966; Bakun and Lindh, 1985). However, the event happened 11 years late in 2004 and is consistent with a Gutenberg-Richter relationship derived for the area defined by Michael and Jones (1998), suggesting that the probability of $M_w\sim6$ 'characteristic' earthquakes was significantly overestimated by the CE model (Jackson and Kagan, 2006). Thus, despite arguably performing better in some places, for example, Japan (Ishibe and Shimazaki, 2012), the commonly used CE model is under increasing scrutiny (e.g. Wesnousky, 1994; Parsons, 2000). The CE assumption is also problematic as it blinds models to larger events by assuming earthquakes cannot propagate across segment boundaries; for example, the 2011 Great East Japan 'Tohoku' and 2004 Sumatra earthquakes ripped across several previously hypothesized 'segment boundaries' (cf. Sparkes *et al.*, 2010; Kagan, Jackson and Geller, 2012).

Despite current unknowns and inadequacies in earthquake forecasting leading to large uncertainties in risk calculations, a probabilistic approach of calculating annual risk is of primary importance to stakeholders in order to understand the risk and for planning purposes.

### 3.7.2 Damage Caused by the Peril

Annual economic losses from earthquakes are close to US$50 billion (CRED, 2014) and projected to average about US$100 billion (Daniell *et al.*, 2015) in the coming century, with

**Figure 3.22** Illustrative types of damage: (a) concrete columns can shear to case 'pancake' collapse; (b) exterior walls falling can result in partial or total collapse of the roof system; (c) buildings can slide off their foundations; (d) 'X' cracks in exterior; (e) freeway collapse; (f) broken furniture.
*Sources:* (a) 21 May 2003. Boumerdes and Algiers. Algeria. earthquake, http://www.ngdc.noaa.gov/hazardimages/picture/show/1109; (b) 2 May 1983. Coalinga, USA, earthquake, http://www.ngdc.noaa.gov/hazardimages/picture/show/187; (c) 25 April 1992. Cape Mendocino. USA. earthquake, http://www.ngdc.noaa.gov/hazardimages/picture/show/337; (d), A typical 'X' crack from L'Aquila City following the 2009 earthquake, courtesy Faure Walker; (e) 18 October 1989, Loma Prieta, USA, earthquake and tsunami, http://www.ngdc.noaa.gov/hazardimages/picture/show/99; (f) 2 May 1983, Coalinga, USA, earthquake, http://www.ngdc.noaa.gov/hazardimages/picture/show/194.

76% in Asia, 8% in Europe, 9% in America, and 5% in Latin America (UNISDR, 2013). Earthquakes have caused three of the ten largest insurance losses since 1970 (i.e. Tohoku in, 2011, Northridge in 1994, and Christchurch in 2011; Aon Benfield, 2015), with Japan, the United States, New Zealand, Chile, and Italy having recently suffered heavy insured losses (i.e., >$15 billion).

Damage caused by earthquakes is typically assumed to be driven by the strength of ground shaking at its maximum intensity, quantified either as peak ground acceleration (PGA) or peak ground velocity (PGV), the latter being recognized as a better indicator of damage potential for large structures (Earthquake Engineering Research Institute, 1994). However, a longer duration of shaking caused by large earthquake also worsens the damage (Trifunac and Brady, 1975). Buildings also differ in their response. Low-rise buildings are sensitive to the dominant high frequencies from small earthquakes, while tall buildings respond to the lower frequencies created by larger ones. And therefore the spectral acceleration (SA) is also an indicator of potential damage. The SA is the description of the maximum amplitudes of modestly damped resonant responses of single-degree-of-freedom oscillators (an idealization of simple building responses) to a particular ground motion time history, as a function of natural period or natural frequency (Kramer, 1996; Scawthorn and Chen, 2002). Further damage can come from temporary or permanent ground deformation, like the landslides observed during the 1995 Kobe event, which might cause especially high losses if they are near the coast and result in flooding, as some argue happened during the 1908 Messina event (Billi *et al.*, 2008).

Damage can be structural or not. Structural damage is varied (Figure 3.22), costly, and a threat to life. It can result in partial or total collapse of a building and can require any remaining parts to be destroyed and rebuilt. Non-structural damage can be equally economically significant, consisting of broken furniture or equipment, broken water, ruptured gas lines, and extensive cracking of non-structural interior walls. These can all result in major costs, extended vacancies, and loss of productive use of the building. Sometimes, non-structural damage to vital emergency facilities can be catastrophic. For example, following the 1994 $M_w$6.7 Northridge Earthquake, failure of non-structural walls ruptured water lines in Santa Monica's St John's Hospital and forced an evacuation (Pickett, 1997). Infrastructure can also be heavily affected. The 1987 $M_w$7.1 Ecuador event damaged the Trans-Ecuadorian oil pipeline and generated a US$815 million **indirect loss** by ceasing to work, in comparison to the US$185 million **direct loss** cost to repair it (GEM-ECD, 2015). Damage to critical infrastructure such as road and water networks can also compound losses by hindering emergency responders, perhaps preventing the fire department from attacking fires and slowing reconstruction due to access restrictions.

### 3.7.3 Forecasting Ability and Mitigation

It is not currently possible to **predict** earthquakes, that is, give an exact location, date, and time of future events. Foreshocks (e.g. Papazachos, 1975; Felzer, Abercrombie and Ekstrom, 2004), changes in strain accumulation (see Main *et al.*, 1996, for a review), ground deformation, changes in radon gas levels (Kulachi *et al.*, 2009), and animal behaviour (e.g. Kirschvink, 2000) continue to be investigated, but these are not reliable precursors (International Commission on Earthquake Forecasting for Civil Protection, 2011). Only Coulomb Stress Transfer calculations following major events have shown any predictive power or statistical significance (e.g. McCloskey, Nalbant and Steacy, 2005).

Earthquake forecasts are probabilistic, and for a given area, size range and time frame (e.g. Parkfield; see Section 3.7.1) are more reliable at lower spatial resolutions and when data are collected across larger areas or timeframes. Current catastrophe models typically use hazard that is based on historical rates of seismicity in regions. In regions considered to be seismically homogeneous, *a* and *b* of the Gutenberg-Richter law are typically derived from instrumentally recorded seismicity, and historical records (e.g. Aki, 1965). However, even relatively long historical records often cover an insufficient time to capture the full hazard because earthquake recurrence intervals are longer than historical records (see Stein, Geller and Liu, 2012). For example, in Italy, typical recurrence intervals range from hundreds to thousands of years on individual faults but the historical record (considered one of the most extensive in the world) is thought to be complete for damaging earthquakes (>$M_w$5.5) only since about 1349. Furthermore, comparisons of long-term strain rates measured from geological observations across faults and the historical record show that there are areas which have experienced no large earthquakes within the historical record but are at risk in the future as strain is accumulating along mapped faults (Faure Walker *et al.*, 2010, 2012). Regarding the estimation of maximum magnitude, the limitations of using historical records when not supplemented by geological observations or kinematic models (e.g. Manaker *et al.*, 2008) were highlighted in the 2011 Great East Japan Earthquake; because such a large event was absent in the historical record, it was not included in catastrophe models (AIR Worldwide, 2011; EEFIT, 2013). However, such events were evident in the geological record (e.g. tsunami deposits dated at 869 AD; Minoura *et al.*, 2001). Probabilistic forecasts can also be made on the basis of individually mapped faults (e.g. Parsons, 2004). Long-term deformation rates across faults can be measured from geological data, while GPS geodesy can be used to calculate short-term deformation rates (e.g. Friedrich *et al.*, 2003). The earthquake history can be derived from seismic catalogues, historical records of earthquake damage, palaeoseismic records (e.g. trenches across faults or cosmogenic exposure dating; Grosse and Phillips, 2001), and geological records, including evidence of landslides and tsunami (e.g. see Chapter 4.3.3.5). However, observational limitations make the evaluation of earthquake forecasts inherently difficult. The Collaboration for the Study of Earthquake Prediction (CSEP) has ongoing investigations testing the quality of a forecasting method against actual earthquake occurrences for particular regions around the world (Jordan, 2006; International Commission on Earthquake Forecasting, 2011). Meanwhile, unpredicted earthquakes will likely continue to surprise us.

A key technique to **mitigate** loss of life, not financial loss, is to apply stringent building codes to new structures and retrofitting of older ones. Public preparedness is also a mitigation tool, as are relief operations and recovery plans, including seeking financial protection through insurance mechanisms (Scawthorn and Chen, 2002). Examples of retrofitting measures include: adding base isolation systems to decouple the ground motion from the building; adding supplementary dampers to absorb additional energy; adding cross-braces to walls or additional frames or walls to lower floors to provide extra strength; using steel to reinforce concrete buildings to make them less brittle; and column jacketing to increase concrete confinement, shear strength, and flexural strength (see FEMA, 2010). Innovative potential options may also include 'superplastic alloy' (Omori *et al.*, 2011) and 'seismic wallpaper' (e.g. POLYTECT, 2012). Mitigation measures for lifeline systems may include emergency operations, equipment/non-structural anchorage, emergency power acquisition, strategy for alternative supplies, added flexibility in parts of the pipeline network, and automatic lock-down/shut down above certain ground motion threshold (nuclear facilities, transport or water systems; Scawthorn and Chen, 2002).

### 3.7.4 Representation in Industry Catastrophe Models

**HAZARD DESCRIPTION AND EVENT SETS:** Earthquake **events** in catastrophe models are modelled stochastically using large (e.g. $\sim$10,000–1,000,000) ensembles or sets of synthetic events (see Chapter 4.3.1). These event sets are constructed to be representative of all possible earthquake scenarios that could occur in the region, each with a probability of occurrence. **Area sources** are used where earthquakes are distributed uniformly across regions considered by scientists as homogeneous, sometimes augmented by **mapped faults** when their locations are considered reliably known. Most models use faults mapped by national agencies.

Most simply, for both area and mapped sources, earthquake sizes are set according to a local historically calibrated Gutenberg-Richter relationship. However, because fault systems are finite in size, typically an upper truncated power law is used to set a maximum possible size for earthquakes (Marzocchi and Sandri, 2003). Alternatively, the characteristic earthquake model may be introduced in some regions to model worst-case magnitude scenarios and their likelihood, with the potential difficulties discussed above (see also Chapter 4.3.6.3).

Currently, simpler models use earthquake probabilities that assume earthquakes are random in time (i.e. Poisson; Chapter 1.11.11.1) and with no dependence of the timing of an earthquake on either the elapsed time since the last earthquake on the same fault or activity on other faults within the system. More sophisticated models include time or slip dependence or Coulomb Stress Changes in their earthquake probability calculations (e.g. RMS, 2012). Most peril-regions do not incorporate time dependence in the models, although it has been included in some places (e.g. California) where sufficient data are available. Different scientific opinions regarding the inputs into hazard assessments are starting to be included in the uncertainty of some models (e.g. AIR Worldwide, 2012).

The area expected to be impacted by shaking for each event or footprint is calculated across space (i.e. *x*, *y*) using empirical strong ground motion prediction equations (e.g. Equation (3.3))

$$Shaking = A_1 + A_2 \times M + A_3 \times (M - M_{REF}) + A_4 \times ln\left[\left(R^N + C_{Source}\right)^N\right] + A_S \times R + A_6$$

$$\times F^{source} + A_7 \times F^{site} + A_8 \times F^{HW} + A_9 \times F^{main}, \text{with } \sigma_{shaking} = A_{10} \qquad (3.3)$$

Shaking is the ground motion parameter of interest (peak acceleration, velocity, displacement, response spectral ordinate, etc.; see Kramer, 1996), $M$ is the magnitude, $R$ is a distance measure, $M_{REF}$ and $C_{Source}$ are magnitude and distance terms that define the change in amplitude scaling, and the $F^{(source,\ site,\ HW,\ main)}$ are indicator variables of source type, site type, geometry of the fault, and main shock discriminator. The $A_i$ coefficients are determined by a regression of past observations, and the last term, $\sigma_{shaking}$, represents the estimate of the standard deviation. This functional form of the ground motion has been used for the 'Next Generation Attenuation' (Abrahamson and Silva, 2008; Boore and Atkinson, 2008; Campbell and Bozorgnia, 2008; Chiou and Youngs, 2008; Idriss, 2008) project which is now commonly implemented in the industry catastrophe models. Older forms were limited to a few terms only (Douglas, 2003) and are equally found in the industry catastrophe models. Using the preceding equation, shaking is calculated for locations either on a grid (e.g. regular or multi-resolution) or related to administrative areas such as postcodes. Importantly, the resolution needs to be sufficient to capture the variation of soil $F^{(site)}$ or, as is always the case, uncertainty $\sigma_{shaking}$ should be introduced to account for not knowing the detail. Local site effects that affect the damage are soil deposits, dynamic response, topography or liquefaction and are reflected in the coefficient $F^{(site)}$.

**DAMAGE METRIC AND VULNERABILITY:** Earthquake damage is typically modelled as a damage ratio that is a function of shaking intensity (Figure 3.23). Intensity can be MMI, or more usually PGA or PGV. More sophisticated models use **spectral accelerations (SA)**, though these typically require building height to be known. These **vulnerability functions** (also known as **vulnerability curves**) express damage to buildings as a mean ratio, but actual monetary loss estimates in models do account for uncertainties to handle the nonlinear effects of policy mechanisms in the calculation chain.

Vulnerability curves have typically been developed on the basis of analysis of claims data from past events throughout the world, engineering-based analytical studies, expert opinions, and testing (Rossetto, Ioannou and Grant, 2013). Vulnerability functions have been developed for structural damage to buildings, as well as for business interruption losses and damage to contents (e.g. Figure 3.23).

Damage to a building depends on the characteristics of the earthquake-induced shaking and its vulnerability, which is dependent on a number of building attributes, these include: construction type (e.g. unreinforced masonry, brick or stone buildings have a higher vulnerability than steel frame buildings), height (a structure is most sensitive to ground motions with

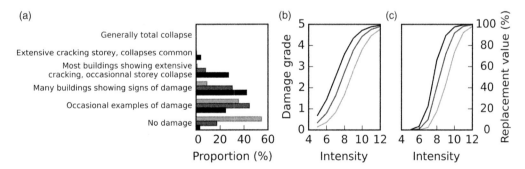

**Figure 3.23** Illustrative vulnerability functions, derived according to Risk-UE methodology (Mouroux and Le Brun, 2006). (a) shows the distribution of damage states for a particular intensity level in a particular area for three different building types; (b) displays the average damage status per intensity value for three different building types in a particular area; (c) displays the resulting replacement cost as a percentage of insured value, which can be computed once the damage grade is known.

frequencies near its natural resonant frequency in Hz, which is approximately 10/number of stories), age (generally older buildings are more vulnerable although there are exceptions), compliance with building codes (generally these are updated to incorporate new findings following events), **building regularity** (horizontal and vertical changes in mass, stiffness, and symmetry will increase building vulnerability, especially a soft or irregular ground floor), any underlying slope (a sloping base will increase vulnerability), retrofitting measures, building use and its contents, engineered compared with nonengineered structure, building orientation relative to direction of motion, and directivity of seismic waves. Models which take these things into account have the potential to perform better, although the onus is on the model user to ensure that accurate building characteristic information on the risks of concern is input into the model.

**ADDITIONAL LOSS DRIVERS INCLUDED:** Earthquake-induced fires have the potential to result in very large losses. This was demonstrated by the large conflagrations in the 1923 Great Kanto earthquake in Tokyo and the 1906 San Francisco earthquake in which 98% of the property damage is estimated to be from the fires (iinc, 2012). Fire following earthquake (FFEQ) losses are usually available within earthquake models either as a FFEQ model explicit modelling based on factors such as building spacing and ignition research in more-developed regions or as a simplified additional loss driver in less sophisticated models (Aon Benfield, 2012; AIR Worldwide, 2012). The FFEQ effect depends on the shaking damage, the building construction type, and building density, as well as the regional supply of water and fire department resources, which are limited. Flooding of properties caused by the fracturing of fire-protection sprinkler systems, or other fire extinguishing systems, by an earthquake are also a source of additional loss, particularly in the United States.

### 3.7.5 Secondary Perils and Non-Modelled Items

Compounding losses due to critical infrastructure damage, resource scarcity, reduced workforce, and secondary perils are generally not modelled explicitly in catastrophe models, but included indirectly under loss amplification, which is typically set as a percentage of the modelled loss (e.g. 5–10%).

Most of the world's most damaging tsunamis have been earthquake-induced, but their representation in catastrophe models was typically absent or rudimentary until recently when it caused an estimated 15–25% of insured losses (AIR, 2013a) in the 2011 Great East Japan Earthquake and Tsunami (GEJET); see Section 3.9. Similarly, earthquakes induce landslides and debris flows (see Section 3.8) that can cause high economic losses, especially following earthquakes in mountainous terrain such as in the Himalayas, China, and central Italy (e.g. Yin, Wong and Sun, 2009). The damages from the GEJET to the Fukushima Daiichi Nuclear Power Plant were a result of the triple hazard of earthquake, tsunami, and local landslide (e.g. Faure Walker *et al.*, 2013).

Soils can behave as a liquid when shaken, a process known as **liquefaction** causing losses such as seen in the eastern suburbs of Christchurch following the February 2011 earthquake; this may or may not be included explicitly within a catastrophe model. This is a particular risk for reclaimed land and infill. Daniell and colleagues (2015) estimates liquefaction caused 15% of economic losses in the last century, compared with 62% from shaking.

Following a main event, there are typically aftershocks that follow a standard frequency-magnitude distribution (i.e. b ~1), meaning that there are ten times as many aftershocks for each integer increase in magnitude (Utsu, 1961). Current probabilistic methods exclude such dependent events as individual entities; this is a material issue as the Emilia-Romagna 2012 earthquake in Italy and the Christchurch earthquake in New Zealand in 2011 caused high insured losses from aftershocks (EQECAT, 2012), which were not explicitly accounted for in the models. Current research in engineering is investigating the effects of repeated shaking on building vulnerability curves (e.g. Polese *et al.*, 2012) and hence whether serious damage could be expected from aftershocks if the main shock has weakened the structure. Such effects are not included in vulnerability estimates within models (EQECAT, 2012).

**Induced seismicity** is earthquake activity raised above the expected natural level by human activity. For instance, the USGS has identified a 10-fold increase in seismicity in the Mid and Eastern United States since 2010 (Ellsworth, 2013), with some $M_w$ >5 events (e.g. $M_w$ 5.3 in Colorado, 2011), potentially because of the injection of waste water from oil and gas production into deep disposal wells. Concerns also exist regarding mining (Gibson and Sandiford, 2013), particularly in Australia (e.g. Albaric *et al.*, 2013) and South Africa. Induced seismicity cannot be modelled with a Poisson distribution because of the role of pore fluid pressure (e.g. McGarr, 2014) and is currently not included explicitly in most catastrophe models.

### 3.7.6 Key Past Events

Historical events are used by model developers and users to justify the need for and validate models, cross-check between them, and identify factors that need to be included. The 1811–1812 New Madrid (in the United States) sequence of large earthquakes of unknown source caused damage over a large area and highlights the role that time dependence may play. Of historical earthquakes in the United States, this event is modelled to have the highest insured loss if it were to occur today (AIR Worldwide, 2012a). The 1906 San Francisco quake is taken as an example of the potential for high value losses from fire following earthquake, although the reduction in buildings constructed of wood and the installation of fire-protection sprinklers in commercial buildings have mitigated this risk to a large degree. The 1994 Northridge earthquake created a large financial impact, insured losses of about US $12.5 billion and total losses of about US$40 billion, justifying the need for earthquake catastrophe models. The 2004 Sumatra and 2011 GEJET 'great' (i.e. $M_w$ >8) quakes each ripped through several previously hypothesized 'segment boundaries' highlighting deficiencies in the segmentation theory for subduction zones (Kagan, Jackson and Geller, 2012). The GEJET has also highlighted the need for explicit and less simplistic modelling of losses from tsunami. The 2011 Christchurch earthquake highlighted the high losses possible from liquefaction and aftershocks.

### 3.7.7 Open Questions/Current Hot Topics/Questions to Ask Your Vendor

When making decisions from the output of earthquake catastrophe models, the following questions could be revealing:

- Which secondary perils (i.e. tsunami, fire following, landslide) are modelled explicitly, included as a generic factor, or not at all? Note that anything included in the historical loss figures to which the model has been calibrated is implicitly factored in.
- Where are fault maps used to define the hazard, how complete are they, and where have area sources been used? Area sources will have a lower resolution and higher associated uncertainties with the ground motion equations.
- How long is the observational period used to assess earthquake rates? Short historical (i.e. < a few seismic cycles) records may miss some possible events (e.g. GEJET).
- Is the model based on one or multiple hazard models? Including alternative scientific views avoids single assumptions dominating, and may allow the discrepancies between models to be considered.

- Since clusters of events will increase worst-case losses, are fault interactions included? Similarly, is time dependence (e.g. the earthquake cycle) included?
- What has been used as the definition for a single event? Are aftershocks associated with the main shock? Are other triggered events (e.g. < 72 hours later) linked? It is important to know what definitions are, especially for the kth event, and what parametric triggering rules are used.
- Have local vulnerability curves been used, or have US ones just been adjusted? An educated guess where information is lacking will be less reliable.
- How is uncertainty in compliance with building regulations assessed and accounted for? Namely, have the variable standards of local building practices been considered?

Probabilistic models of rare events cannot be robustly reality-checked on a human time scale. Indeed, no risk assessment is considered definitive, so comparing several models may be wise. A maximum credible scenario is also a useful tool.

### 3.7.8 Nonproprietary Data Sources

Authoritative hazard data sources of intensity, probabilistic models, location for earthquakes include: Global Earthquake Model (http://www.globalquakemodel.org/), International Seismological Center (http://www.isc.ac.uk/), and the following sources from the United States of Geological Society and Earthquake Research Institute:

GEM, Global Historical Earthquake Catalogues: http://www.globalquakemodel.org/what/seismic-hazard/historical-catalogue
Japan Earthquake Catalogues: https://wwweic.eri.u-tokyo.ac.jp/db/
Japan National Hazard Maps: http://www.eri.u-tokyo.ac.jp/BERI/pdf/IHO81304.pdf
United States Earthquake Hazard Maps: http://earthquake.usgs.gov/hazards/
United States Seismic Hazard Analysis Tools: http://earthquake.usgs.gov/hazards/apps/
Earthquake data catalogues: http://earthquake.usgs.gov/data/?source=sitenav/

Detailed data for calculating vulnerability, that is, contemporaneous loss and exposure data, are normally proprietary and not publicly available. Historical loss/claims and exposure data can be difficult to obtain, but US insured losses are collated and archived by the PCS at the US state and event level. Alternatively, CRED's EM-DAT database (http://www.cred.be/) and GEM (http://gemecd.org/) provide valuable insights. Elsewhere, aggregated loss data (e.g. by country, by event, or by line of business) are generally proprietary and assembled by the insurance and reinsurance industries and described in publications by companies such as Swiss Re, Munich Re, and broker Aon Benfield; some summary statistics, however, are available.

Aon Benfield: http://www.aon.com/impactforecasting/impact-forecasting.jsp
Munich Re: http://www.munichre.com/en/reinsurance/business/non-life/natcatservice/index.html
PCS: www.verisk.com/property-claim-services/pcs-estimates-of-insured-property-loss.html
Swiss Re: http://www.swissre.com/sigma/

---

**Further Reading** Basics are covered by 'Earthquake Hazards 101': http://earthquake.usgs.gov/hazards/about/basics.php or Bolt (2003), and the necessity for improving on hazard maps is introduced by Stein, Geller, and Liu (2012).

## 3.8   Mass Movement

*Tom Dijkstra, Craig Verdon, and John Hillier*

*Reviewer: Joanna Faure-Walker*

### 3.8.1   What Is the Peril?

**DESCRIPTION: Landslides**, or **mass movements (MMs)**, are defined as the downslope movements of rock, earth, or debris and can be grouped into six types: falling, toppling, sliding, spreading, flowing, and slope deformation (Hutchinson, 1989; Cruden, 1991; Hungr, Leroueil and Picarelli, 2014). Generally, landslides take place as a result of shear failure along a **slip surface** when the **shear strength** (friction and cohesion) along this surface is no longer able to withstand **driving forces** (primarily gravity) acting down the slope (Figure 3.24(a)). Material generally originates between **crown** and **foot**, moving to create a **scarp** of exposed material and, in a process called **runout**, can extend beyond its original footprint forming a **toe** (see Figure 3.24 (b)). Landslide velocities vary widely (see Figure 3.24), and volumes vary from a few small boulders to millions of cubic metres (Table 3.7). Landslide occurrence focuses in hotspots around the world (Nadim *et al.*, 2006) and may be associated with volcanoes, earthquakes, or rain (Gill and Malamud, 2014). Most serious consequences occur as a result of large and rapidly moving individual landslides, particularly as those that behave like fluids may travel rapidly and for many kilometres such as Zhouqu in 2010 (Tang *et al.*, 2011) and Elm in 1881 (e.g. Heim 1932; Selby 1993), or when many landslides occur in clusters in time and space (e.g. Wenchuan in, 2008; see Table 3.7).

   **PHYSICAL DRIVERS:** Landslides occur on slopes where the driving forces are greater than the resisting forces (material strength). Natural processes or mismanagement of marginally stable slopes can initiate a landslide by: (i) increasing the driving forces (e.g. seismic shock, slope oversteepening, external loading); (ii) decreasing the resisting forces (e.g. strength loss through weathering or changes in groundwater); or a combination of the two (Bromhead, 2006). Preceding or **antecedent conditions** determine how stable a slope is and affect the magnitude of the trigger needed to initiate a landslide event. Triggers may either be local or spatially correlated across a large footprint within a restricted time window (e.g. storm, earthquake).

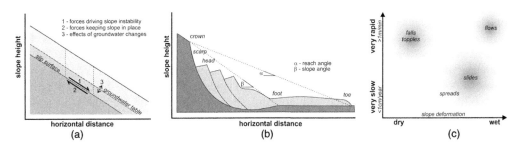

**Figure 3.24**  (a) The main components determining shallow landsliding, that is, for depths to the slip surface of several metres (Lu and Godt, 2013). (b) A section through a rotational slide; sizes range from several metres to hundreds of metres. (c) A characterization of major landslide types.

**KEY BEHAVIOURAL SUBTLETIES:** The likelihood of landslides occurring in a region, or its **susceptibility**, is determined by complex interplays between topographic slope, geology, soil materials, land use, seismic activity, climate, etc. Rainfall can lead to wetter, heavier slopes becoming unstable. Small landslides (slip surface depths of just a few metres) respond very quickly to this **hydro-meteorological triggering** while large landslides respond more slowly, sometimes months later than rainfall events (Lu and Godt, 2013). Many empirical- and process-based models have been developed to account for the relationships between landsliding and rainfall types (e.g. Guzzetti *et al.*, 2008; Montrasio *et al.*, 2014). Triggering by earthquakes generally leads to more events at high mountain peaks due to **topographic amplification** of shaking amplitudes (Meunier, Hovius and Haines, 2008) and where soft ground over hard rock is found due to **soil amplification** (Drosos, Gerolymos and Gazetas, 2012). Human behaviour can mitigate the hazard (e.g. by slope management, drainage, reinforcement; Choi and Cheung, 2013), or severely exacerbate it, for example, by transport infrastructure, dam construction, soil tips, land use change, mining (Bromhead, 2006; Table 3.7). There are further complexities; vegetation sucks water out of the ground and provides additional strength through roots, but any benefit is generally limited to shallow-based landslides and disappears if the vegetation burns (Cannon, Kirkham and Parise, 2001; Cannon *et al.*, 2008). Thus, landslide susceptibility changes over time with its drivers (e.g. climate; Dijkstra and Dixon, 2010). In addition, large triggers (e.g. earthquakes, hurricanes) can enhance the landslide susceptibility of the affected region for many years after the trigger event (Hovius *et al.*, 2011).

### 3.8.2    Damage Caused by the Peril

Although the largest single landslides can result in losses of $1–2 billion (e.g. 2013 Bingham Canyon Mine collapse; Pankow *et al.*, 2014; Table 3.7), these events are rare and their impact from a reinsurance industry (i.e. global) perspective is relatively modest. Landslide clusters, however, can seriously affect regions or countries, contributing to financial losses of significant proportions of GDP when many landslides are triggered by single trigger events (e.g. Hurricane Tomas, 2010, >43% overall St Lucia GDP; Economic Commission for Latin America and the Caribbean, 2011) or through gradual hydro-meteorological triggering causing attritional damage across a financial year (e.g. annually, landslides cost around 25% GDP in Italy (Lunio, 2002; Lu and Godt, 2013). Direct damage is to lives, livelihoods, and infrastructure (e.g. utilities, transport), with indirect effects hard to quantify and including, for example, tsunami-like waves if a landslide enters water (Genevois and Ghirotti, 2005; Schnellmann *et al.*, 2006) or, if a landslide dams a river, through extensive flooding both as it stores water and if the dam fails (e.g. Costa and Schuster, 1988).

### 3.8.3    Forecasting Ability and Mitigation

Complex interactions among processes, incomplete data, and intrinsic uncertainty in triggers still make it very difficult to accurately forecast time and location of specific landslides. For pragmatic risk assessment, maps of landslide susceptibility are useful (e.g. Nadim *et al.*, 2006; Günther *et al.*, 2014), but challenges remain in their construction (van Westen, Van Asch and Soeters, 2006; van den Eeckhaut and Hervás, 2012). Typically, they combine first-order drivers of instability (e.g. topographic slope, soil moisture, shallow geology) and are calibrated against catalogues of events, that is, **landslide inventories** (Malamud *et al.*, 2004a, 2004b). In addition to extensive fieldwork, these inventories are increasingly based on observations from satellite images (Guzzetti *et al.*, 2012; Jordan *et al.*, 2015,) satellite multi-temporal interferometry (Wasowski and Bovenga, 2014) and aerial photographs (Metternicht, Hurni and Gogu, 2005). Analysis of these inventories can yield size-frequency relationships (Malamud *et al.*, 2004a) and precipitation intensity-duration (I-D) landslide trigger thresholds (Guzzetti *et al.*, 2008). However, because susceptibility is now widely recognized to vary through time (Section 3.8.1; Dijkstra and Dixon, 2010; Hovius *et al.*, 2011), models based on physical process representation are increasingly being used to describe how, in time and space, landscapes respond to changing stresses by generating landslides. Also, probabilistic projections of landslide occurrence (e.g. AIR, 2012) are replacing deterministic ones. Mitigation can be cost-effective; in Hong Kong, for example, disasters in 1972 (138 deaths) and 1976 (18 deaths) contributed to the establishment of the Landslip Preventive Measure (LPM) Programme where, through investment in research and successful retrofitting of substandard man-made slopes, landslide risk is now effectively managed (Choi and Cheung, 2013).

### 3.8.4 Representation in Industry Catastrophe Models

**HAZARD DESCRIPTION AND EVENT SETS:** Landslides are generally regarded as a triggered or **secondary peril** in industry catastrophe models. There are no stand-alone commercial landslide models and the peril is excluded from weather-related models (e.g. hurricane). Even within earthquake models the approach taken is basic (see Corominas *et al.*, 2014), and damage due to landslides is not explicitly modelled; that is size, speed, and mode of failure are not considered for individual slides. Models assign a susceptibility rating and landslide initiation thresholds based on topographic slope, substrate (i.e. soil or rock) material strength and seasonal water saturation (e.g. Wilson and Keefer, 1985; Jibson, Harp and Michel, 2000; AIR, 2012); this may use physics-based models such as Newmark's displacement (Newmark, 1965). Then, a set of footprints comprising landslide-influenced areas can be created by determining where these thresholds are exceeded due to modelled seismic ground acceleration (i.e. for each of an earthquake event set). Although useful as a first approximation, this kind of approach can significantly deviate from observed co-seismic landslide populations (BGS/Durham, 2015). There is therefore a need for better models of the landslide peril, and not just for earthquake triggered landslides, but also for precipitation induced events.

**DAMAGE METRIC AND VULNERABILITY:** Industry models do not normally include landslide-specific vulnerability curves that consider the exposure at risk, type of landslide, or mechanism by which damage is caused (see Corominas, 2014). This is changing slowly, with some vendors introducing curves that vary by construction type (AIR, 2012), although at this stage confidence in these curves depends on the amount of detailed information provided about how they were formulated. In addition, vulnerability curves require substantial amounts of data to construct and need to be updated regularly (e.g. Papathoma-Koehle *et al.*, 2015). Loss as the fraction of the repair value of the assets at risk, or damage ratio (DR), is determined from both the susceptibility of the site and a metric quantifying calculated seismic ground motion for each event impacting that site. Specifically, co-seismic landslide damage might be calculated using vulnerability functions including either a relative measure of slope performance (i.e. resistance to failure) such as Newark Displacement, or on the degree to which a trigger shaking intensity is exceeded (e.g. Newmark, 1965; AIR, 2012).

### 3.8.5 Secondary Perils and Non-Modelled Items

Most initiating and contributory factors to landslides (e.g. weather-induced landslides, failure of man-made slopes and enhanced susceptibility due to wildfires) are not modelled. Secondary perils such as landslide lakes, increased fire risk, or landslide-induced tsunamis are also not considered. Indirect impacts (e.g. contingent business interruption) and effects such as severed transport links are not considered.

### 3.8.6 Key Past Events

Examples of some significant landslides since 1900, including the largest recorded submarine and subaerial landslides, are shown in Table 3.7.

Table 3.7 A small selection of significant landslide events since 1900, and the largest recorded submarine and sub-aerial landslides.

| Event | Year | Type | Volume | Impact | Comment | References |
|---|---|---|---|---|---|---|
| Storegga, Norway | 8200BP | Submarine landslide | 2500 to 3500 km$^3$ | Tsunami impact in GB/ Norway | A very large submarine landslide affecting a large coastal zone | Bryn *et al.* 2005 |
| Heart mountain landslide | Eocene | landslide | Area covered: 3400 km$^2$ | Rapid, catastrophic sliding >50 km | The largest known subaerial landslide (56–33.9 My BP) | Goren, Aharonov and Anders 2010 |
| Frank Slide, Alberta, Canada | 1903 | rock avalanche | 30 Mm$^3$ | 70 deaths | mining related | Cruden and Krahn 2012 |
| Haiyuan landslides, China | 1920 | slides, flows | Area covered >50000 km$^2$ | >100000 landslide deaths | Earthquake triggered landslides | Zhang and Wang 2007 |
| Vaiont landslide, Italy | 1963 | landslide | 270 Mm$^3$ | 2,500 deaths | Destroyed the village of Vaiont. Dam construction. | Genevois and Ghirotti 2005 |
| Aberfan landslide, Wales | 1966 | Flowslide | Travel distance >600 m | 144 | Rainfall trigger spoil tip failure. 115 deaths were school children | Hutchinson 1986 |
| Vargas, Venezuela | 1999 | Flash floods and debris flows | ~1.9 Mm$^3$ | >30000 deaths; US $1.8billion | Death toll approaching 10% of total population. | Wieczorek *et al.* 2001 |
| Chi-Chi landslides, Taiwan | 1999 | Landslides and debris flows | >20000 events. Area covered: 113 km$^2$ | 2,415 deaths; US$14 billion overall loss; US750 million insured loss | Co-seismic landslides and heightened susceptibility for post-seismic rainfall-triggered events | Lin *et al.* 2006 Munich Re 2011 |
| Kashmir landslides, Pakistan | 2005 | Landslides, rockfalls, debris flows, debris avalanche | Several thousand events, from small to ~80 Mm$^3$ | >25000 deaths due to co-seismic landslides | Earthquake triggered landslides | Owen *et al.* 2008 Dunning *et al.* 2007 |
| Wenchuan landslides, China | 2008 | Avalanches, flows, slides | 197,481 events affecting a of >100000 km$^2$ | 84000 deaths; US$85 billion overall loss; US$300 million insured loss | Earthquake trigger, e.g. rockslide in Beichuan destroyed school, >1600 deaths. | Xu *et al.* 2013 Yin *et al.* 2011 Munich RE 2011 |

Table 3.7 (*Continued*)

| Event | Year | Type | Volume | Impact | Comment | References |
|---|---|---|---|---|---|---|
| Zhouqu debris flow, China | 2010 | debris flow | 2.2 Mm$^3$ | 1,765 deaths | rainfall triggered | Tang *et al.* 2011 Dijkstra *et al.* 2012 |
| Hurricane Tomas landslide, St Lucia | 2010 | Landslides, debris flows | >1,200 events | >40% GDP | Landslide swarm triggered by hurricane, enhancing landscape sensitivity for many years thereafter | ECLAC 2011 Jordan *et al.* 2015 |

### 3.8.7 Open Questions/Current Hot Topics/Questions to Ask Your Vendor

Landslides result from a complex interplay of processes, and are crudely represented in catastrophe models. Current hot topics/questions therefore include:

- Is landsliding explicitly modelled? If not, is it implicitly included by calibration to historic losses that include landslide damage? Alternatively, for explicit modelling, how were landsliding losses separated out in the historic losses?
- In vulnerability curves, what is the metric that drives damage? What exactly is the calculation used to incorporate landslide losses?
- Slope steepness is key to susceptibility mapping, but how does the model allow for areas below an unstable slope, for example, how does it include runout pathways (Pareschi *et al.*, 2000; Toyos *et al.*, 2007)?
- How do different mechanisms (geophysical, meteorological, hydrological) manifest themselves in slope damage?
- How significant are landslides in hazard cascades?
- How are time-varying processes that affect annual susceptibility (e.g. wild fires causing deforestation, or earthquakes weakening slopes, e.g. Hovius *et al.*, 2011) accounted for, if at all?
- How can climate change be pragmatically accounted for (Dijkstra and Dixon, 2010; Dijkstra *et al.*, 2014)?

### 3.8.8 Nonproprietary Data Sources

Yearly landslide fatalities (2004–2010) have been collated (Petley, 2012). National landslide databases may be available through Web portals (e.g. BGS, NWRED); disasters are databased on GLIDE, with summary statistics on the PreventionWeb pages.

British Geological Survey: www.bgs.ac.uk/landslides/NLD.html
GLIDE: www.glidenumber.net
Norwegian Water Resources and Energy Directorate: www.xgeo.no
PreventionWeb: http://www.preventionweb.net/english/hazards/statistics/?hid=65

> **Further Reading** Smith and Petley (2009) give an overview of the hazard of mass movment. Lu and Godt (2013) provide a substantial analysis of hillslope hydrology and stability. Wilson and Keefer (1985) and Jibson and colleagues (2000) provide reviews of the computational stages required to make a model. Corominas and colleagues (2014) provide recommendations for the quantitative analysis of landslide risk.

## 3.9 Tsunami

*Anawat Suppasri and Yo Fukutani*

*Reviewer: Tiziana Rossetto*

### 3.9.1 What Is the Peril

**DESCRIPTION: Tsunami** are a series of long-duration waves that are generated by an impulsive, rapidly occurring disturbance of a large water mass. Their **wave period**, or time between successive waves, can range between 10 minutes and 2 hours. Their **wavelength** ($\lambda$), or distance between successive waves (Figure 3.25a), is much larger than the water depth ($d$) in which they propagate so they are **shallow-water waves**. Because the total energy of a tsunami remains approximately constant, its speed decreases as it enters shallow water and its amplitude grows; amplitude is its maximum height above sea level in offshore areas before it breaks. For example, in water 7 km deep, a tsunami can travel as fast as 943 km/h ($\lambda$ of 280 km), while in shallow water of 10 m, the same tsunami is expected to slow down to 36 km/h ($\lambda$ of 11 km) (Edward, 2008: 27–29). A tsunami travelling inland causes flooding or **inundation**, which is measured as **inundation height**; this is defined as the maximum elevation above mean sea level, at any given point, that gets submerged during the wave's passage. **Inundation depth** is, at any point, the extent to which the inundation height is above the land surface (see Figure 3.25 (a)). **Runup height** is the maximum height above sea level reached by the inundation. Tsunami inundation is the main cause of damage, however, the return flow to the sea because of gravity, or **back wash**, also causes significant damage in some cases.

**PHYSICAL DRIVERS:** Tsunami are mainly generated by submarine earthquakes (EQs). Other drivers are impacts of objects from outer space (Goto *et al.*, 2004), submarine landslides, or those into water including lakes and reservoirs (Schnellmann *et al.*, 2004; Salamon *et al.*, 2007; Fritz, Mohammed, and Yoo, 2009) and volcanic eruptions (e.g. Latter, 1981). All these can, but do not necessarily, generate tsunamis. The Pacific Ocean's rim is notably active due to its many subduction zones and EQs (e.g. Keller and Blodgett, 2008: 83). EQ magnitude and frequency vary by geographic region. Medium to large EQs (M < 8) that generate small to moderate tsunami (< 3 m height) recur approximately every 50–100 years while 'great' EQs (M > 8) capable of causing major tsunami (> 3 m) may be separated on average by 500–1,000 years (e.g. Minoura *et al.*, 2001; Jankaew *et al.*, 2008; Sparkes *et al.*, 2010); however, see Chapter 3.7.1 regarding earthquake clustering.

**KEY BEHAVIOURAL SUBTLETIES:** Tsunami are typically generated by large earthquakes (M > 7.0) at shallow depths (<50 km) but in deep sea (>1 km; Suppasri, Koshimura and

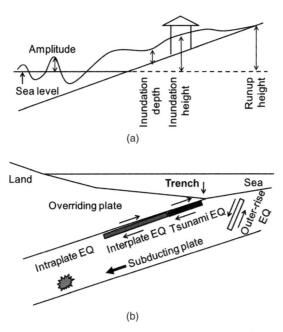

Figure 3.25 (a) Basic terminology associated with a tsunami, and (b) illustration of tsunamigenic earthquakes in a subduction zone (Fowler, 2005: 6) setting.

Imamura, 2011). Most tsunami are generated by **interplate earthquakes** (see Figure 3.25 (b)) between plates at plate boundaries along subduction zones (e.g. Crete in 1530 and Indian Ocean in 2004; Fujii and Satake, 2007) and Japan in 2011 (Satake and Fujii, 2014). Dangerous passive margin intraplate EQs are, however, also possible, for example, Lisbon in 1755 (Chester, 2008). Tsunami generated by **intraplate earthquakes**, for example, South Carolina in 1886 (Bollinger, 1972) and Sumatra in, 2012 (Satriano *et al.*, 2012) and **outer-rise earthquakes**, for example, Sanriku in 1933 (Geist *et al.*, 2009) are relatively rare. Abnormal earthquakes that cause tsunami waves larger than expected for their magnitude are called **tsunami earthquake** (Kanamori, 1972), exemplified by Niaragua 1992. They are typically explained by invoking rupture at slow velocities and very shallow depths, and enhance risk in unexpected locations (Bell *et al.*, 2014). Tsunami activity can also cluster, such as for the Indian Ocean after the 2004 event where nearby parts of the margin ruptured (Henstock *et al.*, 2010). Tsunami interact with coastal topography giving significant spatial variability to their impact (e.g. Fraser *et al.*, 2013), most obviously penetrating further (e.g. 2–5 km) into low-lying areas. Tsunami can also be amplified significantly (e.g. height doubled) by **focusing** in bays, and can propagate many tens of km from the sea up rivers (Fritz, Mohammed and You, 2009).

### 3.9.2  Damaged Caused by the Peril

EQ and subsequent tsunami can generate very large losses, with Japan 2011 causing US$210 billion of damage of which US$40 billion was insured (Insurance Information Institute, 2013). Because tsunami have high flow velocity (Fritz *et al.*, 2012), flows as shallow as 0.5–2 m can cause serious damage to some structures (e.g. wooden houses). It is, however, not only the hydrodynamic force of the water that drives the damage. Damage also results from impacts by floating debris such as marine vessels, cars, containers, trees, and destroyed coastal defence structures/housing (Koshimura *et al.*, 2009; Suppasri *et al.*, 2011, Fraser *et al.*, 2013; Macabuag and Rossetto, 2014). In addition, tsunami can trigger beach erosion, saline intrusion that causes damage to farmland, liquefaction and scouring that damage buildings' foundations destabilizing them, and fires that are perhaps caused by an electronic short circuit of a power line or car's battery igniting nearby items (e.g. gas tanks).

### 3.9.3  Forecasting Ability and Mitigation

Tsunami occurrence is less predictable than their triggers because tsunami generation is not guaranteed. However, warnings of a few minutes to a day in advance are possible depending on the distance between the tsunami source and its impact area. Warnings can be given within 3 minutes of the rupture for nearby EQ (Japan Meteorological Agency, 2015). Prior modelling can aid mitigation planning such as suggesting the height of seawalls, elevation of industrial plants and storage facilities, evacuation routes and emergency buildings and the designation of non-residential zones (Muhari *et al.*, 2011; Iuchi, Johnson and Olshansky, 2013). Currently, it is possible to give advance warning for the arrival of tsunami generated by EQ and landslides (Vanneste *et al.*, 2011).

### 3.9.4  Representation in Industry Catastrophe Models

**HAZARD DESCRIPTION AND EVENT SETS:** Catastrophe models historically tended to include simplistic tsunami footprints for high-risk regions based on a 'bathtub' model, which assumes a horizontal water surface from the coast inland. Typically, these are for historical events and runup distributions, although variants for lower magnitudes and high or low tides may be present. This is sufficient for a first-order assessment of how losses may be compounded with the EQ peril. As of 2014, some models include sets of probabilistic tsunami footprints (e.g. see Thio, 2009) generated by numerical simulations that explicitly consider the physical mechanisms of tsunami generation (Okada, 1985), propagation (Titov and González, 1997), and inundation (Titov and Synolakis, 1998). This physical modelling is often still limited to a set of historical great EQs, but sometimes catastrophe model event sets are built using stochastic synthetic tsunami (e.g. AIR, 2013). In these, tsunami are typically calculated by changing earthquakes depth, strike, dip, rake, and heterogeneity of slip distribution (e.g. Sørensen *et al.*, 2012; Løvholt *et al.*, 2012; Fukutani, Suppasri and Imamura, 2014) randomly resampled from the probability distributions derived from observed regional data. Non-earthquake-driven tsunami are not considered.

**DAMAGE METRIC AND VULNERABILITY:** To a first approximation, the fraction of damage caused by a tsunami or damage ratio (DR) is taken as 100% in areas inundated, and otherwise it is considered as driven by inundation depth. Based on damage data from the 2004 Indian Ocean and the 2011 Japanese tsunami (e.g. Ministry of Land, Infrastructure Transportation and Tourism, 2011), vulnerability (fragility) functions for building damage prediction have also been proposed (e.g. Koshimura *et* al., 2009; Suppasri *et al.*, 2013). Some catastrophe models use these results to modify their loss assessments. Tsunami flow velocity is known from the 2011 Japan tsunami to affect DRs (Kosa, 2012; Fraser, 2013) but is not currently included in any model.

### 3.9.5  Secondary Perils and Non-Modelled Items

Fires following a tsunami have caused vast amounts of damage, but are not currently modelled, and their generation mechanism is poorly understood (e.g. Hokugo, Nishino and Inada, 2013). Casualties from tsunami are not currently in catastrophe models.

### 3.9.6 Key Past Events

Some events typically used as scenarios are: Kamchatka in 1952, Chile Valdivia in 1960, and Alaska in 1964 (Tsunami laboratory, 2015). The Indian Ocean 2004 event is notable for its US$1 billion losses in an area of low insurance penetration (Stein and Okal, 2005; Fujii and Satake, 2007; Insurance Information Institute, 2013) and because it drove the development of tsunami warning services in this region together with education programs (Siripong, 2010; Pacific Tsunami Warning Center, 2015). Neither the Japan 2011 M9.0 earthquake, the $40 billion in loss, nor that such a large tsunami would be caused by it were expected (Suppasri, Imamura and Koshimura, 2012; Satake and Fujii, 2014; Tappin *et al.*, 2014), again highlighting the need to model tsunami and to correctly estimate the worst EQ magnitude (Wesnousky, 1994; Parsons, 2009), location, and its recurrence (Parsons, 2004).

### 3.9.7 Open Questions/Current Hot Topics/Questions to Ask Your Vendor

Differences in modelled losses result from differing tsunami source assumptions, simulation technique, and vulnerability data. Key current questions are:

- Does the model have fully probabilistic tsunami footprints?
- If so, how are the effects of earthquake source parameters (e.g. slip distribution, slip amount, strike, dip, rake) on tsunami impact accounted for (Fukutani, Suppasri and Imamura, 2014)?
- Are non-earthquake tsunami sources such as the 1958 Lituya landslide (Fritz, Mohammed and Yoo, 2009) and the 1883 Krakatoa volcanic eruption (Latter, 1981) accounted for, even as scenarios?
- How are secondary impacts such as floating debris and fires quantified? Implicitly in vulnerability curves, or as optional loading factors (Hokugo, Nishino and Inada, 2013)?
- There are the different types of tides (e.g. meteorological tide, sea tide). In calculating tsunami inundation, how is tidal level set, and is it included statically or included dynamically in the numerical modelling calculation (e.g. Mofjeld *et al.*, 2007)?
- Typical vulnerability curves were largely developed after the 2004 Indian Ocean tsunami but will differ between peril-regions. Are differing building materials and construction methods accounted for, and if so by observation (Suppasri *et al.*, 2012) or generalizing functions analytically or experimentally (Macabuag and Rossetto, 2014)?

### 3.9.8 Nonproprietary Data Sources

The Global Historical Tsunami Database (National Geophysical Data Center/World Data Service, 2014) contains global information (e.g. date, location, validity of the source, maximum height) on tsunami events from 2100 BC to the present in the Atlantic, Indian, and Pacific Oceans and the Mediterranean and the Caribbean Seas. The Japan Tsunami Trace Database (Japan Nuclear Energy Safety Organization and Tohoku University, 2014) includes data on tsunami traces, literature-based geographic assessments of impact, recorded around Japan. Underlying scientific papers and documents for many past tsunami are in the database of the International Tsunami Information Center (United Nations Educational, Scientific and Cultural Organization, 2014). Web links to the data are in the bibliography.

---

**Further Reading** Keller and Blodgett (2008) give a good introduction to tsunami hazard. Boris and Mikhail (2009) detail the up-to-date physics in models of tsunami waves.

## 3.10 Volcanoes

*Sue Loughlin, Rashmin Gunasekera, and John Hillier*

*Reviewer: Robin Spence*

### 3.10.1 What Is the Peril?

**DESCRIPTION:** There are 1551 recognized, recently active (i.e., <11.7 ka) volcanoes on Earth (Figure 3.26). Most are associated with tectonic plate boundaries, with a few exceptions over 'hot-spots' such as Hawaii (Cottrell, 2014). Volcanoes produce multiple potentially hazardous phenomena (Figure 3.27). **Lava flows** (Connor *et al.*, 2012), **pyroclastic density currents** (pyroclastic flows and surges; e.g. Charbonnier *et al.*, 2013), **lahars** (volcanic mud flows; Pistolesi *et al.*, 2014), **jökulhlaups** (glacial outburst floods), **debris flows** (landslides), and **debris avalanches** cause extensive damage and may extend to tens of kilometres. **Volcanic bombs** and earthquakes may cause proximal damage before and during eruptions. **Volcanic ash**, **volcanic gas**, and **volcanic aerosol** may cause distal (i.e. at a distance from the edifice), hemispheric, or even global impacts during large magnitude eruptions. Volcanic ash fall has multiple impacts (Wilson *et al.*, 2012; Bonadonna and Costa, 2013). Eruptions may be **explosive**, or **effusive** where lava is extruded, or combine these behaviours (see Figure 7 of Simkin, Siebert and Kimberly, 2010). Volcanic hazards vary in intensity and type over time during an eruption and may occur before or after the main eruption.

Eruption intensity is quantified by the rate of eruption or mass flux and magnitude (*M*) is calculated from erupted mass (Pyle, 2015). In total, about 50–60 volcanoes erupt each year (Brown *et al.*, 2015). The global frequency of explosive eruptions is inversely proportional to *M* (Deligne, Coles and Sparks, 2010; Caricchi *et al.*, 2014), and average return periods increase steadily from *M*4 (~2.5 years) to *M*6.5 (~380 years), but then become much greater as *M* increases (Brown *et al.*, 2015; Table 2.1). Impacts of *M* >6 eruptions could last for years and

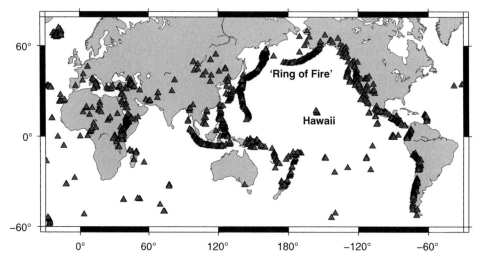

**Figure 3.26** Location map of the Earth's volcanoes active in the Holocene (i.e., last ~11,700 years. *Source:* Smithsonian Institution (2013).

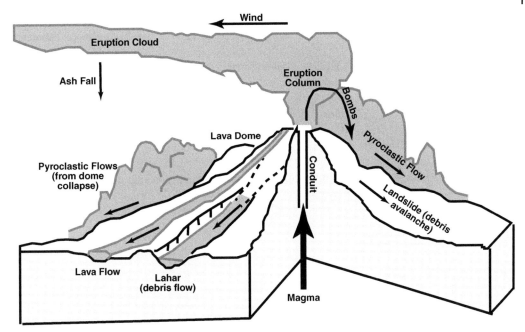

Figure 3.27 Volcanic hazards products. Also see Fagents, Gregg, and Lopes (2013) and Sigurdsson and colleagues (2015). *Source:* Adapted from USGS (http://volcanoes.usgs.gov/hazards/).

be global in extent (Oppenheimer, 2011) correlating risk between peril-regions. 'Super-eruptions' $M8+$ are rare events (Mason, Pyle and Oppenheimer, 2004; Rampino, 2008) with return periods of ~130,000 years (Crosweller *et al.*, 2012).

The most common measure of eruption magnitude is the **Volcanic Explosivity Index (VEI)**, a logarithmic scale 0–8 for estimating the relative size of explosive eruptions; VEI is based on the total volume of ejected **tephra** (explosively erupted fragments of any size including ash), and sometimes the eruption column height (Newhall and Self, 1982). However, VEI only applies to explosive eruptions.

The median duration of historical eruptions is seven weeks but eruptions may last for hours to decades (Siebert, Simkin and Kimberly, 2010), usually following a detectable episode of unrest as magma moves upwards. Roughly half the episodes of unrest at stratovolcanoes lead to an eruption (Phillipson, Sobradelo and Gottsman, 2013). (See Figure 2.5 in Brown *et al.*, 2015, for common volcano types.) This may allow stakeholders to re-evaluate annually renewed risks after an event has started but before it necessarily impacts.

**PHYSICAL DRIVERS:** Eruptive style, which governs hazard, changes over time and is primarily dictated by the composition, volatile content, and rate of supply of the rising molten rock or **magma** (Parfitt and Wilson, 2007). Vent location and topography also influence the character and distribution of different hazards and their impacts. Different hazards may interact, for example, intense rainfall may trigger lahars (i.e. volcanic mud flows), cause edifice instability or lava dome collapse (e.g. Matthews *et al.*, 2002). Large magnitude explosive '**Plinian**' eruptions are driven by violent expansion of gas bubbles that fragment magma, creating a large (i.e., $>10$ km) buoyant, convective eruption column of ash and gas, which may reach the stratosphere, for example, Pinatubo in 1991 (Newhall and Punongbayan, 1996). Empirically, the column height of such eruptions relates to mass eruption rate, a key parameter for ash dispersal models (Mastin *et al.*, 2009; Mastin, 2014). Gases, aerosols, and ash particles that reach the stratosphere may have both cooling and warming effects on global climate. In the troposphere, wind speed and direction vary with altitude and largely control which areas will be affected by volcanic ash fall (Jenkins *et al.*, 2012a, 2012b, 2015), volcanic gases, and volcanic aerosol; patterns of dispersal over time may be complex.

**KEY BEHAVIOURAL SUBTLETIES:** Globally, eruptions are not purely random but cluster in time, with similar sizes tending to occur together and the largest eruptions occurring during the most volcanically active periods (Gusev, 2006). Marzocchi and Zaccarelli (2006) identified two volcanic regimes (open and closed conduit systems) that relate to inter-event times, with some volcanoes erupting frequently and persistently for decades (Ogburn, Loughlin and Calder, 2015). Pyroclastic density currents may form when eruption columns or lava domes collapse and the low density of these currents make them highly mobile. Lahars and floods are generated when intense rainfall, or ice/snow melt entrains unconsolidated volcanic debris. Volcanic gas (that includes sulphur dioxide) may convert to sulphate aerosols, severely affecting air quality (e.g. Laki in 1783) and causing acid rain (e.g. Thordarson and Self, 2003).

### 3.10.2 Damage Caused by the Peril

Damage can be to property, infrastructure, life (e.g. Jenkins *et al.*, 2013), livelihoods, e.g. such as crops (Wilson *et al.*, 2007), and disruption to business and supply chains (e.g. Picquout *et al.*, 2013). Economic losses can be significant (EM-DAT; http://www.emdat.be/) and are typically driven by ash fall such as US$960 million for Mt. Etna in 1980 (Munich Re, 2007) or disruption to air traffic, although pyroclastic density currents can have severe impacts locally (e.g. Montserrat).

The modest Eyjafjallajökull eruption in Iceland in 2010 caused US$5 billion losses globally due to business and supply chain disruption (Ragona, Hannstein and Mazzocchi, 2011). Small eruptions (VEI <3) have caused one-fifth of the documented ash-plane incidents, and moderate-size ones (VEI 3–6) have caused nearly two-thirds (Prata, 2009), because their ash is at a height disruptive to aviation. This said, during the large 1991 Pinatubo eruption, a number of jets flying far to the west of the Philippines encountered ash that was dispersed by intense storm winds, causing costs to aviation that were estimated at over US$100 million (Casadevall, Reyes and Schneider, 1995).

Damage to buildings (i.e. their roofs) is caused by the weight of volcanic tephra/ash 'fall' deposited on them (Jenkins *et al.*, 2014). Tephra fall also affects power and telecommunications infrastructure, crops and livestock, water quality and supply systems, drains, sewage, and other critical infrastructure (Wilson *et al.*, 2012, 2014). Wet ash has much higher bulk density and behaves more like concrete, sticking to vegetation and other surfaces. Volcanic ash is easily remobilized causing lahars, and may persist for weeks or months affecting critical systems (see Section 3.8). Volcanic earthquakes can also damage buildings (see Section 3.7).

### 3.10.3 Forecasting Ability and Mitigation

Volcanic eruptions are usually preceded by detectable unrest which may last from years or months to hours (Potter, Scott and Jolly, 2012; Phillipson, Sobradelo and Gottsman, 2013). If a volcano is monitored (e.g. by a **volcano observatory**) and effective emergency management systems are in place, early warnings and short-term forecasts can be issued before and during eruptions allowing for evacuations (Surono *et al.*, 2012). The impacts of ash fall can be mitigated in the short term by rapid and effective clean-up and wearing ash masks and in the long term through building and infrastructure design. There are examples of adaptation of critical infrastructure systems to persistent ash fall (Sword-Daniels, 2011). There are also nine Volcanic Ash Advisory Centers (VAACs) worldwide that forecast dispersal of ash clouds for the aviation industry enabling avoidance and re-routing (Lechner *et al.*, 2015).

#### 3.10.4    Representation in Catastrophe Models

**HAZARD DESCRIPTION AND EVENT SETS:**  Volcano catastrophe modelling is in its early stages. Eruption scenarios exist, such as for Vesuvius (Baxter *et al.*, 2008; Spence, Gunasekera and Zuccaro, 2009; Zuccaroa *et al.*, 2008). Probabilistic modelling has been done for pyroclastic flows, for example, Montserrat (Wadge and Issacs, 1987). For ash fall, probabilistic modelling has been used to create maps of the chance of exceeding some loading or thickness threshold over a given period of time and may be applied at local to global scales (Macedonio, Costa and Folch, 2008; Jenkins *et al.*, 2012a, 2012b, 2015). These do not include variability in eruption style, frequency, magnitude, and location and their associated probabilities of occurrence (Jenkins *et al.*, 2015). But, methodologies are being created to address this using a Bayesian Event Tree Volcanic Hazard tool. So, although the basis for creating stochastic events sets exists, these have not been implemented in exposure-based catastrophe models. A caution is that statistical studies of existing data (e.g. Deligne, Coles and Sparks, 2010; Furlan, 2010; Brown *et al.*, 2014) show under-reporting of eruptions before 1950, giving apparent long return periods at many individual volcanoes so it is likely critical to address this incompleteness in creating event sets (Whelley, Newhall and Bradley, 2015, Deligne, Coles and Sparks, 2010; Bebbington, 2010).

  **DAMAGE METRIC AND VULNERABILITY:** Crudely, damage from all flows can be treated as binary, with total loss if an asset is inundated. This is true of lava flows, but damage to infrastructure due to pyroclastic density currents can be partial if the flow is dilute and also depends upon flow velocity and temperature (e.g. Wilson *et al.*, 2007; Jenkins *et al.*, 2013). Damage by lahars may also be partial (e.g. Bélizal *et al.*, 2013). Dry ash fall thickness can be used as a hazard intensity proxy relating to damage and functionality (Jenkins *et al.*, 2015), with 25 cm of dry ash sufficient to cause the collapse of some roofs (Spence, Gunasekera and Zuccaro, 2009). Aviation risk footprints are typically expressed as a function of concentration of ash, with damage relating to the length of flight path affected. For earthquakes, see Section 3.7.

#### 3.10.5    Secondary Perils and Non-Modelled Items

Secondary perils such as lahars (see Figure 3.26) may result decades after a large eruption (e.g. Pinatubo), but are not usually modelled for the insurance industry. Interacting and cascading perils are not modelled; neither is the clustering of eruptions or global impacts (e.g. temperature on crop yield). The long-term health effects of exposure to ash are increasingly well understood (Bonadonna, Macedonio and Sparks, 2002; Horwell and Baxter, 2006) but are not modelled.

#### 3.10.6    Key Past Events

Vesuvius eruptions (AD 79 and AD 1661) showcased the impacts of ash fall and pyroclastic density currents; the 1883 Krakatoa eruption highlighted the impacts of dilute pyroclastic density currents and tsunamis (Self, 2006); and Tambora (1815) and Laki (1783–1784) demonstrated the devastating impacts of ash fall and air pollution (Thordarsson and Self, 2003). The modest (VEI 3) eruption of Nevado del Ruiz, Colombia, in 1985 resulted in the loss of a town and more than 23,000 lives due to lahars (Voight, 1996), demonstrating that smaller events should not be ignored. The long-lived eruption of Soufriere Hills volcano, Montserrat, which began in 1995, has highlighted the use and applicability of probabilistic volcanic risk models (Aspinall, 2010) for emergency management. The 2010 Eyjafjallajökull eruption caused a focus on ash dispersal modelling by causing €3.3 billion losses to nine European airlines alone (Mazzocchi, Hansstein and Ragona, 2010).

### 3.10.7 Open Questions/Current Hot Topics/Questions to Ask Your Vendor

Most aspects of creating a probabilistic catastrophe model are hot topics or open questions.

### 3.10.8 Non-Proprietary Data Sources

Some of the key data sources and databases include:

BET_VH: (Marzocchi, Hansstein and Ragona *et al.*, 2010); available at vhub.org/resources/betvh

Global Volcano Model (GVM): http://globalvolcanomodel.org/

Smithsonian Institution – Global Volcano Program: http://www.volcano.si.edu/

US Geological Survey: http://pubs.usgs.gov/gip/volc/eruptions.html

VHUB: https://vhub.org/ (Palma *et al.*, 2014)

Volcano Global Risk Identification and Analysis Project (VOGRIPA): http://www.bgs.ac.uk/vogripa/index.cfm

WOVOdat: http://www.wovodat.org/

---

**Further Reading** The *Encyclopedia of Volcanic Hazards* (Sigurdsson *et al.*, 2015) gives clear introductory explanations of the types of volcanic hazard and the physical processes driving them. Loughlin (2015) gives a summary of global hazards and threats.

---

## References

**Overview**

Aon Benfield (2015) Global insured loss by peril. http://catastropheinsight.aonbenfield.com/pages/insuredglobal.aspx?region=globalandlosstype=insured (accessed 27 August 2015).

Corti, T., Muccione, V., Köllner-Heck, P., Bresch, D. and Seneviratne, S.I. (2009) Simulating past droughts and associated building damages in France. *Hydrology Earth System Sciences*, **13**, 1739–1747.

Donat, M.G., Leckebusch, G.C., Wild, S., and Ulbrich, U. (2011) Future changes in European winter storm losses and extreme wind speeds inferred from GCM and RCM multi-model simulations. *Natural Hazards and Earth System Sciences*, **11**, 1351–1370.

Gill, M. and Malamud, B.D. (2014) Reviewing and visualizing the interactions of natural hazards. *Reviews of Geophysics*, **52**, 680–722.

Hillier, J.K., Macdonald, N., Leckebusch, G. and Stavrinides A. (2015) Interactions between apparently 'primary' weather-driven hazards and their cost. *Environmental Research Letters*, **10**, http://dx.doi.org/10.1088/1748-9326/10/10/104003

Hovius, N., Meunier, P., Ching-Weei, L., *et al.* (2011) Prolonged seismically induced erosion and the mass balance of a large earthquake. *Earth and Planetary Science Letters*, **304**, 347–355.

Jansa, A., Alpert, P., Arbogast, P., *et al.* (2014) MEDEX: A general overview. *Natural Hazards and Earth System Sciences*, **14**, 1965–1984.

Kappes, M.S., Keiler, M., von Elverfeldt, K. and Glade, T. (2012) Challenges of analyzing mulit-hazard risk: A review. *Natural Hazards*, **64**, 1925–1958.

Marzocchi, W., Mastellone, M.L., Di Ruocco, A., *et al.* (2007) Principles of multi-risk assessment. *European Commission Project Report*, EUR23615.

Munich Re (2008) Topics geo natural hazards 2007. NatCatSERVICE. http://www.munichre.com/natcatservice (accessed 28 August 2015).

Munich Re (2012) Topics geo 2011. NatCatSERVICE. http://www.munichre.com/natcatservice (accessed 28 August 2015).

Munich Re (2016) World map of natural disasters, 2015. NatCatSERVICE, http://www.munichre.com/en/reinsurance/business/non-life/natcatservice/annual-statistics/index.html (accessed 28 August 2015).

*Swiss Re* (2014) Natural catastrophes and disasters in 2013. *Sigma*, no 1/2014., http://media.swissre.com/documents/sigma1_2014_en.pdf (accessed 27 August 2015).

*Swiss Re* (2015) Natural catastrophes and man-made disasters in 2008: North America and Asia suffer heavy losses. *Sigma* No. 2/2015. http://media.swissre.com/documents/sigma2_2015_en.pdf (accessed 23 March 2016).

The Royal Society (2014) Resilience to extreme weather, *The Royal Society Science Policy Centre Report*, 02/14 DES3400.

Trapero, L., Bech, J., Duffourg, F., Esteban, P. and Lorente, J. (2013) Mesoscale numerical analysis of the historical November 1982 heavy precipitation event over Andorra (Eastern Pyrenees). *Natural Hazards and Earth System Sciences*, **13**, 2969–2990.

Vitolo, R., Stephenson, D.S., Cook, I. and Mitchell-Wallace, K. (2009) Serial clustering of intense European storms. *Meteorologische Zeitschrift*, **18**, 411–424.

Wang, C. and Lee, S. (2009) Co-variability of tropical cyclones in the N Atlantic and NE Pacific. *Geophysical Research Letters*, **36**, L24702.

## Tropical Cyclones

Aguado, E. and Burt, J. (2012) *Understanding Weather and Climate*, 6th edn, Prentice Hall, Hoboken, NJ.

AIR (2012) AIR worldwide updates loss estimates for post-tropical cyclone sandy, Press Release. http://www.air-worldwide.com/Press-Releases/AIR-Worldwide-Updates-Loss-Estimates-for-Post-Tropical-Cyclone-Sandy/ (accessed 26 November 2012).

Belanger, J.I., Curry, J.A. and Hoyos, C.D. (2009) Variability in tornado frequency associated with U.S. landfalling tropical cyclones. *Geophysical Research Letters*, **36**, L17805.

Bonazzi, A., Dobbin, A., Turner, J.K., *et al.* (2014) A simulation approach for estimating hurricane risk over a 5-year horizon. *Weather, Climate and Society*, **6**, 77–90.

Brunkard, J., Namulanda, G. and Ratard, R. (2008) Hurricane Katrina deaths, Louisiana 2005. *Disaster Medicine Public Health Preparedness*, **2**, 215–223.

Bove, M.C., O'Brien, J., Elsner, J.B., Landsea, C.W. and Niu, X. (1998) Effect of El Niño on U.S. landfalling hurricanes, revisited. *Bulletin of the American Meteorological Society*, **79**, 2477–2482.

Caron, L-P., Boudreault, M. and Bruyère, C.L. (2014) Changes in large-scale controls of Atlantic tropical cyclone activity with the phases of the Atlantic multidecadal oscillation. *Climate Dynamics*, DOI: 10.1007/s00382-014-2186-5.

Chan, J.C.L. (2008) Decadal variations of intense typhoon occurrence in the western North Pacific. *Proceedings of the Royal Society*, **464A**, 249–272.

Chang, E.K.M. and Guo, Y. (2007) Is the number of North Atlantic tropical cyclones significantly underestimated prior to the availability of satellite observations? *Geophysical Research Letters*, **34**, L14801.

Chang, P., Ji, L. and Li, H. (1997) A decadal climate variation in the tropical Atlantic Ocean from thermodynamic air-sea interactions. *Nature*, **385**, 516–518.

Czajkowski, J. and Done, J.M. (2014) As the wind blows? Understanding hurricane damages at the local level through a case study analysis. *Weather Climate and Society*, **6**, 202–217.

Dailey, P.S., Zuba, G., Ljung, G., Dima, I.M. and Guin, J. (2009) On the relationship between North Atlantic sea surface temperatures and U.S. hurricane landfall risk. *Journal of Applied Meteorology and Climatology*, **48**, 111–129.

Daley, W.M., Baker, J. and Kelly, J.J. Jr. (2000) Hurricane Floyd Floods of September 1999. Service Assessment. U.S. Dept of Commerce. http://www.nws.noaa.gov/os/assessments/pdfs/floyd.pdf (accessed 13 November 2016).

Dehring, C.A. and Halek, M. (2013) Coastal building codes and hurricane damage, *Land Economics*, **80**, 597–613.

DeMaria, M. (1996) The effect of vertical shear on tropical cyclone intensity change. *Journal of the Atmospheric Sciences*, **53**, 2076–2087.

Demuth, J.L., Morss, R.E., Morrow, B.H. and Lazo, J.K. (2012) Creation and communication of hurricane risk information. *Bulletin of the American Meteorological Society*, **93**, 1133–1145.

Demuth, J.L., DeMaria, M. and Knaff, J.A. (2006) Improvement of advanced microwave sounder unit tropical cyclone intensity and size estimation algorithms. *Journal of Applied Meteorology*, **45**, 1573–1581.

Dinesh, A. (2012) Modeling hurricane-induced storm surge. http://www.eqecat.com/pdfs/ hurricane-storm-surge-modeling-dinesh-2012-08-30.pdf (accessed 1 December 2014).

Done, J.M., Bruyère, C.L., Jaye, A. and Ge, M. (2014) Internal variability of North Atlantic tropical cyclones. *JGR-Atmospheres*. DOI: 10.1002/2014JD021542.

Dube, S.K., Chittibabu, P., Sinha, P.C., Rao, A.D. and Murty, T.S. (2004) Numerical modelling of storm surge in the Head Bay of Bengal using location-specific models. *Natural Hazards*, **31**, 437–453.

Elsner, J.B. and Jagger, T.H. (2006) Prediction models for annual U.S. hurricane counts. *Journal of Climate*, **19**, 2935–2952.

Elsner, J.B., Liu, K.B. and Kocher, B. (2000) Spatial variations in major US hurricane activity: Statistics and a physical mechanism. *Journal of Climate*, **13**, 2293–2305.

Emanuel, K.A. (1986) An air-sea interaction theory for tropical cyclones. Part I: Steady-state maintenance. *Journal of the Atmospheric Sciences*, **43**, 585–604.

Emanuel, K.A. (1988) The maximum intensity of hurricanes. *Journal of the Atmospheric Sciences*, **45**, 1143–1155.

Emanuel, K.A. (2003) Tropical cyclones. *Annual Review of Earth and Planetary Sciences*, **31**, 75–104.

Emanuel, K.A. (2005) *Divine Wind: The History and Science of Hurricanes*, Oxford University Press, Oxford.

Emanuel, K.A. (2011) Global warming effects on U.S. hurricane damage. *Weather Climate and Society*, **3**, 261–268.

Emanuel, K.A., Sundararajan, R. and Williams, J. (2008) Hurricanes and global warming: Results from downscaling IPCC AR4 simulations. *Bulletin of the American Meteorological Society*, **89**, 347–367.

EQE (2012) Superstorm Sandy post-landfall estimates, *Catwatch Reports.*, http://www.eqecat.com/ catwatch/post-landfall-loss-estimates-superstorm-sandy-released-2012-11-01/ (accessed 1 November 2012).

Evans, D.L., Gudes, S.B. and Kelly, J.J. Jr. (2001) Tropical storm Allison heavy rains and floods Texas and Louisiana, June 2001. Service Assessment. U.S. Dept of Commerce. http://www.nws .noaa.gov/om/assessments/pdfs/allison.pdf (accessed 13 November 2016).

Fink, A.H., Schrage, J.M. and Kotthaus, S. (2010) On the potential causes of the nonstationary correlations between West African precipitation and Atlantic hurricane activity. *Journal of Climate*, **23**, 5437–5456.

Fiorino, M. and Elsberry, R.L. (1989) Contributions to tropical cyclone motion by small, medium and large scales in the initial vortex. *Monthly Weather Review*, **117**, 721–727.

Florida Building Code (2014) *Building*, 5th edn. http://codes.iccsafe.org/app/book/content/2014_Florida/Building%20Code/Chapter%2016.html (accessed 13 November 2016).

Forbes, C., Rhome, J., Mattocks, C. and Taylor, A. (2014) Predicting the storm surge threat of hurricane Sandy with the National Weather Service SLOSH Model. *Journal of Marine Science and Engineering*, **2**, 437–476.

Frank, W. and Ritchie, E. (2001) Effects of vertical wind shear on the intensity and structure of numerically simulated hurricanes. *Monthly Weather Review*, **129**, 2249–2269.

Frank, W. and Roundy, P.E. (2006) The role of tropical waves in tropical cyclogenesis. *Monthly Weather Review*, **134**, 2397–2417.

Frank, W. and Young, G.S. (2007) The interannual variability of tropical cyclones. *Monthly Weather Review*, **135**, 3587–3598.

Galarneau, T.J., Jr. and Davis, C.A. (2013) Diagnosing forecast errors in tropical cyclone motion. *Monthly Weather Review*, **141**, 405–430.

Goh, A.Z–C. and Chan, J.C.L. (2010) Interannual and interdecadal variations of tropical cyclone activity in the South China Sea. *International Journal of Climatology*, **30**, 827–843.

Goldenberg, S.B., Landsea, C.W., Mestas-Nunez, A.M. and Gray, W.M. (2001) The recent increase in Atlantic hurricane activity: Causes and implications. *Science*, **293**, 474–479.

Gray, W.M. (1979) Hurricanes: Their formation, structure and likely role in the tropical circulation, in *Meteorology over Tropical Oceans* (ed. D.B. Shaw), Royal Meteorological Society, James Glaisher House, Bracknell, Berkshire.

Grieser, J. and Jewson, S. (2012) The RMS TC-Rain model. *Meteorologische Zeitschrift*, **21**, 79–88.

Hall, J.D., Matthews, A.J. and Karoly, D.J. (2001) The modulation of tropical cyclone activity in the Australian region by the Madden–Julian oscillation. *Monthly Weather Review*, **129**, 2970–2982.

Hall, T.M. and Jewson, S. (2007) Statistical modeling of North Atlantic tropical cyclone tracks. *Tellus*, **59A**, 486–498.

Hall, T.M. and Jewson, S. (2008) Comparison of local and basin-wide methods for risk assessment of tropical cyclone landfall. *Journal of Applied Meteorology and Climatology*, **47**, *361-367*. DOI: 10.1175/2007JAMC1720.1.

Hart, R.E. and Evans, J.L. (2001) A climatology of the extratropical transition of Atlantic tropical cyclones. *Journal of Climate*, **14**, 546–564.

Henry, D., Cooke-Hull, S., Savukinas, J., *et al.* (2013) Economic impact of hurricane Sandy, potential economic activity lost and gained in New Jersey and New York. U.S. Department of Commerce. http://www.esa.gov/sites/default/files/sandyfinal101713.pdf (accessed 13 November 2016).

Holland, G.J. (1983) Tropical cyclone motion: Environmental interaction plus a beta effect. *Journal of the Atmospheric Sciences*, **40**, 68–75.

Holland, G.J. (1997) The maximum potential intensity of tropical cyclones *Journal of the Atmospheric Sciences*, **54**, 2519.

Holland, G.J. (2007) Misuse of landfall as a proxy for Atlantic tropical cyclone activity. *Eos Transactions American Geophysical Union*, **88**, 349–350.

Holland, G.J., Belanger, J. and Fritz, A. (2010) A revised model for radial profiles of Hurricane winds. *Monthly Weather Review*, **138**, 4393–4401.

Holland, G.J. and Bruyère, C.L. (2013) Recent intense hurricane response to global climate change. *Climate Dynamics*, **42**, 617–627.

Holland, G.J. and Webster, P.J. (2007) Heightened tropical cyclone activity in the North Atlantic: Natural variability or climate trend? *Philosophical Transactions of the Royal Society A*, **365**, 2695–2716.

Holton, J.R. and Hakim, G.J. (2012) *An Introduction to Dynamic Meteorology*, 5th edn, Academic Press, Oxford.

Insurance Information Institute (2013) http://www.iii-insurancematters.org/insurance-and-disasters/facts/index.cfm.

Intergovernmental Panel on Climate Change (2012) Summary for policymakers, in *Managing the Risks of Extreme Events and Disasters to Advance Climate Change Adaptation: A Special Report of Working Groups I and II of the Intergovernmental Panel on Climate Change* (eds C.B. Field, V. Barros, T.F. Stocker, *et al.*), Cambridge University Press, Cambridge.

Jagger, T.H. and Elsner, J.B. (2012) Hurricane clusters in the vicinity of Florida. *Journal of Applied Meteorology and Climatology*, **51**, 869–877.

Jagger, T.H., Elsner, J.B. and Burch, R.K. (2011) Climate and solar signals in property damage losses from hurricanes affecting the United States. *Natural Hazards*, **58**, 541–557.

Jagger, T.H., Elsner, J.B. and Saunders, M.A. (2008) Forecasting U.S. insured hurricane losses, in *Climate Extremes and Society* (eds H.F. Diaz and R.J. Murnane), Cambridge University Press, Cambridge.

Jain, V. (2010) The role of wind duration in damage estimation. *AIR Currents*. http://www.air-worldwide.com/Publications/AIR-Currents/2010/The-Role-of-Wind-Duration-in-Damage-Estimation/ (accessed 1 December 2014).

Jarvinen, B.R., Neumann, C.J. and Davis, M.A.S. (1984) A tropical cyclone data tape for the North Atlantic Basin, 1886–1983: Contents, limitations, and uses. NOAA Tech. Memo. NWS NHC 22. http://www.nhc.noaa.gov/pdf/NWS-NHC-1988-22.pdf (accessed 13 November 2016).

Jewson, S., Bellone, E., Khare, S., *et al.* (2008) 5-year prediction of the number of hurricanes which make U.S. landfall, in *Hurricanes and Climate Change* (eds J.B. Elsner and T.H. Jagger), Springer, New York.

Jonkman, S.N., Maaskant, B., Boyd, E. and Levitan, M.L. (2009) Loss of life caused by the flooding of New Orleans after Hurricane Katrina: Analysis of the relationship between flood characteristics and mortality. *Risk Analysis*, **29**, 676–698.

Leith, C. and Nolan, B.D. (2010) Using mesoscale simulations to train statistical models of tropical cyclone intensity over land. *Monthly Weather Review*, **138**, 2058–2073.

Lonfat, M., Boissonnade, A. and Muir-Wood, R. (2007) Atlantic basin, U. S. and Caribbean landfall rates over the 2006–2010 period: An insurance industry perspective. *Tellus A*, **59**, 499–510.

Kaplan, J. and DeMaria, M. (1995) A simple empirical model for predicting the decay of tropical cyclone winds after landfall. *Journal of Applied Meteorology*, **34**, 2499–2512.

Karen Clark and Company (2012) Historical hurricanes that would cause $10 billion or more of insured losses today. http://www.karenclarkandco.com/pdf/HistoricalHurricanes_Brochure.pdf (accessed 2 December 2014).

Knaff, J., Longmore, S. and Molenar, D. (2014) An objective satellite-based tropical cyclone size climatology. *Journal of Climate*, **27**, 455–476.

Knapp, K., Kruk, M., Levinson, D., Diamond, H. and Neumann, C. (2010) The international best track archive for climate stewardship (IBTrACS). *Bulletin of the American Meteorological Society*, **91**, 363–376.

Knudsen, M.F., Seidenkrantz, M., Jacobsen, B.H. and Kuijpers, A. (2011) Tracking the Atlantic Multidecadal Oscillation through the last 8,000 years. *Nature Communications*, **2**, 178.

Knutson, T.R., McBride, J., Chan, J., *et al.*, (2010) Tropical cyclones and climate change. *Nature Geoscience*, **3**, 157–163.

Kossin, J.P., Camargo, S.J. and Sitkowski, M. (2010) Climate modulation of North Atlantic hurricane tracks. *Journal of Climate*, **23**, 3057–3076.

Kossin, J.P., Olander, T.L. and Knapp, K.R. (2013) Trend analysis with a new global record of tropical cyclone intensity. *Journal of Climate*, **26**, 9960–9976.

Landsea, C.W., Harper, A.A., Hoarau, K. and Knaff, J.A. (2006) Can we detect trends in extreme tropical cyclones? *Science*, **313**, 452–454.

Langousis, A. and Veneziano, D. (2009) Theoretical model of rainfall in tropical cyclones for the assessment of long-term risk. *Journal Geophysical Research*, **114**, D02106.

Lecomte, E. and Gahagan, K. (1998) Hurricane insurance protection in Florida, in *Paying the Price: The Status and Role of Insurance against Natural Disasters in the United States* (eds H. Kunreuther and R. Roth Sr), Joseph Henry Press, Washington, DC.

Li, Y. and Ellingwood, B.R. (2006) Hurricane damage to residential construction in the US: Importance of uncertainty modeling in risk assessment. *Engineering Structures*, **28**, 1009–1018.

Loridan, T., Khare, S., Scherer, E., Dixon, M. and Bellone, E. (2015) Parametric modeling of transitioning cyclone's wind fields for risk assessment studies in the western North Pacific. *Journal of Applied Meteorology and Climatology*, **54**, 624–640.

Maloney, E.D. and Hartmann, D.L. (2000) Modulation of hurricane activity in the Gulf of Mexico by the Madden-Julian oscillation. *Science*, **287**, 2002–2004.

Mori, N., Kato, M., Kim, S. *et al.,* (2014) Local amplification of storm surge by Super Typhoon Haiyan in Leyte Gulf. *Geophysical Research Letters*, **41**, 5106–5113.

Mumby, P.J., Vitolo, R. and Stephenson, D.B. (2011) Temporal clustering of tropical cyclones and its ecosystem implications. *Proceedings of the National Academy of Sciences*, **108**, 17626–17630.

National Climate Data Center (2009) *U.S. Storm Events Database, 1950–2008*, National Climate Data Center, Asheville, NC.

National Hurricane Center (2014) Forecast Verification. http://www.nhc.noaa.gov/verification/ (accessed 13 November 2016).

Nordhaus, W. (2010) The economics of hurricanes and implications of global warming. *Climate Change Economics*, **1**, 1–20.

Olsen, A. and Porter, K. (2011) What we know about demand surge: Brief summary. *Natural Hazards Review*, **12**, 62–71.

Owens, B.F. and Landsea, C.W. (2003) Assessing the skill of operational Atlantic seasonal tropical cyclone forecasts. *Weather and Forecasting*, **18**, 45–54.

Palmen, E. (1948) On the formation and structure of the tropical hurricane. *Geophysica*, **3**, 26–38.

Pielke, R.A., Jr, Gratz, J., Landsea, C.W., Collins, D., Saunders, M.A. and Musulin, R. (2008) Normalized hurricane damage in the United States: 1900–2005. *Natural Hazards Review*, **9**, 29–42.

Pielke, R.A., Jr. and Landsea, C.W. (1999) La Niña, El Niño, and Atlantic hurricane damages in the United States. *Bulletin of the American Meteorological Society*, **80**, 2027–2033.

Pinelli, J., Simiu, E., Gurley, K., *et al.* (2004) Hurricane damage prediction model for residential structures. *Journal of Structural Engineering*, **130**, 1685–1691.

Pita, G., Pinelli, J-P., Gurley, K. and Mitrani-Reiser, J. (2015) State of the art hurricane vulnerability estimation methods: A review. *Natural Hazards Review*, **16**, 04014022.

Powell, M., Soukup, G., Cocke, S., *et al.* (2005) State of Florida hurricane loss projection model: Atmospheric science component. *Journal of Wind Engineering and Industrial Aerodynamics*, **93**, 651–674.

Rappaport, E.N. (2014) Fatalities in the United States from Atlantic tropical cyclones: New data and interpretation. *Bulletin of the American Meteorological Society*, **95**, 341–346.

Rappin, E.D., Nolan, D.S. and Emanuel, K.A. (2010) Thermodynamic control of tropical cyclogenesis in environments of radiative–convective equilibrium with shear. *Quarterly Journal of the Royal Meteorological Society*, **136**, 1954–1971.

Risk Management Solutions, Inc. (2013) Modelling Sandy: A high resolution approach to storm surge. *RMS White Paper* Feb 2013. http://forms2.rms.com/rs/729-DJX-565/images/tc_2013_rms_modeling_sandy_storm_surge.pdf (accessed 4 May 2016).

Rogers, R., Marks, F. and Marchok, T. (2009) Tropical cyclone rainfall, in *Encyclopedia of Hydrological Sciences*. DOI: 10.1002/0470848944.hsa030.

Saji, N.H., Goswami, B.N., Vinayachandran, P.N. and Yamagata, T. (1999) A dipole in the tropical Indian Ocean. *Nature*, **401**, 360–363.

Schmidt, S., Kemfert, C. and Hoppe, P. (2010) The impact of socio-economics and climate change on tropical cyclone losses in the USA. *Regional Environmental Change*, **10**, 13–26.

SCOR (2013) SCOR Global P&C Guide to Hurricanes: An introduction to quantifying the hazard and managing the peril. https://www.scor.com/images/stories/pdf/library/newsletter/pc_nl_hurricanes_en.PDF (accessed 5 May 2016).

Swiss Re (2013) Natural catastrophes and man-made disasters in 2012. *Sigma*, 2/2013. http://www.swissre.com/media/news_releases/nr_20130327_sigma_natcat_2012.html (accessed 20 April 2016).

Takahashi, H., Fujinami, H., Yasunari, T., Matsumoto, J. and Baimoung, S. (2014) Role of tropical cyclones along the monsoon trough in the 2011 Thai flood and interannual variability. *Journal of Climate*. DOI: 10.1175/JCLI-D-14-00147.1.

Tan, F., Lim, H.A. and Abdullah, K. (2012) The effects of orography in Indochina on wind, cloud, and rainfall patterns during Typhoon Ketsana (2009). *Asia-Pacific Journal of Atmospheric Sciences*, **48**, 295–314.

Tang, B. and Emanuel, K. (2010) Midlevel ventilation's constraint on tropical cyclone intensity. *Journal of the Atmospheric Sciences*, **67**, 1817–1830.

Tsai, F., Hwang, J-H., Chen, L-C. and Lin, T.H. (2010) Post-disaster assessment of landslides in southern Taiwan after 2009 Typhoon Morakot using remote sensing and spatial analysis. *Natural Hazards and Earth System Science*, **10**, 2179–2190.

Unanwa, C.O., McDonald, J.R., Mehta, K.C. and Smith, D.A. (2000) The development of wind damage bands for buildings. *Journal of Wind Engineering and Industrial Aerodynamics*, **84**, 119–149.

Vecchi, G.A. and Knutson, T.R. (2011) Estimating annual numbers of Atlantic hurricanes missing from the HURDAT database (1878–1965) using ship track density. *Journal of Climate*, **24**, 1736–1746.

Vickery, P.J. (2005) Simple empirical models for estimating the increase in the central pressure of tropical cyclones after landfall along the coastline of the United States. *Journal of Applied Meteorology*, **44**, 1807–1826.

Vickery, P.J., Masters, F.J., Powell, M.D. and Wadhera, D. (2009) Hurricane hazard modeling: The past, present, and future. *Journal of Wind Engineering and Industrial Aerodynamics*, **97**, 392–405.

Vickery, P.J., Skerlj, P.F. and Twisdale, L.A. (2000) Simulation of hurricane risk in the US using empirical track model. *Journal of Structural Engineers*, **126**, 1222–1238.

Villarini, G., Vecchi, G.A. and Smith, J.A. (2012) U.S. Landfalling and North Atlantic Hurricanes: Statistical modeling of their frequencies and ratios. *Monthly Weather Review*, **140**, 44–65.

Vimont, D.J. and Kossin, J.P. (2007) The Atlantic meridional mode and hurricane activity. *Geophysical Research Letters*, **34**, L07709.

Wang, C., Lee, S-K. and Enfield, D.B. (2008) Atlantic warm pool acting as a link between Atlantic multidecadal oscillation and Atlantic tropical cyclone activity. *Geochemistry, Geophysics, Geosystems*, **9**, Q05V03.

Wang, C., Liu, H., Lee, S-K. and Atlas, R. (2011) Impact of the Atlantic warm pool on United States landfalling hurricanes. *Geophysical Research Letters*, **38**, L19702.

Watson, A.I., Jamski, M.A., Turnage, T.J., Bowen, J.R. and Kelley, J.C. (2005) The tornado outbreak across the North Florida Panhandle in association with Hurricane Ivan. Paper presented at 32nd Conference on Radar Meteorology, 22–29 Oct., American Meteorological Society, Albuquerque http://www.srh.noaa.gov/images/tae/pdf/research/AMS_RadarConf.pdf (accessed 1 December 2014).

Webster, P.J., Holland, G.J., Curry, J.A. and Chang, H-R. (2005) Changes in tropical cyclone number, duration, and intensity in a warming environment. *Science*, **309**, 1844–1846.

Willoughby, H.E. and Black, P.G. (1996) Hurricane Andrew in Florida: Dynamics of a disaster. *Bulletin of the American Meteorological Society*, **77**, 543–549.

WMO (2008) Guidelines for converting between various wind averaging periods in tropical cyclone conditions (eds B. A. Harper, J.D. Kepert, and J.D. Ginger) https://www.wmo.int/pages/prog/www/tcp/Meetings/HC31/documents/Doc.3.part2.pdf (accessed 13 November 2016).

Wu, L. and Braun, S. (2004) Effects of environmentally induced asymmetries on hurricane intensity: A numerical study. *Journal of the Atmospheric Sciences*, **61**, 3065–3081.

Zhai, A.R. and Jiang, J.H. (2014) Dependence of US hurricane economic loss on maximum wind speed and storm size. *Environmental Research Letters*, **9**, 064019.

Zhang, F. (2011) The future of hurricane prediction. *Computing in Science and Engineering*, **13**, 9–12.

Zhao, H., Wu, L. and Wang, R. (2014) Decadal variations of intense tropical cyclones over the Western North Pacific during 1948–2010. *Advances in Atmospheric Sciences*, **31**, 57–65.

Zhu, P. (2008) Impact of land-surface roughness on surface winds during hurricane landfall. *Quarterly Journal of the Royal Meteorological Society*, **134**, 1051–1057, DOI: 10.1002/qj.265.

**Extra-Tropical Cyclones**

AIR (2009) Looking back, looking forward: Anatol, Lothar, and Martin ten years later. *AIR CURRENTS*. http://www.air-worldwide.com/Publications/AIR-Currents/Looking-Back,-Looking-Forward--Anatol,-Lothar-and-Martin-Ten-Years-Later/ (accessed 13 November 2016).

AIR (2010) European windstorms: Implications of storm clustering on definitions of occurrence losses. http://www.air-worldwide.com/Publications/AIR-Currents/2010/European-Windstorms--Implications-of-Storm-Clustering-on-Definitions-of-Occurrence-Losses/ (accessed 14 November 2016).

AIR (2013) Coastal flooding in the United Kingdom, 1953 and now. http://www.air-worldwide.com/Publications/AIR-Currents/2013/Coastal-Flooding-in-the-United-Kingdom,-1953-and-Now/. (accessed 6 May 2016).

Alexandersson, H., Schmith, T., Iden, K., and Tuomenvirta, H. (1998) Long-term variations of the storm climate over NW Europe. *Global Atmospheric Ocean Systems*, **6**, 97–120.

Barredo, J.I. (2010) No upward trend in normalised windstorm losses in Europe (2010) 1970–2008. *Natural Hazards and Earth Systems Sciences*, **10**, 97–104.

Beaufort, F. (1805) Beaufort scale. http://www.spc.noaa.gov/faq/tornado/beaufort.html (accessed 13 November 2016).

Berliner, M.L. (2001) Monte Carlo based ensemble forecasting. *Statistics and Computing*, **10**, 269–275.

Browning, K.A. (1997) The dry intrusion perspective of extra-tropical cyclone development. *Meteorological Applications*, **4**, 317–324.

Browning, K.A. (2004) The sting at the end of the tail: Damaging winds associated with extratropical cyclones. *Quarterly Journal of the Royal Meteorological Society*, **130**, 375–399.

Burt, S.D. and Mansfield, D.A. (1988) The Great Storm of 15–16 October 1987. *Weather*, **43**, 90–110.

Cook, N.J. (1985) *The Designer's Guide to Wind Loading of Building Structures, Part 1.* Butterworths for Building Research Establishment, London.

Dacre, H.F., Hawcroft, M.K., Stringer, M.A., and Hodges, K.I. (2012) An extratropical cyclone atlas: A tool for illustrating cyclone structure and evolution characteristics. *Bulletin of the American Meteorological Society*, **93**, 1497–1502. DOI: http://dx.doi.org/10.1175/BAMS-D-11-00164.1

Davies-Jones R., Burgess. D. and Foster, M. (1990) Test of helicity as a tornado forecast parameter. In *Proceedings of the 16th Conference on Severe Local Storms*, Kananaskis Park, AB, Canada, pp. 588–592.

Della-Marta, P.M., Liniger, M.A., Appenzeller, C., *et al.* (2010) Improved estimates of the European winter wind storm climate and the risk of reinsurance loss using climate model data. *Journal of Applied Meteorology and Climatology*, **49**, 2092–2120.

Della-Marta, P.M., Mathis, H., Frei, C., *et al.* (2008) The return period of wind storms over Europe. *International Journal of Climatology*, **29**, 437–459.

Della-Marta, P.M. and Pinto, J.G. (2009) Statistical uncertainty of changes in winter storms over the North Atlantic and Europe in an ensemble of transient climate simulations. *Geophysical Research Letters*, **36**, L14703, DOI: 10.1029/2009GL038557.

Fujita, T. T. (1981) Tornadoes and downbursts in the context of generalized planetary scales, *Journal of Atmospheric Sciences*, **1981**, 1511–1534.

Gómara, I., Pinto, J.G., Woollings, T., *et al.* (2014) Rossby wave-breaking analysis of explosive cyclones in the Euro-Atlantic sector. *Quarterly Journal of the Royal Meteorological Society*, **140**, 738–753. DOI: 10.1002/qj.2190.

Guy Carpenter (2010) Windstorm Xynthia Update 2. http://www.gccapitalideas.com/2010/03/08/update-2-windstorm-xynthia/ (accessed 13 November 2016).

Haas, R. and Pinto, J.G. (2012) A combined statistical and dynamical approach for downscaling large-scale footprints of European windstorms. *Geophysical Research Letters*, **39**, L23804, DOI: 10.1029/2012GL054014.

Hillier, J.K., Macdonald, N., Leckebusch, G., and Stavrinides, A. (2015) Interactions between apparently 'primary' weather-driven hazards and their cost. *Environmental Research Letters*, **10**, 104003, DOI: 10.1088/1748-9326/10/10/104003.

Hoerling, M.P. and Ting, M. (1994) Organization of extratropical transients during El Niño. *Journal of Climate*, **7**, 745–766. DOI: http://dx.doi.org/10.1175/1520-0442(1994)007<0745:OOETDE>2.0.CO;2.

Hohl, R. (2001) Relationship between hailfall intensity and hail damage on ground, determined by radar and lightning observations, PhD thesis, University of Fribourg, Switzerland.

Holton, J. (2004) Introduction. in *Dynamic Meteorology*, Elsevier Academic Press, London.

Hoskins, B.J. and Valdes, P.J. (1990) On the existence of storm-tracks. *Journal of the Atmospheric Sciences*, **47**, 1854–1864.

Hunter, A. (2014) Quantifying and understanding the aggregate risk of natural hazards, PhD thesis, University of Exeter. http://ethos.bl.uk

Huntingford, C., Marsh, T., Scaife, A.A. and Kendon, E.J. (2014) Potential influences on the United Kingdom's floods of winter 2013/14. *Natural Climate Change*, **4**, 769–777.

Hurrell, J.W. (1995) Decadal trends in the North Atlantic Oscillation: Regional temperatures and precipitation, *Science*, **269**, 67–679. DOI: 10.1126/science.269.5224.676.

Innes, P.M. and Dorling, S. (2013) *Operational Weather Forecasting*, Wiley-Blackwell, Hoboken, NJ.

Jain, V. (2010) The role of wind duration in damage estimation. http://www.air-worldwide.com/Publications/AIR-Currents/2010/The-Role-of-Wind-Duration-in-Damage-Estimation/ (accessed 13 November 2016).

Jones, S.C., Harr, P.A., Abraham, J., *et al.* (2003) The extratropical transition of tropical cyclones: Forecast challenges, current understanding, and future directions. *Weather and Forecasting*, **18**. DOI: http://dx.doi.org/10.1175/1520-0434(2003)018<1052:TETOTC>2.0.CO;2

Karremann, M.K., Pinto, J.G., von Bomhard, P.J. and Klawa, M. (2014) On the clustering of winter storm loss events over Germany. *Natural Hazards and Earth System Sciences*, **14**, 2041–2052.

Keller, J.L., Dailey, P.S., and Fischer, M.D. (2004) AMS Annual Meeting–CD-ROM edition, 84; 5.1 Applied Climatology, 14th Conference.

Kendon, M., and McCarthy, M. (2015) The UK's wet and stormy winter of 2013/2014. *Weather*, **7**, 40–47. DOI: 10.1002/wea.2465.

Khare, S., Bonazzi, A., Mitas, C., and Jewson, S. (2015) Modelling clustering of natural hazard phenomena and the effect on re/insurance loss perspectives. *Natural Hazards and Earth System Sciences*, **15**, 1357–1370, DOI: 10.5194/nhess-15-1357-2015.

Kistler, R., Kalnay, E., Collins, W. *et al.* (2001) The NCEP– NCAR 50–year reanalysis. *Bulletin of the American Meteorological Society*, **82**, 247–268.

Kolen, B., Slomp, R., and Jonkman, S.N. (2012) The impacts of storm Xynthia, February 27–28 2010 in France: Lessons for flood risk management. *Journal of Flood Risk Management*, **6**, 261–278.

Lamb, H.H. (1991) *Historic Storms of the North Sea, British Isles and Northwest Europe*, Cambridge University Press, Cambridge.

Lambert, S.J. and Fyfe, J.C. (2006) Changes in winter cyclone frequencies and strengths simulated in enhanced greenhouse warming experiments: Results from the models participating in the IPCC diagnostic exercise. *Climate Dynamics*, **26**, 713–728.

Lavers, D.A., Allan, R.P., Wood, E.F., *et al.* (2011) Winter floods in Britain are connected to atmospheric rivers. *Geophysical Research Letters*, **38**, L23803. DOI: 10.1029/2011GL049783.

Lowe, J.A. and Gregory, J.M. (2005) The effects of climate change on storm surges around the United Kingdom, *Philosophical Transactions of the Royal Society of London*, **363**, 1313–1328.

Mailier, P.J., Stephenson, D.B., Ferro, C.A.T., and Hodges, K.I. (2006) Serial clustering of extratropical cyclones. *Monthly Weather Review*, **134**, 2224–2240, DOI: 10.1175/MWR3160.1.

Matthews, T., Murphy, C., Wilby, R.L. and Harrigan, S. (2014) Stormiest winter on record for Ireland and UK. *Natural Climate Change*, **4**, 738–740.

McCallum, E. (1990) The Burn's Day storm, 25 January 1990. *Weather*, **45**, 166–173.

McIlveen, R. (2010) *Fundamentals of Weather and Climate*, Chapman and Hall, Oxford.

Meteorologisk Institutt (2008) http://met.no/?module=Articles;action=Article.publicShow; ID=1080 (accessed 21 May 2015).

Munich Re (2002) Winter storms in Europe (II). Analysis of 1999 losses and loss potentials. http://www.planat.ch/fileadmin/PLANAT/planat_pdf/alle_2012/2001-2005/Munich_Re_Group_2002_-_Winter_storms_in_Europe_II.pdf (accessed 13 November 2016).

NOAA (2015) Superstorm, http://www.erh.noaa.gov/ilm/archive/Superstorm93/

Nordhaus, W. (2010) The economics of hurricanes and implications of global warming. *Climate Change Economics*, **1**, 1–20.

NSSL (2016) Severe weather 101 – thunderstorms, https://www.nssl.noaa.gov/education/svrwx101/thunderstorms (accessed 16 November 2016).

Petterssen, S. and Smebye, S.J. (1971) On the development of extratropical cyclones. *Quarterly Journal of the Royal Meteorological Society*, **97**, 457–448.

Renfrew, I.A., Thorpe, A.J. and Bishop, C.H. (1997) The role of the environmental flow in the development of secondary frontal cyclones. *Quarterly Journal of the Royal Meteorological Society*, **123**, 1653–1675. DOI: 10.1002/qj.49712354210.

RMS (2000) Windstorms Lothar and Martin. http://ipcc-wg2.gov/njlite_download.php?id=6144 (accessed 13 November 2016).

RMS (2003) 1953 UK floods. http://static.rms.com/email/documents/fl_1953_uk_floods_50_retrospective.pdf (accessed 13 November 2016).

RMS (2008) The 1993 Superstorm: 15-year retrospective. *RMS Special Report* http://resilientriskmanagement.com/Publications/1993_SuperStorm.pdf.

Roberts, J.F., Champion, A.J., Dawkins, L.C., *et al.* (2014) The XWS open access catalogue of extreme European windstorms from 1979 to 2012. *Natural Hazards and Earth System Sciences*, **14**, 2487–2501. DOI: 10.5194/nhess-14-2487-2014.

Rogers, J.C. (1997) North Atlantic storm track variability and its association to the North Atlantic Oscillation and climate variability of Northern Europe. *Journal of Climate*, **10**, 1635–1647. DOI: http://dx.doi.org/10.1175/1520-0442(1997)010<1635:NASTVA>2.0.CO;2

Roth, D.M. (2015) Tropical cyclone averages and maxima per duration. *Tropical Cyclone Rainfall Data*. http://www.wpc.ncep.noaa.gov/tropical/rain/tcrainfall.html (accessed 23 March 2016).

Scaife, A.A., Arribas, A., Blockley, E., *et al.* (2014) Skillful long-range prediction of European and North American winters. *Geophysical Research Letters*, **41**, 2514–2519, DOI: 10.1002/2014GL059637.

Schmidt, S., Kemfert, C., and Hoppe, P. (2010) The impact of socio-economics and climate change on tropical cyclone losses in the USA. *Regional Environmental Change*, **10**, 13–26.

Sibley, A., Cox, D., and Titley, H. (2015) Coastal flooding in England and Wales from Atlantic and North Sea storms during the 2013/2014 winter, *Weather*, **70**, 62–70.

Svillo, J.K., Ahlquist, J.E. and Toth, Z. (1997) An ensemble forecasting primer. *Weather and Forecasting*, **12**, 809–818.

Swiss Re (2015) *Sigma* No 2/2015. http://media.swissre.com/documents/sigma2_2015_en_final.pdf (accessed 13 November 2016).

Ulbrich, U., Fink, A.H., Klawa, M., and Pinto, J.G. (2001) Three extreme storms over Europe in December 1999. *Weather*, **56**, 71–80.

Vitolo, R., Stephenson, D.B., Cook, I.M. and Mitchell-Wallace, K. (2009) Serial clustering of intense European storms. *Meteorologische Zeitschrift*, **18**, 411–424, DOI: 10.1127/0941-2948/2009/0393.

Wang, X.L., Wan, H., Zwiers, F.W., *et al.* (2011) Trends and low-frequency variability of storminess over western Europe, 1878–2007. *Climate Dynamics*, **37**, 2355–2371. DOI: 10.1007/s00382-011-1107-0.

Waisman, F. (2015) European windstorm vendor model comparison. Paper presented ate International Underwriting Association of London (IUA) Conference. www.iua.co.uk (accessed 23 March 2016).

Wernli, H., Dirren, S., Liniger, M.A. and Zillig, M. (2002) Dynamical aspects of the life cycle of the winterstorm Lothar (24–26 December 1999). *Quarterly Journal of the Royal Meteorological Society*, **128**, 405.

Willis Re (2015) Extratropical cyclones Lothar and Martin, 1999 15-year anniversary review and alternative scenarios. http://www.willisresearchnetwork.com/assets/templates/wrn/files/ETC_Lothar_Martin_15thAnniversaryReport_v7_Blog.pdf (accessed 13 November 2016).

Zappa. G., Shaffrey, L.C., Hodges, K.I., Sansom, P., and Stephenson, D. (2013) A multimodel assessment of future projections of North Atlantic and European extratropical cyclones in the CMIP5 climate models. *Journal of Climate*, **26**, 5846–5862.

**Severe Convective Storms**

ABC (2015) Brisbane super storm damage bill tops $1 billion. http://www.abc.net.au/news/2015-02-14/brisbane-super-storm-damage-bill-tops-1-billion/6092338 (accessed 28 February 2015).

Akaeda, K., Reisner, J., and Parsons, D. (1995) The role of mesoscale and topographically induced circulations in initiating a flash flood observed during the TAMEX Project. *Monthly Weather Review*, **123**, 1720–1739.

Allen, J.T., Tippett, M.K. and Sobel A.H. (2015) An empirical model relating U.S. monthly hail occurrence to large-scale meteorological environment. *Journal of Advances in Modeling Earth Systems*, **7**, 226–243.

AMS (2013) *Glossary of Meteorology*, 2nd edn, American Meteorological Society, glossary.ametsoc. org/(accessed 14 November 2016).

Atkins, N., Bouchard, C., Przybylinski, R., Trapp, R., and Schmocker, G. (2005) Damaging surface wind mechanisms within the 10 June 2003 Saint Louis bow echo during BAMEX. *Monthly Weather Review*, **133**, 2275–2296.

Barthlott, C., Burton, R., Kirshbaum, D., *et al.* (2011) Initiation of deep convection at marginal instability in an ensemble of mesoscale models: A case study from COPS. *Quarterly Journal of the Royal Meteorological Society*, **137**, 118–136.

Beatty, K., Rasmussen, E., and Straka, J. (2008) The supercell spectrum. Part I: A review of research related to supercell precipitation morphology. *E-Journal of Severe Storms Meteorology*, **3**, 1–21.

Berthet, C., Dessens, J., and Sanchez, J.L. (2011) Regional and yearly variations of hail frequency and intensity in France. *Atmospheric Research*, **100**, 391–400.

Bissolli, P., Grieser, J., Dotzek, N., and Welsch, M. (2007) Tornadoes in Germany 1950–2003 and their relation to particular weather conditions. *Global Planetary Change*, **57**, 124–138.

Bluestein, H.B. (2013) *Severe Convective Storms And Tornadoes: Observations and Dynamics*, Springer, Heidelberg.

Brooks, H.E. (2013) Severe thunderstorms and climate change. *Atmospheric Research*, **123**, 129–138.

Brooks, H.E. and Dotzek, N. (2008) The spatial distribution of severe convective storms and an analysis of their secular changes. In *Climate Extremes and Society* (eds H.F. Diaz and R. Murnane), Cambridge University Press, Cambridge.

Brooks, H.E., Lee, J.W., and Craven, J. (2003) The spatial distribution of severe thunderstorm and tornado environments from global reanalysis data. *Atmospheric Research*, **67**, 73–94.

Brown, A., Milton, S., Cullen, M., Golding, B., Mitchell, J., and Shelly, A. (2012) Unified modeling and prediction of weather and climate: A 25-year journey. *Bulletin of the American Meteorological Society*, **93**, 1865–1877.

Browning, K.A. and Foote, G.B. (1976) Airflow and hail growth in supercell storms and some implications for hail suppression. *Quarterly Journal of the Royal Meteorological Society*, **102**, 499–533.

Bunkers, M.J. and Zeitler, J.W. (2000) On the nature of highly deviant supercell motion. Paper presented at 20th Conference on Severe Local Storms, American Meteorological Society, 15–20 Sept. 2000, Orlando, FL.

Cecil, D.J. and Blankenship, C.B. (2012) Toward a global climatology of severe hailstorms as estimated by satellite passive microwave imagers. *Journal of Climate*, **25**, 687–703.

Cecil, D.J., Buechler, D.E., and Blakeslee, R.J. (2014) Gridded lightning climatology from TRMM-LIS and OTD: Dataset description. *Atmospheric Research*, **135–136**, 404–414.

Changnon, S.A. (1970) Hailstreaks. *Journal of the Atmospheric Sciences*, **27**, 109–125.

Changnon, S.A. (1977) The scales of hail. *Journal of Applied Meteorology*, **16**, 626–648.

Chisholm, A.J. and Renick, J.H. (1972) The kinematics of multicell and supercell Alberta hailstorms. *Alberta Hail Studies*, **72**, 24–31.

Crenshaw, V. and Koontz, J.D. (2001) Simulated hail damage and impact resistance test procedures for roof coverings and membranes. *Interface,* http://www.jdkoontz.com/articles/simulated.pdf (accessed 23 February 2017).

Davies-Jones, R., Trapp, R.J., and Bluestein, H. (2001) Tornadoes and tornadic storms, severe convective storms. *Meteorological Monographs*, **28**, 167–221.

Deepen, J. (2006) Schadenmodellierung extremer Hagelereignisse in Deutschland. Master's thesis, Universität Münster.

Diffenbaugh, N.S., Trapp, R.J. and Brooks, H. (2008) Does global warming influence tornado activity? *Eos Transactions*, **89**, 553.

Doswell III C.A. (2001) Severe convective storms: An overview. *Meteorological Monographs*, **28**, 1–26.

Doswell III C.A. (2005) Progress toward developing a practical societal response to severe convection. *Natural Hazards and Earth System Sciences*, **5**, 691–702.

Dotzek, N. (2003) An updated estimate of tornado occurrence in Europe. *Atmospheric Research*, **67–68**, 153–161.

Dotzek, N., Kurgansky, M.V., Grieser, J., Feuerstein, B., and Nevir, P. (2005) Observational evidence for exponential tornado intensity distributions over specific kinetic energy. *Geophysical Research Letters*, **32**, L24813.

Eccel, E., Cau, P., Riemann-Campe, K., and Biasioli, F. (2012) Quantitative hail monitoring in an alpine area: 35-year climatology and links with atmospheric variables. *International Journal of Climatology*, **32**, 503–517.

Fujita, T.T. (1971) Proposed characterization of tornadoes and hurricanes by area and intensity. SMRP research paper, 91, University of Chicago, Chicago.

Fujita, T.T. (1973) Tornadoes around the world. *Weatherwise*, **26**, 56–62.

Fujita, T.T. (1981) Tornadoes and downbursts in the context of generalized planetary scales. *Journal of the Atmospheric Sciences*, **38**, 1511–1534.

Fujita, T. and Caracena, F. (1977) An analysis of three weather-related aircraft accidents. *Bulletin of the American Meteorological Society*, **58**, 1164–1181.

Gessler, S. and Petty, S. (2013) Hail fundamentals and general hailstrike damage assessment methodology, in *Forensic Engineering* (ed. S. Petty), CRC Press, Boca Raton, FL.

Glass, F.H. and Britt, M. (2002) The historic Missouri-Illinois high precipitation supercell of 10 April 2001. Preprints, 21st Conference on Severe Local Storms, San Antonio, Texas, American Meteorological Society.

Groenemeijer, P.H. and van Delden, A. (2007) Sounding-derived parameters associated with large hail and tornadoes in the Netherlands. *Atmospheric Research*, **83**, 473–487.

Guy Carpenter (2014) Guy Carpenter launches probabilistic European hail model. *News Release*. http://www.guycarp.com/content/dam/guycarp/en/documents/PressRelease/2014/Guy%20Carpenter%20Launches%20Probabilistic%20European%20Hail%20Model.pdf (accessed 14 November 2016).

Hand, W.H. and Cappelluti, G. (2010) A global hail climatology using the UK Met Office convection diagnosis procedure (CDP) and model analyses. *Meteorological Applications*, **18**, 446–458.

Houze, R.A. (2004) Mesoscale convective systems. *Review of Geophysics*, **42**, RG4003.

Houze, R.A. (2014) *Cloud Dynamics*, 2nd edn, Academic Press, Cambridge, MA.

Houze, R.A and Hobbs, PV. (1982) Organization and structure of precipitating cloud systems. *Advances in Geophysics*, **24**, 225–315.

Houze, R., Rutledge, S.A., Biggerstaff, M.I. and Smull, B.F. (1989) Interpretation of Doppler weather radar displays of midlatitude mesoscale convective systems. *Bulletin of the American Meteorological Society*, **70**, 608–619.

Huntrieser, H., Schiesser, H.H., Schmid, W., and Waldvogl, A. (1997) Comparison of traditional and newly developed thunderstorm indices for Switzerland. *Weather and Forecasting*, **12**, 108–125.

IPCC. (2012) *Managing the Risks of Extreme Events and Disasters to Advance Climate Change Adaptation. A Special Report of Working Groups I and II of the Intergovernmental Panel on Climate Change*, Cambridge University Press, Cambridge.

IPCC. (2013) *Climate Change 2013: The Physical Science Basis. Contribution of Working Group I to the Fifth Assessment Report of the Intergovernmental Panel on Climate Change*, Cambridge University Press, Cambridge.

Johns, R. and Hirt, W. (1987) Derechos: Widespread convectively induced windstorms. *Weather and Forecasting*, **2**, 32–49.

Johnson, A.W. and Sugden, K.E. (2014) Evaluation of sounding-derived thermodynamic and wind-related parameters associated with large hail events. *E-Journal of Severe Storms Meteorology*, http://www.ejssm.org/ojs/index.php/ejssm/article/viewArticle/137 (accessed 14 November 2016).

Kantonale Gebäudeversicherungen. (2012) Ereignisanalyse Hagel, 2009. Untersuchung der Hagelunwetter vom 26. Mai und 23. Juli, 2009, *Interkantonaler Rückversicherungsverband IRV*.

Kapsch, M-L., Kunz, M., Vitolo, R. and Economou, T. (2012) Long-term trends of hail-related weather types in an ensemble of regional climate models using a Bayesian approach. *Journal of Geophysical Research*, **117**, D15107.

Khain, A., Rosenfeld, D., Pokrovsky, A., Blahak, U., and Ryzhkov, A. (2011) The role of CCN in precipitation and hail in a mid-latitude storm as seen in simulations using a spectral (bin) microphysics model in a 2D dynamic frame. *Atmospheric Research*, **99**, 129–146.

Knight, N.C. and Heymsfield, A.J. (1983) Measurement and interpretation of hailstone density and terminal velocity. *Journal of the Atmospheric Sciences*, **40**, 1510–1516.

Kottmeier, C., Kalthoff, N., Barthlott, C., *et al.* (2008) Mechanisms initiating deep convection over complex terrain during COPS. *Meteorologische Zeitschrift*, **17**, 931–948.

Kunkel, K.E., Bromirski, P.D., Brooks, H.E., *et al.* (2008) Observed changes in weather and climate extremes. In *Weather and Climate Extremes in a Changing Climate. Regions of Focus: North America, Hawaii, Caribbean, and U.S. Pacific Islands* (eds T.R. Karl, G.A. Meehl, D.M. Christopher, *et al.*), U.S. Climate Change Science Program and the Subcommittee on Global Change Research, Washington, DC.

Kunz, M. (2007) The skill of convective parameters and indices to predict isolated and severe thunderstorms. *Natural Hazards and Earth System Sciences*, **7**, 327–342.

Kunz, M. and Puskeiler, M. (2010) High-resolution assessment of the hail hazard over complex terrain from radar and insurance data. *Meteorologische Zeitschrift*, **19**, 427–439.

Kunz, M., Sander, J. and Kottmeier, C. (2009) Recent trends of thunderstorm and hailstorm frequency and their relation to atmospheric characteristics in southwest Germany. *International Journal of Climatology*, **29**, 2283–2297.

Lakshmanan, V., Smith, T., Stumpf, G., and Hondl, K. (2007) The warning decision support system–integrated information. *Weather and Forecasting*, **22**, 596–612.

Lemon, L.R. (1998) The radar three-body scatter spike: An operational large-hail signature. *Weather and Forecasting*, **13**, 327–340.

Lemon, L.R. and Doswell III C.A. (1979) Severe thunderstorm evolution and mesocyclone structure as related to tornadogenesis. *Monthly Weather Review*, **107**, 1184–1197.

Leslie, L.M., Leplastrier, M., and Buckley, B.W. (2008) Estimating future trends in severe hailstorms over the Sydney Basin: A climate modelling study. *Atmospheric Research*, **8**, 37–57.

Maddox, R.A. (1980) Mesoscale convective complexes. *Bulletin of the American Meteorological Society*, **61**, 1374–1387.

Markowski, P. and Richardson, Y. (2010) *Mesoscale Meteorology in Midlatitudes*. Wiley-Blackwell, Chichester.

Marshall, T.P. and Herzog, R.F. (2006) Protocol for assessment of hail-damaged roofing. Proceedings of the North American Conference on Roofing Technology, Haag Engineering, Carrollton, Texas.

Martinet, P., Fourrié, N., Bouteloup, Y., Bazile, E., and Rabier, F. (2014) Towards the improvement of short-range forecasts by the analysis of cloud variables from IASI radiances. *Atmospheric Science Letters*, **15**, 342–347.

Mohr, S. and Kunz, M. (2013) Recent trends and variabilities of convective parameters relevant for hail events in Germany and Europe. *Atmospheric Research*, **123**, 211–228.

Mohr, S., Kunz, M., and Keuler, K. (2015) Development and application of a logistic model to estimate the past and future hail potential in Germany. *Journal of Geophysical Research*, **120**, 3939–3956.

Munich Re (2011) NATHAN: World map of natural hazards, http://www.munichre.com/site/corporate/get/documents/mr/assetpool.shared/Documents/0_Corporate%20Website/_Publications/302-05972_en.pdf (accessed 28 February 2015).

Munich Re (2014) Natural Catastrophes, 2013, *Topics Geo,* http://www.munichre.com/en/reinsurance/magazine/publications/index.html (accessed 23 March 2016).

NOAA (2015a) Storm prediction center: Climatological or past storm information, http://www.spc.noaa.gov/climo/historical.html (accessed 12 May 2015).

NOAA. (2015b) Section 3: Thunderstorms and severe weather spotting, http://www.erh.noaa.gov/lwx/swep/Spotting.html (accessed 28 February 2015).

Otto, M. (2009) Modellierung von Hagelschäden in der Pkw-Kaskoversicherung in Deutschland. Master's thesis, Technische Universität Dresden.

Potter, S. (2007) Fine-tuning Fujita. *Weatherwise*, **60**, 64–71.

Pruppacher, H.R. and Klett, J.D. (2010) *Microphysics of Clouds and Precipitation*, 2nd edn, Springer, Dordrecht.

Przybylinski, R. (1995) The bow echo: Observations, numerical simulations, and severe weather detection methods. *Weather and Forecasting*, **10**, 203–218.

Punge, H., Werner, A., Bedka, K., and Kunz, M. (2014) A new physically based stochastic event catalogue for hail in Europe. *Natural Hazards*, **73**, 1625–1645.

Rauber, R.M., Walsh, J.E. and Charlevoix, D.J. (2014) *Severe and Hazardous Weather: An Introduction to High Impact Meteorology*, Kendall and Hunt Publishers, Iowa.

Reinhold, T., Reinolds, R., and Morrison, M. (2014) Attached structures high wind research, insurance institute for business and home safety, https://www.disastersafety.org/wp-content/uploads/attached-structure-high-wind-research_IBHS.pdf (accessed 16 June 2015).

RMS (2007) HailCalc Europe, http://riskinc.com/publications/HailCalc.pdf (accessed 15 May 2015).

RMS (2008) U.S. and Canada severe convective storm, http://riskinc.com/Publications/US_Canada_Severe_Convective_Storm.pdf (accessed 16 June 2015).

RMS (2015) Severe thunderstorm risk: What you don't know can hurt you, http://www.rms.com/blog/2014/02/27/severe-thunderstorm-risk/ (accessed 28 February 2015).

Romero, R., Gaya, M., and Doswell III C.A. (2007) European climatology of severe convective storm environmental parameters: A test for significant tornado events. *Atmospheric Research*, **83**, 389–404.

Sánchez, J.L., Gil-Robles, B., Dessens, J., *et al.* (2009) Characterization of hailstone size spectra in hailpad networks in France, Spain, and Argentina. *Atmospheric Research*, **93**, 641–654.

Sanderson, M.G., Hand, W.H., Groenemeijer, P., *et al.* (2014) Projected changes in hailstorms during the 21st century over the UK. *International Journal of Climatology*, **35**, 15–24.

Schuster, S.S., Blong, R.J., Leigh, R.J., and McAneney, K.J. (2005) Characteristics of the 14 April 1999 Sydney hailstorm based on ground observations, weather radar, insurance data and emergency calls. *Natural Hazards and Earth System Sciences*, **5**, 613–620.

Schuster, S.S., Blong, R.J. and McAneney, K.J. (2006) Relationship between radar-derived hail kinetic energy and damage to insured buildings for severe hailstorms in Eastern Australia. *Atmospheric Research*, **81**, 215–235.

Schuster, S.S., Blong, R.J. and Speer, M.S. (2005) A hail climatology of the greater Sydney area and New South Wales. Australia, *International Journal of Climatology*, **25**, 1633–1650.

Smith, P.L., Johnson, L.R., and Priegnitz, D.L. (1997) An exploratory analysis of crop hail insurance data for evidence of cloud seeding effects in North Dakota. *Journal of Applied Meteorology*, **36**, 463–473.

Straka, J.M., Rasmussen, E.N., Davies-Jones, R.P., and Markowski, P.M. (2007) An observational and idealized numerical examination of low-level counter-rotating vortices in the rear flank of supercells. *E-Journal of Severe Storms Meteorology*, **2**, 1–22.

Straka, J.M., Zrnic, D.S., and Ryzhkov, A.V. (2000) Bulk hydrometeor classification and quantification using polarimetric radar data: Synthesis of relations. *Journal of Applied Meteorology*, **39**, 1341–1372.

Swiss Re (2012) Natural catastrophes and man-made disasters in 2011, *Sigma* No. 2, http://media .swissre.com/documents/sigma2_2012_en.pdf (accessed 23 March 2016).

Swiss Re (2014) Natural catastrophes and man-made disasters in 2013, *Sigma* No. 1, http://media .swissre.com/documents/sigma1_2014_en.pdf (accessed 23 March 2016).

Swiss Re (2015) Natural catastrophes and man-made disasters in 2014, *Sigma* No. 2, http://media .swissre.com/documents/sigma2_2015_en_final.pdf (accessed 23 March 2016).

Wakimoto, R. (1982) The life cycle of thunderstorm gust fronts as viewed with Doppler radar and rawinsonde data. *Monthly Weather Review*, **110**, 1060–1082.

Wakimoto, R. (2001) Convectively driven high wind events. *Meteorological Monographs*, **28**, 255–298.

Wakimoto, R., Kessinger, C. and Kingsmill, D. (1994) Kinematic, thermodynamic, and visual structure of low-reflectivity microbursts. *Monthly Weather Review*, **122**, 72–92.

Weisman, M.L. and Klemp, J.B. (1982) The dependence of numerically simulated convective storms on vertical wind shear and buoyancy. *Monthly Weather Review*, **110**, 504–520.

Weisman, M.L. and Rotunno, R. (2004) A theory for strong long-lived squall lines revisited. *Journal of the Atmospheric Sciences*, **61**, 361–382.

Xie, B., Zhang, Q., and Wang, Y. (2008) Trends in hail in China during 1960–2005. *Geophysical Research Letters*, **35**, L13801.

## Inland Flooding

Aerts, J.C.J.H., Botzen, W.J.W., de Moel, H. and Bowman, M. (2013) Cost estimates for flood resilience and protection strategies in New York City. *Annals of the New York Academy of Sciences*, **1294**, 1–104. DOI: 10.1111/nyas.12200.

AIR Worldwide (2011) AIR Worldwide launches inland flood model for Germany, http://www.air-worldwide.com/Press-Releases/AIR-Worldwide-Launches-Inland-Flood-Model-for-Germany (accessed 14 November 2016).

AIR Worldwide (2013) Introducing the AIR inland flood model for the United States, http://www .air-worldwide.com/Publications/AIR-Currents/2013/Introducing-the-AIR-Inland-Flood-Model-for-the-United-States/ (accessed 14 November 2016).

Alberta Transportation (2004) *Guidelines on Extreme Flood Analysis*. Alberta Transportation, Transportation and Civil Engineering Division, Civil Projects Branch.

Alexander, L.V., Zhang, X., Peterson, T.C., *et al.* (2006) Global observed changes in daily climate extremes of temperature and precipitation. *Journal of Geophysical Research*, **111**, D05109.

Alfieri, L., Burek, P., Dutra, E., *et al.* (2013) GloFAS: global ensemble streamflow forecasting and flood early warning. *Hydrology and Earth System Sciences*, **17**, 1161–1175. DOI: 10.5194/hess-17-1161-2013.

Allan, R.P. and Soden, B.J. (2008) Atmospheric warming and the amplification of precipitation extremes. *Science*, **321**, 1481–1484. DOI: 10.1126/science.1160787.

American Academy of Actuaries. (2001) *Insurance Industry Catastrophe Management Practices*, American Academy of Actuaries, Washington, DC.

Arakawa, A. and Lamb, V.R. (1977) Computational design of the basic dynamical processes of the UCLA general circulation model, *Methods of Computational Physics, 17*. Academic Press, New York, pp. 173–265.

Arthur, S., Crow, H., and Karikas, N. (2009) Including public perception data in the evaluation of the consequences of sewerage derived urban flooding. *Water Science and Technology*, **60**, 231–242.

Australian Government (1984) Insurance Contract Act 1984 (Cth) and amendments. https://www.comlaw.gov.au/Details/C2014C00310 (accessed 14 November 2016).

Aviva (2004) Norwich Union's revolutionary flood map begins roll-out, https://www.aviva.co.uk/media-centre/story/1684/norwich-unions-revolutionary-flood-map-begins-roll/ (accessed 14 November 2016).

Baker, V., Kochel, R.C. and Patton, P.C. (1988) *Flood Geomorphology*, John Wiley and Sons, Inc., Hoboken, NJ.

Barredo, J. (2007) Major flood disasters in Europe: 1950–2005. *Natural Hazards*, **42**, 125–148.

Barry, R.G. and Chorley, R.J. (2009) Atmosphere, Weather and Climate, 9th edition, *Routledge*, pp. 536.

Beck, H.E., de Roo, A. and van Dijk, A.I.J.M. (2015) Global maps of streamflow characteristics based on observations from several thousand catchments. *Journal of Hydrometeorology*. DOI: 10.1175/JHM-D-14-0155.1.

Beven, K. (2009) *Environmental Modelling: An Uncertain Future? An Introduction to Techniques for Uncertainty Estimation in Environmental Prediction*, Routledge, London.

Beven, K. (2011) *Rainfall-Runoff Modelling: The Primer*, 2nd edn, John, & Sons, Inc, Hoboken, NJ.

Beven, K. and Hall, J. (2014) *Applied Uncertainty Analysis for Flood Risk Management*, World Scientific Publishers, Hackensack, NJ.

Beven, K., Young, P., Romanowicz, R., *et al.* (2008) *Analysis of Historical Data Sets to Look for Impacts of Land Use and Management Change on Flood Generation*. R&D Technical Report FD2120/TR, Department for Environment, Food and Rural Affairs, London.

Botzen, W. (2013) *Managing Extreme Climate Change Risks Through Insurance*, Cambridge University Press, Cambridge.

Boughton, W. and Droop, O. (2003) Continuous simulation for design flood estimation—a review. *Environmental Modelling and Software*, **18**, 309–318.

Buijs, F., Hall, J., Sayers, P., and Van Gelder, P. (2009) Time-dependent reliability analysis of flood defences. *Reliability Engineering and System Safety*, **94**, 1942–1953.

*Business Wire* (2011) AIR Worldwide launches inland flood model for Germany.

Bye, P. and Horner, M. (1998) *Easter 1998 Floods: Report by the Independent Review Team to the Board of the Environment Agency*, Environment Agency, Bristol.

Cabinet Office. (2008) *Learning Lessons from the 2007 Floods: The Pitt Review* (Final report), Cabinet Office, London.

Cameron, D., Beven, K., Tawn, J., Blazkova, S., and Naden, P. (1999) Flood frequency estimation by continuous simulation for a gauged upland catchment (with uncertainty). *Journal of Hydrology*, **219**, 169–187.

Castellarin, A., Kohnova, S., Gaal, L., *et al.* (2012) *Review of Applied-Statistical Methods for Flood-Frequency Analysis in Europe*. WG2 Milestone Report, COST Action ES0901, NERC, Centre for Ecology & Hydrology.

Česká Asociace Pojišťoven. (2003) Flood maps, http://www.cap.cz/en/calculators-and-applications/flood-maps (accessed 14 November 2016).

Chow, V., Maidment, D.R., and Mays, L.W. (1988) *Applied Hydrology*, McGraw-Hill, New York.

Coiffier, J. (2011) *Fundamentals of Numerical Weather Prediction*, Cambridge University Press, New York.

Coles, S. (2001) *An Introduction to Statistical Modeling of Extreme Values*, Springer-Verlag, London.

Contestabile, M. (2012) Economic impacts: global flood risk. *Nature Climate Change*, **2**, 644.

Costa, J.E. and Schuester, R.L. (1988) The formation and failure of natural dams. *Geological Society of America Bulletin* **100**, 1054–1068.

Davis, S.A. (1985) *Business Depth-Damage Analysis Procedures*, Research Report 85-R-5, Institute for Water Resources, US Army Corps of Engineers, Ft. Belvoir, VA.

Davis, S.A. and Skaggs, L.L. (1992) *Catalog of Residential Depth-Damage Functions: Used by the Army Corps of Engineers in Flood Damage Estimation*, IWR Report 92-R-3, Institute for Water Resources, US Army Corps of Engineers, Ft. Belvoir, VA.

Delgado, J.M., Merz, B., and Apel, H. (2012) A climate-flood link for the lower Mekong River. *Hydrology and Earth Systems Science*, **16**, 1533–1541, DOI: 10.5194/hess-16-1533-2012.

De Waal, D., Van Gelder, P., and Nel, A. (2007) Estimating joint tail probabilities of river discharges through the logistic copula. *Environmetrics*, **18**, 621–631.

DeWalle, D. and Rango, A. (2008) *Principles of Snow Hydrology*, Cambridge University Press, Cambridge.

Di Mauro, M. and De Bruijn, K.M. (2012) Application and validation of mortality functions to assess the consequences of flooding to people. *Journal of Flood Risk Management*, **5**, 92–110.

Eastoe, E. and Tawn, J. (2012) Modelling the distribution of the cluster maxima of exceedances of subasymptotic thresholds. *Biometrika*, **99**, 43–55.

ECMWF (2015) ECMWF | World leader in global medium-range numerical weather prediction. Available at: http://www.ecmwf.int/en/forecasts/datasets/set-i#I-i-a (accessed 28 August 2015).

Engineers Australia (2015) Peak discharge estimation. Book 3, Australian Rainfall and Runoff. Draft, March, 2015. Engineers Australia.

Environment Agency (1998) *Environment Agency Response to the Independent Report on the Easter 1998 Floods: Action Plan*. Environment Agency, Bristol.

Environment Agency (2008) *The Improved FEH Statistical Method*. Environment Agency &D report SC050050/TR, Environment Agency, Bristol, https://www.gov.uk/government/publications/benchmarking-the-latest-generation-of-2d-hydraulic-flood-modelling-packages (accessed 14 November 2016).

Environment Agency (2013) *Benchmarking the Latest Generation of 2D Hydraulic Flood Modelling Packages*, Environment Agency Report SC120002, Ref LIT 8570, Environment Agency Bristol.

Environment Agency (2014) *Flood and Coastal Erosion Risk Management: Long-Term Investment Scenarios (LTIS)*, Report LIT 10045, Environment Agency, Bristol.

EQECAT. (2005) Euroflood™—EQECAT's Europe flood model, http://www.eqecat.com/catastrophe-models/flood/europe/ (accessed 14 November 2016).

European Parliament. (2007) *Directive, 2007/60/EC of the European Parliament and of the Council of 23 October, 2007 on the Assessment and Management of Flood Risks*, 23 October, 2007.

Farr, T.G., Rosen, P.A., Caro, E., *et al.*, and the Shuttle Radar Topography Mission (2001) http://www2.jpl.nasa.gov/srtm/SRTM_paper.pdf (accessed 14 November 2016).

Faulkner, D., Warren, S. and Burn, D. (2016) Design floods for all of Canada. *Canadian Journal of Water Resources* http://dx.doi.org/10.1080/07011784.2016.1141665 (accessed 22 February 2017).

Federal Emergency Management Agency (2003) *Guidelines and Specifications for Flood Hazard Mapping Partners, Volume 1: Flood Studies and Mapping*, http://www.fema.gov

Federal Emergency Management Agency (2008) *HAZUS-MH MR4 Technical Manual*, http://www.fema.gov

Federal Emergency Management Agency (2015) *Flood Zones*, http://www.fema.gov/flood-zones (accessed 6 August 2015).

Feng, L. and Luo, G. (2010) Proposal for a quantitative index of flood disasters. *Disasters*, **34**, 695–704.

Few, R. and Matthies, F. (eds) (2006) *Flood Hazards and Health: Responding to Present and Future Risks*, Earthscan, London.

Fleming, G. (2002) How can we learn to live with rivers? The findings of the Institution of Civil Engineers Presidential Commission on flood-risk management. *Philosophical Transactions of the Royal Society A: Mathematical Physical and Engineering Sciences*, **360**, 1527–1530.

Gallus, W. A., Jankov, I., and Correia Jr. J. (2005). The 4 June 1999 Derecho event: A particularly difficult challenge for numerical weather prediction. *Weather and Forecasting*, **20** (5), 705–728.

Gumbel, E.J. (1941) The return period of flood flows. *Annals of the Mathematical Statistics*, **12**, 163–190. DOI: 10.1214/aoms/1177731747.

Gupta V.K. (2004) Emergence of statistical scaling in floods on channel networks from complex runoff dynamics. *Chaos, Solitons, and Fractals*, **19**, 357–365.

Gupta, V.K. and Dawdy, D.R. (1995) Physical interpretations of regional variations in the scaling exponents of flood quantiles. *Hydrological Processes*, **9**, 347–361.

Hartmann, D. (1994) *Global Physical Climatology*, Academic Press, San Diego.

Haseldine, L., Baxter, S., Wheeler, P., and Thomson, T. (2014) Developing a Malaysia flood model. *Geophysical Research Abstracts*, 16 EGU2014–10040.

Haven, J. (2013) A comparison of actual fluvial embankment flood defence performance to RASP estimated performance. MSc Dissertation, University of Bristol.

Hornberger, G. (1998) *Elements of Physical Hydrology*, Johns Hopkins University Press, Baltimore, MD.

Hunter, N., Bates, P., Neelz, S., *et al.* (2008) Benchmarking 2D hydraulic models for urban flooding. *Proceedings of the Institution Of Civil Engineers, Water Management*, **161**, 13–30.

Institute of Hydrology (1999) *Flood Estimation Handbook* (5 vols), Institute of Hydrology, Wallingford.

*Insurance Journal* (2005) RMS Releases Germany River Flood Model. http://www.insurancejournal.com/news/international/2006/09/12/72374.htm.

Jarvis, A., Reuter, H.I., Nelson, A., and Guevara, E. (2008) Hole-filled SRTM for the globe Version 4, available from the CGIAR-CSI SRTM 90 m Database http://srtm.csi.cgiar.org (accessed 14 November 2016).

JBA Risk Management Ltd (2012a) JBA releases Thailand Flood model, *Actuarial Post*, http://www.actuarialpost.co.uk/article/jba-releases-thailand-flood-model-2256.htm (accessed 14 November 2016).

JBA Risk Management Ltd (2012b) Photograph: Internal wrack marks in main Rojana industrial site, 3 Jan, 2012. By permission.

JBA Risk Management Ltd (2012c) Photograph: Water contamination in Lat Karabang, Thailand, 2 Jan 2012. By permission.

JBA Risk Management Ltd (2015a) JBA risk to model Flood Re risk, *Intelligent Insurer*, http://www.intelligentinsurer.com/news/jba-risk-to-model-flood-re-risk-5872 (accessed 14 November 2016).

JBA Risk Management Ltd (2015b) Europe river flood (accessed 14 November 2016, http://www.jbarisk.com/europe-river-flood.

JBA Risk Management Ltd (2015c) Probabilistic model (accessed 14 November 2016, http://www.jbarisk.com/our-datasets/probabilistic-models.

JBA Trust (2014) How well do flood defence models match reality? http://www.jbatrust.org/howwe-help/research/infrastructure/how-well-do-flood-defence-models-match-reality/

Jones, M., Fowler, H., Kilsby, C., and Blenkinsop, S. (2013) An assessment of changes in seasonal and annual extreme rainfall in the UK between 1961 and 2009. *International Journal of Climatology*, **33**, 1178–1194.

Jonkman, S.N. (2005) Global perspectives of loss of human life caused by floods. *Natural Hazards*, **34**, 151–175.

Jonkman, S.N., Maaskant, B., Boyd, E., and Levitan, M.L. (2009) Loss of life caused by the flooding of New Orleans after Hurricane Katrina: Analysis of the relationship between flood characteristics and mortality. *Risk Analysis*, **29**, 676–698. DOI: 10.1111/j.1539-6924.2008.01190.x.

Jothityangkoon, C. and Sivapalan, M. (2001) Temporal scales of rainfall-runoff processes and spatial scaling of flood peaks: space-time connection through catchment water balance. *Advances in Water Resources*, **24**, 1015–1036.

Kabat, P., Claussen, M., Dirmeyer, P.A., *et al.* (2004) *Vegetation, Water, Humans and the Climate: A New Perspective on an Interactive System*, Springer, Berlin.

Kiefer, J.C. and Willett, J.S. (1996) *Analysis of Nonresidential Content Value and Depth-Damage Data for Flood Damage Reduction Studies*, IWR Report 96-R-12, Institute for Water Resources, US Army Corps of Engineers, Alexandria, VA.

Kilsby, C.G., Jones, P.D., Burton, A.A., *et al.* (2007) A daily Weather Generator for use in climate change studies. *Environmental Modelling and Software*, **22**, 1705–1719.

Korup, K. (2002) Recent research on landslide dams: A literature review with special attention to New Zealand. *Progress in Physical Geography*, **26**, 206–235.

Kreibich, H., Christenberger, S., and Schwarze, R. (2011) Economic motivation of households to undertake private precautionary measures against floods. *Natural Hazards and Earth System Sciences*, **11**, 309–321.

Lamb, H. H. (1991) *Historic Storms of the North Sea, British Isles and Northwest Europe*, Cambridge University Press, Cambridge.

Lamb, R. (2005) Rainfall-runoff modelling for flood frequency estimation, in *Encyclopedia of Hydrological Sciences* (eds M. Anderson and J. McDonnell), John Wiley and Sons, Ltd, Chichester.

Lamb, R., Crossley, M., and Waller, S. (2009) A fast two-dimensional floodplain inundation model. *Proceedings of the ICE-Water Management*, **162**, 363–370.

Lamb, R., Keef, C., Tawn, J.A., *et al.* (2010) A new method to assess the risk of local and widespread flooding on rivers and coasts. *Journal of Flood Risk Management*, **3**, 323–336.

Landsea, C.W., Franklin, J.L., Blake, E.S., and Tanabe, R. (2013) The revised Northeast and North Central Pacific hurricane database (HURDAT2). US National Oceanic and Atmospheric Administration's National Weather Service, http://www.nhc.noaa.gov/data/hurdat/hurdat2-format-nencpac.pdf (accessed 28 July 2013).

LeVeque, R.J. (2002) *Finite Volume Methods for Hyperbolic Problems*, Cambridge University Press, Cambridge.

Liu, B., Henderson, M., Xu, M., and Zhang, Y. (2011) Observed changes in precipitation on the wettest days of the year in China, 1960–2000) *International Journal of Climatology*, **31**, 487–503.

Lloyd's (2014) Catastrophe modelling and climate change, https://www.lloyds.com/~/media/lloyds/reports/emerging%20risk%20reports/cc%20and%20modelling%20template%20v6.pdf (accessed 14 November 2016).

Lloyd's (2015) Realistic disaster scenarios: Scenario specification, http://www.lloyds.com/The-Market/Tools-and-Resources/Research/Exposure-Management/Realistic-Disaster-Scenarios (accessed 14 November 2016).

Lohmann, D., Eppert, S., Hilberts, A., Honegger, C., and Steward-Menteth, A. (2009) Correlation in time and space: Economic assessment of flood risk with Risk Management Solution (RMS) UK River Flood Model, in *Flood Risk Management Research and Practice* (eds P. Samuels, S. Huntington, W. Allsop, and J. Harrop), CRC Press, Boca Raton, FL.

Lohmann, D. and Yue, F. (2011) Correlation, simulation and uncertainty in catastrophe modelling. in *Proceedings of the 2011 Winter Simulation Conference*. 11–14 Dec, IEEE, Phoenix, AZ. DOI: 10.1109/WSC.2011.6147746.

Luo, T., Maddocks, A., Iceland, C., Ward, P., and Winsemius, H. (2015) World's 15 countries with the most people exposed to river floods, http://www.wri.org/blog/2015/03/world%E2%80%99s-15-countries-most-people-exposed-river-floods (accessed 7 July 2015).

Mascarenhas, F. (2005) *Flood Risk Simulation*, WIT, Southampton.

May. B., Clark, P., Cooper, A. *et al.* (2004) *Numerical Weather Prediction: Flooding at Boscastle, Cornwall on 16 August 2004: A Study of Met Office Forecasting Systems.* Met Office Forecasting Research Technical Report 429.

McGuffie, K. and Henderson-Sellers, A. (2005) *A Climate Modelling Primer*, John Wiley & Sons, Ltd., Chichester.

Mens, M., Klijn, F., and Schielen, R. (2015) Enhancing flood risk system robustness in practice: insights from two river valleys. *International Journal of River Basin Management*, **13**, 297–304.

Merz, R. and Blöschl, G. (2003) A process typology of regional floods. *Water Resources Research*, **39**, 1340, DOI: 10.1029/2002WR001952.

Metropolitan Transportation Authority of the State of New York. (2014) Photograph: Flooding at LIRR station, s https://www.flickr.com/photos/mtaphotos/14906782465/in/album-72157646010676347/ (accessed 14 November 2016).

Michel-Kerjan, E.O. (2010) Catastrophe economics: The national flood insurance program. *Journal of Economic Perspective*, **24**, 165–186.

Munich Re (2005) *What is a Flood? Defining Flood Loss Occurrences for Reinsurance Purposes.* Munich Reinsurance Company, Munich, Germany.

Munich Re (2010) *Topics Geo: Natural Catastrophes, 2009: Analyses, Assessments, Positions,* Munich Reinsurance Company, Munich, Germany.

Murphy, C., Cunnane, C., Das, S. and Mandal, U. (2014) *Flood Frequency Estimation. Vol. II, Flood Studies Update Technical Research Report*, Office of Public Works, Dublin, Ireland.

Oasis Loss Modelling Framework (2014) *Model Developer Checklist R1.1*, Oasis Loss Modelling Framework, London.

O'Connell, P.E., Ewen, J., O'Donnell, G. and Quinn, P. (2007) Is there a link between agricultural land-use management and flooding? *Hydrology and Earth System Sciences*, **11**, 96–107.

Onof, C., Chandler, R.E., Kakou, A. *et al.* (2000) Rainfall modelling using Poisson-cluster processes: a review of developments, *Stochastic Environmental Research and Risk Assessment*, **14**, 384–411.

Owusu, S., Wright, G., and Arthur, S. (2015) Public attitudes towards flooding and property-level flood protection measures. *Natural Hazards*, **77**, 1963–1978.

Pagano, T.C. and Sorooshian, S. (2005) Global water cycle (fundamental, theory, mechanisms), in *Encyclopedia of Hydrological Sciences* (ed. M.G. Anderson), John Wiley and Sons, Hoboken, NJ.

Pender, G. and Faulkner, H. (2010) *Flood Risk Science and Management*, John Wiley and Sons, Inc., Hoboken, NJ.

Penning-Rowsell, E., Johnson, C., Tunstall, S., *et al.* (2005). The benefits of flood and coastal risk management: a manual of assessment techniques Middlesex University Press. http://www.mcm-online.co.uk/about/

Penning-Rowsell, E., Priest, S., Parker, D., *et al.* (2013) *Flood and Coastal Erosion Risk Management: A Manual for Economic Appraisal*, Routledge, London.

Ploughmann, L. (2015) Photograph: The Shoal Creek effect, https://www.flickr.com/photos/criminalintent/18621159399 (accessed 14 November 2016).

Pritchard, J., Lipski, V., Wanna, J., and Boston, J. (2014) *Future-Proofing the State*, ANU Press, Canberra, Australia.

Rijkswaterstaat. (2006) Lessons learned from flood defence in the Netherlands. *Irrigation and Drainage*, **55**, 121–132. DOI: 10.1002/ird.242.

RMSI (2014) Development news: RMSI launches India flood model, http://www.rmsi.com/ri/newsletter/In_Focus.html

Robinson, J.S. and Sivapalan, M. (1997) Temporal scales and hydrological regimes: Implications for flood frequency scaling. *Water Resources Research*, **33**, 2981–2999.

Rosbjerg, D., Bloschl, G., Burn, D., *et al.* (2013) (eds) *Runoff Prediction in Ungauged Basins: Synthesis across Processes, Places and Scales*, Cambridge University Press, Cambridge.

Rose, S. and Peters, N.E. (2001) Effects of urbanization on streamflow in the Atlanta area (Georgia, USA): A comparative hydrological approach. *Hydrological Processes*, **15**, 1441–1457.

Saha, S., Moorthi, S. H., Pan, X., *et al.* (2010). The NCEP Climate Forecast System Reanalysis, *Bull. Amer. Meteor. Soc.*, **91**, 1015–1057, doi: 10.1175/2010BAMS3001.1.

Sampson, C.C., Fewtrell, T.J., O'Loughlin, F., *et al.* (2014) The impact of uncertain precipitation data on insurance loss estimates using a flood catastrophe model. *Hydrology and Earth Systems Science*, **18**, 2305–2324. DOI: 10.5194/hess-18-2305-2014.

Schaake, J., Hamill, T., Buizza, R., and Clark, M. (2007) HEPEX: The hydrological ensemble prediction experiment. *Bulletin of the American Meteorological Society*, **88**, 1541–1547.

Schanze, J., Zeman, E., and Marsalek, J. (2006) *Flood Risk Management: Hazards, Vulnerability and Mitigation Measures*, Springer, Dordrecht.

Schumann, A. (2011) *Flood Risk Assessment and Management: How to Specify Hydrological Loads, Their Consequences and Uncertainties*, Springer, Dordrecht.

Shaw, E., Beven, K., Chappell, N., and Lamb, R. (2010) *Hydrology in Practice* (4th edn), Spon Press, New York.

Singh, V. (1995) *Computer Models of Watershed Hydrology* (rev. edn), Water Resources Publications, Highlands Ranch, CO.

Sithole, G. and Vosselman, G. (2004) Experimental comparison of filter algorithms for bare-earth extraction from airborne laser scanning point clouds. *ISPRS Journal of Photogrammetry and Remote Sensing*, **59**, 85–101.

Sleigh, P.A., Gaskell, P.H., Berzins, M., and Wright, N.G. (1998) An unstructured finite-volume algorithm for predicting flow in rivers and estuaries. *Computers and Fluids*, **27**, 479–508.

SMHI (2015) Large-scale multi-basin modelling and comparative hydrology, http://www.smhi.se/en/research/research-departments/hydrology/hype-model-systems-eng-1.21603 (accessed 24 July 2015).

Smith, A., Sampson, C., and Bates, P. (2015) Regional flood frequency analysis at the global scale. *Water Resources Research*, **51** (1), 539–553.

Stedinger, J.R., Vogel, R.M., and Foufoula-Georgiou, E. (1993) Frequency analysis of extreme events, in *Handbook of Hydrology* (ed D. Maidment), McGraw-Hill, New York.

Stephenson, D.B. (2008) Definition, diagnosis, and origin of extreme weather and climate events, in *Climate Extremes and Society* (eds R. Murnane and H. Diaz), Cambridge University Press, Cambridge.

Veldhuis, J.T., Clemens, F., Sterk, G., and Berends, B. (2010) Microbial risks associated with exposure to pathogens in contaminated urban flood water. *Water Research*, **44**, 2910–2918.

Toothill, J. (2002) Central European flooding, August, 2002. EQECAT Technical Report. http://www.absconsulting.com/es/resources/Catastrophe_Reports/flood_rept.pdf (accessed 14 November 2016).

Trenberth, K.E., Jones, P.D., Ambenje, P., *et al.* (2007) Observations: Surface and atmospheric climate change, in *Climate Change, 2007: The Physical Science Basis. Contribution of Working Group I to the Fourth Assessment Report of the Intergovernmental Panel on Climate Change, 2007* (eds S. Solomon, D. Qin, D. Manning *et al.*), Cambridge University Press, Cambridge.

Uppala, S. M., KÅllberg, P. W., Simmons, A. J., *et al.* (2005) The ERA-40 re-analysis. *Q.J.R. Meteorol. Soc.*, **131**, 2961–3012. doi:10.1256/qj.04.176

United Nations International Strategy for Disaster Reduction. (2009) *Global Assessment Report on Disaster Risk Reduction, Risk and Poverty in a Changing Climate*. United Nations International Strategy for Disaster Reduction Secretariat, Geneva.

US Army Africa (2012) Photograph: U.S. Army Africa engineer visits river-flood damaged areas in Tanzania, 201002, https://www.flickr.com/photos/usarmyafrica/4348611150 (accessed 14 November 2016).

US Army Corps of Engineers (1999) *Risk-Based Analysis in Geotechnical Engineering for Support of Planning Studies*. Publication No. ETL 1110-2-556. US Army Corps of Engineers, Washington, DC.

US Army Corps of Engineers (2003) Economic guidance memorandum (EGM) 04-01, generic depth-damage relationships for residential structures with basements.

US Department of Agriculture (1986) *Urban Hydrology for Small Watersheds*. Technical Release 55, Natural Resources Conservation Service, Conservation Engineering Division, Washington, DC.

Van Heerden, I.L.L. (2007) The failure of the New Orleans levee system following Hurricane Katrina and the pathway forward. *Public Administration Review*, **67**, 24–35. DOI: 10.1111/j.1540-6210.2007.00810.x.

Verstraeten, G. and Posen, J. (1999) The nature of small-scale flooding, muddy floods and retention pond sedimentation in central Belgium. *Geomorphology*, **29**, 275–292. DOI: 10.1016/S0169-555X(99)00020-3.

Vrijling, J.K. (2001) Probabilistic design of water defense systems in the Netherlands. *Reliability Engineering and System Safety*, **74**, 337–344.

Ward, P., Jongman, B., Weiland, F., *et al.* (2013) Assessing flood risk at the global scale: Model setup, results, and sensitivity, *Environmental Research Letters*, **8**, 1748–9326. 044019, DOI: 10.1088/1748-9326/8/4/044019.

Ward, R. and Robinson, M. (2000) *Principles of Hydrology* (4th edn), McGraw-Hill, New York.

Westra, S., Alexander, L.V., and Zwiers, F.W. (2013) Global increasing trends in annual maximum daily precipitation. *Journal of Climate*, **26**, 3904–3918.

Westra, S., Fowler, H., Evans, J., *et al.* (2014) Future changes to the intensity and frequency of short-duration extreme rainfall. *Reviews of Geophysics*, **52**, 522–555.

Wilby, R.L. and Wigley, T.M.L. (1997) Downscaling general circulation model output: A review of methods and limitations. *Progress in Physical Geography* **21**, 530–548.

Winsemius, H.C., Van Beek, L.P.H., Jongman, B., Ward, P.J., and Bouwman, A. (2013) A framework for global river flood risk assessments. *Hydrology and Earth Systems Science*, **17**, 1871–1892. DOI: 10.5194/hess-17-1871-2013.

Woollings, T. (2010) Dynamical influences on European climate: An uncertain future. *Philosophical Transactions of the Royal Society A*, **368**, 3733–3756. DOI: 10.1098/rsta.2010.0040

Worni, R., Huggel, C., and Stoffel, M. (2013) Glacial lakes in the Indian Himalayas: From an area-wide glacial lake inventory to on-site and modelling-based risk assessment of critical glacial lakes. *Science of the Total Environment*, **567**, 468–469.

Xia, J., Falconer, R.A., Lin, B., and Tan, G. (2011) Numerical assessment of flood hazard risk to people and vehicles in flash floods. *Environmental Modelling and Software*, **26**, 987–998.

## Shrink-Swell Subsidence

Brady, N.C. and Weil, R.R. (2002) *The Nature and Properties of Soils*, 13th edn, Prentice Hall, Upper Saddle River, NJ.

BRE (1993) Lowrise building on shrinkable clay soil. *Building Research Establishment (BRE) Digest*, 240-2, CRC, London.

Brink, A.B.A., Partridge, T.C., and Williams, A.A.B. (1982) *Soil Survey for Engineering*. Clarendon Press, Oxford.

BGS (2012) Ground shrinking and swelling. *UK Geohazard Note*, www.bgs.ac.uk/downloads/start.cfm?id=2499 (accessed 19 January 2015).

Cheetham, R. (2008) Subsidence: A gradual catastrophe. *Insurance Times*, 10 Dec.

Corti, T., Muccione, V., Köllner-Heck, P., Bresch, D., and Seneviratne, S.I. (2009) Simulating past droughts and associated building damages in France. *Hydrology and Earth Systems Science*, **13**, 1739–1747.

Hao, Z., AghaKouchak, A., Nakhiiri, N., and Farahmand, A. (2014) Global integrated drought monitoring and prediction system. *Scientific Data*, DOI: 10.1038/sdata.2014.1.

ISE (1994) *Subsidence of Low Rise Buildings*, ISE, London.

Jones, L.D. (2004) Cracking open the property market. *Planet Earth*, **Autumn,** 30–31.

Lawson, M. and O'Callaghan, D. (1995) A critical analysis of the role of trees in damage to low rise buildings. *Journal of Arboriculture*, **21**, 90–97.

Page, R.C.J. (1998) Reducing the cost of subsidence damage despite global warming. *Structural Survey*, **16**, 67–75.

Pritchard, O.G., Hallett, S.H. and Farewell, T.S. (2014) Soil impacts on UK infrastructure: Current and future climate. *Engineering Stability*, **160**, 170–184. DOI: 10.1680/ensu.13.00035.

Pugh, R.S. (2002) Some observations of the influence of recent climate change on the subsidence of shallow foundations. *Geotechnical Engineering*, **155**, 23–25.

Salagnac, J.L. (2007) Lessons from the 2003 heat wave: A French perspective. *Building Research and Information*, **35**, 450–457.

Swiss Re (2011) The hidden risks from climate change: An increased risk of property damage from soil subsidence in Europe, http://www.swissre.com/media/news_releases/nr_20110704_soil_subsidence.html (accessed 19 January 2015.

## Earthquakes

Abrahamson, N.A. and Silva, W.J. (2008) Summary of the Abrahamson and Silva NGA ground-motion relations. *Earthquake Spectra*, **24**, 67–97.

AIR Worldwide (2011) *Rethinking the Unthinkable: Modeling Unprecedented Ruptures Like the Great Tohoku Earthquake*, http://www.air-worldwide.com/Publications/AIR-Currents/2010/Rethinking-the-Unthinkable--Modeling-Unprecedented-Ruptures-Like-the-Great-Tohoku-Earthquake/ (accessed 13 May 2015).

AIR Worldwide (2012) *Top 10 Historical Hurricanes and Earthquakes in the US: What Would They Cost Today*? http://www.air-worldwide.com/Publications/AIR-Currents/2012/Top-10-Historical-Hurricanes-and-Earthquakes-in-the-U-S---What-Would-They-Cost-Today/ (accessed 14 May 2015).

AIR Worldwide. (2013a) *Introducing the Industry's First Fully Probabilistic Tsunami Model of March 26 2013*, http://www.air-worldwide.com/Publications/AIR-Currents/2013/Introducing-The-Industry%E2%80%99s-First-Fully-Probabilistic-Tsunami-Model/ (accessed 6 May 2016).

Air Worldwide (2013b) *Modeling Fire Following Earthquake for the 2013 Japan Model Update, 23 May 2013*, http://www.air-worldwide.com/Publications/AIR-Currents/2013/Modeling-Fire-Following-Earthquake-for-the-2013-Japan-Model-Update/ (accessed 20 May 2015).

Aki, K. (1965) Maximum likelihood estimate of b in the formula log (N) = a − bM and its confidence limits. *Bulletin of the Earthquake Research Institute Tokyo University*, **43**, 237–239.

Akkar, S. and Bommer, J.J. (2007) Prediction of elastic displacement response spectra in Europe and the Middle East. *Earthquake Engineering and Structural Dynamics*, **36**, 1275–1301.

Albaric, J., Oye, V., Langet, N., et al. (2013) Monitoring of induced seismicity during the first geothermal reservoir stimulation at Paralana, *Australia. Geothermics*. DOI: 10.1016/j.geothermics.2013.10.013.

Ameri, G., Massa, M., Bindi, D., et al. (2009) The 6 April 2009 Mw 6.3 L'Aquila (central Italy) earthquake: Strong ground motion observations. *Seismological Research Letters*, **80** 951–966.

Aon Benfield (2012) *Earthquake Catastrophe Models in Disaster Response Planning, Risk Mitigation and Financing in Developing Countries in Asia, Winspear, Musulin and Sharma*, http://www.aon.com/attachments/reinsurance/201202_eq_cat_models_drp.pdf (accessed 20 May 2015).

Aon Benfield (2015) *Top 10 Global Economic Insured Loss Events*, http://catastropheinsight. aonbenfield.com/Top10/Global-Economic-Insured-Loss-Events.pdf (accessed 17 May 2015).

Armijo, R., Pondard, N., Meyer, B., *et al.* (2005) Submarine fault scarps in the Sea of Marmara pull-apart (North Anatolian Fault): Implications for seismic hazard in Istanbul. *G3*, **6**. DOI: 10.1029/ 2004GC000896.

Bakun, W.H. and Lindh, A.G. (1985) The Parkfield, California, earthquake prediction experiment. *Science*, **229**, 619–624.

Belardinelli, M.E., Bizzari, A., and Cocco, M. (2003) Earthquake triggering by static and dynamic stress changes. *Journal of Geophysical Research*, **108**. DOI: 10.1029/2002JB001779.

Billi, A., Funiciello, R., Minelli, L., *et al.* (2008) On the cause of the 1908 Messina tsunami, southern Italy. *Geophysical Research Letters*, **35**, L06301.

Bolt, B.A. (2003) *Earthquakes* (5th edn), W. H. Freeman, New York.

Bonilla, L.F. (2000) Computation of linear and nonlinear site response for near-field ground-motion, PhD thesis, University of California, Santa Barbara.

Boore, D.M. and Atkinson, G.M. (2008) Ground-motion prediction equations for the average horizontal component of PGA, PGV, and 5%-damped PSA at spectral periods between 0.01s and 10.0 s. *Earthquake Spectra*, **24**, 99–138.

Bull, J.M., Barnes, P.M., Lamarche, G., *et al.* (2006) High-resolution record of displacement accumulation on an active normal fault: Implications for models of slip accumulation during repeated earthquakes. *Journal of Structural Geology*, **28**, 1146–1166.

Campbell, K.W. and Bozorgnia, Y. (2008) NGA ground motion model for the geometric mean horizontal component of PGA, PGV, PGD and 5% damped linear elastic response spectra for periods ranging from 0. 01 to 10 s. *Earthquake Spectra*, **24**, 139–171.

Chiou, B.S.J. and Youngs, R.R. (2008) An NGA model for the average horizontal component of peak ground motion and response spectra, *Earthquake Spectra*, **24**, 173–215.

CRED (2014) Annual disaster statistical review, 2013: The numbers and trends, www.cred.be/sites/ default/files/ADSR_2013.pdf (accessed 14 November 2016).

Daniell, J.E., Skapski, J-U., Vervaeck, A., Wenzel, F., and Schaefer, A. (2015) Global earthquake and volcanic eruption economic losses and costs from 1900–2014: 115 years of the CATDAT database. Trends, normalisation and visualisation. EGU General Assembly.

Delouis, B., Nocquet, J-M., and Vallee, M. (2011) Slip distribution of the February 27, 2010 Mw = 8.8 Maule Earthquake, central Chile, from static and high-rate GPS, InSAR, and broadband teleseismic data. *Geophysical Research Letters*, **37**, L17305.

dePolo, C.M. and Slemmonds, D.B. (1990) Estimation of earthquake size for seismic hazards, in *Neotectonics in Earthquake Evaluation* (E.L. Krinitzsky and D. B. Slemmons eds), Geological Society of America Reviews in Engineering Geology, Boulder, CO, 8, 1–28.

Douglas, J. (2003) Earthquake ground motion estimation using strong-motion records: A review of equations for the estimation of peak ground acceleration and response spectral ordinates, *Earth Science Reviews*, **61**, 43–104.

Durukal, E., Erdik, M., and Sesetyan, K. (2006) Expected earthquake losses to buildings in Istanbul and implications for the performance of the Turkish catastrophe insurance pool, http://dc. engconfintl.org/geohazards/34 (accessed 6 March 2015).

Earthquake Engineering Research Institute (1994) EERI Special Earthquake report. *EERI Newsletter*, **26**, 1–12.

Earthquake Engineering Research Institute (2010) EERI Special Earthquake Report, April, 2010, http://escweb.wr.usgs.gov/share/mooney/138.pdf (accessed 14 November 2016).

EEFIT (2013) Financial management and Japan earthquake insurance, Recovery two years after the 2011 Tohoku earthquake and tsunami: A return mission report by EEFIT, http://www.istructe .org/webtest/files/23/23e209f3-bb39-42cc-a5e5-844100afb938.pdf (accessed 14 November 2016).

Ellsworth, W.L. (2013) Injection-induced earthquakes. *Science*, **341**.

EQECAT (2012) Spatial and temporal earthquake clustering: Part 2 earthquake aftershocks, EQECAT White Paper, http://www.eqecat.com/pdfs/global-earthquake-clustering-whitepaper-part-2-2012-02.pdf (accessed 10 May 2015).

Faure Walker, J.P., Pomonis, A., Monk-Steel, B., Goda, K., and Smith, R. (2013) UK-Japan Workshop on disaster risk reduction, learning from the 2011Great East Japan Earthquake. IRDR Special Report, 2013-01, http://www.ucl.ac.uk/rdr/documents/docs-publications-folder/IRDR-Special-Report-2013-01 (accessed 13 May 2015).

Faure Walker, J.P., Roberts, G.P., Cowie, P.A., *et al.* (2012) Relationship between topography, rates of extension and mantle dynamics in the actively-extending Italian Apennines. *Earth and Planetary Science Letters*, **325–326**, 76–84.

Faure Walker, J.P., Roberts, G.P., Sammonds, P., and Cowie, P.A. (2010) Comparison of earthquake strains over 100 and 10,000 year timescales: Insights into variability in the seismic cycle in the central Apennines, Italy. *Journal of Geophysical Research*, **115** (B10418). DOI: 10.1029/2009JB006462.

Felzer, K.R., Abercrombie, R.E., and Ekstrom, G. (2004) A common origin for aftershocks, foreshocks, and multiplets. *Bulletin of the Seismological Society of America*, **94**, 88–98.

FEMA (2010) Earthquake-resistant design concepts, FEMA P-749, https://c.ymcdn.com/sites/www.nibs.org/resource/resmgr/BSSC/FEMA_P-749.pdf (accessed 6 May 2016).

Forsyth, D.W. and Uyeda, S. (1975) On the relative importance of the driving forces of plate motion. *Geophysical Journal of the Royal Astronomical Society*, **43**, 163–200.

Fowler, C.M.R. (1990) *The Solid Earth: An Introduction to Global Geophysics*, Cambridge University Press, Cambridge.

Friedrich, A.M., Wernicke, B.P., Niemi, N.A., Bennett, R.A., and Davis, J.L. (2003) Comparison of geodetic and geologic data from the Wasatch region, Utah, and implications for the spectral character of Earth deformation at periods of 10 to 10 million years. *Journal of Geophysical Research*, **108**, 2199–2222.

GEM-ECD (2015) GEM earthquake consequences database, http://gemecd.org/event/92 (accessed 14 November 2016).

Gibson, G. and Sandiford, M. (2013) Seismicity and induced earthquakes, A Background Paper to the Office of the NSW Chief Scientist and Engineer (OCSE) providing information and a discussion about induced seismicity, microseismic monitoring and natural seismic impacts, in relation to CSG activities, http://www.chiefscientist.nsw.gov.au/__data/assets/pdf_file/0017/31616/Seismicity-and-induced-earthquakes_Gibson-and-Sandiford.pdf (accessed 22 November 2016).

Grosse, J.C. and Phillips, F.M. (2001) Terrestrial in situ cosmogenic nuclides: Theory and application. *Quarternary Science Reviews*, **20**, 1475–1560.

Gutenberg, B. and Richter, C. (1954) *Seismicity of the Earth and Associated Phenomena* (2nd edn), Princeton University Press, Princeton, NJ.

Henstock, T., McNeill, L., Dean, S., *et al.* (2010) Exploring structural control on Sumatran earthquakes. *Eos*, **91**, 405.

Housner, G.W. and Thiel, C.C. Jr. (1990) Competing against time: Report to the Governor, Governor's Board of Enquiry, State of California. http://authors.library.caltech.edu/5013/1/HOUes90.pdf

Idriss, I.M. (2008) An NGA empirical model for estimating the horizontal spectral values generated by shallow crustal earthquakes. *Earthquake Spectra*, **24**, 217–242.

International Commission on Earthquake Forecasting (2011) *Annals of Geophysics*, **54**, 4. DOI: 10.4401/ag-5350.

iinc (2012) *The 1906 Earthquake and Fire*, http://www.iinc.org/articles/345/1/The-1906-Earthquake-and-Fire/Page1.html

Ishibe, T. and Shimazaki, K. (2012) Characteristic earthquake model and seismicity around late quaternary active faults in Japan. *Bulletin of the Seismological Society of America*, **102**, 1041–1058.

Jackson, D.D. and Kagan, Y.Y. (2006) The 2004 Parkfield earthquake, the 1985 prediction, and characteristic earthquakes: Lessons for the future. *Bulletin of the Seismological Society of America*, **96**, S397–S409.

Jordan, T.H. (2006) Earthquake predictability, brick by brick. *Seismological Research Letters*, **77**, 3–6.

Kagan, Y.K. and Jackson, D.D. (1991) Seismic gap hypothesis: Ten years after. *Journal of Geophysical Research*, **96** (B13), 419–421.

Kagan, Y.Y., Jackson, D.D. and Geller, R.J. (2012) Opinion, characteristic earthquake model, 1884-2011, R.I.P. *Seismological Research Letters*, **83**.

Kanamori, H. (1977) The energy release in great earthquakes. *Journal of Geophysical Research*, **82**, 2981–2987.

Kanamori, H. (1983) Magnitude scale and quantification of earthquakes. *Tectonophysics*, **93**, 185–199.

King, G.C.P., Stein, R.S., and Lin, J. (1994) Static stress changes and the triggering of earthquakes. *Bulletin of the Seismological Society of America*, **84**, 935–953.

Kirschvink, J.L. (2000) Earthquake prediction by animals: Evolution and sensory perception. *Bulletin of the Seismological Society America*, **90**, 312–323.

Kramer, S.L. (1996) *Geotechnical Earthquake Engineering*, Prentice Hall, New York.

Kulachi, F., Inceoz, M., Dogru, M., Aksoy, E., and Baykara, O. (2009) Artificial neural network model for earthquake prediction with radon monitoring. *Applied Radiation and Isotopes*, **67**, 212–219.

Loveless, J.P. and Meade, B.J. (2011) Spatial correlation of interseismic coupling and coseismic rupture extent of the 2011 MW = 9.0 Tohoku-oki earthquake. *Geophysics Research Letters*, **38**, L17306.

Main, I.G. (1996) Statistical physics, seismogenesis, and seismic hazard. *Review of Geophysics*, **34**, 433–462.

Manaker, D.M., Calais, E., Freed, A.M., *et al.* (2008) Interseismic plate coupling and strain partitioning in the Northeastern Caribbean. *Geophysical Journal International*, **174**, 889–903.

Marzocchi, W. and Sandri, L. (2003) A review and new insights on the estimation of the b-value and its uncertainty. *Annals of Geophysics*, **46**.

McCann, W.R., Nishenko, S.P., Sykes, L.R. and Krause, J. (1979) Seismic gaps and plate tectonics: Seismic potential for major boundaries. *Pure Applied Geophysics*, **117**, 1082–1147.

McCloskey, J., Nalbant, S.S. and Steacy, S. (2005) Earthquake risk from co-seismic stress. *Nature*, **434**, 291.

McGarr, A. (2014) Maximum magnitude earthquakes induced by fluid injection. *Journal of Geophysical Research Solid Earth*. DOI: 10.1002/2013JB010597.

Michael, A.J. and Jones, L.M. (1998) Seismicity alert probabilities at Parkfield, California, revisited. *Bulletin of the Seismological Society of America*, **88**, 117–130.

Mouroux, P. and Le Brun, B. (2006) Presentation of RISK-UE Project. *Bulletin of Earthquake Engineering*, **4**, 323–339.

O'Connell, D.R.H. and Ake J.P. (2002) *Earthquake Ground Motion Estimation*, Routledge, Ltd./Academic Press, Amsterdam.

Omori, T., Ando, K., Okano, M., *et al.* (2011) Superelastic effect in polycrystalline ferrous alloys. *Science*, **333**, 68–71. DOI: 10.1126/science.1202232

Ozawa, S., Murakami, M., Kaidzu, M. *et al.* (2011) Detection and monitoring of ongoing aseismic slip in the Tokai region, central Japan, *Science*, **298**, 1009–1012. DOI: 10.1126/science.1076780.

Papazachos, B.C. (1975) Foreshocks and earthquake prediction. *Tectonophysics*, **28**.

Parsons, T. (2004) Recalculated probability of M>7 earthquakes beneath the Sea of Marmara, Turkey. *Journal of Geophysical Research*, **109**. DOI: 10.1029/2003JB002667

Parsons, T. (2009) Is there a basis for preferring characteristic earthquakes over a Guttenberg-Richter distribution in probabilistic earthquake forecasting?' *Bulletin of the Seismological Society of America*, **99**, 2012–2019.

Parsons, T., Toda, S., Stein, R.S., Barka, A., and Dieterich, J.H. (2000) Heightened odds of large earthquakes near Istanbul: An interaction-based probability calculation. *Science*, **288**, 661–665.

Pickett, M. (1997) *Northridge Earthquake: Lifeline Performance and Post-Earthquake Response*, Report No. NIST GCR 97–712, http://fire.nist.gov/bfrlpubs/build97/PDF/b97112.pdf (accessed 22 November 2016).

Polese, M., Di Ludovico, M., Prota, A., and Manfredi, G. (2012) Damage-dependent vulnerability curves for existing buildings. *Earthquake Engineering and Structural Dynamics*, **42**, 853–870.

Pollitz, F., Stein, R.S., Sevilgen, V.M. and Burgmann, R. (2012) The 11 April 2012 east Indian Ocean earthquake triggered large aftershocks worldwide. *Nature*, **290**, 250–253.

POLYTECT. (2012) Final Report of the POLYTECT project, http://cordis.europa.eu/publication/rcn/13401_en.html (accessed 6 May 2016).

Pousse, G. (2005) Analysis of K-net and Kik-net data: implications for ground motion prediction: Acceleration time histories, response spectra and nonlinear site response. PhD thesis, University of Grenoble, France.

Protti, M., Gonzalez, V., Newman, A., *et al.* (2014) Nicoya earthquake rupture anticipated by geodetic measurement of the locked plate interface. *Nature Geoscience*, **7**, 117–121.

RMS (2011) *The 2010 Maule, Chile Earthquake: Lessons and Future Challenges*. RMS, http://forms2.rms.com/rs/729-DJX-565/images/eq_2010_chile_eq.pdf (accessed 22 November 2016).

RMS (2012) *The M9.0 Tohoku, Japan Earthquake: Short-Term Changes in Seismic Risk RMS Special Report*, http://www.riskinc.com/Publications/2011_Tohoku_Seismic_Risk.pdf (accessed 14 November 2016).

Rossetto, T., Ioannou, I., and Grant, D.N. (2013) Existing empirical vulnerability and fragility functions: Compendium and guide for selection GEM Technical Report, 2013-X, GEM Foundation, Pavia, Italy.

Sammonds, P.R., Meredith, P.G. and Main, I.G. (1992) Role of pore fluids in the generation of seismic precursors to shear fracture. *Nature*, **359**, 228–230.

Scawthorn, C. and Chen, W-F. (2002) *Earthquake Engineering Handbook*, CRC Press, Boca Raton, FL.

Scholz, C.H. (1982) Scaling laws for large earthquakes: Consequences for physical models. *Bulletin of the Seismological Society of America*, **72**, 1–14.

Scholz, C.H., Aviles, C.A., and Wesnousky, S.G. (1986) Scaling differences between large interpolate and intraplate earthquakes. *Bulletin of the Seismological Society of America*, **76**, 65–70.

Scholz, C.H., Wyss, S., and Smith, W. (1969) Seismic and aseismic slip on the San Andreas Fault. *Journal of Geophysical Research*, **74**, 2049–2069.

Schwartz, D.P. and Coppersmith, K.J. (1984) Fault behavior and characteristic earthquakes: Examples from Wasatch and San Andreas fault zones. *Journal of Geophysical Research*, **89**, 5681–5698.

Somerville, P. and Pitarka, A. (2006) Differences in earthquake source and ground motion characteristics between surface and buried earthquakes. In *Proceedings, Eighth National Conference on Earthquake Engineering*, Proceedings CD-ROM, Paper No. 977. Earthquake Engineering Research Institute, Oakland, CA.

Sparkes, R., Tilmann, F., Hovius, N., and Hillier, J.K. (2010) Subducted seafloor relief stops rupture in South American great earthquakes: Implications for rupture behaviour in the 2010 Maule,

Chile earthquake. *Earth and Planetary Science Letters*, **298**, 89–94. DOI: 10.1016/j.epsl.2010.07.029.

Staecy, S., Gomberg, J., and Cocco, M. (2005) Introduction to special section: Stress transfer, earthquake triggering, and time-dependent seismic hazard. *Journal of Geophysical Research*, **110**, B05S01. DOI: 10.1029/2005JB003692.

Stein, R.S. (1997) Progressive failure on the North Anatolian fault since 1939 by earthquake stress triggering. *Geophysical Journal International*, **128**, 594–604.

Stein, R.S. (1999) The role of stress transfer in earthquake occurrence. *Nature*, **402**, 605–609.

Stein, S., Geller, R.J., and Liu, M. (2012) Why earthquake hazard maps often fail and what to do about it. *Tectonophysics*, **562–563**, 1–25.

Stirling, M., Goded, T., Berryman, K., and Litchfield, N. (2013) Selection of earthquake scaling relationships for seismic-hazard analysis. *Bulletin of the Seismological Society of America*, **103**, 2993–3011.

Swiss Re (2000) *Random Occurrence or Predictable Disaster? New Models in Earthquake Probability Assessment*. Swiss Re, Zurich.

Thatcher, W. (1984) The earthquake deformation cycle, recurrence, and the time-predictable model. *Journal of Geophysical Research*, **89**, 5674–5680.

Toda, S. and Stein, R.S. (2013) The 2011 M=9.0 Tohoku-Oki earthquake more than doubled the probability of large shocks beneath Tokyo. *Geophysical Research Letters*, **40**, 1–5, DOI: 10.1002/grl.50524.

Toksoz, M.N. (1979) Space-time migration of earthquakes along the North Anatolian fault zone and seismic gaps. *Pageoph*, **117**, 1258–1270.

Trifunac, M.D. and Brady, A.G. (1975) A study on the duration of strong earthquake ground motion, *Bulletin of the Seismological Society of America*, **65**, 581–626.

UNISDR. (2013) *From Shared Risk to Shared Value: The Business Case for Disaster Risk Reduction. Global Assessment Report on Disaster Risk Reduction*. United Nations Office for Disaster Risk Reduction, Geneva.

USGS (1996) USGS response to an urban earthquake, Northridge '94, Open-File Report 96–263, http://pubs.usgs.gov/of/1996/0263/report.pdf (accessed 14 November 2016).

USGS (2006) ShakeMap Manual. 20 May 2015, http://pubs.usgs.gov/tm/2005/12A01/pdf/508TM12-A1.pdf.

USGS (2012) Earthquake facts and statistics,, http://earthquake.usgs.gov/earthquakes/eqarchives/year/eqstats.php (accessed 12 May 2015).

USGS (2014) Magnitude/Intensity comparison, http://earthquake.usgs.gov/learn/topics/mag_vs_int.php (accessed 12 May 2015).

Utsu, T. (1961) A statistical study on the occurrence of aftershocks, *Geophysical Magazine*, **30**, 521–605.

Wald, D.J., Quitoriano, V., Heaton, T.H., and Kanamori, H. (1999) Relationship between peak ground acceleration, peak ground velocity, and Modified Mercalli Intensity in California. *Earthquake Spectra*, **15**, 557–564.

Wells, D.L. and Coppersmith, K.J. (1994) New empirical relationships among magnitude, rupture length, rupture width, rupture area, and surface displacement. *Bulletin of the Seismological Society of America*, **84**, 974–1002.

Wesnousky S.G. (1994) The Guttenberg-Richter or characteristic earthquake distribution, which is it?' *Bulletin of the Seismological Society of America*, **84**, 1940–1959.

Yin, Y., Wang, F., and Sun, P. (2009) Landslide hazards triggered by the 2008 Wenchun earthquake, Sichuan, *China. Landslides*, **6**, 139–152.

**Mass Movement**

AIR (2012) The AIR Earthquake Model for Canada. Brochure, http://www.air-worldwide.com/Facet-Search/Search-Results/ (accessed 2 July 2015).

BGS/Durham (2015) *Landslide Inventory following 25 April and 12 May Nepal Earthquakes.* http://ewf.nerc.ac.uk/2015/06/30/updated-30-june-landslide-inventory-following-25-april-and-12-may-nepal-earthquakes/ (accessed 6 June 2015).

Bromhead, E. (2006) *The Stability of Slopes.* CRC Press, Boca Raton, FL.

Bryn, P., Berg, K., Forsberg, C.F., Solheim, A., and Kvalstad, T. (2005) Explaining the Storegga slide. *Marine and Petroleum Geology*, **22**, 11–19.

Cannon, S.H., Gartner, J.E., Wilson, R.C., Bowers, J.C., and Layber, J.L. (2008) Storm rainfall conditions for floods and debris flows from recently burned areas in southwestern Colorado and southern California. *Geomorphology*, **96**, 250–269.

Cannon, S.H., Kirkham, R., and Parise, M. (2001) Wildfire-related debris flow initiation processes, Storm King Mountain, Colorado. *Geomorphology*, **39**, 171–188.

Choi, K.Y. and Cheung, R.W. (2013) Landslide disaster prevention and mitigation through works in Hong Kong. *Journal of Rock Mechanics and Geotechnical Engineering*, **5**, 354–365.

Corominas, J., Van Westen, C., Frattini, P., *et al.* (2014) Recommendations for the quantitative analysis of landslide risk. *Bulletin of Engineering Geology and the Environment*, **73**, 209–263.

Costa, J.E. and Schuster, R.L. (1988) The formation and failure of natural dams. *Geological Society of America Bulletin*, **100**, 1054–1068.

Cruden, D.M. (1991) A simple definition of a landslide. *Bulletin of the International Association of Engineering Geology*, **41**, 27–29.

Cruden, D.M. and Krahn, J. (2012) Frank rockslide, Alberta, Canada. *Rockslides and Avalanches*, **1**, 97–112.

Dijkstra, T.A., Chandler, J., Wackrow, R., *et al.* (2012) Geomorphic controls and debris flows: The 2010 Zhouqu disaster, China, in *Landslides and Engineered Slopes*, Proceedings of the 11th International Symposium on Landslides (ISL) and the 2nd North American Symposium on Landslides, Banff, Canada (eds E. Eberhardt, C. Froese, A.K. Turner, and S. Leroueil), CRC Press/Balkema, Leiden, The Netherlands.

Dijkstra, T.A. and Dixon, N. (2010) Climate change and slope stability in the UK: Challenges and approaches. *Quarterly Journal of Engineering Geology and Hydrogeology*, **43**, 371–385.

Dijkstra, T., Dixon, N., Crosby, C., *et al.* (2014) Forecasting infrastructure resilience to climate change. *Proceedings of the ICE-Transport*, **167**, 269–280.

Drosos, V.A., Gerolymos, N., and Gazetas, G. (2012) Constitutive model for soil amplification of ground shaking: Parameter calibration, comparisons, validation. *Soil Dynamics and Earthquake Engineering*, **42**, 255–274.

Dunning, S.A., Mitchell, W.A., Rosser, N.J., and Petley, D.N. (2007) The Hattian Bala rock avalanche and associated landslides triggered by the Kashmir Earthquake of 8 October, 2005. *Engineering Geology*, **93**, 130–144.

Economic Commission for Latin America and the Caribbean. (2011) *Saint Lucia: Macro Socio-Economic and Environmental Assessment of the Damage and Losses Caused by Hurricane Tomas: A Geo-Environmental Disaster.* Towards Resilience. Economic Commission for Latin America and the Caribbean, Santiago, Chile.

Genevois, R. and Ghirotti, M. (2005) The 1963 Vaiont landslide, *Giornale di Gelogia Applicata*, **1**, 41–52.

Gill, M. and Malamud, B.D. (2014) Reviewing and visualizing the interactions of natural hazards. *Review of Geophysics*, **52**, 680–722.

Goren, L., Aharonov, E., and Anders, M.H. (2010) The long runout of the Heart Mountain landslide: Heating, pressurization, and carbonate decomposition. *Journal of Geophysical Research*, **115**, B10210. DOI: 10.1029/2009JB007113.

Günther, A., Van Den Eeckhaut, M., Malet, J-P., Reichenbach, P., and Hervás, J. (2014) Climate-physiographically differentiated Pan-European landslide susceptibility assessment using spatial multi-criteria evaluation and transnational landslide information. *Geomorphology*, **224**, 69–85.

Guzzetti, F., Mondini, A.C., Cardinali, M., *et al.* (2012) Landslide inventory maps: New tools for an old problem. *Earth-Science Reviews*, **112**, 42–66.

Guzzetti, F., Peruccacci, S., Rossi, M., and Stark, C.P. (2008) The rainfall intensity–duration control of shallow landslides and debris flows: an update. *Landslides*, **5**, 3–17.

Heim, A. (1932) *Landslides and Human Lives*, trans. N Skermer, BiTech Publishers, Vancouver.

Hovius, N,. Meunier, P., Ching-Weei, L., *et al.* (2011) Prolonged seismically induced erosion and the mass balance of a large earthquake. *Earth and Planetary Science Letters*, **304**, 347–355.

Hungr, O., Leroueil, S., and Picarelli, L. (2014) The Varnes classification of landslide types, an update. *Landslides*, **1**, 167–194.

Hutchinson, J.N. (1986) A sliding-consolidation model for flow slides. *Canadian Geotechnical Journal*, **23**, 115–126.

Hutchinson, J.N. (1989) General report: Morphological and geotechnical parameters of landslides in relation to geology and hydrogeology. Proceedings of the 5th International Symposium on Landslides, Lausanne, 10–15 July 1988. Rotterdam: AA Balkema. In *International Journal of Rock Mechanics and Mining Sciences and Geomechanics Abstracts*, **26** (1988).

Jibson, R.W., Harp, E.L., and Michel, J.A. (2000) A method for producing digital probabilistic seismic landslide hazard maps. *Engineering Geology*, **58**, 271–289.

Jordan, C.J.J., Grebby, S., Dijkstra, T., Dashwood, C., and Cigna, F. (2015) Risk information services for disaster risk management (DRM) in the Caribbean: Operational documentation. BGS Report OR/15/001. British Geological Survey, Keyworth, Nottingham.

Lin, C.W., Liu, S.H., Lee, S.Y., and Liu, C.C. (2006) Impacts of the Chi-Chi earthquake on subsequent rainfall-induced landslides in central Taiwan. *Engineering Geology*, **86**, 87–101.

Lu, N. and Godt, J.W. (2013) *Hillslope Hydrology and Stability*, Cambridge University Press, Cambridge.

Lunio, F. (2002) Sequence of instability processes triggered by heavy rainfall in the northern Italy. *Geomorphology*, **66**, 13–39.

Malamud, B.D., Turcotte, D.L., Guzzetti, F., and Reichenbach, P. (2004a) Landslide inventories and their statistical properties. *Earth Surface Processes and Landforms*, **29**, 687–711.

Malamud, B.D., Turcotte, D.L., Guzzetti, F., and Reichenbach, P. (2004b) Landslides, earthquakes, and erosion. *Earth and Planetary Science Letters*, **29**, 45–59.

Metternicht, G., Hurni, L., and Gogu, R. (2005) Remote sensing of landslides: An analysis of the potential contribution to geo-spatial systems for hazard assessment in mountainous environments. *Remote Sensing of Environment*, **98**, 284–303.

Meunier, P., Hovius, N., and Haines, J.A. (2008) Topographic site effects and the location of earthquake induced landslides. *Earth and Planetary Science Letters*, **275**, 221–232.

Montrasio, L., Valentino, R., Corina, A., Rossi, L., and Rudari, R. (2014) A prototype system for space–time assessment of rainfall-induced shallow landslides in Italy. *Natural Hazards*, **74**, 1263–1290.

Munich Re. (2011) Significant earthquakes worldwide 1980–2011. Münchener Rückversicherungs-Gesellschaft, Geo Risks Research, NatCatSERVICE, https://www.munichre.com/site/corporate/get/documents_E245007932/mr/assetpool.shared/Documents/0_Corporate Website/6_Media Relations/Press Dossiers/New Madrid Seismic Zone/significant-earthquakes-worldwide-en.pdf

Nadim, F., Kjekstad, O., Peduzzi, P., Herold, C., and Jaedicke, C. (2006) Global landslide and avalanche hotspots. *Landslides*, **3**, 159–173.

Newmark, N.M. (1965) Effects of earthquakes on dams and embankments. *Geotechnique*, **15**, 139–160.

Owen, L.A., Kamp, U., Khattak, G.A., *et al.* (2008) Landslides triggered by the 8 October 2005 Kashmir earthquake. *Geomorphology*, **94**, 1–9.

Pankow, K.L., Moore, J.R., Hale, J.M., *et al.* (2014) Massive landslide at Utah copper mine generates wealth of geophysical data. *GSA Today*, **24**, 4–9.

Papathoma-Köhle, M., Zischg, A., Fuchs, S., Glade, T., and Keiler, M. (2015) Loss estimation for landslides in mountain areas: An integrated toolbox for vulnerability assessment and damage documentation. *Environmental Modelling and Software*, **63**, 156–169.

Pareschi, M.T., Favalli, M., Giannini, F., *et al.* (2000) May 5 1998, debris flow in circum-Vesuvian areas (Southern Italy): Insights for hazard assessments. *Geology*, **28**, 639–642.

Petley, D. (2012) Global patterns of loss of life from landslides. *Geology*, **40**, 927–930.

Schnellmann, M., Anselmetti, F.S., Giardini, D., and McKenzie, J.A. (2006) 15,000 Years of mass-movement history in Lake Lucerne: Implications for seismic and tsunami hazards. *Eclogae Geologicae Helvetiae*, **99**, 409–428.

Selby, M.J. (1993) *Hillslope Materials and Processes* (2nd edn), Oxford University Press, Oxford.

Smith, K., and Petley, D. (2009) *Environmental Hazards: Assessing Risk and Reducing Disaster* (5th edn), Routledge, New York.

Tang, C., Rengers, N., van Asch, T.W., *et al.* (2011) Triggering conditions and depositional characteristics of a disastrous debris flow event in Zhouqu city, Gansu Province, northwestern China. *Natural Hazards and Earth System Sciences*, **11**, 2903–2912.

Toyos, G., Dorta, D.O., Oppenheimer, C., *et al.* (2007) GIS-assisted modelling for debris flow hazard assessment based on the events of May 1998 in the area of Sarno, Southern Italy: Part 1. Maximum run-out. *Earth Surface Process and Landforms*, **32**, 1491–1502.

Van Den Eeckhaut, M., and Hervás, J. (2012) State of the art of national landslide databases in Europe and their potential for assessing susceptibility, hazard and risk. *Geomorphology*, **139–140**, 545–558. DOI: 10.1016/j.geomorph.2011.12.006.

Van Westen, C.J., Van Asch, T.W., and Soeters, R. (2006) Landslide hazard and risk zonation—why is it still so difficult? *Bulletin of Engineering geology and the Environment*, **65**, 167–184.

Wasowski, J., and Bovenga, F. (2014) Investigating landslides and unstable slopes with satellite Multi Temporal Interferometry: Current issues and future perspectives. *Engineering Geology*, **174**, 103–138.

Wieczorek, G.F., Larsen, M.C., Eaton, L.S., Morgan, B.A., and Blair, J.L. (2001) *Debris-Flow and Flooding Hazards Associated with the December 1999 Storm in Coastal Venezuela and Strategies for Mitigation.* USGS Open File Report 01–0144 pubs.usgs.gov/of/2001/ofr-01-0144/(accessed 14 November 2016).

Wilson, R.C. and Keefer, D.K. (1985) Predicting areal limits of earthquake-induced landsliding, in *Evaluating Earthquake Hazards in the Los Angeles Region: An Earth-Science Perspective* (J.I. Ziony, ed.), U.S. Geological Survey professional paper, 1360, 316–345.

Xu, C., Xu, X.W., Yao, Q., and Wang, Y.Y. (2013) GIS-based bivariate statistical modelling for earthquake-triggered landslides susceptibility mapping related to the 2008 Wenchuan earthquake, China. *Quarterly Journal of Engineering Geology and Hydrogeology*, **46**, 221–236, http://dx.doi.org/10.1144/qjegh2012-006.

Yin, Y., Zheng, W., Li, X., Sun, P., and Li, B. (2011) Catastrophic landslides associated with the M8.0 Wenchuan earthquake. *Bulletin of Engineering Geology and the Environment*, **70**, 15–32.

Zhang, D and Wang, G. (2007) Study of the 1920 Haiyuan earthquake-induced landslides in loess (China). *Engineering Geology*, **94**, 76–88.

## Tsunami

AIR Worldwide (2013) Introducing the industry's first fully probabilistic tsunami model of March 26 2013, http://www.air-worldwide.com/Publications/AIR-Currents/2013/Introducing-The-Industry%E2%80%99s-First-Fully-Probabilistic-Tsunami-Model. (accessed 6 May 2016).

Bell, R., Holden, C., Power, W., Wang, X., and Downes, G. (2014) Hikurangi margin tsunami earthquake generated by slow seismic rupture over a subducted seamount. *Earth and Planetary Science Letters*, **397**, 1–9. DOI: 10.1016/j.epsl.2014.04.005.

Bollinger, G.A. (1972) Historical and recent seismic activity in South Carolina. *Bulletin of the Seismological Society of America*, **62**, 851–864.

Boris, L. and Mikhai, N. (2009) *Physics of Tsunamis*, Springer, Berlin.

Chester, D.K. (2008) The effects of the 1755 Lisbon earthquake and the tsunami on the Algarve region, southern Portugal. *Geography*, **93**, 78–90.

Edward, B. (2008) *Tsunami: The Underrated Hazard* (2nd edn), Springer, New York.

Fritz, H.M., Mohammed, F. and Yoo, J. (2009) Lituya bay landslide impact generated mega-tsunami 50th anniversary, tsunami science four years after the 2004 Indian Ocean tsunami. *Pure and Applied Geophysics*, **166**, 153–175.

Fritz, H.M., Phillips, D.A., Okayasu, A., *et al.* (2012) The 2011 Japan tsunami current velocity measurements from survivor videos at Kesennuma Bay using LiDAR. *Gephysical Research Letters*, **39**, L00G23.

Fowler, C.M.R. (2005) *The Solid Earth: An Introduction to Global Geophysics*, Cambridge University Press, Cambridge.

Fraser, S., Raby, A., Pomonis, A., *et al.* (2013) Tsunami damage to coastal defences and buildings in the March 11th 2011 Mw 9.0 Great East Japan earthquake and tsunami. *Bulletin of Earthquake Engineering*, **11**, 205–239.

Fujii, Y. and Satake, K. (2007) Tsunami source of the 2004 Sumatra-Andaman earthquake inferred from tide gauge and satellite data. *Bulletin of the Seismological Society of America*, **97** (1A), S192–S207.

Fukutani, Y., Suppasri, A., and Imamura, F. (2014) Stochastic analysis and uncertainty assessment of tsunami wave height using a random source parameter model that targets a Tohoku-type earthquake fault. *Stochastic Environmental Research and Risk Assessment*. DOI: 10.1007/s00477-014-0966-4.

Geist, E., Kirby, S., Ross, S., and Dartnell, P. (2009) Samoa disaster highlights danger of tsunamis generated from outer-rise earthquakes. *Sound Waves: Coastal and Marine Research News from Across the USGS* (accessed 14 November 2016, http://soundwaves.usgs.gov/2009/12/research.html

Goto, K., Tajika, E., Tada, R., and Matsui, T. (2004) Formation of a large oceanic impact crater and generation of impact tsunamis at the Cretaceous/Tertiary boundary. *Journal of Japan Society for Planetary Sciences*, **13**, 241–248.

Henstock, T., McNeill, L., Dean, S., *et al.* (2010) Exploring structural control on Sumatran earthquakes. *Eos*, **91**, 405.

Hokugo, A., Nishino, T. and Inada, T. (2013) Tsunami fires after the Great East Japan Earthquake. *Journal of Disaster Research*, **8**, 584–593.

Insurance Information Institute. (2013) Earthquake and tsunamis, http://www.iii.org/fact-statistic/earthquakes-and-tsunamis (accessed 14 November 2016).

Iuchi, K., Johnson, L.A. and Olshansky, R.B. (2013) Securing Tohoku's future: Planning for rebuilding in the first year following the Tohoku-Oki earthquake and tsunami, *Earthquake Spectra*, **29**, S479–S499.

Jankaew, K., Atwater, B.F., Sawai, Y. *et al.* (2008) Medieval forewarning of the 2004 Indian Ocean tsunami in Thailand. *Nature*, **455**, 1228–1231.

Japan Meteorological Agency. (2015) Monitoring of earthquakes, tsunamis and volcanic activity, http://www.jma.go.jp/jma/en/Activities/earthquake.html (accessed 23 January 2015).

Japan Nuclear Energy Safety Organization and Tohoku University (2014) Japan tsunami trace database, http://tsunami3.civil.tohoku.ac.jp (accessed 5 February 2015).

Kanamori, H. (1972) Mechanism of tsunami earthquakes. *Physics of the Earth and Planetary Interiors*, **6**, 346–359.

Keller, E.A. and Blodgett, R.H. (2008) *Natural Hazards: Earth's Processes as Hazards, Disasters and Catastrophes*. Pearson, London.

Kosa, K. (2012) Damage analysis of bridges affected by tsunami due to Great East Japan Earthquake, in Proceedings of the International Symposium on Engineering Lessons Learned from the 2011 Great East Japan Earthquake, March 1–4, 2012, Tokyo, Japan.

Koshimura, S., Takayuki, O., Hideaki, Y. and Imamura, F. (2009) Developing fragility functions for tsunami damage estimation using numerical model and post-tsunami data from Banda Aceh, Indonesia. *Coastal Engineering Journal*, **51**, 243. DOI: 10.1142/S0578563409002004.

Latter, J.H. (1981) Tsunamis of volcanic origin: Summary of cause, with particular reference to Krakatoa, 1883. *Bulletin Volcanologique*, **44**, 467–490.

Løvholt, F., Pedersen, G., Bazin, S., *et al.* (2012) Stochastic analysis of tsunami runup due to heterogeneous coseismic slip and dispersion. *Journal of Geophysical Research*, **117**, c3047. DOI: 10.1029/2011JC007616.

Macabuag, J. and Rossetto, T. (2014) Towards the development of a method for generating analytical tsunami fragility functions. Proceedings of 2nd European Conference on Earthquake Engineering and Seismology (2ECEES).

Ministry of Land, Infrastructure, Transportation and Tourism (2011) Survey of tsunami damage condition, http://www.mlit.go.jp/toshi/toshi-hukkou-arkaibu.html (accessed 14 November 2016).

Minoura, K., Imamura, F., Sugawara, D., Kono, Y. and Iwashita, T. (2001) The 869 Jogan tsunami deposit and recurrence interval of large-scale tsunami on the Pacific coast of northeast Japan. *Journal of Natural Disaster Science*, **23**, 83–88.

Mofjeld, H.O., González F.I., Titov, V.V., Venturato, A.J., and Newman, J.C. (2009) Effects of tides on maximum tsunami wave heights: Probability distributions, *Journal of Atmospheric and Oceanic Technology*, **24**, 117–123. DOI: http://dx.doi.org/10.1175/JTECH1955.1.

Muhari, A., Imamura, F., Koshimura, S., and Post, J. (2011) Examination of three practical run-up models for assessing tsunami impact on highly populated areas. *Natural Hazards and Earth System Sciences*, **11**, 3107–3123.

National Geophysical Data Center/World Data Service (2014) Global historical tsunami database. DOI: 10.7289/V5PN93H7, http://www.ngdc.noaa.gov/hazard/tsu_db.shtml (accessed 14 November 2016).

Okada, Y. (1985) Surface deformation due to shear and tensile faults in a half-space. *Bulletin of the Seismological Society of America*, **75**, 1435–1454.

Pacific Tsunami Warning Center (2015) PTWC responsibilities, http://ptwc.weather.gov/ptwc/responsibilities.php (accessed 25 January 2015).

Parsons, T. (2004) Recalculated probability of M>7 earthquakes beneath the Sea of Marmara, Turkey. *Journal of Geophysical Research*, **109**. DOI: 10.1029/2003JB002667

Parsons, T. (2009) Is there a basis for preferring characteristic earthquakes over a Guttenberg-Richter distribution in probabilistic earthquake forecasting? *Bulletin of the Seismological Society of America*, **99**, 2012–2019.

Salamon, A., Rockwell, T., Ward, S.N., Guidoboni, E., and Comastri, A. (2007) Tsunami hazard evaluation of the Eastern Mediterranean: Historical analysis and selected modeling. *Bulletin of the Seismological Society of America*, **97**, 1–20. DOI: DOI: 10.1785/0120060147.

Satake, K. and Fujii, Y. (2014) Review: Source models of the 2011 Tohoku earthquake and long-term forecast of large earthquakes. *Journal of Disaster Research*, **9**, 272–280.

Satriano, C., Kiraly, E., Bernard, P., and Vilotte, J.P. (2012) The 2012 Mw 8.6 Sumatra earthquake: Evidence of westward sequential seismic ruptures associated to the reactivation of a N-S ocean fabric. *Geophysical Research Letters*, **39**, L15302.

Schnellmann, M., Anselmetti, F.S., Giardini, D., McKenzie, J., and Ward, S.N. (2004) A recent survey of the sediments beneath a Swiss lake reveals a series of prehistoric tremblors. *American Scientist*, **92**, 38–45.

Siripong, A. (2010) Education for disaster risk reduction in Thailand. *Journal of Earthquake and Tsunami*, **4**, 61.

Sørensen, M., Spada, M., Babeyko, A., Wiemer, S., and Grünthal, G. (2012) Probabilistic tsunami hazard in the Mediterranean Sea. *Journal of Geophysical Research*, **117**, B01305.

Sparkes, R., Tilmann, F., Hovius, N., and Hillier, J.K. (2010) Subducted seafloor relief stops rupture in South American great earthquakes: Implications for rupture behaviour in the 2010 Maule, Chile earthquake. *Earth and Planetary Sciences Letters*, **298**, 89–94. DOI: 10.1016/j.epsl.2010.07.029.

Stein, S. and Okal, E.A. (2005) The 2004 Sumatra earthquake and Indian Ocean tsunami: What happened and why? *Vista Geoscience*, 1–6, http://www.earth.northwestern.edu/people/seth/Texts/virtgeo.pdf. (accessed 14 November 2016).

Suppasri, A., Imamura, F., and Koshimura, S. (2012) Tsunamigenic ratio of the Pacific Ocean earthquakes and a proposal for a tsunami index. *Natural Hazards and Earth System Sciences*, **12**, 175–185.

Suppasri, A., Koshimura, S., and Imamura, F. (2011) Developing tsunami fragility curves based on the satellite remote sensing and the numerical modeling of the 2004 Indian Ocean tsunami in Thailand. *Natural Hazards and Earth System Sciences*, **11**, 173–189.

Suppasri, A., Mas, E., Charvet, I., *et al.* (2013) Building damage characteristics based on surveyed data and fragility curves of the 2011 Great East Japan tsunami. *Natural Hazards*, **66**, 319–341.

Suppasri, A., Muhari, A., Ranasinghe, P., *et al.* (2012) Damage and reconstruction after the 2004 Indian Ocean tsunami and the 2011 Great East Japan tsunami. *Journal of Natural Disaster Science*, **34**, 19–39.

Tappin, D.R., Grilli, S.T., Harris, J.C., *et al.* (2014) Did a submarine landslide contribute to the 2011 Tohoku tsunami?' *Marine Geology*, **357**, 344–361.

Thio, H.K. (2009) Tsunami hazard in Israel. *Report for the Geological Survey of Israel.* http://www.gsi.gov.il/Eng/_Uploads/264Israel-Tsunami-Hazard.pdf (accessed 21 January 2015).

Titov, V.V. and González, F.I. (1997) Implementation and testing of the Method of Splitting Tsunami (MOST) model. *NOAA Technical Memorandum ERL PMEL-112*, NOAA/Pacific Marine Environmental Laboratory, Seattle, Washington.

Titov, V.V. and Synolakis, C.S. (1998) Numerical modeling of tidal wave runup. *Journal of Waterway, Port, Coastal, and Ocean Engineering*, **124**, 157–171.

Tsunami Laboratory (2015) Analysis of the tsunami travel time maps for damaging tsunamis in the world ocean. http://tsun.sscc.ru/ttt_rep.htm (accessed 5 February 2015).

United Nations Educational, Scientific and Cultural Organization International Tsunami Information Center (2014) *List of Tsunamis*, http://itic.ioc-unesco.org/index.php?option=com_contentandview=categoryandlayout=blogandid=1160andItemid=1077andlang=en (accessed 4 September 2015).

Vanneste, M., Forsberg, C.F., Glimsdal, S., *et al.* (2013) Submarine landslides and their consequences: What do we know, what can we do? In *Landslide Science and Practice*, Springer, Berlin.

Wesnousky, S.G. (1994) The Guttenberg-Richter or characteristic earthquake distribution, which is it? *Bulletin of the Seismological Society of America*, **84**, 1940–1959.

**Volcanoes**

Aspinall, W.P. (2010) A route to more tractable expert advice. *Nature*, **463**, 294–295. DOI: 10.1038/463294a.

Baxter, P.J., Aspinall, W.P., Neri, A., *et al.* (2008) Emergency planning and mitigation at Vesuvius: A new evidence-based approach. *Journal of Volcanology and Geothermal Research*, **178**, 454–473.

Bebbington, M.S. (2010) Trends and clustering in the onsets of volcanic eruptions. *Journal of Geophysical Research (Solid Earth)*, **115**, B1. DOI: 10.1029/2009JB006581.

Bélizal, E., Lavigne, F., Hadmoko, D., *et al.* (2013) Rain-triggered lahars following the 2010 eruption of Merapi volcano, Indonesia: A major risk. *Journal of Volcanology and Geothermal Research*, **261**, 330–347.

Bonadonna, C. and Costa, A. (2013) Modeling of tephra sedimentation from volcanic plumes, in *Modeling Volcanic Processes: The Physics and Mathematics of Volcanism* (eds S.A. Fagents, T.K.P. Gregg, and R.M.C. Lopes), Cambridge University Press, Cambridge.

Bonadonna, C., Macedonio, G., and Sparks, R.S.J. (2002) Numerical modelling of tephra fallout associated with dome collapses and vulcanian explosions: Application to hazard assessment on Montserrat, in *The Eruption of Soufrière Hills Volcano, Montserrat from 1995 to 1999* (eds T.H. Druitt and B.P. Kokelaar), Memoir No 21, Geological Society of London, London.

Brown, S.K., Crosweller, S.K., Sparks, R.S.J., *et al.* (2014) Characterisation of the Quaternary eruption record: Analysis of the Large Magnitude Explosive Volcanic Eruptions (LaMEVE) database. *Journal of Applied Volcanology*, **3**, 5.

Brown, S.K., Loughlin, S.C., Sparks, R.S.J., *et al.* (eds) (2015) *Global Volcanic Hazard and Risk*, Cambridge University Press, Cambridge.

Caricchi, L., Annen, C., Blundy, J., Simpson, G., and Pinel, V. (2014) Frequency and magnitude of volcanic eruptions controlled by magma injection and buoyancy. *Nature Geoscience*, **7**, 126–130.

Casadevall, T.J., Reyes, P.J.D. and Schneider, D.J. (1995) The 1991 Pinatubo eruptions and their effects on aircraft operations, in *Fire and Mud* (C.G. Newhall and R.S. Punongbayan eds), http://pubs.usgs.gov/pinatubo/contents.html (accessed 14 November 2016).

Charbonnier, S.J., Germa, A., Connor, C.B., *et al.*, (2013) Evaluation of the impact of the 2010 pyroclastic density currents at Merapi volcano from high-resolution satellite imagery, field investigations and numerical simulations. *Journal of Volcanology and Geothermal Research*, **261**, 295–315, http://dx.doi.org/10.1016/j.jvolgeores.2012.12.021

Connor, L.J., Connor, C.B., Meliksetian, K., and Savov, I. (2012) Probabilistic approach to modeling lava flow inundation: A lava flow hazard assessment for a nuclear facility in Armenia. *Journal of Applied Volcanology*, **1**, 3. DOI: 10.1186/2191-5040-1-3.

Cottrell, E. (2014) Global distribution of active volcanoes, in P. Papale and JF Shroeder (eds.), *Volcanic Hazards, Risks and Disasters*, Elsevier, Oxford.

Crosweller, H.S., Arora, B., Brown, S.K., *et al.* (2012) Global database on large magnitude explosive volcanic eruptions (LaMEVE). *Journal of Applied Volcanology*, **1**, 13.

Deligne, N.I., Coles, S.G., and Sparks, R.S.J. (2010) Recurrence rates of large explosive eruptions. *Journal of Geophysical Research*, **115**, B06203.

Fagents, S., Gregg, T., and Lopes, R. (eds) (2013) *Modelling Volcanic Processes: The Physics and Maths of Volcanism*, Cambridge University Press, Cambridge.

Furlan, C. (2010) Extreme value methods for modelling historical series of large volcanic magnitudes. *Statistical Modelling*, **10**, 113–132.

Gusev, A.A. (2008) Temporal structure of the global sequence of volcanic eruptions: clustering and intermittent discharge rate. *Physics of the Earth and Planetary Interiors*, **166** (3–4), 203–218. DOI: 10.1016/j.pepi.2008.01.004.

Horwell, C.J. and Baxter, P.J. (2006) The respiratory health hazards of volcanic ash: A review for volcanic risk mitigation. *Bulletin of Volcanology*, **69**, 1v24.

Jenkins, S., Komorowski, J.C., Baxter, P.J., *et al.* (2013) The Merapi 2010 eruption: An interdisciplinary impact assessment methodology for studying pyroclastic density current dynamics. *Journal of Volcanology and Geothermal Research*, **261**, 316–329.

Jenkins, S., Magill, C., McAneney, K., and Blong, R. (2012a) Regional ash fall hazard I: A probabilistic assessment methodology. *Bulletin of Volcanology*, **74**, 699–712.

Jenkins, S., McAneney, K., Magill, C., and Blong, R. (2012b) Regional ash fall hazard II: Asia-Pacific modelling results and implications. *Bulletin of Volcanology*, **74**, 713–727.

Jenkins, S., Spence, R., Fonseca, J., Solidum, R. and Wilson, T. (2014) Volcanic risk assessment: Quantifying physical vulnerability in the built environment. *Journal of Volcanology and Geothermal Research*, **276**, 105. DOI: 10.1016/j.jvolgeores.2014.03.002.

Jenkins, S., Wilson, T.M., Magill, C., *et al.* (2015) Volcanic ash fall hazard and risk, in *Global Volcanic Hazard and Risk* (eds S.C. Loughlin, S. Brown, R.S.J. Sparks, *et al.*), Cambridge University Press, Cambridge.

Lechner, P,. Tupper, A,. Guffanti, M,. Loughlin, SC,. and Casadevall, T. (2015) Volcanic ash and aviation—the challenges of real-time, global communication of a natural hazard, in *Observing the Volcano World* (eds C. Fearnley *et al.*), Springer, Berlin.

Loughlin, S.C. (2015) *Global Volcanic Hazards and Risk*. Cambridge University Press, Cambridge.

Macedonio, G., Costa, A,. and Folch, A. (2008) Ash fallout scenarios at Vesuvius: Numerical simulations and implications for hazard assessment. *Journal of Volcanology and Geothermal Research*, **178**, 366–377.

Marzocchi, W. and Zaccarelli, L. (2006) A quantitative model for the time-size distribution of eruptions. *Journal of Geophysical Research*, **111**. DOI: 10.1029/2005JB003709

Mason, B.G., Pyle, D.M., and Oppenheimer, C. (2004) The size and frequency of the largest explosive eruptions on Earth. *Bulletin of Volcanology*, **66**, 735–748.

Mastin, L. (2014) Testing the accuracy of a 1-D volcanic plume model in estimating mass eruption rate. *Journal of Geophysical Research (ATM)*, **119**, 2474–2495. DOI: 10.1002/2013JD020604.

Mastin, L.G., Guffanti, M., Servranckx, R. *et al.*, (2009) A multidisciplinary effort to assign realistic source parameters to models of volcanic ash-cloud transport and dispersion during eruptions. *Journal of Volcanology and Geothermal Research*, **186**, 10–21.

Matthews, A.J., Barclay, J., Carn, S., *et al.* (2002) Rainfall-induced volcanic activity on Montserrat. *Geophysical Research Letters*, **29**, 1644. DOI: 10.1029/2002GL014863.

Mazzocchi, M,. Hansstein, F,. and Ragona, M. (2010) The 2010 volcanic ash cloud and its financial impact on the European airline industry. *CESifo Forum* 2/2010, **11**, 92–100.

Munich Re (2007) Volcanism: Recent findings on the risk of volcanic eruptions. *Schadenspiegel*, **1/2007**, 34–39.

Newhall, C.G. and Punongbayan, A.S. (1996) *Fire and Mud* http://pubs.usgs.gov/pinatubo/contents.html (accessed 14 November 2016).

Newhall, C.G. and Self, S. (1982) The Volcanic Explosivity Index (VEI): An estimate of explosive magnitude for historical volcanism. *Journal of Geophysical Research*, **87**, 1231–1238.

Ogburn, S,. Loughlin, S.C. and Calder, E. (2015) The association of lava dome growth with major explosive activity (VEI ≥ 4): DomeHaz, a global dataset. *Bulletin of Volcanology*, **77**, 40. DOI: 10.1007/s00445-015-0919-x.

Oppenheimer, C. (2011) *Eruptions That Shook The World*, Cambridge University Press, Cambridge.

Palma, J.L., Courtland, L., Charbonnier, S., Tortini, R., and Valentine, G.A. (2014) Vhub: a knowledge management system to facilitate online collaborative volcano modeling and research. *Journal of Applied Volcanology*, **3**, 2. DOI: 10.1186/2191-5040-3-2.

Parfitt, L. and Wilson, L. (2007) *Principles of Physical Volcanology*, Wiley-Blackwell, Hoboken, NJ.

Phillipson, G., Sobradelo, R. and Gottsman, J. (2013) Global volcanic unrest in the 21st century: An analysis of the first decade. *Journal of Volcanology and Geothermal Research*, **264**, 183–196.

Picquout, A., Lavigne, F., Mei, E.T.W., *et al.* (2013) Air traffic disturbance due to the 2010 Merapi volcano eruption. *Journal of Volcanology and Geothermal Research*, **261**, 366–375.

Pistolesi, M., Cioni, R., Rosi, M., and Aguilera, E. (2014) Lahar hazard assessment in the southern drainage system of Cotopaxi volcano, Ecuador: Results from multiscale lahar simulations. *Geomorphology*, **207**, 51–63.

Potter, S.H., Scott, B.J. and Jolly, G.E. (2012, *Caldera Unrest Management Sourcebook*. GNS Science Report *2012/12*, Lower Hutt, NZ: GNS Science.

Prata, A.J. (2009) Satellite detection of hazardous volcanic clouds and the risk to global air traffic. *Natural Hazards*, **51**, 303–324.

Pyle, D.M. (2015) Sizes of volcanic eruptions, in *Encyclopedia of Volcanoes* (eds H. Sigurdsson, B. Houghton, S. McNutt, H. Rymer and J. Stix), (2nd edn), Elsevier, Oxford.

Ragona, M., Hannstein, F., and Mazzocchi, M. (2011) The impact of volcanic ash crisis on the European airline industry. in *Governing Disasters: The Challenges of Emergency Risk Regulations* (ed A. Alemanno), Edward Elgar Publishing, Cheltenham.

Rampino, M.R. (2008) Super-volcanism and other geophysical processes of catastrophic import, in *Global Catastrophic Risks* (eds N. Bostrom and M. M. Cirkovic), Oxford University Press, Oxford.

Self, S. (2006) The effects and consequences of very large explosive volcanic eruptions. *Philosophical Transactions of the Royal Society A*, **364**, 2073–2097. DOI: 10.1098/rsta.2006.1814.

Siebert, L., Simkin, T., and Kimberly, P. (2010) *Volcanoes of the World* (3rd edn), University of California Press, Berkeley.

Sigurdsson, H., Houghton, B., McNutt, S., Rymer, H., and Stix, J. (2015) *Encyclopedia of volcanoes* (2nd edn), Elsevier, Oxford.

Smithsonian Institution (2013) Volcanoes of the world database version 4.40, doi.org/10.5479/si. GVP.VOTW4-2013(accessed 27 July 2015).

Spence, R., Gunasekera, R., and Zuccaro, G. (2009) Insurance risks from volcanic eruptions in Europe, http://www.willisresearchnetwork.com/assets/templates/wrn/files/WRN%20-% 20Insurance%20Risks%20from%20Volcanic%20Eruptions_Final.pdf (accessed 9 February 2015).

Surono, J.P., Pallister, J., Boichu, M., *et al.* (2012) The 2010 explosive eruption of Java's Merapi volcano-A '100-year' event. *Journal of Volcanology and Geothermal Research*, **241–242**, 21–135.

Sword-Daniels, V. (2011) Living with volcanic risk: The consequences of, and response to, ongoing volcanic ashfall from a social infrastructure systems perspective on Montserrat. *New Zealand Journal of Psychology*, **40**, 131–138.

Thordarson, T and Self, S. (2003) Atmospheric and environmental effects of the 1783–1784 Laki eruption: A review and reassessment. *Journal of Geophysical Research*, **108**, 4011. DOI: 10.1029/ 2001JD002042.

Voight, B. (1996) The management of volcano emergencies: Nevado del Ruiz, in *Monitoring and Mitigation of Volcano Hazards* (eds R. Scarpa and R. I. Tilling), Springer, Berlin.

Wadge, G. and Isaacs, M.C. (1987) Volcanic hazards from Soufriere Hills volcano, Montserrat, West Indies. A report to the government of Montserrat and the Pan Caribbean Preparedness and Prevention Project.

Wilson, G., Wilson, T.M., Deligne, N.I., and Cole, J.W. (2014) Volcanic hazard impacts to critical infrastructure: A review. *Journal of Volcanology and Geothermal Research*, **286**, 148–182.

Wilson, T., Kaye, G., Stewart, C., and Cole, J. (2007, Impacts of the 2006 eruption of Merapi volcano, Indonesia, on agriculture and infrastructure. GNS Science Report, 2007/07. https:// www.massey.ac.nz/massey/fms/Colleges/College%20of%20Humanities%20and%20Social% 20Sciences/Psychology/Disasters/pubs/GNS/2007/SR%202007-07%20Impacts%20of%20the% 202006%20eruption%20of%20Merapi%20volcano.pdf?4F36D8B6AE5CE1E8F79AFF0C1275AC58

Wilson, T.M., Stewart, C., Sword-Daniels, V., *et al.* (2012) Volcanic ash impacts on critical infrastructure. *Physics and Chemistry of the Earth, Parts A/B/C*, **45**, 5–23.

Whelley, P., Newhall, C.G., and Bradley, K.E. (2015) The frequency of explosive eruptions in Southeast Asia. *Bulletin of Volcanology*, **77**, 1.

Zuccaroa, G., Cacacea, F., Spence, R.J.S., and Baxter P.J. (2008) Impact of explosive eruption scenarios at Vesuvius. *Journal of Volcanology and Geothermal Research*, **178**, 416–453.

# 4

# Building Catastrophe Models

*Editors: Matthew Foote, Kirsten Mitchell-Wallace, Matthew Jones, and John Hillier*

*Contributing authors: Barbara Page, Adam Podlaha, Claire Souch, Rashmin Gunasekera, Shane Latchman, Milan Simic, Renato Vitolo, Nilesh Shome, Alexandros Georgiadis, Petr Puncochar, Goran Trendafiloski, Chris Ewing, Radovan Drinka, Radek Solnicky, Sarka Cerna, Marc Hill*

## 4.1 Overview

### 4.1.1    What Is Included

This chapter provides a more detailed description of the framework of catastrophe models described in Chapter 1.8, and outlines, in a series of studies, how they are developed. It describes the major steps commonly undertaken in the construction of catastrophe models from the viewpoint of the model developer. Descriptions of the types of data, methods as well as the limitations faced in catastrophe model development and methods commonly employed to handle them are included. It illustrates the concepts with a number of case study examples, drawn from across the industry. While the chapter attempts wherever possible to be 'model agnostic' with respect to any particular development method or platform design, the authors have provided examples specific to their own approaches.

*Natural Catastrophe Risk Management and Modelling: A Practitioner's Guide*, First Edition.
Edited by Kirsten Mitchell-Wallace, Matthew Jones, John Hillier and Matthew Foote.
© 2017 John Wiley & Sons Ltd. Published 2017 by John Wiley & Sons Ltd.

#### 4.1.2    What Is Not Included

This chapter does not attempt to cover the model development process in sufficient detail to render it exactly reproducible; many model development methodologies are the intellectual property of the companies who develop them, and they differ between companies. Indeed, much of the literature pertaining to specific models is restricted to those who have commercial access to these models. This chapter cannot, and does not, attempt to reproduce detail on any particular model methodology.

It specifically excludes:

- detailed descriptions of catastrophe model component designs for all natural catastrophe perils;
- details of specific modelling company approaches;
- details of mathematical or statistical methods;
- specific uncertainty calculation or detailed financial model methodologies;
- model platform software or output considerations.

#### 4.1.3    Why Read This Chapter?

This chapter provides a descriptive overview of the typical framework used in the development of catastrophe models. It uses examples from a range of modelling projects, to illustrate how catastrophe models are designed, constructed and validated. It also covers the common sources of data and model processes; these are also discussed in Chapters 1, 2, 3 and 5, to which the reader is directed as appropriate. As a whole, the chapter should provide insight into the main challenges faced by model developers and the range of solutions commonly applied.

## 4.2    Introduction

As described in Chapter 1, catastrophe models incorporate a wide range of scientific, statistical and financial risk knowledge within complex computational engines, producing financial loss estimates to assist risk decision-making and planning. An effective catastrophe model has to produce quantitative information that enables decisions to be taken with an appropriate level of confidence. Each catastrophe model represents a view of the most appropriate quantification of a risk at a particular time, synthesizing many inter-related and often inter-dependent factors. Each model is constructed from a range of information sources (e.g. historical loss records, or other models of key processes, such as Global Circulation Models, GCMs, see Section 4.3.6.6) combined with an understanding of the processes that determine damage and eventual loss.

Apart from a relatively basic representation of uncertainty, the details of a particular model development approach, its validation, and the epistemic and other uncertainties inherent in the process (see Chapter 2.16), are usually not available publicly or in the final loss estimates produced by the model. Furthermore, the *fundamental scope* of each model, for example, its geographic region or inclusion of secondary perils (e.g. Sections 4.3.1 or 4.3.7), is a pragmatic set of decisions made by the modelling developers. Ultimately, such decisions rest with the party commissioning the model, known as its **sponsor**. Projects to build catastrophe models will generally be designed and specified by representatives of the model's sponsors to an agreed requirement. This requirement may often be based on a specific and materially significant risk identified by a national agency, business or market sector, or by a commercial organization based on their market intelligence and business plans. To other, later users of the models, these original requirements may not always be obvious or completely explicable.

There are a number of challenges for the model developer in the choice of input data, scientific theories, necessary or pragmatic assumptions, and the structure of the computational approaches. The ultimate intention is to produce 'fit for purpose' (e.g. accurate and timely) loss estimates but the choices made by the model developer also present challenges for the model user in terms of best use of the model for their own specific requirements. The approaches taken by model developers, especially decisions that involve trade-offs between computational limitations, data availability and model parameterizations, will determine the suitability of a model for a particular use. It is therefore important that users are familiar with the approaches used to construct a particular model, e.g. to inform their own view of risk (see Chapter 5). This familiarity can (in part) be gained by detailed review of model documentation, supplemented by discussion with model developers, but will be more robust if grounded in an understanding of the physical processes characterizing the hazard (e.g. Chapter 3).

Figure 4.1 illustrates the structure of a typical catastrophe model from a development perspective. Generally, the development process is performed by the construction of a series of blocks of knowledge or computer code that perform one or more of the key tasks required. In catastrophe modelling, these blocks are known as **components** (see Grossi and Kunreuther, 2005). In environmental modelling terms, this is equivalent to the concept of **modularization**. As a useful background to the general concepts of mathematical modelling of natural processes, which mirrors many of the approaches described here, including modularization, the reader is directed to Wainright and Mulligan (2013).

Catastrophe models may be considered in terms of the computational structure of each component, as well as their conceptual basis. Both are equally critical to understanding the provenance and applicability of the final model to risk assessment. The development process is also critically dependent on the interconnections between each component, as defined by the inputs and outputs of each.

Figure 4.1 emphasizes the importance of calibration and validation for the different modules (i.e. components) of the model, but also specifically notes the requirement for overall calibration and validation of losses once the modules are combined (see also Chapter 5). Multiple data sources from different agencies and disciplines are required not only in the model building itself but also for the subsequent calibration and validation. The framework shown is generally applicable in that it can be scaled up to represent any catastrophe model, however complex.

**Calibration** and **validation** have a tendency to be used somewhat interchangeably, and sometimes incorrectly. However, they are different and distinct. Calibration can be considered 'fine tuning' of the key model parameters, for example, refining a vulnerability function based on generic engineering assessments of a particular type of insured asset or occupancy. Validation is the testing of the model outputs and parameters to ensure that they reflect 'reality', checking these against available external data, independent to the data used in calibration, for example, modelled wind gust speeds against observations. Calibration and validation are generally undertaken at both component and overall model levels, although calibration is usually part of the model development process, while validation is undertaken by both model developers and by model users (as described in Chapter 5.2). A common type of validation, which illustrates this, is the comparison of modelled 'as if' industry losses with industry losses reported by industry bodies; this is undertaken both by model developers, and by model users (see e.g. Chapter 2.6.2.1 and Chapter 2.6.4.2). Further validation using company-specific losses can be an important tool for model sense-checking against an individual organization's own portfolio as discussed in Chapter 5.4.3.

Model design and development projects often operate using the framework shown in Figure 4.1, with specialist groups responsible for each different component and the overall model construction. Specifically, developers typically frame their workflow in terms of the four interrelated but relatively discrete elements: hazard (and event), exposure, vulnerability, and financial calculations. They define these elements with varying degrees of complexity, depending on the nature of the hazard, the availability of data, and their relative impact. The structure of the chapter follows these blocks, with sections sequentially discussing each of these elements in more detail, together with examples illustrating particular concepts and areas of uncertainty. In

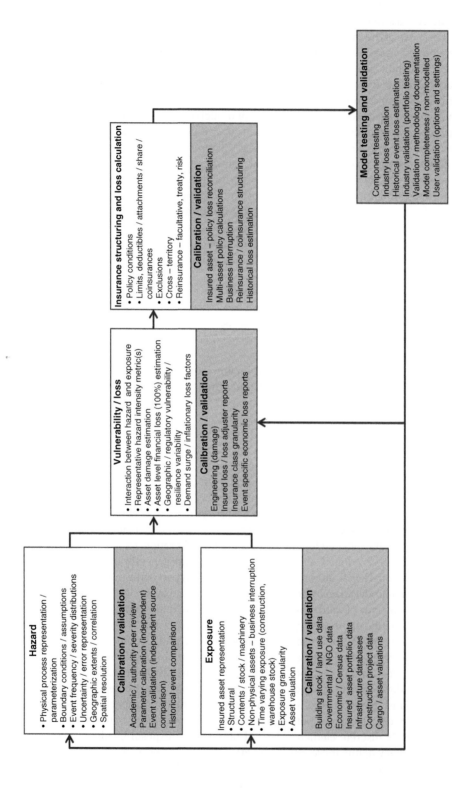

**Figure 4.1** Generic Catastrophe Model Development Framework. Common titles for the components (i.e. hazard, exposure, vulnerability, financial) are given in bold. The white regions highlight key elements of the components, while the shaded zones list approaches to calibrating or validating the component.

addition, the important role of geographical frameworks and associated geospatial concepts as a key integration framework for catastrophe model data and development processes is discussed in Section 4.6. There are many different ways in which different models represent event footprint intensity and associated damage. Some just provide mean values (simply mean damage models); others provide full probability distributions for both intensity and vulnerability. Many use point values for intensity coupled with damage (vulnerability) distributions (OASIS, 2015).

## 4.3   Hazard

A **peril** is a natural phenomenon with the potential to cause loss or damage (e.g. earthquake), and **hazard** is the danger (e.g. ground shaking) resulting from the peril (Chapter 1.2). The foundation of all natural catastrophe models is the hazard component, which is sometimes referred to as a model or sub-model in its own right.

The initial, key decision when building a catastrophe model is its scope: specifically what it includes, and the **domain** (both spatial and temporal) to be modelled. This is, or at least should be, strongly influenced by a view of the hazard. To date, domain definition has placed limits both on the geographic extent and perils to be included, leading to the concept of a **peril-region**. The extent of the geographic domain needs to be computationally tractable, both for the user and during model development, i.e. a single main peril is typically chosen. This leads to models of peril-regions such as 'UK flood', 'European windstorm' or 'US hurricane', which are usually assumed to be statistically independent of each other in terms of the causal physical processes and thus losses expected over the time period of concern. This may or may not be true in actual terms, for instance, many hydro-meteorological perils, such as storms and floods, are considered to be driven by large-scale ocean and atmosphere processes at various spatial and temporal scales. The extent to which some peril-regions are independent is a subject for ongoing debate, but is beyond the scope of this chapter (see Chapters 3.1.1 and 6.2.2) and instead illustrates the complexity of the underlying physical processes and the challenges of representing them effectively within computationally efficient models.

Hazards are the local, physical manifestation of a natural phenomenon that provides extreme conditions capable of causing damage or loss. Aspects in describing a hazard include its likelihood, its **intensity**, its geographical extent and how intensity varies *within* this extent. This raises questions of the appropriate **resolution** of intensity representation, and how this varies with the peril and hazard (see Chapter 3). The (pragmatic) choices made to answer these questions also depend on the information available for construction, which differs markedly by hazard and geography. In addition, choices are not isolated from the required use of the model.

Hazard model development teams typically include specialists in fields such as: geophysics, seismology, meteorology, climatology, atmospheric physics, hydrology, computational fluid dynamics and applied mathematics. These specialists will usually work in collaboration with data, analytics and risk specialists, as well as computational geo-statistical and geographical experts. The specialists may be members of the core catastrophe model development team, but equally they may be representatives of independent entities, such as an academic or governmental research team, producing data or approaches useful to the overall catastrophe model construction.

For a **probabilistic** risk model (Section 4.3.6), some organizations may further split the hazard components into separate 'event' and 'hazard' components. This differentiates the representation of the range and spatial-temporal distribution of peril events (the causal processes, e.g. hurricane tracks) and their associated likelihoods from the spatial impact of each event (e.g. wind fields).

Section 4.3.1 starts with some fundamental questions: what is the difference between a probabilistic model and a deterministic one, and why is it crucial to use events rather than other probabilistic representations?

With the need for a probabilistic event-set based model established, the chapter continues with discussion of how event-based models are constructed. First, choices about how exactly to quantify the severity of a hazard are outlined (Section 4.3.2). Once this is determined, historical hazard is considered (Section 4.3.3), starting with the **interpolation** step required to estimate a spatial intensity distribution for a single event. Discussion moves on to the construction of **catalogues** of hazardous historical events. This is the stage at which a deterministic event-based model is possible (Section 4.3.4). The next step considered is how to extend that which has been experienced to what is *likely or possible to be experienced* in the future (e.g. in the next 12 months, or five years). Approaches can be statistical (e.g. Extreme Value Analysis – EVA) or physical/computational (e.g. Global Circulation Models, or GCMs), illustrated in a single geographic area (Section 4.3.5) before full, spatially distributed event set construction is illustrated using a number of examples (Section 4.3.6).

### 4.3.1 Deterministic Versus Probabilistic Hazard Models

There are numerous ways of quantifying hazard and risk (e.g. Kappes *et al.*, 2012). It is therefore natural to question why the insurance industry uses the particular form of an event set-based probabilistic model known as a catastrophe model. This can be answered by considering the two basic types of hazard model:

- **Deterministic models:** These quantify the impact of an event or collection of events, but do not attempt to describe the full range of events that might happen in the near future (e.g. next year). They are often referred to as 'scenario' models. They usually do not describe the events' likelihood. Examples of deterministic models often used include historical hazard events (see Section 4.3.3), 'what-if' scenarios (e.g. Chapter 2.7.1.3), and accumulation zones (Chapter 2.7.1.1). In pure terms, deterministic models will be fully described by the initial conditions set, and will contain no inherent randomness.
- **Probabilistic models:** These generally contain ensembles of a large number of synthetic events (e.g., $n > 10,000$), with associated estimates of event probability, to reflect the whole range of possible events. Each of the events is created by combining the historical record with physical, theoretical and statistical models that allow the record to be **extrapolated** beyond what has been observed. These are also known as **stochastic** event sets and models, and both terms are in common usage. The probabilistic model therefore can be considered at one level, to be a stochastically structured collection of multiple, individual deterministic (hazard event) models, defined in terms of their frequency-severity, and which include an element of randomness in their likelihood of occurrence.

This classification, however, is simplistic. There are other types of hazard representation which straddle this deterministic-probabilistic model classification, which we will call **hybrid** models. These products may have a probabilistic component, in defining a certain peril by its frequency, and consider the risk geographically, but do not contain any concept of an *event*. Deterministic and hybrid hazard models are also sometimes used in the probabilistic model validation process (also see Chapter 5.3.5). For example, a modeller may compare a flood map of given hydrological probability (e.g. 1-in-100 year) derived from their model against national or other authoritative, independent flood risk mapping sources such as the United States official flood zone maps (FIRM – Flood Insurance Rate Map) issued by FEMA (Federal Emergency Management Agency).

Examples of hybrid models include:

- Hazard zoning maps, such as 1-in-50-year return period flood risk maps (e.g. Chapter 3.5.6). These represent the risk at locations shown across an area, and are often developed for planning purposes where information about a single site in isolation can be useful. They can

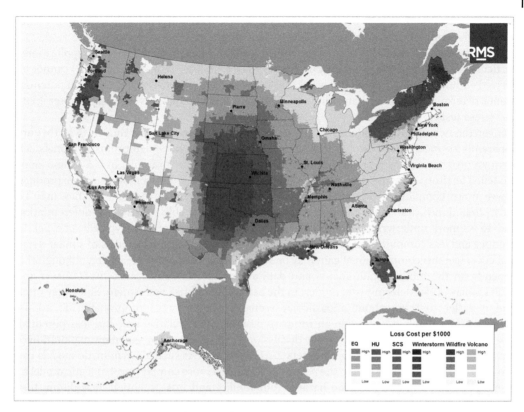

Figure 4.2 'Heat map' showing the relative risk from different perils in the United States. The relative risk from different perils is illustrated in different colours and terms of the loss cost (AAL) per $1000. EQ is earthquake, HU is hurricane, and SCS is severe convective storm. *Source*: Reproduced with permission from Risk Management Solutions Inc.

be created by linking local point estimates of extremes (see e.g. Chapter 3.5.4), or as a by-product of an event-based regional probabilistic model.

- Risk heat maps represent some derived view of the risk (such as the average annual loss, Chapter 1.10.1) that combines measures of hazard, vulnerability and exposure (see Figure 4.2).

Zoning maps and hybrid models can be useful for differential underwriting – by comparing risk between locations to pick which to include in the portfolio or to decide the price (e.g. Chapter 2.6). However, limitations arise when using hybrid models, even those probabilistically generated, in most reinsurance-related applications. Probabilistic events are fundamental in correctly understanding risk due to natural catastrophes because they allow for the **spatial and temporal correlation** of risk for a credibly scaled event. That is, events allow more than one asset at risk (e.g. a house) to be affected by the same adverse episode (e.g. a storm) but are constrained by the particular peril-hazard boundary conditions which determine its severity and extent. Hybrid models do not include this level of event correlation. They cannot be used for occurrence-based reinsurance (e.g. CAT XL) or ILS structures (e.g. Chapter 2.15).

In general, the notion of a probabilistic event is therefore critical to catastrophe re/insurance and to most applications relating to it, particularly when considering accumulation (e.g. Chapters 1.9.3 and 2.7). All downstream processes requiring accumulation, for example, capital modelling (see Chapter 2.10) or capacity setting (see Chapter 2.7.2.1) cannot be completed without consideration of a probabilistic event. It is also notable that public sector risk assessments, such as in Disaster Risk Reduction (DRR), are underpinned by the notion of a 'disaster': an event that impacts multiple sites (e.g. with houses or people). Definition of the event is considered in more detail in Section 4.3.1.1, but as this paragraph shows, the lack of events restricts the applicability of hybrid models to catastrophe risk assessment.

The deterministic models are in some ways equally limited in their potential applications, particularly as many of the commonly used metrics (see Chapter 1.10 and Chapter 2) cannot be derived without the stochastic framework. However, since the scenarios are based on potential events, deterministic models can still be used for simple accumulations (see e.g. Chapter 2.7.1) and stress tests.

Given this, why might a model developer choose to build a deterministic model, given they are apparently so much more limited in scope? For model developers, deterministic models can form an attractive entry point to understanding a new risk, for example, in a new and non-modelled territory. Alternatively, they can support risk assessment for a new insurance product, where more complex models, or historical claims experience do not exist. They may be incorporated into risk rating and decision support tools to enable relative risk differentiation and to support underwriting controls. Deterministic models are also typically cheaper to produce and less computationally resource-intensive than probabilistic models of similar detail and coverage; this computational gain may be several orders of magnitude in size, although this depends on the model requirements and data availability. In some cases, model users and analysts may prefer deterministic models in the early stages of risk assessment. Since they often find it easier to think about a particular scenario than a range of probabilistic events, deterministic models can provide an *intuitive* description of relative risk and loss potential. A scenario-based approach can provoke discussion, help visualization and allow exploration of the impacts of specific events (e.g. Chapters 2.7.1.3 or 5.4.3). As such, deterministic models can provide a useful stepping stone on the path to developing a more complex probabilistic model by gaining detailed knowledge of the hazard, vulnerability and loss potential. In addition, they enable the model developers or sponsors to gauge interest and to foster working relationships with partner companies and local experts.

While development costs for deterministic models do tend to be lower than those required for probabilistic models, as noted above, they can still be expensive projects, requiring significant development specialist expertise and computational capacity. In many cases, deterministic risk maps will be outputs from large-scale, complex and costly national risk assessment and statutory planning support projects, such as the US FIRMs, or the UK Environment Agency's flood risk maps where accuracy and validation are of primary importance.

### 4.3.1.1 Hazard Event Footprints

The geographical extent of the area impacted by an event in (e.g. the wind field of a windstorm) is called a **footprint** (Figure 4.3) and each footprint contains within it the spatial distribution of

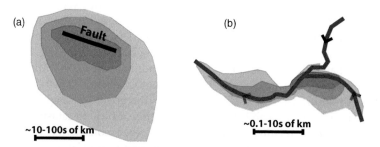

**Figure 4.3** Schematic illustration of the concept of a footprint for an earthquake in (a) and a flood in (b). The view is plan view (i.e. from above), in (b) the river is in dark blue, and scales are broadly indicative. In both cases, the footprint is the entire coloured area, and intensity of the hazard within that area is given by the intensity of the colour: see also Figure 1.3.

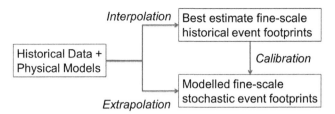

Figure 4.4 Conceptual overview of how physical and historical data are used to build impact footprints for hazardous events. This illustrates the relationship between deterministic modelling (historical footprints) and probabilistic modelling (*synthetic* footprints). It also highlights the role of interpolation, extrapolation and calibration. Note again that 'stochastic' and 'probabilistic' are used somewhat inter-changeably in common usage. *Source*: Reproduced with permission of RMS.

damage-causing intensities (e.g. peak wind gust speed, ground motion or water depth). To summarize the distinction discussed above, deterministic models contain footprints, but in probabilistic models events will have both a footprint and a probability and these are designed to span the entire range of what is thought to be physically plausible.

Figure 4.4 gives a conceptual overview of how physical and historical data are used to build footprints, showing the different ways footprints are developed for deterministic and probabilistic hazard models. In both cases, catastrophe model developers combine historical data with physical models; see 'Hazard Description and Event Sets' in each of the hazard sections in Chapter 3 and examples in Section 4.3.6. In building the footprints, the concepts of interpolation and extrapolation are important, and commonly used in combination. Model developers use statistical and physical methods both to **interpolate** *between* (e.g. Powell *et al.*, 2010) and to **extrapolate** *beyond* the historical observation points (e.g. Macdonald, 2013), respectively.

*Interpolation* is defined as the construction of new data points within the range of known data points. Interpolation is often applied in the combined spatial-temporal domain. In the *spatial* context, interpolation can be used to produce a regularized hazard intensity surface for historical events where observational data are directly recorded at a smaller sample of scattered measurement points; it is also possible to use statistical techniques to add plausible fine-scale geographical structure between the measurement points (e.g. see Section 4.3.6.3). In the combined *spatial-temporal* context, interpolation can be used to create a more complete distribution of events across the whole frequency range, for example, by including additional events to a sparse set of historical earthquakes in given source zones.

*Extrapolation* is, by definition, the use of methods to *extend* and make predictions *beyond* an observed data set, in this case a limited historical record, for example, in the form of site-based extrapolations (see Section 4.3.5.1). In more complex models, both extrapolation and interpolation can be combined, for example, to develop a set of simulated probabilistic event footprints by the use of weather-resolving GCM-derived storm tracks (Section 4.3.6 or specific sections in Chapter 3.3).

Since interpolated, fine-scale historical event footprints for deterministic events can be used to understand and calibrate (e.g. Hall and Jewson, 2007 and in Chapter 3.2.4) the fine structure of modelled probabilistic events, deterministic model development can be seen as an integral part of probabilistic model development.

Storm surge inundation modelling provides an example of how extrapolation and interpolation are combined to produce hazard footprints, for both deterministic and probabilistic models. Section 4.3.7.1 provides a case study example of how storm surge is combined in a tropical cyclone model as a secondary peril. The term storm surge describes the elevation of the ocean's surface in response to meteorological forcing during windstorms, including both tropical and extra-tropical cyclones. Storm surge can lead to catastrophic coastal flooding, as seen historically in 1953 along Europe's North Sea coast, in 1959 from the Isewan Typhoon (Vera) in Japan, and more recently when Hurricane Katrina hit New Orleans in 2005, and in downtown New York following superstorm Sandy (2012).

The hydrodynamic parameter which determines the potential extent of coastal surge inundation is the residual surge elevation above expected Still Water Level (SWL) and its intersection with the landward surface elevation. Historical records of time-varying water height are sparse, rely on tidal gauges at defined coastal locations, and often are short in length. Inundation inland of the shoreline requires hydraulic modelling, since there will be few records of actual inundation available.

First, from a 'hybrid' perspective, the potential inundation to an event with a particular hydrological frequency (e.g. a '1-in-1000-year surge'), can be derived through the extrapolation of peak surge residuals from time series of water levels recorded at the available gauging sites along the coastline in question, removing the effect of tides to determine surge potential over assumed SWL.

Using an appropriate spatial surface generation function (see Section 4.6.5) to interpolate between gauging site levels at the defined frequency can derive estimates of surge height along the coastline between the observed locations. These can then be used to generate an inundation flood risk map via a suitable inundation model, whether via a 'bathtub' flood (a simple intersection of the water elevation surface against the inland terrain to compute a water depth) or a hydraulic computational model, for example, MIKE 21 (Danish Hydraulic Institute, 2015) to compute more accurate water flow, depth and ponding.

A similar approach, but employing both spatial and temporal interpolation to create the deterministic historical surge inundation footprints from Hurricanes Katrina and Rita, is described by Gesch (2009). Here, the surface interpolation from surge data collected during the storms was combined with a **digital terrain model** (DTM) and a series of time-varying water surfaces generated from water levels sampled at different time periods during the Rita event surge, to represent the temporal variability of inundation.

Another way of producing probabilistic storm surge events involves generating each storm surge event by association with a wind event. The advantage of this approach is that it allows the model to assess the correlation between wind and surge losses generated by the same underlying climatic phenomenon, for example, a hurricane.

Such models use the stochastic event wind field to generate a probabilistic surge event or range of events that account for oceanographic wave/tide models (see, e.g. Taylor and Glahn, 2008). The model computes the surge elevations at each point along the affected coastline of each storm through a wave propagation and shoaling model (reflecting the hydrodynamic progression of waves up to the shoreline as they are affected by seabed friction). This approach uses both extrapolation (the frequency-severity distribution of storms, and tidal harmonic models), and interpolation (the generation of the wind-wave energy and momentum interactions and local surge elevations).

The Florida Public Hurricane Loss Model (FPHLM) is a rare example of an openly developed, well-described event-based probabilistic model (e.g. Powell *et al.*, 2005; Pinelli *et al.*, 2006; Chen *et al.*, 2009; Pinelli *et al.*, 2011). The complete hazard development processes is described in Powell *et al.* (2005). This paper highlights, in the context of hurricane modelling, the development stages common to most probabilistic catastrophe hazard models regardless of the hazard, including the use of extrapolation, interpolation and calibration. These steps, expanding on the 'hazard' box in Figure 4.1, can be generalized as follows:

1) definition of the model domain (e.g. region, territory, impact area, modelling time period);
2) determining geographic source zones (e.g. storm genesis locations, fault zones and fault lines, flood catchment areas and river networks);

3) event parameterization and evolution (e.g. storm tracks, central pressure and wind intensities, river hydrology, earthquake generation);
4) localized impact estimation and interpolation (e.g. wind gust speed surfaces based on land cover roughness coefficients, flood depth extents, earthquake ground motion intensity footprints);
5) calibration, refinement of hazard intensity values (e.g. spatial smoothing of hazard intensities, calibration against observational data).

These steps can be related to Figure 4.4, and will be re-visited in Section 4.3.6. Steps (2) and (3) contain *extrapolation*, step (4) involves *interpolation*, while step (5) is the *calibration* stage.

### 4.3.2 Representing the Hazard Severity

Typically, as discussed in Sections 4.3.3.1 and 4.3.4, intensity measurements from real events are sparse, made at point locations and must be interpolated to develop a footprint. Before interpolation or footprints can be considered, however, a critical decision needs to be made, namely, what metric is to be used to quantify the severity of the hazard.

The majority of current catastrophe models use a single, standard, easy-to-measure parameter to represent the peril severity. This is the quantity that is argued to be the main driver of damage (e.g. to buildings) and for which vulnerability curves should be constructed (Section 4.5). Examples of typical parameters used in this way for different perils are listed in Table 4.1.

The advantage of using a single, well-established parameter is that observations are more likely to be generally available, perhaps from a relatively dense network of instrumentation with a degree of temporal homogeneity and over a long time. In some cases (e.g. peak gust speed), these are also metrics used by engineering and design standards organizations, such as the ASCE 7 minimum design loads standards (see Section 4.1.4). However, these single parameters are not typically the only factor influencing damage levels experienced by the elements at risk. Often they are not a complete description of the main hazards, and may miss important elements of them. For instance, the response of a building to earthquake-induced shaking is

Table 4.1 Parameters that are used, or could be used, to quantify peril intensity; these are typically chosen as they are thought to have significant ability in predicting damage.

| Peril | Single-parameter severity predictor (first-generation models) | Additional predictors | Local hazard modifiers | Chapter section |
|---|---|---|---|---|
| Earthquake | Modified Mercalli Intensity (a scale defined by damage) | Peak ground acceleration, spectral acceleration, spectral displacement | Landslide susceptibility, soil type, liquefaction susceptibility, spectral response | 3.7/3.8 |
| Flood, storm surge, tsunami | Maximum water depth | Duration of flooding | Water velocity, wave action | 3.5/3.9; surge included in 3.2/3.3. |
| Windstorm | Maximum 3-second peak gust wind speed | Duration, full spectrum of gusts | Surface roughness, topographic enhancement | 3.2/3.3/3.4 |
| Hail | Peak hailstone size, hail density | Peak kinetic hail energy | Localized downdraft winds | 3.4 |

influenced by the full spectrum of horizontal and vertical motions it experiences, which the single-parameter peak ground acceleration may not entirely describe (see Chapter 3.7). Further, local or event-specific factors may also be important. For example, both soil and slope conditions can determine whether a location affected by a particular level of shaking is susceptible to secondary perils that can increase the damage, such as liquefaction or landslide (see Chapters 3.7.5 and 3.8). This source of **secondary uncertainty** (see Chapter 2.16.1) creates scatter in the data from which empirical vulnerability functions are created (Section 4.5). Another type of local effect was observed in the 1985 Mexico City earthquake. Mexico City is buit on a dried-up lake bed. The frequency of oscillation of the deep lake-bed sediments during the earthquake was close to the resonance frequency of mid-rise buildings, which were consequently more severely damaged than shorter or taller structures during the earthquake (e.g. Flores-Estrella *et al.*, 2007). See Chapter 3 for information specific to the different perils.

### 4.3.3 Understanding the Historical Hazard

For both deterministic and probabilistic models, the first step in understanding the hazard associated with a peril is to examine the historical record. However, there are challenges in interpreting historical hazard data.

The historical record is generally short, at least compared to the typical intervals between events that can cause catastrophic loss. This is a major challenge in catastrophe model construction. This challenge also arises for users forming a view on a model (see Chapter 5.4.1). An example of how plausible events may not have occurred historically is that the 110-year-old New York subway system has only flooded extensively once, when Hurricane Sandy hit the region in October 2012 (Lhota, 2012). This is simply because the typical recurrence interval for a storm-surge of such magnitude in the area is estimated to be greater than the length of the time that the subway system has existed (e.g. Swiss Re, 2013). Indeed, the complete record of historical Atlantic hurricanes, even the number making US landfall, is only considered reliable since 1851 (e.g. Jarrell *et al.*, 1992; Elsner and Jagger, 2006). Likewise, although historical European windstorms can be partly known back to ~120 BC (see Lamb, 1991, p. 3), complete catalogues based entirely on instrumental data are much shorter; for instance, the XWS catalogue covers the period 1979–2012 (Roberts *et al.*, 2014).

Modern-day, dense networks of instrumental gauges (e.g. wind speed, river flow, tides) have only existed for a few decades. International initiatives, such as the UN International Hydrological Programme, which began in the 1970s, have tended to initiate consistent data collection. Data availability and quality vary depending on the type of instrumentation and the region in question. Furthermore, within these networks, records may be affected by inconsistencies and non-stationary data within the recording period, including, for example, data 'drop-outs' and corruptions, or changes to measurement methods (see Section 4.3.3.1). Specifically, the density of observation networks, and the nature of instrumentation used to make measurements of hazard intensity at a particular location, can vary (e.g. as it is upgraded) (see Lamb, 1991, pp. 4–5), so that the historical record is heterogeneous both in space and time. Even the most temporally consistent long duration datasets, for example, the National Oceanographic and Atmospheric Administration (NOAA) National Hurricane Center (NHC) HURDAT 2 hurricane database (Landsea *et al.*, 2015), have missing parameter values and have changed in format over the period of record (see Section 4.3.3.4, for a description of the HURDAT database).

Acknowledging the limitations of the historical record, model developers use assumptions, numerical models, and statistical techniques to help interpret, and interpolate from, the record. However, these techniques in turn introduce additional layers of uncertainty into the modelling process. For example, meteorological agencies use Global Climate Models (GCMs) to interpolate meteorological data and to create 'reanalysis' data products, such as NCEP/NCAR 20CR and

ERA interim (Compo *et al.*, 2011) (see Table 4.1 or Section 4.3.6.3). Section 4.3.5 provides examples of statistical techniques to extend flood time-series. Filling gaps in time-series is non-trivial and model developers need to approach this technique with caution (Schneider, 2001; Alavi *et al.*, 2006; Keef *et al.*, 2009; Dumedah and Coulibaly, 2011; Andersson *et al.*, 2012; see also Kashani and Dinpashoh (2012) for a useful summary). The degree of interpolation depends both on the density of the measured locations and the spatial scale at which the hazard varies. Thus, constructing footprints for some perils such as hail (see Chapter 3.4.4) from single location measurement is particularly challenging given the comparatively small size of the events and the sparse distribution of 'hail pad' sensors (Lozowski and Strong, 1978). Here, other measurement sources, such as radar, are often used to estimate hail intensity. The challenges for larger-scale perils should also not be ignored (see Section 4.3.3.3).

The rest of this section describes in some more detail typical data used to build the hazard component of catastrophe models, and discusses some of the issues associated with appropriately interpreting them. These typical data formats, some of which are pre-processed versions of the others, are: measurements of intensity at a location, regional or broad-scale observations, and catalogues of historical events.

### 4.3.3.1 Measurements of Intensity at a Location

Measurements of the intensity of a hazard at a single geographic location are known as **point observations**. Such point observations provide time series for a specific parameter, or parameters, depending upon the specific equipment or instrumentation at the site. Common physical properties measured as point data are: peak gust wind speed, peak ground acceleration (i.e. ground shaking), or flow rate in a river (i.e. 'discharge' in $m^3 s^{-1}$). When building a model it is important to make sure that the chosen parameters match, or through physical modelling can be related to, the measure of severity that drives the damage and loss for a peril (i.e. wind, earthquake and flood for the examples given, respectively). Such observations may relate to a single event or be a time series of equivalent measurements. For many such parameters, a third party may construct arrays of point data in networks which may be built into spatially distributed maps or products before the model developer has access to them (see Section 4.3.3.2).

Model developers face many issues when interpreting historical intensity data, particularly when validating and standardizing the data. Validating the data involves exploring its *completeness* and the *accuracy of the instrumentation*. For example, a time series of wind speed observations during a storm may be incomplete if the wind speed recorder failed during the storm due to power outages. Developers may need to detect and compensate for incomplete or inaccurate time series using computational or data cleansing techniques (Kashani and Dinpashoh, 2012; Macdonald, 2013). *Standardizing* the data involves understanding how local conditions may make a particular observational site unrepresentative of the surrounding area, then applying corrections to ensure that all observations are adjusted so that they are on a representative, common baseline. For example, wind observations may need to be adjusted for height (e.g. to 10 m), upstream surface roughness conditions, or averaging time (e.g. 1 min) (Hoffman, 2005; Powell *et al.*, 2010). Such discrepancies must be accounted for when combining different sources of data for use in either the model development or validation process. Significant additional, but not insurmountable issues arise when using pre-instrumental data such as ships' records or weather diaries (Lamb, 1991).

Even with standardized data, understanding of the non-stationarity of the measured hazard and its parameters poses a major challenge. A random variable or random process is said to be **stationary** if all of its statistical parameters are independent of time, i.e. it does not change through time (Von Storch and Zweiers, 1999). However, many key perils are thought to be non-stationary (see Chapter 3), for two primary reasons. First, there is natural variability in the

Earth's system (e.g. climate variability). For example, most experts accept that the occurrence of events such as tropical cyclones is influenced by large-scale climatic modes (e.g. the El Niño Southern Oscillation or ENSO, and the Atlantic Meridional Oscillation or AMO – see Chapter 3.2.1) (Elsner and Jagger, 2006). Second, not only the magnitude of losses (Pielke *et al.*, 2008; Barredo, 2010), but the manifestation of hazard itself can be influenced by human (i.e. 'anthropogenic') interactions with the natural processes. Examples of anthropogenic influences that should be considered in model development include the building of river defences, or urbanization and land use change. They can affect the severity, location and probability of future events, making them distinctly differently to those experienced in the past. Thus, model developers must consider the consistency of the time series over its record period.

### 4.3.3.2 Spatially Distributed (Areal) Measurements of Intensity

Spatially distributed data can supplement point observations (i.e. at a single location) to provide information about the spatial distribution of a hazardous phenomenon across a region. Like point observations, these may be at a single time, or depict how a hazard evolves over time (see Section 4.2.4).

Pragmatically, this book categorizes spatially distributed data as those *not explicitly* extracted from a sensor or sensors fixed at and specifically measuring precisely known locations. Thus, spatially distributed data record primarily broad-scale (areal) observations derived from a single sensor capable of sensing hazard across an area; this may be images from weather radar, or images derived from satellite and other remote-sensing platforms (e.g. aircraft). In addition, this book classes areal products derived from point observations, for example, by interpolation as spatially distributed data, especially where the original source data are inaccessible. As well as quantitative data, examples of such data sources also include qualitative data such as the distributions of hazard intensities from eyewitness accounts. An example of the integration of historical and observational data into a combined time series can be seen in Coleman and LaVoie (2012) where pre-1851 hurricane tracks in HURDAT format were derived from varying archival data sources within a GIS to extend the historical record.

In addition to data that measure spatial variations in a hazard, related data describe spatially varying properties that may impact the intensity of the hazard. Such observations include geophysical and geomorphological data, i.e. related to the physical property or form of the Earth and representing in many cases the geophysical extremes of hazards beyond the historical period (see Woodward *et al.* 2010, for an example of paleoflood data). For example, in earthquake modelling, geotechnical factors such as regional geology, slope and soil type data influence ground motion and its attenuation with distance from the earthquake source and a site's susceptibility to liquefaction and landslide (see Chapters 3.7 and 3.8). In hydro-meteorological hazards, physical attributes of the Earth's surface such as topography and surface roughness can modify the severity of winds experienced on the ground, or flood extents and flow paths (see Chapter 3.5 and Powell *et al.* 2005).

### 4.3.3.3 Case Study: Windstorm Gudrun (Erwin) in Sweden, January 2005

An observational network may be insufficiently dense to capture the spatial structure of its most intense impacts. This issue is compounded in some cases when the severity of the peril affects the functioning of, or even destroys, the recording equipment. In such cases, model developers seeking to create an event footprint interpolate between the observations. Windstorm Gudrun (also known as Erwin), which affected Sweden on 8 January 2005, illustrates the potential problems of interpolating measurements to obtain a picture of the fine-scale spatial variations within an event: see Appendix 3 of Winkel *et al.* (2009) for a summary of the event and its aftermath.

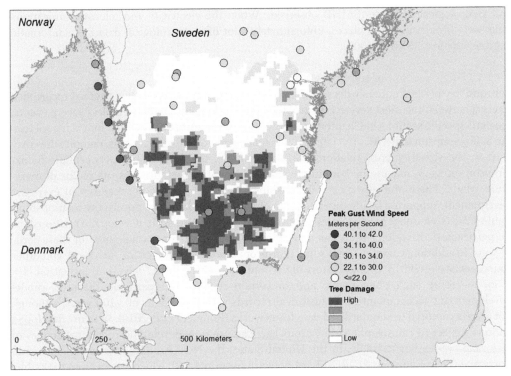

Figure 4.5 Wind observations (dots) and tree damage (green shading) from windstorm Gudrun (Erwin) in Sweden, January 2005. Wind data from SMHI; forest loss data. *Source*: After Claesson and Paulsson (2005), reproduced with permission of the authors.

Gudrun killed 20 people across Europe, and left over 700,000 without power (SMHI, 2005; Winkel *et al.*, 2009). It also felled over 75 million cubic metres of trees. The pattern of damage to the trees offers insights when compared to meteorological data (Figure 4.5). Although Southern Sweden is fairly densely and uniformly forested, the highest recorded wind observations were not in the area with the most forestry damage. Rather, the highest wind speeds were recorded in coastal locations, where tree damage was less severe. Moreover, the spatial scale of the variability in the damage to trees is less than the spacing of the meteorological observations. Several factors contribute to this observation. First, thanks to the wind direction, coastal locations recorded unshielded (over-water) wind speed values, whereas site-specific conditions that modify the strength of the measured gusts affect the inland measurements. Such conditions include the complex frictional impacts of upstream obstacles (including forests) and topographic alterations of the wind speed. This example therefore highlights the importance of the standardization of measurements (Section 4.3.3.1) to obtain values that best represent damage potential. The fine-scale variation in forestry damage likely illustrates small-scale, site-specific changes in wind hazard that cannot be appropriately interpolated by simply using an average. A second possible contributing factor may be that the forestry damage may simply reflect variations in the density of forestry exposure (inventory) or differences in forestry vulnerability across the region. This last point illustrates the difficulty in using proxy data (i.e. indirect measures) to estimate hazard. Nevertheless, the tree damage intrinsically records the element of hazard of most interest (i.e. the part that ultimately causes damage), so there are insights to be gained through examining loss data (Hillier *et al.*, 2015).

Gudrun also provides an example whereby the passage of an event itself compromised the instrumentation network's capability to record and transmit data. The Swedish Meteorological and Hydrological Institute (SMHI) observed: 'When the electricity and telecommunications links were broken in many places, unfortunately a lot of meteorological data from automatic stations was lost' (SMHI, 2005).

#### 4.3.3.4 Catalogues of Historical Events

For most key perils, frequency decreases with increasing event size, i.e. the observed recurrence interval between the most severe catastrophic events is longer than between less severe events; see peril-specific sections in Chapter 3 for how 'size' or **magnitude** is defined for each peril (e.g. the Saffir-Simpson scale, $M_w$, the Fujita scale). Of course, for individual perils, complexities may exist in the size-frequency relationships, for example, the relationship between the Saffir-Simpson category, maximum 1-minute wind speed, central pressure and the radius of maximum winds. These and other key relationships form the basis for the majority of event set development, supported by an understanding of the physical laws that govern the relationships. Table 4.2 lists a number of catalogues favoured by model developers to develop the relationships for particular peril-regions; these can almost be considered reference catalogues for these peril-regions. In addition, Chapter 3 lists some non-proprietary data sources for each peril, and other sources exist e.g. GEM's 'Global Historical Earthquake Archive' (http://www.emidius.eu/GEH/).

So, for some such phenomena, and in certain regions and particular hazards, model developers can access quantitative historical records with time spans of >100 years, although not always captured in standardized data frameworks (Rougier *et al.*, 2010). In many developed nations subject to catastrophe risk, various agencies are now developing robust catalogues of historical events. For example, in the United States, the NOAA/NHC database of historical hurricane tracks known as HURDAT (as noted in Section 4.3.3), which has been in the public domain since 1982 (Jarvinen *et al.*, 1984) has been maintained annually, and was updated to HURDAT 2 through a reanalysis project in 2015 to include additional parameters related to storm track and intensity (Landsea *et al.*, 2015). These datasets form the basis for the majority of catastrophe models that evaluate hurricane risk in the North Atlantic basin (see Section 4.3.6.4).

Table 4.2 Some favoured sources of historical events for building catastrophe models. Figures in brackets indicate the associated chapter describing the peril in more detail.

| Peril | Source (provider) | Description |
|---|---|---|
| Earthquake (3.7) | USGS Centennial Catalogue | Global catalogue of locations and magnitudes of instrumentally recorded earthquakes from 1900 to 2008 |
| European windstorm (3.3) | ERA40 (ECMWF) | Global atmospheric reanalysis (Sept. 1957–Aug. 2002) |
| | ERA-interim (ECMWF) | Global atmospheric reanalysis from 1979 |
| | Reanalysis (NCEP/NCAR) | Global atmospheric reanalysis (1948–2010) |
| Tropical cyclone (3.2) | HURDAT (NOAA) | Catalogue of tropical cyclone (Atlantic, East Pacific) tracks since 1851 |
| | IBTrACS (NOAA) | Catalogue of tropical cyclone (worldwide) tracks from many sources |
| Inland flood (3.5) | Copernicus (ESA, EU) | Remote sensing data of ongoing catastrophes including historical catalogue |
| | Dartmouth Flood Observatory | Global and regional analyses of flood |

HURDAT 2 includes a range of parameters that can be used to define storm characteristics, including central pressure location (latitude-longitude), central pressure, radius of maximum winds (varying speed bands), and the status of a storm. For non-Atlantic tropical cyclone, the international best track archive, IBTrACS has become a standard source (Schreck, 2013), see Chapter 3.2.8 for more detail.

When compiling an event catalogue for catastrophe models, developers need to decide which summary parameters to use. Summary parameters commonly underpin the event generation process, or at least are used to validate its outcomes. The model developer can describe each event's impact, and summarize its complex geographical pattern using the summary parameters. These summary parameters can then be used to predict the consequences of other potential events. When selecting these parameters, model developers must apply various criteria to ensure fitness for purpose. Such criteria typically include longevity and ease of measurement. Thus, events today need to be comparable to those of hundred years ago, on a single, calibrated scale. The parameters chosen to represent the event should be easy to measure, and their measurement should not be affected by changes over time, for example, with the introduction of new instrumentation.

For simplicity, and to optimize the length of the event record, model developers often use pragmatic parametric definitions that give equivalence to instrumental and pre-instrumental observations (e.g. km/h and damage). In addition, they may adopt well-established and widely recognized classification systems such as the Saffir-Simpson scale for hurricanes (see Chapter 3.2). Other metrics such as a 'Storm severity index' (Lamb, 1991) are widely used in science. These summary metrics capture extremeness from the point of view of the whole event, and *not from the experience of an observer at a particular location*, as is described in Section 4.3.3.1. For instance, the moment magnitude ($M_w$) summary parameter quantifies the total energy released by an earthquake, in contrast to the Modified Mercalli Index (MMI) which categorizes the impact of shaking at a given location: see Chapter 3.7 for definitions and details.

Capturing a complete record of events in the instrumented era is relatively easy, but this time period is relatively short. As such, historical accounts are used where possible to extend catalogues and improve understanding of a particular phenomenon, especially when employed as evidence in support of an appropriate, peer-reviewed theoretical framework. Lamb (1991), for example, uses a reassuringly robust and transparent method of data collation and verification. Bayliss and Reed (2001) reviewed the approaches for collation and interpretation of historical flood records for the United Kingdom to both interpolate and extrapolate flood time series to include pre-instrumentation information. In forecasting a $41 \pm 14\%$ chance of a $M > 7$ earthquake impacting Istanbul in the next 30 years, Parsons *et al.* (2000) and Parsons (2004) illustrate how geophysics can augment historical evidence. However, the recurrence intervals of the most extreme events may exceed any written record of the local civilization; in these cases archaeological, geomorphological or geological evidence may be used, as illustrated for earthquake risk in the Pacific North-West of the United States in Section 4.3.3.5.

### 4.3.3.5  Case Study: Using Other Evidence to Supplement the Written Record – Cascadia Earthquake

The potential may exist for a very large (i.e. $M > 9$) 'megathrust' earthquake (Gregor *et al.*, 2002) along the Cascadia subduction zone (Figure 4.6). This possibility and its initial investigation are authoritatively documented by Atwater *et al.* (2005). An overview is given below, and for background information on earthquakes, see Chapter 3.7.

The subduction zone of the Juan de Fuca tectonic plate extends for 1,000 km north to south. The subducting slab is moving under the Cascades region of the Pacific North West of America (Figure 4.6); and this gives the zone its name (i.e. Cascadia).

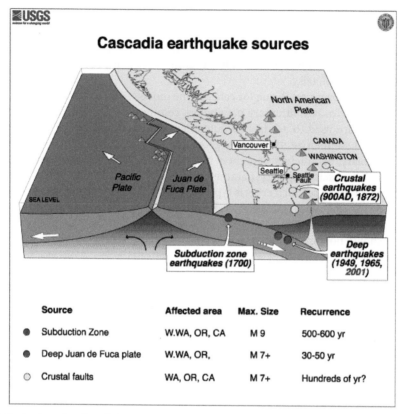

**Figure 4.6** Illustration of earthquake sources in the Cascadia Subduction Zone. There are three types, including subduction zone quakes shown as a red dot, which are of most concern as they may be up to magnitude (denoted M) 9 in size. The affected area abbreviations refer to US states. *Source*: USGS.

Large subduction zone earthquakes, such as the 2004 Sumatra-Andaman earthquake in the Indian Ocean (Henstock *et al.*, 2010) and the 2011 Tohoku event in Japan (Loveless and Meade, 2011; Fukutina *et al.*, 2014), can create extreme shaking and tsunami hazard (see Chapter 3.9). Although no written records describe such an earthquake in the Cascadia region, aboriginal histories refer to flooding from the sea, which has been taken to represent **tsunamigenic** events (see Chapter 3). This interpretation is consistent with the occurrence of a large subduction earthquake (see Chapter 3.7.1); from such events, tsunami comprise an expected behaviour along with ground shaking and widespread (e.g. across ~1000 km) subsidence of the coast. Despite the evidence from aboriginal communities, and until as recently as the early 1980s, the scientific community regarded this kind of catastrophe event in the Cascadia region as unlikely. However, this assumption has since been challenged, through research into the region's geological record.

Researchers reported sand sheets, commonly buried under a single band of estuarine muds and silts, were found along ~1000 km of coastline. These sheets require a high-energy wave to deposit them. The sheets thinned inland and contained microscopic siliceous shells of marine diatoms, demonstrating that they must have come from the sea. Numerous examples of apparently recent soil layers found beneath 1–2 m of mud and silts deposited in estuaries demonstrate, as global sea level has not risen to that extent over the period, that the land's surface has moved down (i.e. subsided) in this region. Furthermore, this coincidence of the subsidence and the sand sheets is not explained by a storm or remotely generated tsunami.

Researchers also found direct sedimentary evidence of local ground shaking, including sand dikes indicating liquefaction, and offshore deposits (i.e. **turbidites**) revealing where shaking and dislodged material was deposited. Radiocarbon dating constrained the age of the deposits to 1695–1720. Together, these sources of paleoseismic evidence point to the occurrence of at least one severe ($M > 8$) earthquake in that time frame. However, this evidence does not distinguish between a single $M > 9$ earthquake and a cluster (see Chapter 3.7.1) of $M > 8$ events (Henstock *et al.*, 2010).

In the case of Cascadia, tree-ring evidence gathered from widespread areas of submerged tree stumps indicated that parts of the coastline and inland mud creeks that had previously comprised mature forests flooded suddenly and catastrophically between August 1699 and May 1700. Wide rings up to their edge showed that trees were healthy until the moment that they died: there was no evidence for a slow death by gradual immersion. This evidence ties in very strongly with records of an 'orphan' tsunami with an unknown earthquake source that hit Japan on 26 January 1700; computer simulations and comparison with other tsunami of distant sources suggest a single large earthquake in Cascadia as the most likely source. This conclusion also fits with archaeological evidence of destroyed campsites in Cascadia in the upper levels of the buried soil layer.

Evidence for one $M > 9$ earthquake in Cascadia raises the question of the recurrence interval of such events. A stratigraphic sequence of buried soils such as the 1700 one, each overlain in turn by their muds and silts, suggests approximately one such event every 500 years. Recent geophysical studies of the offshore deposits (i.e. turbidites) have helped to improve the accuracy of recurrence interval estimates for a full megathrust rupture (Goldfinger *et al.*, 2012). Current estimates of the recurrence interval for a full megathrust magnitude 9+ rupture in this region is of the order 500–550 years (Petersen *et al.*, 2014).

Thus, this example illustrates how multi-disciplinary geological, geomorphological, paleo-seismic and tree-ring data have been combined to understand the recurrence interval for a rare and potentially highly damaging event. Catastrophe model developers use such research findings when deciding on the range of magnitudes and recurrence frequencies to apply to the events included in their models. Such decisions can have significant implications for the relative seismicity and therefore relative hazard in such regions (see Petersen *et al.*, 2014). They may therefore provide a key differentiator between models. Equally, updates to the scientific consensus on extreme, long-period events such as those generated by the Cascadia Subduction Zone may also influence the magnitude of change between model versions.

### 4.3.4 Deterministic Hazard Models: Historical Reconstructions

This section discusses the process for developing a deterministic hazard model from the data sources described in preceding sections. As noted in Section 4.3.1.1, deterministic hazard models may use synthetic scenarios (see Section 2.7.1.3) or be based entirely on historical events, but in either case historical reconstructions are a good place for development to start.

Historical reconstructions of individual events allow hazard to be related to damage (e.g. Figure 4.5), and provide insights and case studies to examine the assumptions used to create stochastic hazard footprints. A set of historical reconstructions spanning a specified time period (e.g. footprints for the events in the HURDAT catalogue – Table 4.1) can also provide a validation benchmark, perhaps through calculating 'as if' losses for the event (see Chapters 2.6.4, 5.4.3, and 5.4.5). Simplistically, probabilities may be assigned to the event as $1/Y$ where $Y$ is the number of years in the catalogue, although this has disadvantages especially if the time-span of the catalogue is short, when the rate can be different from 1/Y because of the expected infrequency of the event, clustering or time dependency of events. Alternatively, statistical approaches exist (Della-Marta *et al.*, 2009).

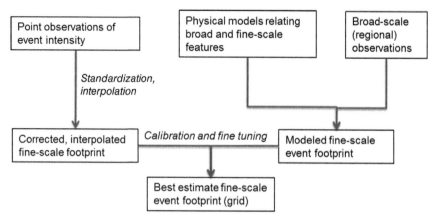

**Figure 4.7** Schematic illustration of a suggested process for reconstructing historical events from point observations of the event's intensity (Section 4.3.3.1), augmented by the use of physical models and spatially distributed observations (Section 4.3.3.2) to assist interpolation. See text for details.

When reconstructing a catastrophe event, a mixture of data types may be used as a starting point, selected appropriate to the peril (Section 4.3.3). This may be augmented by physical and theoretical models and/or statistical techniques such as down-scaling (e.g. Section 4.3.6.3) to better interpolate fine-scale (i.e. small-scale) structure (Figure 4.7) shows one plausible schematic framework, which is typical of that used by catastrophe model developers to combine the differing observations to recreate a footprint of the event.

First, a footprint is obtained by standardizing (e.g. wind speed to 10 m altitude) and interpolating spatially and temporally from point observations (Section 4.3.3.1). This process generates corrected data at the best-constrained locations, and a visualization for the modeller. Second, a physical computational model generates a broad-scale footprint. Further refinements use spatially distributed information to infill the phenomenon's fine-scale features (e.g. maps of topographic roughness to modulate wind). By iteratively adjusting to the modelled footprint to minimize the misfit between it and the point observations, the model developer generates a calibrated best estimate fine-scale event footprint for use in loss calculations. During this process, the model developer usually explores the sensitivity of the final footprint to small changes in the assumptions used in the physical model and any interpolation procedures used. The exercise attempts to retain an accurate representation of the non-homogeneity within the hazard's extent, reduce uncertainty of the intensity estimates between the observed points, and attain some accuracy over the whole area impacted.

### 4.3.5  Site-Based Extrapolation: A Local Solution

This section describes a site-based assessment of probabilistic hazard, i.e. that which seeks to describe *all likely future events* as well as historical ones. In this example the process uses extrapolation. Site-based extrapolation provides a first illustration of the probabilistic approach. It is not an event-based approach, but serves to illustrate some of the issues in extrapolating limited observations to a longer time-frame.

**Frequency analysis** (see also Chapter 3.5.1) refers to an assessment of the probability of an occurrence at a site, defined by a relevant measure of its magnitude such as peak flow rate (i.e. $m^3s^{-1}$) for flooding. An occurrence that has a $1/T$ chance of occurring in any year is said to have a $T$-year return period. Probabilities may be empirically determined from historically observed maxima and analysed statistically using extreme value theory (Gumbel, 1941; Stedinger *et al.*, 1993, Coles, 2001). Since the intensity of any phenomenon can be quantified in a variety of ways (see Section 4.3.2), there is potentially an array of return period values for any occurrence. All metrics will vary spatially, leading to differing return period values between sites of any given event.

Such methods are used in creating zoning maps for planning (British Standards Institute, 2011). They can also evaluate risk for *specific* insured properties. That is, model developers can create site-specific hazard models as a stepping stone to building a regional event-based model. Conversely, site-specific analyses can serve as calibration or validation points for catastrophe models (i.e. locations to compare predictions to data). Decisions on the reasonable timeframe or shape (i.e. statistical distribution used) of the extrapolation require expert judgement to avoid unrealistic exaggeration or underplaying of the risk.

Use of a single site, however, *cannot* capture the spatially distributed footprint of an event. This limits their use in (re)insurance for accumulation or reinsurance pricing. However, such models can be used to understand the relativity of risk at different locations, for example, for primary insurance pricing (see Chapter 2.6). To be of significant use for catastrophe risk assessment, methods must consider the spatial correlation of risk between sites, the underlying drivers of the hazard's severity, and the complex factors that may lead to geospatial and temporal variability between and beyond sites. Hazards experienced at multiple locations, or from multiple sources, require consideration of joint probabilities. This outcome can be achieved using computer modelling, to simulate many realizations of possible weather events (Cameron *et al.*, 1999). Alternatively, records of instrumental observations (e.g. river flows) can be modelled using copula functions (Chen *et al.*, 2012) or conditional probability models (Lamb *et al.*, 2010) (see Chapter 3.5.1).

### 4.3.5.1 Case Study: Extrapolation of Discharge to High Return Periods in a Single-Location Flood Risk Model

This case study illustrates a probabilistic flood risk assessment from a time-series of measurements at a single site of a parameter relevant for estimating insurance loss (i.e. **discharge** in $m^3s^{-1}$). Local or regional institutes such as hydrological or meteorological government departments, usually provide this kind of instrumental recorded data in tabular form for each observation point (e.g. a river gauging station).

Although agencies make considerable attempts to maintain a large, representative network of long-term, complete and homogeneous measurements, time series are typically limited both in space and time. In particular, they typically lack historical observations of extreme, catastrophic events which are required for probabilistic flood modelling. Therefore, the developer must use interpolation and extrapolation to provide a spatially complete picture while exceeding the time range of observations. This process, however, introduces new sources of uncertainty, particularly the lack of knowledge of how to define the relationship between severity and frequency beyond the range of observations, including a potential maximum event scale.

Figure 4.8 illustrates a selection of methods to extrapolate an annual maxima discharge data series instrumentally recorded between 1958–2008 at Colwick (near Nottingham) on the River Trent in the United Kingdom (black triangles). Macdonald (2013) uses four approaches in the flood frequency analysis, the first two of which are considered conventional.

1) A single-site analysis of the gauged flow data, fitting a generalized logistic distribution by L-moments (black).
2) A pooled analysis following UK common practice (i.e. Kjedsen and Jones, 2009). Since only 13 of the 51 years have flows over $500\,m^3\,s^{-1}$, the authors used data from equivalent sites (i.e. interpolation) to assemble 1405 annual maxima data points (black dashed).
3) Augmenting the gauged data with historical annual maxima flow estimates (1795–2008), fitting data above a threshold ($733\,m^3\,s^{-1}$) following the method of Bayliss and Reed (2001) (grey).

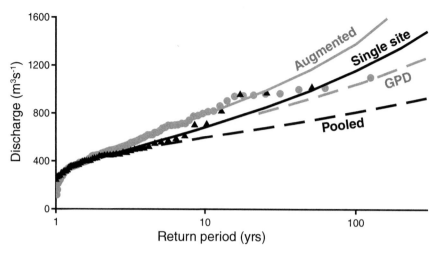

**Figure 4.8** Example of extrapolating an annual maxima discharge data series recorded at 1958–2008 at Colwick on the River Trent in the United Kingdom (black triangles). Different approaches are used by Macdonald (2013), see the text for details. Note that the pre-instrumental annual maxima series shown (grey circles) spans 1884–2008, a shorter period than the 1795–2008 'augmented' series of Macdonald: Annual maxima time-series supplied by N. Macdonald.

4) Fitting a **Generalized Pareto Distribution** (GPD) (see Chapter 1.11.2.2) to data above a threshold ($733 \, \text{m}^3 \, \text{s}^{-1}$) by probability weighted moments (grey dashed).

The return period estimates of discharge for Nottingham diverge significantly for less frequent events. The historically augmented data estimate a 100-year return period flow of $1,386 \, \text{m}^3 \, \text{s}^{-1}$, which would be ranked second in the observational series of Macdonald (2013) and thus appears credible. The GPD, fitted to the same data, illustrates how the distribution chosen can affect estimates. The apparent visual underestimation of the GPD, fitted to 1795–2008 data, to 1884–2008 data plotted on Figure 4.8 may be explained by the fact that three of the four largest floods in Nottingham since 1400 happened between 1795 and 1875; this illustrates the influence of a few very severe events.

The $1,161 \, \text{m}^3 \, \text{s}^{-1}$ 100-year return period value estimated by single site analysis, using 1958–2008 data, would be the third largest in the 214 years since 1795. It is thus credible. The same cannot be said for the interpolated, pooled estimate. Its 100-year return period estimate of $814 \, \text{m}^3 \, \text{s}^{-1}$ has been exceeded 14 times since 1795 and four times since 1958 in the gauged record; this is probably caused by the lack of other suitable sites for the pool as few UK sites have similar-sized flows and this makes it difficult to derive a comparable pooling regression from available data. Overall, Macdonald (2013) highlights the utility of a longer historical record for validation, and the uncertainty associated with both interpolation and extrapolation.

Statistically, a much wider range of different types of distribution could be used, leading to wide uncertainties. As a practical solution to the wide range of outcomes predicted in this way, the model developer could choose a distribution producing estimates somewhere in the middle of the spectrum and build stochastic events with that assumption. However, this approach disregards the uncertainty due to lack of knowledge (see Chapter 2.16). The ideal solution, although often impractical, would be to account for all reasonable distributions when creating the stochastic event set and measure the effect of different options on the modelled losses from the catastrophe model (see Chapter 1.10). Nevertheless, a stress test which considers the effects on modelled losses of choosing different discharge distributions should at least be considered (e.g. see Chapter 5.3.5).

### 4.3.6 Building a Probabilistic Event-Set

This section describes the development of a probabilistic event set-based hazard model, illustrated with case studies. An event-based probabilistic risk model is the 'traditional' form of model in insurance (Chapter 1.8), and is what is most often meant when industry practitioners refer to a catastrophe model. A probabilistic hazard model contains ensembles of a large number of synthetic events (e.g. $n > 10,000$), with associated appropriate probability, that reflect the whole range of what is possible in reality in its domain (e.g. its spatial extent). The events are created by combining the historical record with physical, theoretical and statistical models that allow the record to be extrapolated beyond what has been observed. Together, these are also known as a stochastic (or probabilistic) event set.

Figure 4.9 illustrates the conceptualization of the process of model creation described by Powell *et al.* (2005) and introduced in Section 4.3.1. The stages are described briefly in turn below, although *Events* and *Hazard* cannot be readily distinguished as separate stages for some methodologies.

First, *Geography* is the spatial extent (i.e. domain) of the model. This may have been specified by the model sponsors, but domain may be influenced by market needs, available data, and the scientific and financial plausibility of developing a model for a particular peril and region. The choice of modelling domain can therefore be a trade-off between the scientific desire to correctly understand the regional risk posed by a peril and a combination of practical and commercial considerations. For instance, a model sponsor may have commissioned a study with built-in geographical extent definitions. In such cases, it is often necessary for the model domain to extend beyond these boundaries of the study, considering events that originate outside the domain but have spatial scales which influence the nature of the hazard within it, for example, tropical cyclone genesis regions and their landfalling areas may be separated by hundreds of kilometres.

Developers also choose the temporal domain of the model. How far in extremeness beyond the short measured record should the stochastic catalogue project? For all event sets, this decision should be governed by the limit of confidence in the extrapolation method (i.e. statistical or computer model) to correctly represent extreme events. For developers using computer models, and a particular simulation period (see below), there is a trade-off between runtime and reliability of the output. A longer simulation period will more fully include the whole range of modelled outcomes, but requires more computational time extending development timelines and the larger resulting event set may be unwieldy for end users. In this case, statistical techniques may be used to reduce the size of the event set while maintaining its geospatial and probabilistic integrity. This process is sometimes known as 'boiling down' (see Section 4.3.6.1).

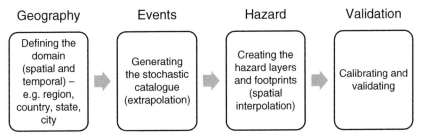

**Figure 4.9** An expansion upon the 'Hazard' box in Figure 4.1, highlighting key decisions in hazard model design. Matched to the list in Section 4.3.1, 'Geography' is step (1); 'Events' combines (2) and (3), 'Hazard' is (4), and 'Validation' is (5).

Second, how plausible events and their hazard are generated varies widely. Construction of a probabilistic event set varies by peril and various model development methodologies exist even for the same peril. Events may be generated by parameter-based statistical approaches, approaches based on the results of other models, for example, a climate model, or combinations of these approaches. Table 4.3 summarizes some of the common questions encountered that require decisions.

For simpler earthquake models, events are statistically created as independent (i.e. Poisson, see Chapter 1.11.1.1) entities, with abundances appropriate to their key attributes (e.g. size, spatial location); see Chapter 3.7.4 and Section 4.3.6.1. Similarly, tropical cyclones can be generated using stochastic combinations of **cyclogenesis**, track, size and intensity (Chapter 3.2.4) (Powell *et al.*, 2005). As an example, distributions for some parameters sometimes used to define tropical cyclone are shown in Figure 4.10.

In Powell *et al.* (2005), this catalogue is transformed into an event set using the following steps:

1) The number of storms per year is selected from the distribution of annual hurricane rate.
2) The genesis time and location are selected from temporal and geographical distributions.
3) Storm track and intensity are modelled as a step-wise process at 1 hour intervals using geographic motion and intensity distributions (see Chapter 3.2.4 and Section 4.3.6.4).
4) Storm parameters are determined to define initial wind field characteristics – these are pressure difference, radius to maximum winds, cyclone system forward speed, and the surrounding pressure profile.

### 4.3.6.1 Stochastic Event Set Resampling ('Boiling Down')

Some modelling development processes will add a further step in the development of the probabilistic event set, by re-sampling the stochastic catalogue of synthetic events into a smaller set with the same statistical properties. Perhaps the most important aspect of this is that it does not simply consist of deleting events. Very simplistically, consider a stretch of river affected by 1-in-100-year flood events, represented by a model. If all conceivable 1-in-100 years were identical, one event footprint with a probability of 0.01 in the model would be sufficient to describe the flooding. However, perhaps variants on the flooding are possible. If two different footprints are modelled to better explain what is possible, each must have a probability of 0.005 in the model such that the probability of the river flooding is correct (i.e. $0.005 + 0.005 = 0.01$). If ten variants were used, the probability needed would be 0.001 for each of them. Now, consider hydrological modelling that has produced many variants (e.g. 437) for the 1-in-100-year flood, but a number are very similar. It would be more computationally efficient to only use one of each similar set; thus if three are similar, one could be maintained but assigned the probability of the others (i.e. this becomes three times greater). The difficult issue here when merging footprints is: What is the 'same'? Same return period, same location, or the same magnitude of loss? Pragmatically, a large number of possible approaches exist to boiling down. One might be to randomly pair up events, deleting and re-assigning probability only if the OEP curve and AAL for a standard distribution remained within pre-defined tolerances.

Another approach to optimize the event set size might be to remove or reduce the recurrence period for smaller, unimportant events in the catalogue, to balance out the addition of possible events with recurrence periods that fall outside the length of the simulation period. For example, with earthquake models, when adding larger events to specific faults, given the uncertainty in relation to characteristic magnitude, model developers apply appropriate physical constraints such as the **moment balance** by source (see Abrahamsen and Blanpied, 2003), and adjust the probabilities of the less severe events in the event catalogue. In this way, the event catalogue then becomes representative of a wider simulation period.

Table 4.3 Questions commonly encountered when constructing a probabilistic hazard event set.

| Issue | Impact | Limitations |
|---|---|---|
| What is a suitable simulation period? | Influences the ability to capture the 'full range' of events | Computational power<br>Adequate scientific knowledge of the potential range and frequency of events<br>Statistical convergence |
| What are the likely maximum intensity events to be considered? | Determines the maximum possible event severity within the range produced<br>May over- or under-estimate possible event maxima and add error or bias to the event severity-frequency distribution | Lack of appropriate scientific knowledge<br>Unverified assumptions or views of likely maxima<br>Geographic variability of event magnitudes (e.g. earthquake) |
| How is an 'event' to be defined? | Determines the physical representation of the loss-making event, range of causal processes and its extent | Ability to isolate an event from interdependent factors (e.g. riverine flooding versus drainage surcharge, European wind storm 'sting jets')<br>Ability to capture the impact from a succession of events (e.g. flood resulting from cumulative impact of multiple rainfall events on a catchment)<br>Ability to assign loss potential from available information (e.g. landslide claims from earthquake or rainfall causes)<br>Ability to identify changes to the event character due to varying geophysical or hydro-meteorological processes (e.g. 'transitioning' tropical cyclones) |
| How should the event set distribution be represented as a sample? | Affects event distributions, particularly in the tail | Computational power<br>Ability to assign an appropriate sampling mechanism |
| How should spatial and temporal dependency be incorporated in the event frequency model? | Can significantly influence occurrence frequencies and loss exceedance probabilities | Lack of scientific knowledge or consensus (e.g. European wind storm clustering) |
| How should event genesis locations and sources be defined? | Can influence spatial distributions of events, and potentially introduce bias in event distributions | Computational power<br>Ability to define and integrate appropriate source zones (e.g. earthquake areal vs fault sources) |
| Is there a temporal variability (non-stationary) trend or cycle that will influence the estimation of event frequencies? | Can influence event rate | Lack of scientific consensus<br>Lack of comprehensive data to represent trends and cycles accurately (time windows) – e.g. hurricane land falling frequencies<br>Uncertainty or bias in results<br>Volatility of results due to re-appraisal or changes in assumptions |

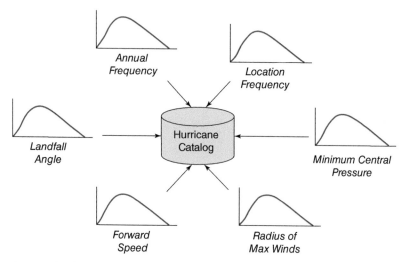

Figure 4.10 An example of the distributions of parameters used within a particular hurricane model to generate a stochastic catalogue. *Source*: Reproduced with permission of AIR Worldwide Corporation.

Regardless of the chosen approach, model developers perform calculations to measure statistical convergence as they optimize the sample size. Statistical convergence represents the degree to which one event set differs from reality by virtue of the chosen event sample. To measure statistical convergence during boiling down or during any other process that involves using randomization to choose parameter values, the model developers produce multiple simulations and measure the degree to which the results converge as the sample size increases. This convergence testing may focus on a single parameter (e.g. Average Annual Loss - see Chapter 1.10.1) or some range of parameters (e.g. specific return-period loss, hurricane landfall rates, peak-gust wind speed, shaking intensity, flood depth, etc.). In general, convergence reduces as the sample size decreases—and the magnitude of this decrease becomes more extreme with increasing geographical resolution.

The following simplistic example considers a c. 250,000-event original event set for Europe windstorm loss. For practical reasons, this full event set is too large to include in a standard desktop realization of the model. Therefore, the developers reduce the event set size as much as possible. First, using criteria that measure the size and strength of the hazard footprint, they discard events generated by the model that have negligible impact on the territories of concern. Next, using an industry exposure database, the developers perform loss analyses using the remaining event set, and then perform the boiling-down process using these loss results. Finally, using key tolerance criteria, they discard 'similar' events, i.e. defined here as those with similar spatial loss footprints, and adjust the probabilities of the remaining storm events accordingly.

The process requires the following criteria to be met:

1) Event rates must remain consistent with the statistical properties of the original event set. Thus, the occurrence-magnitude relationship must be preserved.
2) Loss return periods must remain consistent with the statistical properties of the original event set. During the process of boiling down the event set, the developers measure the average annual loss and key return-period loss metrics both at country resolution and at CRESTA zone resolution. They set different tolerance levels for divergence from these metrics, e.g., maximum 1% at country resolution, and 5% at CRESTA resolution.
3) Spatial correlations must remain consistent with the statistical properties of the original event set. Large windstorms in Europe can affect several territories. For example, windstorm

Daria (1990) caused catastrophe loss in the United Kingdom, the Netherlands, Belgium, and Germany, and significant additional losses in other territories, including France and Denmark. Thus, the spatial correlation of loss can be a key determinant of the overall impact of an event. The developers measure spatial correlations by return period for key parameters including peak-gust wind speed and loss, and measure divergence from those correlated metrics as the event set size decreases.

Thus, the final event set reduces to a more manageable number (e.g. some tens of thousands of events), while still retaining the overall loss characteristics presented by the full 250,000-year stochastic event set.

### 4.3.6.2 Event-Hazard Transformation

In general, creating a hazard footprint for an event involves describing the spatial variation of severity, and the local impacts of particular site conditions such as topography, land use and local geology. The framework shown in Figure 4.11 illustrates the relationship between event generation and hazard. The 'regional' event models generate events that feed into 'local' impact models that explore each event's potential consequences. See Chapter 3, and Sections 4.3.6.1 to 4.3.6.4 for examples.

To transform events to hazard, a hazard footprint is calculated. To generate the hazard footprint, the model **downscales** (Haas and Pinto, 2012; Della-Marta *et al.*, 2010) the initial event severity as defined by the summary parameters to a higher resolution representation of the hazard, that describes the geospatial and temporal evolution of the intensity parameter (or parameters). This is sometimes termed the 'fine mesh'. The final downscaled hazard intensity parameter relates directly to the damage. It provides the means of determining the expected damage within the vulnerability function (see Section 4.5.1).

Each hazard type will require a specific approach to generate an appropriate 'local impact' footprint (see Table 4.4 for a summary). For example, both storm surge and inland river flood footprints will generally be constructed using a hydraulic model, based on hydrographic flow peaks at each point along a river or coastal network with local depths estimated from the propagation of water (by estimated volume) on a cell-by-cell basis (e.g. Sampson *et al.*, 2014). The computational overheads for generating high-resolution footprints may be prohibitive, depending on the processing and storage capability of the model, and the scale of events being modelled. This can lead to a trade-off decision by the model developer – whether to reduce spatial resolution of the underlying footprint to enable larger spatial extents to be modelled, or

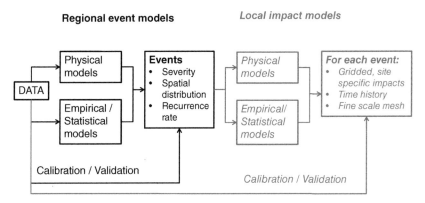

**Figure 4.11** The relationship between the stochastic event set creation, also sometimes called the 'regional' event model, and hazard generation sometimes called the 'local' impact model. *Source*: Reproduced with permission of RMS.

Table 4.4 Common event footprint generation methods and datasets.

| Hazard | Intensity | Commonly used data | Commonly applied methods | Resolution drivers | Common footprint resolution |
|---|---|---|---|---|---|
| Flood | Water depth (velocity, duration) | Digital Terrain Model (DTM) Land cover (surface roughness proxy) Defence data River network and channel geometry (parameterized) Hydrological model (rainfall, flow, surge) | Computational Hydraulic Model Simple water height ('bathtub') | DTM (horizontal and vertical) Defence data (location, type, fragility) Roughness (parameter) uncertainty and impact on flow | 30–1000 m horizontal (but may be lower resolution or postcode-based) |
| Wind | Peak gust speed | 'Over water' (low roughness wind speeds Terrain Land cover (surface roughness proxy) Prevailing wind direction | 'Open terrain' downscaling Wind direction and relative upstream roughness Surface smoothing | Terrain/ roughness data | c.100m–10km |
| Earthquake | Ground motion (Spectral Acceleration, PGA, MMI) | Surface geology/soil classification Terrain (slope) | Ground Motion Attenuation models Slope proxies (e.g. Vs30 – Allen and Wald, 2009) | Soil/geology data resolution | Attenuation models variable Slope proxies (e.g. Vs30–1 km horizontal |

increase the spatial resolution but reduce the size of the computational extents of the model. Many modellers will attempt to optimize the spatial resolution of the underlying footprint mesh. For example, they may apply a spatially varying mesh density, either on a regular grid, or a node-vortex structured vector network (see Section 4.6), to gain detail where such detail is warranted. For example, the mesh may have higher resolution in areas of greater potential exposure, or where representation of hazard processes are considered to require greater detail, for example, at the coastline where detailed sea defences are key to localized flood propagation. This approach, while potentially improving computational efficiency and resolution, does require consideration of the suitability of the chosen hazard downscaling approach on a variable scaled spatial framework, particularly if spatial length scale can have an influence on the relative size of event magnitude calculated (as can be the case, for example, with hydraulic modelling).

The Powell *et al.* (2005) approach for modelling tropical cyclone footprints involves the calculation of a wind field (spatially structured and localized peak wind estimates) every 15 minutes from the wind parameters, themselves varying to reflect the expected storm life cycle (including inland decay of the storm).

From this, the wind data can be extracted and applied to high-resolution surface roughness data as per Powell *et al.* (2005), to generate wind hazard footprints for each event, accounting for the known biases in the initial event generator such as a General Circulation Model (GCM; Chapter 3.3.4) and for the defined simulation period.

Lastly, this is also the point at which developers calibrate and validate the model by comparing the modelled frequency and severity to the historical record (Section 4.3.3.4). Specifically, they may compare observed site-based exceedance probability curves with modelled ones. One example of this applied within catastrophe model validation (especially wind hazards) is in the use of quantile-quantile ('Q-Q') plots, which are graphs to visualize the level to which two datasets are derived from the same common distribution (see, e.g. Figure 4 of Vitolo *et al.*, 2009). To assist in ensuring that their model closely represents actual risk, model developers typically iterate to check the sensitivity of the predicted relationship between severity and probability to key modelling assumptions; large sensitivities to small changes in key assumptions are undesirable. Users also sometimes replicate this stage of modelling to justify their view of risk (Chapter 5).

In summary, developing a stochastic event catalogue that adequately represents the credible range of possible events, their severity, frequency and geographic extents is highly challenging. The extreme and infrequent nature of these catastrophes means their character is poorly constrained, modelling choices require judgement, and scientific understanding may completely change in the light of new evidence. Computational data storage and processing barriers present a challenge, as do short historical data records, and the need for a process with relatively simplistic parameterization. These will continue to influence the methods employed.

The following examples illustrate a variety of approaches adopted by catastrophe model developers for different perils.

### 4.3.6.3 Case Study: A Regional Probabilistic Earthquake Hazard Model

This section describes one simplified example of the implementation of a probabilistic earthquake model for Japan and highlights some of the development issues commonly encountered; see Chapter 3.7 for alternatives, terminology, and technical detail.

In terms of domain, the model considers three types of seismic source. It is set up as a 'regional' model driving 'local' models (Figure 4.12). The model both defines events as occurring on known faults (i.e. mapped at the surface), and attributes known areas of focused seismicity to specific synthetic faults even if the location of the actual faults is unknown. It also uses **area sources** for background seismicity, and randomly and uniformly assigns **epicentre** locations within a region. Then, with locations determined, for each source, the

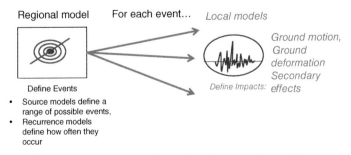

**Figure 4.12** Diagram illustrating the structure of an earthquake hazard model for Japan. A 'regional' model defines source locations, the events that may occur on them, and how often these occur. This drives local intensity models that define ground motion and other hazardous effects. *Source*: Reproduced with permission of RMS.

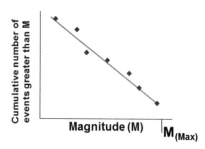

Figure 4.13 Gutenberg-Richter relationship. The graph is plotted on a logarithmic scale. See Chapter 3.7.1. *Source*: Reproduced with permission of RMS.

model defines the size range of possible events; this will depend upon the type of fault and its tectonic setting (Chapter 3.7.1).

Next, a 'recurrence model' calibrated for each seismic source (primarily for sources defined by areas rather than fault lines, see Chapter 3.7.4) is used to define a probability of occurrence for each event via an appropriate size-frequency relationship. The empirical power-law distribution $N = 10^{a-bM}$ known as the Gutenberg-Richter relationship is used where $N$ is the number of earthquakes greater than magnitude $M$, and $a$ and $b$ are constants (Gutenberg and Richter, 1954; Kramer, 1996). Figure 4.13 illustrates this relationship. The recurrence model for fault sources is, however, somewhat different. The recurrence rate is calculated based on the slip rate of faults (Chapter 3.7.1) and the magnitudes (typically defined by the characteristic magnitude, see Petersen *et al.* 2008, for an example) of earthquakes from the concept of moment balancing of fault-rupture models (see Section 4.3.6.1). The maximum magnitude of a fault source, shown as $M_{(max)}$ is estimated based on historic data and on expert judgement that accounts for the underlying geophysical constraints such as the slip observed along a fault in past earthquakes, the fault geometry, and if relevant, the subduction rate along a section of a plate boundary. However, the whole concept of constraining the size of the largest earthquakes by a strict cap at a specific value of $M$ can be controversial (see Chapter 3.7.1). The 2011 $M_w$ 9.0 Tohoku earth quake illustrated the importance of this debate by exceeding many previous $M_{(max)}$ values applied along this section of the Japan Trench. Cautious model developers using $M_{(max)}$ cut-offs may therefore may provide additional 'what if' scenarios outside the stochastic event set that describe the most extreme plausible earthquakes, for example, based on paleo-seismic data.

As noted in Chapter 3.7.1, most models assume that earthquake occurrences on a particular source are independent (i.e. Poisson). However, this assumption may not properly represent the actual nature of the seismic hazard, and regions with good historical catalogues or appropriate paleoseismic data may have sufficient information to use a time-dependent fault rupture model, based on elastic rebound theory (Reid, 1910). See discussion in Chapter 3.7.1, and Section 4.3.7 for context with respect to other perils.

For each of the events produced by the regional event model, the ground motion intensity model determines their local impacts. These have two functions:

1) They use **ground motion prediction equations** (GMPEs) to estimate how much shaking there will be in terms of a metric such as Peak Ground Acceleration (PGA) (Figure 4.14), or other intensity measures. The prediction model uses parameters such as the distance of a site from the fault rupture plane (see Chapter 3.7.1, Figure 3.18) of the quake, its geology, regional characteristics (e.g. active regions where earthquakes occur frequently such as California) and soil type (see Chapter 3.7.4).
2) They estimate other potentially destructive consequences of the earthquake such as landslides (see Chapter 3.8) and **liquefaction**.

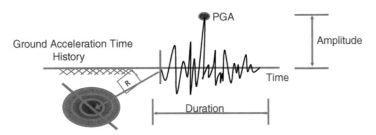

**Figure 4.14** A depiction of measured ground motion during an earthquake by a detector at distance R away from the earthquake's hypocentre (red). The illustrative seismic trace is shown with time increasing left to right, vibrating around the horizontal line that also acts as the Earth's surface. PGA is 'Peak Ground Acceleration' (see Chapter 3.7.2). *Source*: Reproduced with permission of RMS.

There are many GMPEs, and their choice for a region forms a key uncertainty (Chapter 2.16) in earthquake hazard models. GMPEs are generally developed by fitting a simple model that describes different characteristics of ground-motion attenuation to the observed data collected from real earthquakes. However, data may not be available for the modelled region. In such cases, model developers may use GMPEs calibrated with data from a different region or develop GMPEs from simulated ground motion data from some physical models.

### 4.3.6.4 Case Study: High-Definition Regional Precipitation-Induced Inland Flood Model
This short case study presents an outline description of one framework that might be used to create high-definition flood events for a catastrophe model. It uses a region-wide precipitation model, and therefore follows pathway (b) on Figure 3.15. This generic overview is framed in terms of 'regional' and 'local' models; for definitions and detail, such as on how the **hydrological** and **hydraulic models** are used and merged, see Chapter 3.5.

This framework first creates probabilistic precipitation (i.e. rainfall) events in a 'regional' model. It does this using continuous time-series output by a climate simulation (i.e. GCM) at relatively low resolution (i.e. every 1 hour in ~60×60 km cells) that is then statistically **downscaled** to finer resolutions (Figure 4.15). This is done to approximate a realistic temporal and spatial correlation in the precipitation inputs driving flooding. The rainfall events pass into a 'local' hazard model, which first uses a **rainfall-runoff** model (RRM) to simulate the flow across the land's surface allowing a portion of the water to enter river catchments. The other rainwater may cause 'off floodplain' (or **pluvial**) flooding through surface runoff and **flash flooding**, and by entering minor river systems. The water entering rivers then gets directed through the channels within the **catchment** by a routing model.

By being combined with the events' river flows, the outputs of the hydraulic (i.e. inundation) model are used to reflect the hazard associated with floodwater building and cascading across the floodplain over the hours and days following the precipitation events (see Chapter 3.5.4 for details). Pragmatically, the model differentiates between 'major' and 'minor' fluvial flood sources and processes as a consequence of data availability and computational limitations, i.e. different modelling and calibration approaches are used for different scales of floodplain. In some cases, a differentiation may be driven by planning and regulation, such as in the United Kingdom where the Environment Agency defines 'main rivers' as ones it has statutory powers to manage. Additional methods are employed to represent flood defences. The incorporation of local defence responses to flood levels are particularly challenging in a stochastic modelling environment, both in terms of the availability of data to properly represent their spatial and physical characteristics, as well as the uncertainty related to the level of protection they can provide. As a result, it is usual for defence mitigation to be parameterized to some level within catastrophe

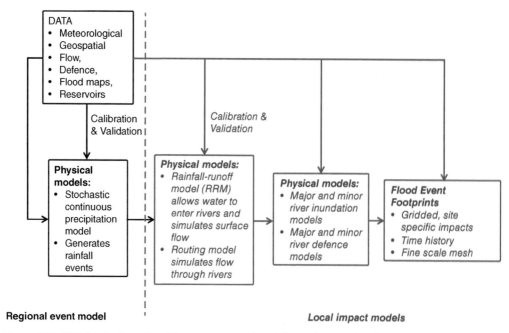

**Figure 4.15** High-level schematic of the processes and models used to simulate flood hazard. See text for a description. *Source*: Reproduced with permission of RMS.

models using general assumptions and approximations, for example, by assuming a given level of protection within urban areas.

The model also includes 'off floodplain' events, including urban drainage and surface runoff type floods, but groundwater flooding could potentially be included here, although groundwater flooding requires additional, time-varying sub-surface considerations in the model formulation. Again, this differentiation is a function of methodological trade-off, reflecting the lower ability to model and calibrate the complexities of urban drainage floods, and surface flooding.

Ultimately, the output hazard maps consist of grids of spatially varying impacts in terms of flood depth; see Chapter 3.5.4 for alternatives. The events have a time stamp, determining their temporal frequency and duration, and the multiple-process technique used allows model developers to attempt to evaluate all identified drivers of flooding in a catchment, including urban flood risk driven by exceeding drainage capacity, not just flooding along large river floodplains and streams. This is important because, as seen in floods that occurred in the United Kingdom in 2012, up to 50% of the insured loss can occur off major flood plains (RMS, 2013).

From the end user's perspective, spatially and temporally correlated approaches better meet the (re)insurance industry's needs (see Chapter 2) than simpler, site-specific extrapolation models (e.g. Section 4.3.5.1). However, each individual model component needs to be independently evaluated and validated, and development is resource-intensive in terms of time, computational resources and expertise, still often requiring decisions that require a trade-off in respect to computational and methodological approaches.

### 4.3.6.5 Case Study: Building a Stochastic Event Set for European Windstorm Using Storms Generated by Climate and Weather Models

The **windstorms** that affect Europe are a sub-set of **extra-tropical cyclones** (ETCs); see Chapter 3.3 for details. ETCs have more complex wind fields than **tropical cyclones** (TCs, e.g. hurricanes). As such, approaches to define events using numerical Global Circulation Models

(GCMs) and NWP models are argued (Keller *et al.*, 2004) to improve upon the time-stepping wind field models used for TCs (see Section 4.3.6.4).

This case study illustrates one means (see Chapter 3.3.4) of constructing a regional, event-based hazard dataset using extrapolated synthetic events extracted from the output of climate simulations.

If known biases (Pinto *et al.*, 2013; Zappa *et al.*, 2013) are accounted for, GCMs and their underlying physics may sufficiently capture the spatial and temporal structure, as well as the intensity ranking of the storms and their seasonal clustering (see Chapter 3.3.1) (Schiwierz *et al.*, 2010). Representation of storm clustering is based on observations in reanalysis data (i.e. NCEP-NCAR 1985–2003), and is characterized by a measure called **over-dispersion** (Chapter 1.11.1.2, Section 4.3.8) (Mailier *et al.*, 2006; Vitolo, 2009; Raschke, 2015), and largely explained by large-scale atmospheric flows quantified by teleconnection indices such as the North Atlantic Oscillation (NAO) (Vitolo *et al.*, 2009).

The resolution of the output events is typically coarse (i.e. 200 km), thus, the extracted storms need to be refined via downscaling to reconstruct the wind speeds at high resolution and there are two basic approaches:

- *Dynamic downscaling* uses the coarse resolution output from the climate model output as boundary conditions to initialize a run of a high-resolution regional numerical weather prediction (NWP) model. The details of NWP modelling are beyond the scope of this book, but a comprehensive summary can be found in Warner (2011). The physics and dynamics of the NWP perform the downscaling to produce the high-resolution event footprints.
- *Statistical downscaling* uses statistical (transfer) functions (e.g. Kirchmeier *et al.*, 2014) that link the coarse and high-resolution wind speed values. An extensive observational sample of corresponding data at both resolutions (low and high) of sufficient temporal and spatial resolution/granularity is necessary to 'train' (i.e. calibrate) these transfer functions. Often the sample of high-resolution wind speeds used in the statistical transfer functions training are actually generated with dynamical downscaling approaches, creating a hybrid approach.

Following the downscaling and reconstruction of the high-resolution event footprints, the final stage calls for detailed validation and calibration of the event set in terms of intensity, frequency and spatial patterns against the existing historical hazard record (historical event set). For European ETCs, historical data such as that produced by national weather bureaux, and in combined products from organizations such as EUMETNET, providing site recording networks of high resolution radar and LIDAR wind profilers across Europe, This step aims at assessing and correcting biases due to the original climate model or errors/noise introduced in the dataset from the downscaling approach. Several published studies have used this approach to generate European wind storm events in order to assess insurance losses (Schwierz *et al.*, 2010; Donat *et al.*, 2011; Haylock, 2011).

### 4.3.6.6 Case Study: Creating a Tropical Cyclone Model: North Atlantic Hurricanes

Tropical cyclones originating in the North Atlantic (i.e. hurricanes) are one of the most significant hazards for (re)insurance given both the severity of the peril (Chapter 3.2.2) and the amount of cover offered via insurance, reinsurance and alternative capital providers (see Chapter 2). As noted in Powell *et al.* (2005) in Section 4.3.6, there is more than one possible approach to creating a regional risk model for hurricanes in the North Atlantic; see Chapter 3.2.4. This case study illustrates the variety of statistical and physical models used in one approach that creates a risk model that encompasses all affected territories, including Central America, the Caribbean, the United States and Canada. The model is constructed this way so that it can establish the correlation between losses from a single event incurred in several territories.

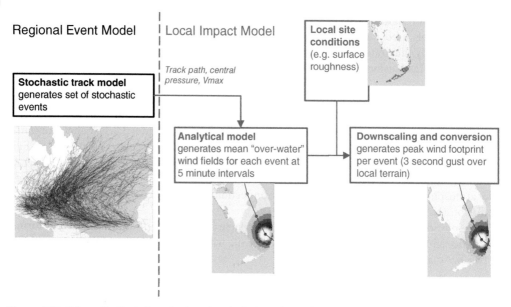

Regional Event Model | Local Impact Model

**Stochastic track model** generates set of stochastic events

*Track path, central pressure, Vmax*

**Local site conditions** (e.g. surface roughness)

**Analytical model** generates mean "over-water" wind fields for each event at 5 minute intervals

**Downscaling and conversion** generates peak wind footprint per event (3 second gust over local terrain)

**Figure 4.16** Schematic illustrating the broad-scale design of a stochastic model for hurricanes in the North Atlantic. Boxes describe actions, and arrows indicate the direction of the computational processes. All stochastic tracks are generated first by the 'regional' event model, then the local impact model is applied to each separately. *Source*: Reproduced with permission of RMS.

First, a 'regional' event model generates thousands of stochastic tracks. Then, a 'local' impact model simulates, for each hurricane, a wind field at regular intervals as time progresses that modulates this hazard to account for local site conditions (Figure 4.16). This is known as a time-stepping wind field approach.

The regional model to generate tracks is semi-parametric and statistical. It uses techniques of Hall and Jewson (2007) (see Chapter 3.2.4). It stochastically generates thousands of tracks for events that represent more than 100,000 years of hurricane activity to extrapolate the HURDAT catalogue (Jarvinen *et al.*, 1984) while preserving similar statistical characteristics (Figure 4.17).

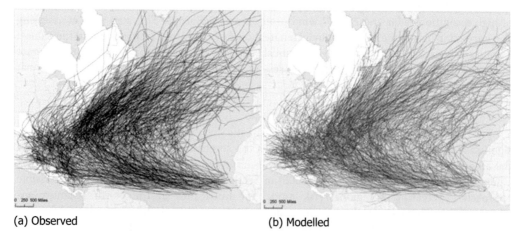

(a) Observed        (b) Modelled

**Figure 4.17** Comparison of observations from 58 years of HURDAT tracks (1950–2007) to One '58-Year','model realization of the Statistical Track Model. *Source*: Reproduced with permission of RMS.

This approach generates stochastic tracks from genesis (starting point) to lysis (end point). Simulated hurricane track parameters provide the key drivers of risk, including landfall intensity, landfall frequency, and landfall correlation.

Then, the local model uses these tracks and their associated parameters to generate a time-stepping, generalized 1-minute mean 'equivalent over-water' wind footprint on a grid. The interval between time steps is 5 minutes.

The model generates the size and shape of the time-stepping wind fields using an analytical wind profile derived from Willoughby *et al.* (2006). Using this framework, at any given point in time and space, the one-minute mean wind (equivalent over water) is entirely prescribed by the position from the storm centre and a set of standard parameters. The required parameters are: maximum wind (*Vmax*), radius of maximum wind (*Rmax*), two shape parameters giving the radial profile inside and outside the eyewall, the angle between the location of the maximum winds and the track, and four additional parameters that reduce the variance between observed and modelled wind fields. As the synthetic modelled tracks do not include all these parameters, the developers derive suitable values from observational data using various statistical methods and assumptions. In this case, the model developers used parameters fitted from the Extended Best Track dataset (Demuth *et al.*, 2006), H*Wind product (Powell *et al.*, 2010) and Weather Research and Forecasting model (WRF) numerical simulations (Skamarock *et al.*, 2005).

The WRF numerical simulations provide an opportunity to understand **extratropical transitioning** (see Chapter 3.2) (Hart and Evans, 2001). This meteorological process occurs as a tropical cyclone such as a hurricane interacts with upper-atmospheric steering currents as it migrates away from the tropics. The fundamental dynamics of the wind field change when it transitions. Transitioning causes distinct changes in the wind field and development of the storm that affect its overall impact on landfall. By using high-resolution (~1 km) numerical simulations of hurricanes known to have undergone a transition, the model developers can establish parameters that describe these changes (Colette *et al.*, 2010). Given the importance of transitioning events on the eastern seaboard of the United States and Canada, this computationally intensive process pays dividends in terms of the resolution and accuracy of the model.

The final step in creating the local wind fields involves downscaling the simulated 1-minute mean wind (equivalent over water) at a site to account for local and upstream roughness conditions. This step captures the transition from sea to land or any change in upstream roughness. It uses a wind engineering model (Cook, 1985, 1997) and roughness lengths derived from the 15–30 m resolution ASTER satellite imagery (Advanced Spaceborne Thermal Emission and Reflection Radiometer, http://asterweb.jpl.nasa.gov) with a 2001–2007 vintage. An additional component of the roughness model converts mean winds over local terrain to 3-second gusts over local terrain (see Vickery and Skerlj, 2005).

### 4.3.7 Secondary or Consequent Perils

Many events leading to a catastrophic loss combine the impacts of more than one peril. In some instances, the damage caused by consequent or 'secondary' peril(s) (see Chapter 3) outweighs the direct impact of the primary peril. For example, hurricane risk is primarily modelled as driven by the severity of the wind; however, in superstorm Sandy (2012), storm surge-related coastal flood losses outweighed direct wind losses, comprising 65–70% of the total (Swiss Re, 2013). The impact of the Indian Ocean tsunami on 26 December 2004 vastly exceeded any direct consequences of the earthquake-induced ground motion. Table 4.5 illustrates typical secondary perils that catastrophe models consider in order to attempt to provide a comprehensive coverage of risk.

**Table 4.5** Secondary or consequent perils for key catastrophe event types. The chapter section points to more information on the primary hazard and its associated secondary perils.

| Catastrophe event | Primary hazard | Secondary perils | Section in book |
|---|---|---|---|
| Tropical cyclone (e.g. hurricane, typhoon) | Wind | Storm-surge (coastal) flooding, inland flooding, landslide, tornadoes | 3.2 |
| Earthquake | Ground shaking | Fire-following, tsunami, landslide, liquefaction, floods, dam failure | 3.7 |
| North America winterstorm | Wind | Snow, ice, freeze | 3.3 |
| Europe windstorm | Wind | Storm-surge, inland flooding | 3.3 |
| Severe convective storm | Tornado | Straight-line wind, hail | 3.4 |

#### 4.3.7.1 Case Study: Secondary Peril Modelling: Rainfall-Induced and Coastal Flooding Associated with Typhoon

Figure 4.18 shows a framework for evaluating inland flood risk (see Chapter 3.5) as a consequent peril associated with typhoons in Japan. In this case, inland flooding can form a more important part of the overall risk than the high winds associated with the typhoon. This framework involves using numerical weather prediction models (NWP) to create a coupled wind and rainfall model for typhoons in the region. The flood peril model can then take the rainfall from each typhoon and associate it with a rainfall runoff model that evaluates the degree to which the rainfall becomes runoff, causing pluvial flooding (runoff, minor streams, drains), or is routed via the catchment into rivers, causing riverine (fluvial, major river) flooding.

Typhoons are also associated with coastal flooding due to storm surges. Thus, to build up a complete view of risk from typhoons in Japan involves also evaluating storm surges associated with each event. The amount of flooding will depend on the way that the high-water levels associated with the typhoon interact with the astronomical tides, on the waves generated by each event, and on the level of protection afforded by sea defences (Figure 4.19). See Section 4.3.1.1 for a description of storm surge hazard development.

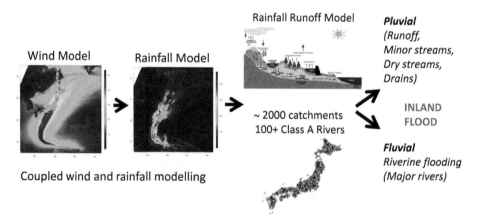

**Figure 4.18** Typhoons are tropical cyclones, and so usually modelled with wind damage as the primary peril. This figure illustrates a modelling framework for including the secondary peril of rainfall-induced flooding. *Source*: Reproduced with permission of RMS.

| Surge | Waves | Tides |
|:---:|:---:|:---:|

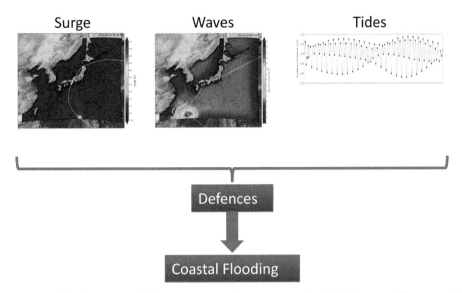

**Figure 4.19** High-level overview of the computational process associated with the modelling coastal flood risk associated with wind-driven storm surge. See text for details. *Source*: Reproduced with permission of RMS.

Such models are computationally expensive. Heavy computational effort is required both to create a full meteorological model and to explore its possible local consequences at a high enough resolution to be meaningful in determining flood risk. However, the key advantage of such an approach for understanding typhoon risk in Japan is that, when combined with a similarly complex linked coastal flood model, it creates a complete picture of correlated risk from a single catastrophic wind and flood event.

### 4.3.8  Time-Dependent Hazard Modelling and Clustering of Catastrophe Events

It is common in catastrophe models to assume stationarity in the hazard-generating processes. Namely, it is commonly assumed that the catastrophic events are independent and are represented through a Poisson process (Chapter 1.11.1.1). This assumption makes modelling straightforward and often yields reliable results, particularly over long time periods. However, patterns observed in historical records of both climate-driven (e.g. Chapter 3.3.1) and geological (e.g. Chapter 3.7.1) hazards sometimes indicate that the assumption is not valid over relatively shorter time scales - which can be decades for hurricanes and hundreds of years for earthquakes (see Section 4.3.6.1 for an earthquake case study example). Therefore, model developers are increasingly using time dependent approaches to model this non-stationarity, rather than applying non-parametric post-processing (e.g. Monte Carlo) to represent clustering of event losses.

A broad category of approaches falls under the umbrella of timeline simulation, i.e. where stochastic events are associated with a synthetic time stamp (i.e. a date) explicitly including temporal correlation in the event set. Model developers are increasingly employing such tactics to define the accumulated risk over some time period of interest, for example, to ensure that the occurrence of the events can be related to the timeframe of contract duration; see discussion of the loss occurrence or 'hours' clause in Chapter 2). Timeline simulations are computationally very intensive, a situation compounded by the high spatial resolution required to model some hazards such as flood. Despite these challenges, timeline simulation models are becoming increasingly possible as computational possibilities evolve. The examples of flood and earthquake are discussed below.

The issue of timeframe definition is particularly relevant in the case of event-based flood risk, where the definition of an 'event' is problematic. Floods can occur on a timescale of days or even weeks, often representing the accumulated impact of several weeks of a persistent weather. Additionally, depending on the size of the catchment and the length of the river system, there can be a lag of several days between a rainfall event and the wave of floodwater passing into the river system and downstream onto floodplains. This provides a substantial impetus to overcome the computational challenges.

The earthquake peril is an example of temporal variability and dependence in event recurrence that results from the underlying physical event-generation mechanisms. Considering only a single fault, time-dependent models can be seen as intuitively appealing as they capture elastic rebound theory (Reid, 1910), i.e. the earthquake cycle. However, dependencies can also exist between faults. For example, when an earthquake occurs, it releases the accumulated strain along the ruptured fault segment and alters the geographical distribution of strain in the near vicinity. This strain transfer may be significant enough to alter the probability of future earthquakes nearby, causing a triggered earthquake, reducing it in some areas or increasing it in others (Stein, 1997; Parsons, 2004) (see Chapter 3.7.1 for more details). The 1811–1812 earthquake events in New Madrid, the cluster of large quakes on the Sumatra margin (Henstock, 2010), and the Canterbury earthquake sequence in New Zealand (Deloitte, 2015) serve as reminders that clusters of earthquakes may occur. In such cases, adopting a timeline simulation approach provides additional flexibility in modelling earthquake sequences.

## 4.4 Exposure Models and Databases

An event only becomes a catastrophe when it affects something and causes loss to property or life. A critical part of catastrophe modelling therefore involves characterizing the properties or infrastructure or assets and their values potentially at risk. Model developers typically refer to this part of the catastrophe model as the **exposure** component. Exposure data specific to the end user's risk is an important, operational input to the model (see Chapter 1) but, as discussed later in this section, development of an exposure model or database is also a necessary part of the catastrophe model development process.

**Exposure modelling** involves quantifying the geographical distribution of affected structures, movable property, and infrastructure, and estimates of the costs associated with the interruption of normal economic activity. Typically this process evaluates the number, replacement cost and type of assets at some granular geographical resolution (e.g. on a grid, or by administrative region, or postal code) appropriate to the region under consideration. For Disaster Risk Reduction (DRR) applications, populations at risk may also need to be considered. See, in addition, Section 4.2.4, for a consideration of the geospatial aspects of the use of such data for risk assignment in models.

Exposure modelling specialists come from various subjects, including geomatics (geospatial technology) and Geographic Information Systems (GIS), imagery and remote sensing, economics, and social and environmental science. This wide set of disciplines reflects the geographical and economic nature of the datasets employed during development, and the specific skills required to interpret them. Many of the datasets originate from national or international agencies such as cartographic (mapping), land use, survey or socio-economic sources with varying quality, age, scale, geographic granularity and resolutions, and often requiring considerable manipulation and augmentation to enable their use in catastrophe modelling. These are used to supplement any available information from (re)insurance companies or associations.

Exposure databases have two roles within the catastrophe modelling process: first, as an end product in themselves with applications for model users, and, second, as a resource used during the model development process. The research into building stock undertaken as part of the exposure model can also feed into the development of vulnerability functions and the mix of buildings it would be appropriate to assume in a particular region, for example, in the production of model building stock inventories (see Chapter 1.8.3).

Catastrophe model developers and other entities often release insured industry exposure databases as a product. For example, insurance companies can use these databases to understand their exposure market share for a particular line of business and geography to understand how their loss from a particular event compares with the industry as a whole. In addition, industry exposure databases are commonly used in the development of loss-based triggers for securitization transactions (see Chapter 1.8.3). However, this section is concerned with the development of exposure data primarily because of its uses within the model development process, see Table 4.6.

Model developers use exposure databases as a primary model design and calibration data source. For example, modellers use exposure databases to create a stationary snapshot of exposure that can be used to assess the relative importance of reconstructed historical events. Comparing reported industry-wide losses for such events can be misleading, given that the exposure affected by historical events can differ widely through time and the problems with on-levelling of historical losses discussed in Chapter 2.6. By reconstructing historical event losses with this static view of current exposure, modellers can compare historical and stochastic **exceedance probability** curves (see Chapter 1.10.2), which may aid model construction. A uniform geographical exposure distribution would neglect the sensitivity of losses to concentrations of exposure in particular geographic regions.

Table 4.6 Uses for industry exposure databases during model development.

| Model | Use |
| --- | --- |
| Hazard modelling | Disaggregating low resolution hazard data to a higher resolution grid, based on assumed distributions of varying asset types and their vulnerability |
| | Estimating event loss distributions and sampling structures, to reduce event sets to a manageable size (i.e. 'boiling down') |
| Vulnerability modelling | Providing an insured inventory that enables the model to automatically apply key vulnerability characteristics as defaults, or choice of composite vulnerability functions when these are not provided by the end user during import |
| | Determining relative loss modifying factors (e.g. business interruption, economic demand surge, regional event resilience) |
| Loss validation | Providing a baseline distribution of insured assets (effectively an 'industry portfolio') for the calibration of loss estimates (e.g. when developing historical event losses) and for model validation (e.g. in comparing historical and stochastic losses on an exceedance probability curve) |
| | Estimating market share when comparing individual portfolio losses |

This section describes some of the approaches adopted by catastrophe modellers to create both economic and industry-wide insured exposure databases.

### 4.4.1 Economic or Insured Exposure Data?

An event's importance to society can be measured in terms of its direct **economic** impact – the total cost of reconstructing and replacing damaged property and infrastructure regardless of whether these elements are insured or not – and the number of casualties. Catastrophe models can assess the economic impact by associating an event-based model with an economic exposure database (e.g. Gunasekera *et al.*, 2015). The economic exposure database represents the cost of replacing all elements at risk from a peril.

For the insurance industry, the impact on insured property (and its derivations for varying assets associated with the structure such as contents and business interruption), tends to be most commonly considered. For this reason, catastrophe models typically include an industry-wide insured exposure database, which represents the total potential pay-out from only the insured elements at risk from a peril. This insured database is therefore a subset of the economic exposure.

Commonly, therefore, model developers construct two different exposure datasets: **economic exposure databases** (EEDs) and **industry exposure databases** (or insured exposure databases, or IEDs).

Model developers express the relationship between the insured exposures and the economic exposures as a **take-up rate** or a **penetration rate**. The take-up rate is the ratio of the number of insured properties to the total number of properties, while the penetration rate is the ratio of the value of insured properties to the total value of properties. This is illustrated in Figure 4.20. Although these terms are sometimes used interchangeably, the distinction is important; the take-up rate in a particular territory is generally lower than the penetration rate, because higher-value properties are more likely to be insured.

Other definitions of penetration rates (such as the ratio of total insured property premium to the total value of GDP in a country) are sometimes used. However, such definitions do not apply in this context.

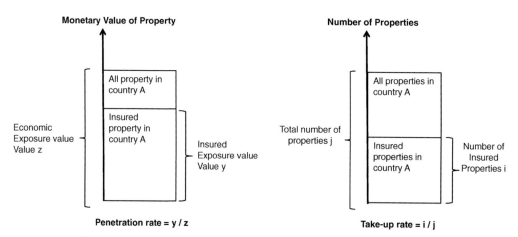

**Figure 4.20** Definition of penetration rates and take-up rates for catastrophe modelling. *Source*: Reproduced with permission of RMS.

---

**Box 4.1 Sources of Global Exposure Data**

Increasingly, global databases and taxonomies for exposure data (most particularly residential building stock or population) are becoming available, based on disaster risk assessment initiatives such as the Global Earthquake Model (GEM). These datasets and methods can provide useful, albeit possibly peril-specific data, which may be used either as a primary data source, or as a validation source. In many cases, these databases will have been developed from varying primary data, i.e. that collected for a specific use and applied in its original form, for example, a Census field, and secondary data, i.e. that derived from one or more sources, for example, a land cover database based on multiple data sets and will apply international taxonomies such as PAGER (Wald *et al.*, 2011), in their formulation.

Currently, many global datasets have been produced to fulfil DRR assessment and modelling requirements, and are therefore most commonly applied to residential and population exposures (Table 4.7). A methodology for exposure data development by Gunasekera *et al.* (2015) has highlighted the challenges associated with effective construction of national and regional exposure models for hazard and risk assessment. DRR-related databases, specific to commercial, industrial, or infrastructure assets, are less commonly available but may be developed in the near future. Future databases are likely to take greater advantage of digital inventories derived from satellite remote sensing data, such as that collected from high-resolution optical sensor platforms. For non-building/infrastructural exposures, including marine hull, construction and cargo exposures, the creation of exposure databases is more problematic, particularly in respect to modelling temporally and spatially varying values at risk (see also Chapter 2.5.4.3). Most currently available catastrophe models are primarily calibrated towards static building inventories.

Table 4.7 List of example global exposure datasets and components (accessibility and commercial use rights will vary).

| Data source | Source/controller | Peril focus | Exposure class/ classes | Resolution/ geography |
|---|---|---|---|---|
| LandScan | Oak Ridge National Laboratories | None | Population/urban proxy | 1 km grid |
| GED4GEM | GEM | Earthquake | Buildings stock (disaggregated) | Varies |
| GED-13 | UNISDR | Earthquake/ Cyclone | Residential Building stock | 5 km grid |
| World Population | GeoData Institute, Southampton University | None | Population | 1 km grid |

### 4.4.2 Economic and Insurance Industry Exposure Database Development Approaches

Catastrophe modelling entities use one or more of the three approaches described here to develop insurance industry and economic exposure databases:

- bottom-up
- top-down
- insurance-aggregate.

Table 4.8  Industry exposure database development approaches.

| Approach | Description | Advantages | Disadvantages |
|---|---|---|---|
| Bottom-up | Evaluates location-level building characteristics for every insurable property and aggregates by geographical region | Detailed, accurate<br>High spatial resolution<br>Uses independent non-insurance data | Time-consuming<br>Data availability and consistency issues, e.g. between adjacent geographies<br>Inconsistent data vintages<br>Requires take-up rate and penetration rate assumptions to derive an insured subset of economic exposure |
| Top-down | Evaluates country-wide exposures and distributes to geographical regions with distribution factors | Quick<br>Easily scalable | Coarse, inaccurate<br>Requires take-up rate and penetration rate assumptions to derive an insured subset of economic exposure |
| Insurance-specific | Collects data from a proportion of market participants and, assuming that they are representative, scales values up to 100% using market share | Accurate market view | Data mapping and data coherence, e.g. double counting of policies<br>Market share by premium different to sum insured<br>Differences between companies in rebuild cost models<br>Requires take-up rate and penetration rate assumptions to derive economic exposure from insured exposure estimate |

All these approaches have their own strengths and weaknesses. Table 4.8 describes each method and outlines its advantages and disadvantages. Given the constraints on time and data, in practice, the final result is often formed by iterations that evaluate and compare the results of all three approaches.

The *bottom-up approach* is based on collecting data at the highest possible geographic resolution. Model developers make assumptions at the finest possible resolution and develop exposure estimates at this resolution, then **aggregate** exposures to postcode, county and state/region resolution.

The *top-down approach* estimates insured exposure value at the national level and then distributes it down to higher resolution geographic frameworks, for example, county or postcode levels. This **disaggregation** process uses income, population, number of workers, numbers of commercial/industrial establishments or any other socio-economic attributes that have a relationship with the asset of interest (usually buildings and structures).

The *insurance-specific approach* attempts to create an accurate market view directly from insurance data, with catastrophe modelling organizations engaging with individual insurance companies to gain access to their data. However, the character of the assets included will often be skewed towards the exposure of the participating companies, and the resolution of the available data depends on how willing participating companies are to disclose details of their exposure data to a third party, and for that data to be used in the creation of a more widely shared dataset. Other disadvantages include the potential for double-counting policies in cases where one large risk (e.g. an industrial facility) is covered by multiple insurers, each covering only part of the same risk. Depending on the granularity of the data shared, this double-counting can be masked by coarse resolution and aggregation.

Both the bottom-up and top-down approaches require the model developer to apply take-up rate and penetration rate assumptions to derive insured exposure estimates, while the insurance-specific approach requires such assumptions to derive economic exposure estimates.

The most granular of these three approaches, bottom-up exposure development, is discussed in more detail in Section 4.4.3.

### 4.4.3 Bottom-Up Industry Exposure Database Development

Building a bottom-up industry-exposure database can be broadly defined as a development process that is based on comprehensive fine-scale data that describe the size and value of all elements at risk. In a series of steps, these fine-scale data are then aggregated to create a database that describes the totality of elements at risk at various scales that can be benchmarked against independent data for verification.

Figure 4.21 illustrates how this process can be applied to the development of an insurance-industry exposure database that covers all the insured properties in a particular territory. The model developer takes datasets that describe the number, size and value of properties at some fine geographical resolution (such as census-tract, postal code, or even individual building resolution). The economic structural exposure in a particular high-resolution geographical area can be calculated by multiplying, across all particular building types, the number of buildings by the total floor area of such buildings and the replacement cost value (per unit area) of such buildings. This economic structural exposure value can then be converted to an insured value by applying various insurance-industry-related assumptions, e.g. regarding the mix of **coverages**, take-up rate assumptions, penetration rates, limits and deductibles for a particular peril.

#### 4.4.3.1 Challenges in Bottom-Up Exposure Modelling and the Role of Trending Models

There are many challenges inherent in creating a coherent industry exposure database. The process involves marrying inconsistently defined secondary datasets of independent demographic, economic, remote-sensing and insurance-company data, none of which are likely to have been designed for this specific purpose. In addition, although the model developer aims to

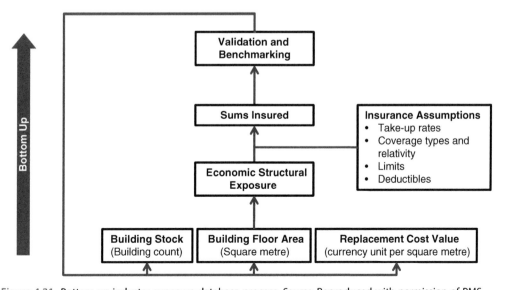

Figure 4.21 Bottom-up industry exposure database process. *Source*: Reproduced with permission of RMS.

provide a snapshot of the insured exposure at some date close to the present day, the data on which this snapshot is based may itself be combined from a wide variety of providers and underlying data vintages.

Model developers resolve these types of data vintage issues by using a variety of **trending** models, based on widely published econometric values, and by performing validation and benchmarking against economic and insurance-industry data.

For example, if the building stock is 2007 vintage, model developers may use government statistics on construction permits, taxation or construction market activity to trend the high-resolution data building-count and floor-area data for different categories of property forward to the present day. Replacement cost values per square metre can be trended by reference to standard construction cost handbooks or information services. In countries where the government and construction-industry data are sparse, modellers typically employ assumptions based on relating such indices to benchmarks such as per-capita GDP and its components (e.g. gross value added). Fluctuations between benchmarking currencies add another dimension of complexity to this issue.

### 4.4.3.2  Case Study: Developing a European Insurance Industry Exposure Database

While exposure databases are valuable components of the modelling process, they are also models of the real world conditions, and so like all sub-models include considerable parameterization and generalization. In many cases, their use will be limited to a particular type of asset class (e.g. residential housing, population) or for a specific region of country.

Thomson and Page (2011) present an example of an industry exposure database development, intended for use in windstorm loss modelling for Europe. As the authors note, this region presents a diverse and heterogeneous range of economies, languages, policy types, insurance penetration and data availability that poses particular challenges in creating a homogenized view of insured exposure.

The first step in the process involved creating a Europe-wide economic exposure database, based on bottom-up assumptions regarding the building stock, floor area and replacement cost value. No single European agency gathers all this information. This case study therefore illustrates the data gathering and assimilation process with examples for individual countries.

The authors gathered building stock data from a variety of sources including national statistical offices and third party market research companies. For example, for residential lines building stock in Germany, GfK GeoMarketing publishes data on number of households, living space, business, population and purchasing power by postal code, while the government statistical office DESTATIS publishes lower-resolution data on vacant and secondary homes and on dwelling type (Figure 4.22).

Similarly, to estimate average floor area for each property type, the authors compiled data from many country-specific sources. For example, in France, the national statistical office INSEE

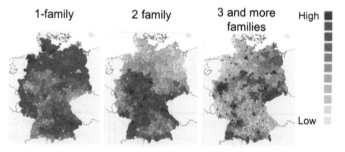

**Figure 4.22** Proportion of dwellings by dwelling type (number of families) in Germany (based on data from DESTATIS and GfK GeoMarketing). *Source*: Reproduced with permission of Geomarketing GfK.

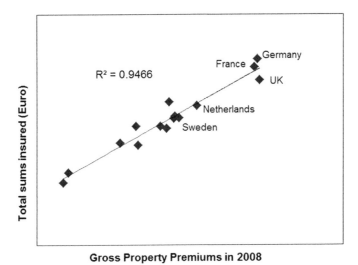

**Gross Property Premiums in 2008**

Figure 4.23 Gross Property Premium versus Total Sum Insured, showing a very close relationship between the two. *Source*: Reproduced with permission of RMS.

published the SITADEL database which included information on new constructed dwellings by number of rooms per department between 1994 and 2007. Additional INSEE data included the average surface area per person by type of tenure.

To estimate the replacement cost value per square metre of floor area, the authors gathered information from a very wide selection of different sources, such as the U.K. Building Cost Information Service, the French Éditions Callons publications, and in Germany entities such as BKI Baukosten, who provide information on building costs per square metre for different types of construction. For some countries, no country-specific information could be sourced, and the rebuild costs per square metre were defined by comparison to other territories. Sometimes the surface areas in these publications are defined differently from the surface areas derived for the building stock, providing additional issues of interpretation.

The model developers needed to create this database at two resolutions: on a grid, suitable for use in model development applications, and on a postal code system, suitable for release onto the market. Postal code resolution data were available for some parameters in some territories. Elsewhere, additional assumptions were required to downscale the data, for example, based on population density, purchasing power per capita, and remote sensing data.

The next step in the process was to adapt the economic exposure data to represent insured exposures for typical policies written.

To account for the many assumptions inherent in this process, the model developers compared the resulting insured exposure databases to a variety of independent metrics, such as gross property premiums, per capita GDP, purchasing power and client insured exposure datasets, to validate the outcome (e.g. see Figure 4.23).

## 4.5 Vulnerability

Vulnerability model development involves quantifying the relationship between the *severity of the hazard at a given location and the resulting loss* (e.g. Rossetto *et al.*, 2013). Vulnerability model developers relate the hazard and the damage via a metric of impact severity, such as

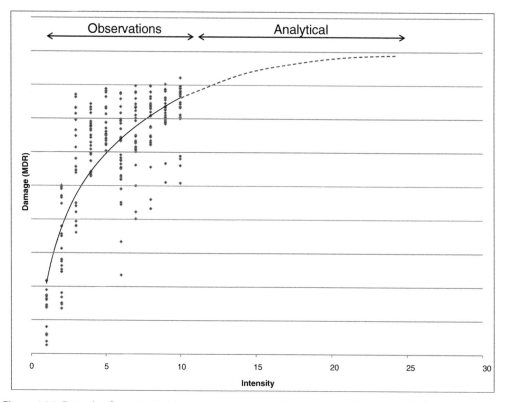

Figure 4.24 Example of a typical hybrid catastrophe vulnerability curve (see Section 4.5.1), illustrating the difference between **empirically** derived (observed) damage/loss (dots and solid line) intensity curves and extrapolation using **analytical** methods (dashed line).

**damage ratios**, also known as damage factors to convert the hazard intensity to a monetized insured loss potential (see Figure 4.24 and Chapter 1.8.2). Vulnerability developers tend to be specialists from the civil engineering, geography, economics and statistical disciplines.

For (re)insurance applications, vulnerability modelling traditionally first considers structural risk, i.e., the structural response of buildings, large facilities, and infrastructure to physical forces such as wind loading, water action and shaking. As well as purely structural considerations, vulnerability modelling then also considers the damage to items housed within the structure ('contents') as well as the consequential loss of profits or income from downtime due to the damage ('business interruption') (see Chapter 1.9.1.2). In addition, some models also deal with other forms of assets, including agricultural (crop, livestock), transportation (automobiles, marine cargo), and population (termed 'casualty' risk in insurance) (see Section 4.5.4).

Regardless of the precise details of the method chosen, by defining the vulnerability as a relationship between value and payout the model defines it in such a way that it can be passed into financial model calculations that express both the expected loss and the associated uncertainty for a portfolio of properties at risk.

Catastrophe models commonly include a range of vulnerability functions, characterized by the exposure primary modifiers (see Chapter 1.9.1.3), enabling model users to select the most appropriate combinations of vulnerability criteria to represent particular asset types. As a result, it is important that model users appreciate the applicability of a particular vulnerability function choice in relation to the asset being modelled, especially when the data available is not specific to that asset. Equally, vulnerability functions are key sources of model uncertainty due to both the aleatory and epistemic considerations of constructing functions to represent the range of possible asset types and damage states (see Section 4.5.4). Capturing this uncertainty within the loss estimation process is one of the primary challenges facing catastrophe modelling. However, for practical reasons (e.g. lack of data), models do not always quantify the full range of epistemic and aleatory uncertainty, even though these uncertainties are most likely to be a large driver of the overall model uncertainty (see Section 4.7.2.1).

### 4.5.1 Vulnerability Function Development

A typical insurance portfolio most likely includes a wide range of exposure types. To characterize the loss potential for such a diversity of exposures, developers need to achieve a level of confidence in the functional relationships between hazard intensity and loss by reviewing a wide range of data sources, and by using a wide range of methods in their construction in the same way as they do when constructing hazard model components. Thus, the same concepts of interpolation, extrapolation, calibration and validation are critical aspects of the vulnerability construction process. Factors such as the relevance, quality and extent of data available to represent, the regional characteristics of asset types, and the peril being considered, all influence the choice of approach in developing vulnerability functions. In many cases, as with hazard component development, developers use data and information to construct the specific components of vulnerability functions sourced from external organizations and other non-catastrophe model projects. Such data sources include catastrophe model end users (e.g. insurance claims and exposure data), information gathered as part of national and international disaster risk resilience initiatives, and engineering data gathered during building code development (see Section 4.5.7). The convention in vulnerability development is to use mean damage ratios by intensity, calculate the mean and standard deviation, and fit distributions to these. A very few cat models are emerging which allow fully histogramic distributions (see for example, the OASIS Financial Module White paper, OASIS, 2015).

In this way, catastrophe model vulnerability functions can be based on engineering studies and concepts, complemented by economic assessments with contributions often provided by peril-specific specialists. The resulting vulnerability functions have an associated uncertainty distribution, which represents the range of damage ratios around the central tendency, usually the mean (or median in older studies), driven by data, sampling and other uncertainties (see Section 4.5.1.1). Given the engineering-based source of many vulnerability functions, it is important to differentiate between the inter-related, but functionally separate concepts of fragility functions, and damage/vulnerability functions.

**Fragility functions** represent the probability that a building component or a whole building will reach or exceed some specified damage state (e.g. slight damage, collapse) as a function of the hazard intensity to which the component or building is subjected (as defined by the GEM project).

Damage/vulnerability functions represent the consequences of damage, often financial loss, from the physical damage experienced. See. for example, Yeo and Cornell (2005).

The Global Earthquake Model (GEM) has also produced a review of seismic vulnerability (available on the GEM website www.globalquakemodel.org), describing the various approaches taken in the construction of both fragility and vulnerability functions. These approaches tend to fall into three general types and are applicable to most perils:

- **Empirical** – usually developers construct empirical vulnerability functions from post-event (historical) databases, and often through regression analysis. In addition, they commonly use ground and aerial survey data. The results may be specific to the types of assets and structures affected. Most model development teams also perform post-event surveys where possible as input to the construction and calibration stages of vulnerability model development.
- **Analytical** – developers construct analytical models, which use computational or physical models, to simulate the expected hazard intensity and to measure the expected structural

---

**Box 4.2 Example of a Damage Function**

---

Developers use various methods to define the relationship between exposure and loss. One commonly used simple measure considers the relationship between the total ground-up repair cost for a structure at a given peril severity, excluding policy terms, and its ground up replacement cost (see Chapter 1.8.4) as shown in Equation (4.1).

In this example, the damage ratio DR $(i, j)$ for a single structure $(i)$ at peril severity $(j)$ relates the cost of repairing the damage to structure $(i)$ at a peril severity $(j)$ to the total cost of replacing structure $(i)$.

$$DR(i, j) = \frac{Cost\ of\ Repair\ (i, j)}{Total\ Replacement\ Cost\ Value\ (i)} \quad (4.1)$$

Extending this concept to a portfolio of (n) similar properties of type $(i)$, all experiencing the designated peril severity $(j)$, provides the following equation defining the mean damage ratio MDR$(i, j)$ for that proper type and peril severity:

$$MDR(i, j) = \frac{\sum {}^n Cost\ of\ Repair(i, j)}{\sum {}^n Total\ Replacement\ Cost\ Value\ (i)} \quad (4.2)$$

In reality, not all properties of type (i) that experience the peril severity (j) will experience the same damage ratio. Therefore, in addition to the mean damage ratio, developers define uncertainty distributions around the mean at a given peril severity.

---

response to physical forces. See D'Ayala *et al.* (2013) for an earthquake-specific review. Commonly applied methods, of varying complexity, include Capacity Spectrum (CSM – e.g. Minas *et al.*, 2015), and performance based earthquake engineering (PBEE) approaches (e.g. Porter *et al.*, 2007) for representing structural response to earthquake ground motion. Further, they might employ experimental data (e.g. from wind-tunnel or shake-table experiments) to measure the precise response of particular building components to certain physical forces.

- **Expert Opinion** – vulnerability functions may require non-analytical or empirical inputs, based on the knowledge of thematic experts. Elicitation methods (see Aspinall and Cooke, 1998) have been employed in many cases, including in support of the ATC-13 design codes (see Section 4.5.4).

Catastrophe modelling teams will often develop or apply **hybrid** models (Figure 4.24) employing a combination of all three methods, to enable both interpolation and extrapolation of functions to represent the expected range of structural responses to varying intensity, and to represent various forms of structural types not necessarily represented from one source. In many cases, independent research projects, such as HAZUS and ATC, will have applied hybrid approaches, and model developers will then adapt and integrate these data sources into their own models.

Pita *et al.* (2012) provide a summary of the history and various forms of vulnerability estimation methods employed for hurricane risk modelling and note that these have generally been based on a range of empirical, insurance loss data, or simulation based curves, taking engineering approaches, or a hybrid approach using both loss data and simulation. In many cases, expert elicitation methods have also been employed to incorporate varying viewpoints or knowledge into the complete functions.

Model developers have several options for choosing how to fit the shape of the vulnerability function and associated uncertainty distribution. For example, they can use **one-step** empirical, statistical relationships between observed hazard impacts and losses, through a **two-step** construction process that considers the chance of having a loss (step 1) together with the conditional vulnerability of a loss (step 2) based on the integration of fragility functions to damage-loss relationships (see, e.g. D'Ayala *et al.*, 2013) and Section 4.5.1.2 for an example or by calculating cumulative distribution functions (OASIS, 2015). In addition, Porter (2016) provides a comprehensive explanation of the mathematical and statistical representation of fragility and vulnerability functions, primarily considering earthquake examples.

### 4.5.1.1 Secondary Uncertainty in Vulnerability Functions

As noted in Section 4.5, the mean damage estimate is only a part of the representation of the vulnerability and functions, however constructed, will incorporate various sources of aleatory and epistemic uncertainty. In the following image (Figure 4.25), buildings are of equal height and similar construction but, from a structural point of view, the buildings experience different levels of damage although we would expect that the buildings have been exposed to a very similar (if not identical) hazard intensity. Vulnerability functions represent this variability as one element of **secondary uncertainty** (see Chapter 2.16.1).

Secondary uncertainty is the uncertainty associated with the damage to physical risks, locations, and facilities should a given event occur. It incorporates several different sources:

Model uncertainty is a source of secondary uncertainty. There is uncertainty in the local intensity (e.g. ground motion or wind speed) of a particular event at a given location. Depending on the underlying assumptions, parameters, and data used, different equations, that is, alternative models, for calculating local intensity are possible.

Translating local intensity to building performance is another source of secondary uncertainty in the model construction process. Since actual damage data are scarce, especially for the most severe events, statistical techniques alone are inadequate for estimating building performance. As a result, modelling agencies construct damage functions based on a combination of historical data, engineering analyses, claims data, post-disaster surveys, and information on the evolution of building codes. The variability of damage for a building of a particular typology is represented by a probability distribution, which represents the intrinsic uncertainty in the estimation of both local intensity and damage.

Catastrophe models must account for such variability, which is represented by a probability distribution of damage for a building given that an event has occurred. The type of probability

Figure 4.25 Variation in damage for buildings in close proximity after Maule 2010 earthquake in Chile. *Source*: Reproduced with permission of AIR Worldwide Corporation.

distribution used varies according to the type of modelled peril and the level of intensity experienced by the structure.

Sometimes a beta distribution (e.g. Sampson *et al.*, 2014, Chapter 1.11.2.1) is chosen for convenience since it is bounded between zero and one and is very flexible. However, other distributions are also becoming increasingly prevalent. For example, some developers are beginning to advocate using a four-parameter distribution (Shome *et al.*, 2012) to capture the full range of possibilities including zero or total loss. In addition, model developers are looking towards non-parametric representations of uncertainty i.e. the ability to sample from a discrete distribution (see Chapter 1.11). Log-normal is commonly applied for earthquake vulnerability distribution modelling, although other parametric and non-parametric approaches can also be used. The overall goodness of fit achieved will vary greatly dependent on the quality of data provided, and the level of parameter uncertainty in the original data and the underlying assumptions (see, e.g., Rossetto *et al.*, 2013 for an overview).

The representation of this uncertainty is very important as it will be critical in the interaction with the financial model and become particularly important in a realistic representation of the application of financial structures (see Section 4.7.2).

The distribution of secondary uncertainty can also be represented by the coefficient of variation around the mean (or in some, often earlier empirical studies, the median), fed into the financial model that calculates losses for a portfolio of exposures (see Chapter 1.11).

#### 4.5.1.2 Case Study: Conditional Vulnerability and Chance of Loss Estimation

This example explores the concept of one-step versus two-step vulnerability functions and is illustrated with an example related to the development of wind vulnerability functions.

- Assume that during a recent windstorm, a particular postcode area was affected by the event. The total loss recorded was $100 and the total value of the sums insured was $400. This gives a damage ratio of 25%.
- A year later, an identical event strikes the same houses in the same area, but this time better resolution loss data is collected. Surprisingly, out of four virtually identical insured locations, two of them exhibit 50% loss ratios while the remaining two experience no loss at all. In other words there is a 50% Chance Of Loss (COL) since two out of four locations have a non-zero loss. This gives the same overall damage ratio of 25% as the year before.

In reality, a wide range of losses will be experienced, ranging from zero to some limit depending on the severity of the wind. This windstorm loss example shows a common feature that occurs when not every insured location in a storm-stricken area reports a loss. There are many variables that may contribute to the manifestation of this effect.

The consequences of using a 25% loss ratio for all buildings versus a 50% loss ratio for 50% of the buildings and a 0% loss ratio for the other 50% may range from being negligible to very serious. The severity of the final effect depends upon how significant the originating factors are and upon the dimensions of the original insurance conditions. Ignoring the originating factors, if high deductibles or low limits are present, this effect will be magnified. If neither is present, there will be no difference in the losses.

Consider the following:

- Four insured locations each with a value of $100.
- Two of the insured locations have a 10% limit ($10) while the other two have no limit (i.e. limit is equal to the insured value of $100).

There are two ways of calculating the loss, either using a one-step approach (loss ratio = 25% for each location) or using a two-step approach (chance of loss = 50%, loss ratio = 50%).

Table 4.9 Comparison of the loss calculation on a one-step and two-step basis.

| Method | Ground Up Loss | Gross Loss |
|--------|----------------|------------|
| One step | $100 | $70 |
| Two step | $100 | $60 |

When the losses are calculated using a one-step basis, the loss for each location is calculated as $25\% \times \$100 = \$25$, leading to a total ground-up loss (before the application of limits and deductibles) of $100. The gross loss (after application of limits and deductibles) will be $25 for the two locations with no limit and $10 for the locations with a 10% limit, leading to an overall gross loss of $70.

When losses are calculated using a two-step basis the conditional (i.e. *given that* a loss has occurred) ground-up loss for each insured location that is impacted will be $50, leading to an overall ground up loss of $50 \times 4 \times 50\% = \$100$ (representing the four locations each with a 50% chance of loss). The gross loss will be $50 for the two locations with no limit and $10 for the two locations with a 10% limit which leads to an overall gross loss of $50\% \times (\$50 + \$50 + \$10 + \$10) = \$60$, which is 14% less than that calculated using the one-step method. Table 4.9 summarizes the losses.

So in order to handle the loss calculation correctly, the concept of conditional vulnerability and chance of loss emerges based on the fact that even if a particular risk is within the storm-stricken area, a loss does not necessarily have to happen. The probability of incurring a loss for a certain type of risk and wind intensity is defined as chance of loss. A scheme of how the conditional vulnerability component works is shown in Figure 4.26. Furthermore, the damage ratio (50%) used after the application of COL (50%) is only applicable (conditionally) on the properties that are defined to suffer a hit (two out of four) by COL. So we can refer to it as conditional mean damage ratio (CMDR). Numerically, multiplication between COL and CMDR gives the MDR (in this example, $50\% \times 50\% = 25\%$).

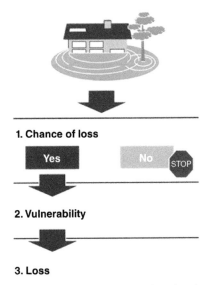

**1. Chance of loss**

Yes — No STOP

**2. Vulnerability**

**3. Loss**

Figure 4.26 Schematic of the application of COL and CMDR. COL describes the probability of a building incurring a loss at a given wind speed. The CMDR is applied on the condition that the building incurs a loss. Courtesy of Impact Forecasting.

### 4.5.2 Empirical Vulnerability Approaches

Most catastrophe vulnerability functions will include some element of an empirical approach, given the need to relate hazard to loss, and the use of post-event field studies as sources for damage information. This will often rely on the availability of claims data at suitable geographical levels.

As a result, geographical accuracy is a key uncertainty in interpreting claims data to develop empirical vulnerability functions for perils with steep hazard gradients. For instance, many perils, such as flood, are heterogeneous over a length scale that is short compared to the precision of locating properties at risk. Properties flooded to depths of several metres may be located only tens of metres from un-flooded properties. In such cases, knowing the location of a property to within even a postal code or street segment may not allow the flood depth to be determined with any degree of confidence. Coupled with uncertainties in evaluating flood depth from historical footprints, this sensitivity means that creating empirical flood vulnerability functions is particularly challenging. In the case of flood, guidance and partnerships with bodies such as the United States Army Corps of Engineers and Middlesex University's Flood Hazard Research Centre are commonly used in developing flood vulnerability functions.

It is also necessary to define a typology or inventory classification as part of the empirical vulnerability development process, since various physical parameters such as construction class, building code and enforcement or building height are known to strongly influence fragility and hence vulnerability.

To develop an empirical vulnerability function from a historical event, the hazard footprint must be related to the precise location of the property so the intensity (e.g. water depth, peak gust wind speed, spectral acceleration) of the severity metric at that location is known. The interpretation of severity in the context of vulnerability is therefore subject to all of the issues and uncertainties associated with interpolation. For example, the practical resolution of depth-damage flood vulnerability functions will be dependent on both the horizontal and vertical resolution of the flood hazard component (and related in large part to the resolution of the digital elevation model), as well as the accuracy of the data used to generate, and critically, calibrate, the damage function (Figure 4.27).

In the case of windstorms, catastrophe modellers typically relate a single measure of wind speed such as the peak (3-second) gust. Although the correlation is reasonably good between

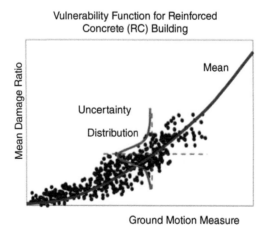

Figure 4.27 An illustration of empirical vulnerability development for earthquake, each point refers to a specific claim. The red line is the relationship between mean damage ratio and intensity. The blue is the hypothecated distribution around this mean. *Source*: Reproduced with permission of RMS.

wind damage and peak gust wind speed, there are other factors that also influence the amount of damage that occurs, including the length of time that the particular hazard is affecting the assets. Although this is often referred to as *duration*, this naming is somewhat of a misnomer, because in reality these factors relate to the sheer number of repeated changes in wind pressure that impact the property. Typically, the available observation data do not pick up the full spectrum of gusts experienced throughout a storm event, and therefore such subtleties are not yet directly included in models. Instead, modellers may attempt to represent them by adding a duration or gust factor.

Where model developers rely on historical loss data to construct empirical vulnerability functions, the resulting relationships are highly dependent on data availability, quality and interpretation.

For a variety of asset classes including fixed structures and moveable contents, catastrophe model developers typically use claims data to assist with estimating the repair cost at a particular peril severity where available. Repair cost can be compared with estimates of the total replacement cost value for the structure in order to obtain a MDR (Figure 4.27). The advantage of this approach is that it derives the vulnerability function from data that are typically captured by insurance companies from their policyholders. However, there are numerous issues associated with interpreting claims and exposure data from ceding insurance companies, which are explored later in the case study (Section 4.5.2.3).

For population vulnerability, developers also typically use empirical data to examine mortality and casualty statistics as a function of the exposed population. Typical data used for population vulnerability assessment include observed population data, death or injury statistics, and an appropriate intensity. For example, Wu *et al.* (2015) constructed empirical mortality curves for earthquake risk in China. In this case, the loss metric was number of deaths, measured against earthquake intensity. The authors applied a range of trend analysis approaches, including S-curves and logistic curves, to the same data. Two regional curves were developed, as well as a national curve to reflect differences in socio-economic development, affecting the overall vulnerability of populations to the hazard. Historic death counts for four earthquakes were used to test the accuracy of each curve and provide a final validation. Financial loss was estimated from a GDP per person calculation.

### 4.5.2.1 Empirical Vulnerability Functions: Property Vulnerability

As mentioned at the beginning of Section 4.5, claims data for assessing property vulnerability are typically more widely available than for the other asset classes. However, issues of heterogeneity between territories, companies and perils can make interpreting property claims data challenging. In addition, the granularity of data is often insufficient to understand the impact of specific taxonomies on the potential damage from an event. Further, the range of historical severities experienced may not extend towards the possible extremes the hazard model reaches. Therefore, developers also use analytical approaches and expert judgment to understand and refine the vulnerability relationships.

Various characteristics of exposed assets can influence their vulnerability to catastrophe risk. For this reason, developers construct asset type taxonomies to differentiate vulnerability within a general catastrophe model. These asset type taxonomies relate directly to the way that a model user can code relevant location-specific details, such as the construction class of a structure, and its occupancy. They also provide users with opportunities to modify the degree of loss at a given level of hazard severity. Therefore, vulnerability development requires the model developers to create damage curves for a range of such property modifiers. A 'family' of vulnerability functions are produced across the asset type taxonomy that consolidate as many of the combinations of modifiers to asset types as possible. In some cases, where empirical or analytical functions cannot be constructed for a particular asset type, for example, due to lack of data for

Table 4.10  List of common property vulnerability modifiers for each peril.

| Modifier/Peril | Flood | Earthquake | Wind |
| --- | --- | --- | --- |
| Occupancy (e.g. ATC class) | X | X | X |
| Construction class (e.g. reinforced concrete, wood frame)- often referring to the primary structural system (lateral load bearing system). | X | X | X |
| Building Height (number of storeys | X | X | X |
| Building Age (e.g. relative to recent wind or snow loading codes that determine building regulations) | X | X | X |
| Presence of basement | X | - | - |
| Roof type (e.g. flat roof, tiled roof) | - | - | X |

construction or validation, a 'best match' function, often generic amalgamations of broader asset or insured line of business, may be employed.

Some modifiers, such as occupancy type, influence damage from all perils, whereas others, such as the presence or absence of basement, are of greatest relevance to a specific peril (flood, in this case). A list of common property modifiers and their importance for different perils is given in Table 4.10. In addition to the modifiers listed, developers are increasingly providing floor area, which has a direct link to exposure, as a modifier.

Typically, developers will create separate vulnerability for each combination of modifier values. For example, a vulnerability curve may represent the potential loss by hazard level for the buildings coverage within a two-storey residential house built between 1980 and 1990. The large number of combinations of different modifier values gives rise to a large number of potential vulnerability curves. This means that there is unlikely to be sufficient claims data spanning the full range of hazard values for all the different combinations of modifier values, and is a key reason why a combination of claims-based and engineering-based approaches are used.

It is common for insurance companies not to have all of the information in Table 4.10; in which case they will be supplying the model with exposure information that does not completely define a unique vulnerability curve for the model to select. Many models make use of an underlying insurance exposure database (IED; see Section 4.4) or building inventory in order to weight together the different unique vulnerability curves into a composite vulnerability curve, which is selected in the absence of exposure information. For example, if the building age is unknown, the model would assume a composite vulnerability curve made up by weighting the unique building age vulnerability curves together, with weights based on the building inventory. These weights could vary by region as the underlying building stock (or building inventory) varies by region. Discrete types of region (from the perspective of building inventory) are often known as inventory regions. These are different to vulnerability regions. See Chapter 5.4.2.5 for further discussion of inventory and vulnerability regions. Some models also increase the uncertainty when composite curves are used, to reflect the fact that an assumption is being made.

In some cases, models provide further options for introducing granularity (i.e. detail in calculations) that users can apply to modulate loss estimates. These options, sometimes termed 'secondary modifiers', include specific non-structural characteristics that influence loss, such as roof type, roof geometry, or the management approach taken by the asset owner or community.

In addition to differentiating vulnerability according to the primary and secondary characteristics of an asset, vulnerability functions also differentiate losses between the types of insurance cover provided. Thus, models provide vulnerability functions that express: structural damage ('buildings' vulnerability – see Chapter 1.9.1); damage to the materials and assets contained within or around the structure ('contents' and 'appurtenant structure' vulnerability – see Chapter 1.9.1); and the additional loss attributable to the inability to continue normal use or operation within the structure (the 'business interruption' or 'time element' vulnerability – see Chapters 1.9.1 and Section 4.5.6). Model developers typically apply empirically-derived multipliers of the structural loss to estimate the relationships between these loss types. Splits of exposed values attributed to buildings, contents and business interruption (sometimes referred to as 'coverages', as noted below – see Chapter 1.8.4) are therefore required to enable correct calculation of loss from the combined property vulnerability function. The relative vulnerability of fixed versus movable contents can be key loss modifier, particularly in respect to flood damages.

Calibrating the mean damage ratio and coefficient of variation across all appropriate peril severities for a particular peril and region requires extensive exposure and loss data, across a range of peril severities. However, in practice, although insurance catastrophe modellers typically have access to some insurance company loss data, the available data may cover only a limited range of peril severities, or may be at insufficiently high resolution to relate to the appropriate historical peril severity.

### 4.5.2.2 Constructing Empirical Property Vulnerability Functions from Per-Location Claims Data

Catastrophe model developers combine insurance claims data, when available, with engineering expertise to construct vulnerability functions calibrated to insurance loss potential. The following examples illustrate some of the key considerations in deriving empirical vulnerability functions from claims data. This process involves relating local historical information about event severity, across a wide range of such severities, to appropriate information about the value of a claim relative to the rebuild cost of the property. In practice, the process requires interpreting a wide variety of data sources including data related to the peril itself, as well as the claims data and associated exposures at the time of the claim.

One of the main challenges is to reconstruct the 'from-ground-up' losses from the claims with insurance and reinsurance conditions including the assumption that the total sum insured equals the building *replacement cost* (see Chapter 1.8.2).

In general, developers construct empirical model vulnerability functions from detailed claims data, if available. The overall appropriateness of the vulnerability function for assessing risk to a particular asset class depends on the representativeness of these data to that asset class in that region. Although detailed data are preferred, nevertheless developers can also use aggregate claims information, (e.g. total losses and exposures at postal code level), for example, as a calibration data set across a particular region.

Although the value of the claim provides a good first estimate of the damage value, this value may itself be impacted by financial structures such as deductibles. In addition, it may include losses from additional coverages such as contents and business interruption, as well as from non-modelled loss and additional expenses, such as claims adjustor's expenses. Developers typically consider and account for these factors in dialog with the entity providing the data, as they impact the calibration. In many cases, this process requires considerable expert judgement.

The claims data are also unlikely to provide the full range of data required to build a vulnerability function. This data can only be derived from recent catastrophe events, which by their nature are rare. There are three main extrapolations from the empirical data:

- Extrapolate to higher-severity MDRs than for which claims data are available.
- Extrapolate to other countries, e.g. claims data may be available for Germany but not Poland.
- Extrapolate to other lines of business/construction or asset types, e.g. claims data are available for wood frame single-story residential properties but not masonry apartment blocks.

An additional challenge is to relate the claim to a replacement cost where this information may not be linked in the database. Often the exposure and claim information may come from different systems and linking this information coherently can be a challenge, for instance the policy numbers may be recorded differently. There are also issues in deriving the appropriate replacement cost: rather than total replacement cost estimates, the insurance company may report the insured fire limit for certain policy types. Again these issues must be carefully considered by the model developer who will need to make a judgement.

Once both the claims and replacement cost data are cleaned, the next step is to collate these data from similar properties to build a picture of the variability of the relationship between the peril severity and the MDR. This stage also encounters issues of claims data interpretation:

- Although the ideal situation would involve constructing different curves that represent different relevant structural characteristics, in practice this information may not be consistently captured within the exposure and claims data.
- The more classes the data is divided into, the smaller the sample size.

Khanduri and Morrow (2003) present a practical general framework is widely adopted in the catastrophe modelling community for overcoming this last difficulty. They recommend an approach that involves creating generic empirical vulnerability functions from claims data. These general vulnerability functions can then be split into vulnerability functions for buildings with specific structural characteristics. The relationship between the general and specific functions is based on engineering judgment, information about the typical inventory of the building stock in the region under study, and the relativity of claims between building classes in neighbouring regions or among similar occupancy classes.

In a recent empirical exercise to test the effectiveness of the FBC (Florida Building Commission) in reducing insured losses in Florida, Simmons *et al.* (2016) analysed Insurance Services Office (ISO) insurance data aggregated to ZIP (postal) code from the period 2001–2010, totalling \$517Bn in claims amounts (2010 values) and nearly 32,000 claims. They developed a regression model from 19 descriptive parameters, characterizing the insurance take up (premium), the number of claims, split by construction type and year, and the proportion of properties constructed after the adoption of the Florida Building Commission (2002). In addition, they used a number of hazard parameters, but reflecting coarse scale hurricane intensity (32 km grid) – distance to coast, maximum wind speed, the frequency of mean speed exceedance over 12 hours, and weightings for category 3 or category 4 storms. Their results suggested that implementation of the codes in Florida have reduced residential losses by as much as 72%.

### 4.5.2.3 Case Study: European Windstorm Vulnerability

The difference between one-step and two-step vulnerability functions was discussed in Section 4.5.1.2 where the concepts of chance of loss and conditional mean damage ratio were introduced. This case study extends this concept further, using the example of European windstorm.

In a 2015 Willis Re study, damage ratios are derived from claims incurred by UK insurers during storms Jeanette in 2002 and Kyrill in 2007. These claims are compared to event footprints from the XWS catalogue (Roberts *et al.*, 2014) to derive claims-based vulnerability functions. Two different horizontal grid resolutions are used: $4 \text{km}^2$ and $25 \text{km}^2$. Thee gusts wind speeds come from the UK Met Office Unified Model at 25 km resolution, which is generated by dynamically downscaling ERA Interim (Dee *et al.*, 2011) re-analysis data, rather than interpolated from point data. This is an example of the process of using models to supplement observations described in Section 4.3.3.5.

Figure 4.28 shows the relationship between MDR calculated using Equation (4.1) and peak gust wind speed for commercial exposures using the two different grids. Neither plot shows a particularly striking relationship between MDR and damage.

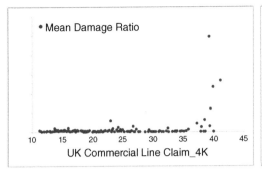
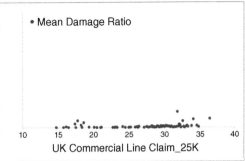

Figure 4.28 Relationship between MDR and 3-second peak gust wind speed (m/s) for commercial exposures using hazard intensity calculated at 4 km$^2$(left) and 25 km$^2$ (right) grid resolution. *Source*: Reproduced with permission of Willis Re.

Figure 4.29 shows a much clearer relationship between claims frequency and wind speed for these same commercial exposures. This analysis is repeated for personal exposures, and given the higher number of claims, also split by coverage in Figure 4.30.

There is, however, much less scatter around the best fit and a visually clearer relationship using the 4 km$^2$ grid.

Damage to homes tends to consist of tile loss, or broken windows so the costs of claims are low, usually the cost of material and a few hours' labour. Contents damage is typically from rain which permeates through the compromised structure. Claims are typically around £1,000 (compared to recent flood claims in the UK of £50,000), but with significant scatter when comparing the average claim by wind speed (see Figure 4.31).

Similar patterns are observed for both resolutions; there is no clear relationship between the peak gust wind speed and the observed average loss. This persists when the hazard is grouped into bins. Furthermore, if the assumed linear relationship between hazard and loss and sum insured held, a positive correlation between claim amount and sum insured is expected. Claims amount here must be the ground up claim amount, as described in Section 4.5.1, not the amount received by the policyholder.

The comparatively low wind speeds associated with European windstorm events may explain why the traditional one-step model is apparently less applicable to this peril and provide an indication whether these findings are applicable more widely.

Figure 4.32 shows the proportions of claims by wind speed for Kyrill, Jeanette and Daria. Kyrill and Jeanette were not particularly extreme events either in terms of loss or wind speed. However, even Daria in 1990[1], which is considered a more extreme event, had 85% of the

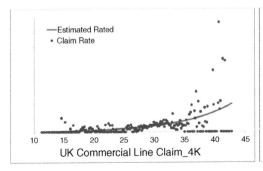
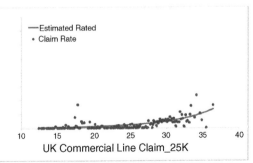

Figure 4.29 Relationship between claims frequency and 3-second peak gust wind speed (m/s) for commercial exposures using hazard intensity calculated at 4 km$^2$ (left) and 25 km$^2$ (right) grid resolution. *Source*: Reproduced with permission of Willis Re.

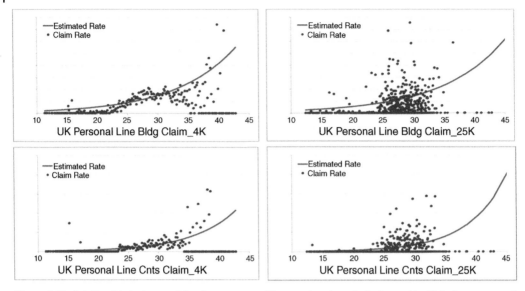

**Figure 4.30** Relationship between claims frequency and 3-second peak gust wind speed (m/s) for homeowners' exposure split by coverage using hazard intensity calculated at 4 km$^2$ (left) and 25 km$^2$ (right) grid resolution. *Source*: Reproduced with permission of Willis Re.

claims between 35 and 45 m/s. This may indicate that the more extreme loss events are not characterized by increasing wind speeds, but by an increased claims frequency and that two-step vulnerability functions are much better at representing this loss. This analysis would indicate that a good representation of these data would be a chance of loss that varies significantly by hazard intensity and an almost constant value for the conditional mean damage ratio.

This case study can be compared to the lessons learned from Superstorm Sandy analysis in Figure 5.8 where the mean damage ratio considering affected properties remains almost flat and the mean damage ratio considering all properties increases with peak gust wind speed. The latter metric includes the claims rate, so the same behaviour is seen in other claims data at similar wind speeds indicating that this may be more broadly applicable to low wind speed events.

This case study illustrates the relevance of two-step vulnerability functions and the importance and influence in the choice of the hazard footprint used in the development of an empirical claims-based vulnerability function.

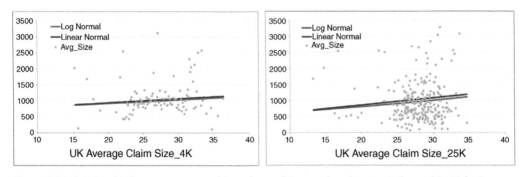

**Figure 4.31** Relationship between average claim value and 3-second peak gust wind speed (m/s) for homeowners' exposure split by coverage using hazard intensity calculated at 4 km$^2$ (left) and 25 km$^2$ (right) grid resolution. *Source*: Reproduced with permission of Willis Re.

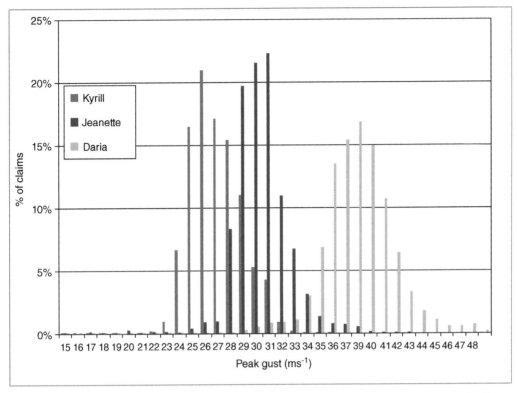

Figure 4.32 Distribution of claims count by peak gust wind speed. *Source*: Reproduced with permission of Willis Re.

### 4.5.3 Analytical Vulnerability Approaches

In contrast to empirical approaches above, analytical and simulation based methods for vulnerability function development, which have their basis in a mix of analytical and probabilistic approaches, examine the theoretical behaviour of specific asset typologies to the structural demands exerted by a particular peril.

Simulation models are based on Monte Carlo simulation of structural models, described either with finite elements, or load path models, or other similar techniques (e.g. Jain and Davidson, 2007). They have been increasingly applied at the individual structural or asset class level within engineering risk assessments, for example to estimate seismic vulnerability as part of the GEM project (Rossetto *et al.*, 2014), attempting to reduce epistemic uncertainties associated with the use of empirical claims and damage data. Another example of the various analytical approaches employed for earthquake structural vulnerability can be found in Causevic and Mitrovic (2010), which highlights the relative differences between non-linear static and dynamic methods, including Capacity Spectrum Method (CSM) and real-time spectral history methods. Both approaches are used in catastrophe vulnerability models.

In simulation models, the loss data are used for validation, not prediction as in empirical models. Expert opinion models exist, but again, in simulation models, expert opinion has a different role, to assist in validation.

Their relative alignment to insured risks, and by consequence uncertainty, will therefore be influenced by the granularity of data available, its geographic coverage as well as its relevance to each type of asset class, or line of business being considered. As with industry exposure database construction, vulnerability data are more readily available for static structural assets, particularly property, and in most cases, less available for non-static exposure.

The specific approach taken will vary with peril and type of risk, for instance, whether to a single, homogeneous structure (e.g. a simple residential dwelling), or a component-based (composite) structure or other asset (e.g. a petrochemical processing site).

One of the reasons that empirical derivation might be preferred is that it offers a potentially less biased ground-truth to analytical approaches such as numerical modelling. Such analytical approaches using, for example, non-linear time-history analysis are only as good as the assumptions fed in and while they can potentially capture a wider range of physical factors, they can also potentially be more biased than the empirical approach.

### 4.5.3.1 Case Study: Analytical Approach of Development of Vulnerability Functions

Since there is limited building damage data from earthquakes, model developers often use analytical approaches to develop vulnerability functions. Jayaram *et al.* (2012) propose the approach summarized in this case study. This approach uses a systematic simulation approach based on the Pacific Earthquake Engineering Research (PEER) Performance-Based Earthquake Engineering (PBEE) loss assessment framework to estimate the total building loss. It uses the building response from detailed nonlinear time-history analysis of buildings to estimate three random variables: (1) repair or demolition cost associated with a damage state; (2) damage state given a demand parameter (EDP); (3) and EDP given an intensity of ground motion. The NIBS & FEMA (2003) *HAZUS Manual* publishes details of these variables.

First, the developers assign building components to either the structural, or non-structural subsystem. The structural subsystem includes professionally engineered building components such as columns and load-bearing walls that resist gravity, earthquake, wind, and other types of loads. In contrast, the non-structural subsystems are not usually designed by engineers. In terms of earthquake vulnerability, non-structural subsystems are categorized as either drift-sensitive (e.g. exterior curtain walls) or acceleration-sensitive (e.g. suspended ceilings). The model development process uses different parameters to predict damage states, i.e. drift, for the structural and the drift-sensitive non-structural subsystems, and acceleration for the acceleration-sensitive non-structural subsystem.

Models predict the damage state for each subsystem based on the relevant fragility functions. HAZUS (2003) defines the following commonly applied damage states: none, minor (DS1), moderate (DS2), extensive (DS3) and complete damage (DS4). For example, moderate structural damage for steel-moment-frame buildings entails the yielding of some steel members, exhibiting observable permanent rotations at connections and major cracks of a few welded connections. In contrast, moderate damage of non-structural components includes damage such as large and extensive cracks of the partition walls, or as an example of moderate damage to acceleration-sensitive components, large numbers of suspended ceiling tiles falling.

The monetary loss from earthquakes reflects the accumulated costs of repair of different subsystems, demolition and collapse. The cost of repair works is generally significantly higher than the cost of new construction, and this difference is more significant for structural subsystems than the non-structural subsystems.

The authors in this study quantify ground-motion hazard using a vector of spectral accelerations. They then predict building response parameters such as storey drifts, floor accelerations and residual drifts under the specified hazard, estimate structural collapse and demolition, and calculate losses by floor based on the responses. The advantage of this method is that it allows the model developer to capture the effects of epistemic and aleatory uncertainties in the random variables, such as ground motions, structural response parameters, loss costs, etc. in order to quantify the overall uncertainty in the building loss results. The repair costs for different subsystems across different damage states are expected to be correlated because of common factors involved in repairs, such as the same contractor hired for the repairs and the same material used for repairs of different subsystems across different stories.

The analytical model then calculates the total loss by summing up the sample losses of different subsystems at each storey, and normalizing the resulting total by the building value to get the damage ratio. The final step in creating the analytical building vulnerability functions is to calculate the mean and coefficient of variation of the damage ratios by grouping data points according to the spectral acceleration vector. By estimating these values a number of times, for example, by first sampling the building parameters from the epistemic uncertainty, the model developer can then derive the overall epistemic uncertainty in building vulnerability.

### 4.5.4  Use of Design Codes in Vulnerability Function Development

As noted by Daniell *et al.* (2014) in relation to seismic risk, many countries have put in place design codes and associated legislation to control the methods and materials used in the construction of buildings and other structures to an appropriate level of resilience. By influencing the structural characteristics of buildings in a given area, design codes are a key development consideration for catastrophe model vulnerability components. International code standards (International Building Code 500-2014) do exist, as do those related to European design (for example, Eurocodes 1 through to 10 as applied in the European region: http:// eurocodes.jrc.ec.europa.eu) and enable formulation of guidelines for structural design for a variety of perils (see Table 4.11). However, most countries will have a range of codes, and will often interpret the generic guidance for their particular hazard and structural characteristics. So, for example, the United Kingdom's British Standards National Annex to Eurocode 1-4 which defines design loads for wind actions on structures (BSI, 2011), is modified to reflect national differences in wind parameters and detailed design wind speed zones.

The proportion of buildings influenced by a particular code varies significantly by the types of structures covered, by region, and also over time. In some countries, codes vary by region and within high risk areas. This variation can influence the geographic structure of model vulnerability regions, and the form of the options available to model users in determining particular vulnerability functions (see Figure 4.33).

Codes tend to be modified for a variety of reasons: in light of new scientific and engineering research findings, changes to building design and materials, and also in the aftermath of catastrophic events. For example, the Florida Building Code (FBC), which is based on the IBC (Simmons *et al.*, 2016), 'high velocity hurricane zone' relates specifically to the Miami-Dade and Broward counties, and was put in place in response to damages from Hurricane Andrew in 1992. Since 2001, the FBC has been updated 12 times, to reflect various changes in design requirement and to come in line with international standards changes.

Design codes will also be related to the particular hazard maps used for legislative and planning purposes. For example, ASCE 7 and FEMA 310 both refer to seismic ground motions from the 2014 updated USGS National Seismic Hazard Maps. Eurocode 8 provides similar seismic hazard maps with zones based on peak ground acceleration, for each country. Interpretation of the design codes will vary with the approach chosen by the modeller. In most cases, the various design code regions are assimilated into vulnerability regions, and the model will select appropriately calibrated vulnerability functions for each. The choice of geographic units or frameworks (see Section 4.6) will often determine the relative vulnerability between modelled regions, and can result in significant differences in loss estimation between relatively near locations.

ASCE 7, which is the US general standard for minimum loading designs for multiple perils, including wind, flood, ice and earthquake, defines structural resilience requirements primarily

**Table 4.11** Some commonly applied design codes within catastrophe model vulnerability functions.

| Code | Natural peril | Country/ Region | Structure(s) covered | Start date | Last revision |
|---|---|---|---|---|---|
| National Earthquake Hazards Reduction Program (NEHRP) | Earthquake | US | Buildings | 1977 | 2004 |
| ATC-3 | Earthquake | US, international | Buildings | 1978 | N/A |
| ATC-14 | Earthquake | US, international | Existing buildings | 1987 | Superseded by ASCE 31 and FEMA 178 |
| ASCE 7 | Multiple perils | US, Worldwide | Buildings and other structures | 1988 | 2010 |
| FEMA 273 | Earthquake | US | Building retrofit | 1991 | superseded by FEMA 310/356 and ASCE 41 |
| FEMA 178 | Earthquake | US | Existing buildings, including educational | 1994 | Superseded by FEMA 310 |
| ATC-40 | Earthquake | US | Concrete buildings | 1996 | N/A |
| International Building Code (IBC) | Multiple perils | Worldwide (national variations) | All buildings | 1997 | 2015 |
| International Residential Code (IRC) | Multiple perils | Worldwide (national variations) | Residential buildings – one/two family dwellings | 1997 | 2015 |
| Eurocode 8 | Earthquake | Europe (national variations) | Buildings and civil engineering works | 2004 | 2013 |
| FEMA 310 | Earthquake | US | Buildings | 1998 | Superseded by ASCE 31 |
| FEMA 356 | Earthquake | US | Structures | 2000 | N/A |
| Florida Building Code | Hurricane, flood | US, Florida | Buildings | 2002 | 2014 |
| ASCE 31 | Earthquake | US | Existing buildings | 2003 | Superseded by ASCE 41 2005 |
| FEMA 440 | Earthquake | US | Structural non-linear static procedures | 2005 | N/A |
| New Zealand NZSEE | Earthquake | New Zealand | Buildings | 2006 | 2015 |
| ASCE 41 | Earthquake | US | Buildings | 2005 | 2007 |
| Eurocode 1 | Wind | Europe (national variations) | Static structures | 2005 | 2014 |

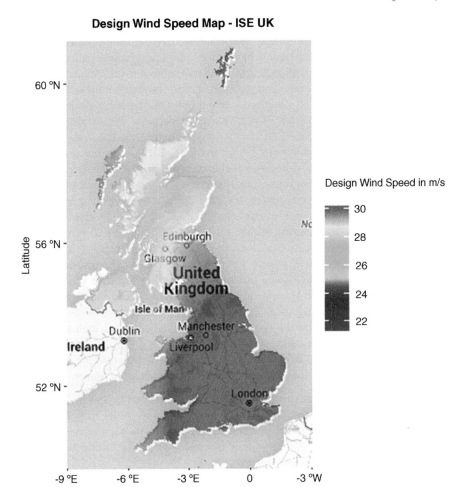

**Figure 4.33** Building Structural Design Wind Speed – UK, *Source*: Reproduced with permission of Willis Re.

in relation to the risk to human life and injury. In the 2016 update, ASCE 7 added tsunami loadings, in line with the 2018 IBC update which included the same. The IBC and its sister code the International Residential Code (IRC) have been adopted by a number of US states, and other nations, as the basis of their own design codes, since 1997.

#### 4.5.4.1 Representing Building Codes in Catastrophe Model Vulnerability Components

One key application of building codes in the design of vulnerability functions involves considering the regional variation in construction quality. The inventory of properties at risk to a peril varies from one place to another. Variations in construction and design include many factors that influence vulnerability. In practice, model developers typically distinguish between two types of regional variation in property vulnerability:

1) *Vulnerability regionalization.* Regional variations in the vulnerability of a particular class and age of building (e.g. 5–7 storey, pre-1939, masonry residential buildings) reflect regional construction practices; these are driven by local architectural practice, design, and by building codes that impose particular standards for design and construction of new buildings, e.g. wind-loading codes, snow-loading codes, and seismic codes. For example,

in seismically active areas, all buildings constructed after a particular seismic code has come into force should follow the appropriate design standards. Based on such variations, models assume that two buildings with identical primary building characteristics – occupancy, construction class, year built, building height – will have different vulnerability depending on the time of construction and on the zoning construction standards imposed where they are built. For example, in the United Kingdom, current building structural design codes ensure that buildings consider the effective 50-year return period peak gust wind speed when new constructions are planned (Figure 4.26). In the United States, ASCE 7 maps of design wind speeds are available in GIS grid formats (Simmons *et al.*, 2016) and can easily be integrated into vulnerability frameworks, although consideration of spatial uncertainty and artefact errors from geospatial data integration must be considered (see Section 4.24).

2) *Inventory regionalization.* In some cases, insurance company portfolios lack information about relevant primary building characteristics such as construction class, year built and building height. In such cases, the model can use an inventory to assess the predominant mix of primary building characteristics that apply to a particular set of aggregated data (such as postcode-resolution aggregate exposure by occupancy class). The model then adjusts vulnerability to account for this typical mix of properties. For example, a model developer may classify postcodes as either urban, rural, semi-urban, or central business district, based on exposure or remote sensing data. Thus, in a European central business district, the building stock might be dominated by post-1980, reinforced concrete, high-rise commercial properties. The general vulnerability of commercial properties to wind damage in such a location will be lower, as a proportion of the overall reconstruction cost of the building, than in a suburban location where low-rise commercial properties predominate.

Both sorts of regional vulnerability variation can be accommodated in a model structure. The first type of regional variability always applies, even when the model user keys in specific information about the primary characteristics of a property. The second type of variability imposes additional geographical differentiation for cases where key information about the property's primary characteristics is unknown.

To develop regional variations in vulnerability, model developers will combine analytical and empirical approaches, by consulting both a range of relevant design codes and supporting data from the engineering literature and, where available, from insurance claims (e.g. Peiris and Hill, 2012).

In contrast, to develop regional variations in the building inventory, developers use data describing the inventory of static assets, such as remote sensing data and census data – in fact, many of the same datasets used in exposure development (see Section 4.4).

As well as defining regional variations in vulnerability that reflect zoning, model developers incorporate building codes into vulnerability functions in a three primary ways expanded upon below.

*Defining Primary Vulnerability Modifiers Related to Specific Code Regions*
Primary vulnerability modifiers relate to the primary structural elements of a building. In particular, the 'year built' (age of building) bands applied in the selection of vulnerability functions, will commonly be derived in relation to building code changes. In South Carolina, for instance, two year bands are often applied – pre-1998 and post-1998.

The frequency of building code updates in a particular region typically reflects a multitude of factors, including incremental political changes (such as the unification of European building codes under the overarching Eurocode umbrella), scientific advances, and the severity of the risk from a particular peril. Thus, the 1944 CP4 UK wind-loading code, after major updates in 1970 and 1997, was superseded in 2005 by the general Eurocode 1 wind loading code, with national parameters and wind maps following in 2008. However, these major overarching updates were accompanied by additional, incremental minor updates to specific codes of practice. For example, the UK CP142 code of practice updates for slating and tiling of houses occurred in 1978, 1990, 1997 and 2003.

In many cases, the occurrence of a catastrophe event can itself act as a catalyst for building code improvements. In the United States, for example, many building code enhancements were made in earthquake-prone regions following the 1994 Northridge earthquake in California. This earthquake highlighted the vulnerability of concrete masonry or tilt-up concrete structures with light-framed wood or steel roofs. A typical structural failure occurred when the connection between the light framed roof and the relatively heavy exterior walls failed. This failure led to partial or full roof collapse. Thus, local building codes including the Uniform Building Code (UBC) enhanced provisions for the design and construction of the roof/wall connections for concrete/masonry buildings with light metal roofs. The 1997 UBC code and subsequent building codes (including the more recent International Code Council International Building Code) incorporated requirements for a higher degree of resistance in these connections.

In some cases, model versions will be in lag to updates of the building codes, and model version changes will often include vulnerability function changes to account for building code changes which have occurred since the last version release. As these changes in the building codes only apply to new or retrofitted buildings – which normally comprise only a small portion of the building stock to start with – they generally have a marginal initial impact on model results for a portfolio that includes mainly older properties. For this reason, model developers may choose to delay incorporation of updated building codes until they can be considered as part of a wider package of more far-reaching model updates.

### Defining and Refining Secondary Modifiers Related to Specific Code Regions

Model developers typically define secondary modifiers as additional, peril-specific building characteristics that impact the vulnerability of a property to a particular peril. These peril-specific modifiers can have a significant impact on losses for an individual property from a particular peril. For example, the presence or absence of a basement can dramatically alter the vulnerability of a property to flood risk, whereas the wind risk remains unaffected. Some secondary modifiers act to mitigate against the risk, in line with design recommendations in a particular area. The secondary modifiers could be building-characteristic specific (e.g., improved roof sheathing or anchors) or external (e.g., storm shutters).

These secondary modifiers modify the base, *average* vulnerability functions according to specific building characteristics or mitigation measures that can be coded in the model as appropriate. In some cases, local regulatory bodies provide reporting opportunities that policyholders and insurance companies can use to codify typical mitigation practice. For example, in Florida, the Florida Office of Insurance Regulation (OIR) 2012 version of the Uniform Mitigation Inspection Forms OIR – B1-1802 introduced the power to capture and inspect the adoption of mitigation measures for pressure-rated doors (garage doors, doors and sliders). Models account for this measure using secondary modifiers that account for retrofitting of garage doors with braces to strengthen resistance to wind pressure.

### Derivation of Performance-Based Design Vulnerability Functions

Design codes contain specific information relating to the performance of building components. With this information, model developers can create theoretical vulnerability curves that account for the behaviour of the specific construction components. One example might be to calculate uplift resistance and restoring moments on roof tiles at different wind speeds. Minor damage to roof tiles dominates wind damage to property in Europe, where wind speeds rarely exceed 40 metres per second in even the most damaging events. A wind suction or uplift force on a tile or roofing element produces an overturning moment. The moment can be calculated as the product of the force and the lever arm at the point of interest. Counterbalancing forces, including the tile's mass and mechanical fixings, resist the uplift force and restore the tile to its initial stationary condition. The British Standards publication BS 55384 (2003) provides a framework for calculating these limiting moments for typical properties in the United Kingdom built since the publication date. Armed with design codes that specify the required minimum mechanical fixing practices, model developers can therefore calculate the impact of these counterbalancing forces for typical properties of different ages. In this way, the developers create theoretical vulnerability curves that can be validated with reference to measurements and observations, such as data collected from wind tunnel experiments and insurance claims.

### 4.5.5   Using Buildings Damage to Determine Other (Non-Structural) Types of Loss

In some cases, the loss is derived from the damage ratio, and usually this is the buildings damage ratio. Typically, fewer sources of data are available to directly calculate the time element (business interruption) losses and these will be derived from the level of buildings damage and the occupancy of the building using weighting factors.

Typical insurance policies cover different types of property and may include losses to contents, or to time-element losses such as additional living expenses or business interruption. The way that each model evaluates losses to different coverages depends on the peril. For example, earthquakes can cause damage to moveable contents in cases where the envelope of the building is not impacted, whereas for climatic hazards such as wind or flood, damage to contents is largely conditional on structural damage to the building in which they are housed.

For some classes of insured cover, catastrophe models derive the loss by directly applying the mean damage ratio (or the chance of loss and conditional mean damage ratio) for that class of property to the total insured exposure input by the model user. This is usually the case for the buildings and contents elements of the loss, where the model user can directly input data on replacement cost.

However, this section describes how models typically apply a more indirect calculation to evaluate other classes of insurance loss, such as time element losses or workers compensation losses. For example, time element losses depend on the level of damage to the facility, and its occupancy class. The models estimate the facility downtime by applying restoration curves that depend on the estimated damage state and the occupancy of the building.

**Business interruption** losses can arise due to the closure of a facility for clean-up and repair. Typical catastrophe model infrastructure takes such losses into account using conditional probabilities based on the structural damage to a facility. 'Downtime' loss tends to be included in catastrophe models as a factor based on the structural MDR by location, assuming that the level of damage will determine the duration and extent of disruption caused to a business. Some models will include secondary modifiers to represent variations in expected BI vulnerability, reflecting differences in likely business restoration. This is often related to the occupancy and construction characteristics of the exposures affected. See Jain and Guin (2009) and Porter and Ramer (2012) for detailed reviews of business interruption and infrastructure downtime modelling methods.

Moreover, when an extreme event occurs, **contingent business interruption (CBI)** loss can also result from loss of lifelines and infrastructure critical for a facility's functioning. In today's modern 'just in time' supply chain paradigm, evaluating such contingent business interruption loss itself forms the basis for complex supply chain management models. Globalization and international networks increase the complexity of the potential from such effects (see Ralph,

2013), and current catastrophe models do not include CBI within their calculations due to this complexity.

Workers' compensation casualty vulnerability also is commonly derived from the building damage. This is particularly problematic, given the variability of overall exposure depending on time of day, or whether a working day, which can significantly affect overall loss potential. Many models will attempt to apply scaled workers' compensation losses depending on the event severity, against average employee numbers and characteristics. Some models will also apply activity distributions, to reflect the likely variation in relative risk with business patterns. In addition, construction type can also be used to modify workers compensation MDRs. Casualty models will use similar damage ratio techniques, reflecting number of people affected to total exposed count. Degree of structural damage and the form of damage (e.g. collapse) will be key drivers. Injury severity classes are employed in most models to structure loss typologies. See Spence *et al.* (2011), for a comprehensive review of earthquake casualty vulnerability assessment).

Sprinkler damage, which can be a significant driver of claims from earthquake, is often applied as a factor to the occupancy type, with damage levels determined by the contents type assumptions. Sprinkler damage estimates will be derived from empirical claims data, and modellers will attempt to vary the impact by occupancy type/use.

**Loss amplification** or **demand surge** factors can also be determined by the level of buildings damage to represent 'sudden and usually temporary increase in the cost of materials, services, and labour due to the increased demand for them following a catastrophe' (Actuarial Standards Board, 2000). A comprehensive description of demand surge can be found in Olsen and Porter (2010). Demand surge is often applied as a loss modifier, commonly a single value. In some cases, demand surge factors will be variable by both region and vulnerability type. This is to represent differences in expected cost inflation, using macro-economic principles, for example, due to the relative ease or difficulty in additional labour sources being made available to a region. If an affected region has relatively porous borders allowing the free movement of labour from other areas (such as continental Europe), supply of labour may be less restricted, and demand surge costs may be lower. Conversely, island states or regions (the commonly cited example is Hawaii) may experience increased demand surge, given relative isolation from potential sources of additional labour or materials. Some models also attempt to represent clustered event impacts on demand surge, for example, if multiple events occur in a particular region, this may increase demand surge costs over and above those estimated for a single event. Models will often apply demand surge on an event-by-event basis, considering the overall portfolio loss rather than attempting to calculate from ground up per location or policy.

Another example, used by Porter *et al.* (2012), applies building damage state as a proxy for estimating casualty rate (both fatal and injury), and hence reflecting the variability of structural damage on likely human casualties.

### 4.5.6 Vulnerabilities for Non-Standard Exposures

Property vulnerability functions depend on building characteristics such as age, height, construction material, occupancy (or use), and peril-specific features, such as, roof type for wind perils, or the presence of a basement for flood perils, or isolation devices for earthquake shaking. The relative importance and impact of these characteristics vary according to the peril being modelled. In addition, vulnerability functions can be constructed for damage to the building envelope or for the contents within that building, or for business interruption due to loss of use of the building.

However, vulnerability functions can also be constructed using other measures for both the numerator and the denominator, for example, economic loss per unit of GDP for a particular hazard intensity.

For risk to population, which is a factor of interest in DRR, parameters such as age, population density, relative poverty levels, education and gender can also assist in defining loss potential. Most population-specific vulnerability functions attempt to represent the impacts of mortality and morbidity (injuries), and have been commonly applied to estimations in Disaster Risk assessment and event monitoring, for example, the USGS *PAGER* project. Most population vulnerability functions are empirically based.

Some of the issues with exposure characterization for non-standard exposures were discussed in Chapter 2.5.4.3; vulnerability issues are discussed here.

Determining appropriate vulnerability functions for Marine Cargo and Specie classes of insurance risk has also been limited to the application of generic property functions, such as warehousing, with associated uncertainties. Until superstorm Sandy in 2012, there was very little claims data available for the development of detailed Marine class risk. In particular, the wide variability in claims size experienced in events such as superstorm Sandy, caused by both difference in peril intensity and causation (storm surge damages, power cuts affecting refrigerated goods, direct damage to warehouse structures), as well as particular damage to given valuables (perishable goods, high value, high vulnerability art and other goods), highlighted the challenges associated with developing vulnerability functions for these classes. The availability of claims data specific to a range of subclasses may improve the granularity of vulnerability functions for cargo in particular, but these are likely to be based on analytical and expert opinion methods as much as empirical loss relationships. An example is the RMS cargo model (RMS, 2015), which has developed a range of vulnerability classes specific to cargo, plus exposure datasets.

Agricultural vulnerability is also problematic, with limited data available for detailed hazard and loss quantification. Agricultural lines exposures are subject to a variety of climate hazards, including drought, freeze, hail, excess rainfall and wind risk. The seasonal nature of crop exposure means that their vulnerability varies during the course of the year – and also with the regional climatic cycles.

Catastrophe models also include, to varying degrees, many other property types such as moveable automobiles and watercraft, marine cargo and aviation risks. Each of these property types poses unique challenges in capturing the key parameters that impact the degree of expected loss from each modelled event.

### 4.5.6.1 Case Study: Understanding Crop Vulnerability to Hail

One example of the challenges associated with agricultural vulnerability function development is that of hail risk (see Chapter 3.4.4).

The measurement of appropriate agricultural damage metrics for comparison against hail intensity is problematic. For crops, damage data are often taken from crop damage reports, and ideally should include:

- type of crop or livestock;
- date, time and location of crops or livestock;
- hail occurrence time and duration;
- extent of damage (as accurately and detailed as possible) as a percentage of crop cover and as a level of yield loss or herd.

In reality, as with other asset types, records are less detailed, and often include errors which reduce the utility and coverage of the data for estimation. Monetary loss will often be based on loss of yield potential, based on the crop or livestock price, and so may also be affected by uncertainties, such as price fluctuation and currency fluctuations, if international data. Trending for such factors will be problematic and complicated by variations in crop prices.

Such factors are also critical to the calibration of damage to loss, given the likely sparsity of additional data for comparison. Calibration of the relationship between intensity and loss is also complicated by a range of secondary considerations, which may well be significant sources of bias or uncertainty. These can be summarized as follows:

- *Crop/livestock types* The effect of hail on a particular crop or livestock type will be highly variable. Soft fruit, for example, will tend to have a higher vulnerability to hail compared to potato plants. There will be considerable variability between crops, and also between crop types, depending on the physiological characteristics of the particular type and relative vulnerability to hail.
- *Wind speed and direction at the ground* Wind-related damages to crops can be a relatively large component of overall impact, but will be very difficult to capture in damage reports. Consequently, this is likely to drive additional uncertainty between a hail intensity metric and damage.
- *Storm duration* The duration of hail fall will have a considerable impact on relative damage. The explicit impact of duration will vary considerably, and will be difficult to isolate from the general intensity factor development.
- *Cropping and livestock husbandry methods* The particular approach taken by a farmer with the management of a crop or livestock herd will also influence vulnerability. For example, crop densities, row lengths and spacing, as well as crop sowing timing, can all influence the relative vulnerability. In particular, methods to increase crop yield may influence the relative vulnerability.
- *Crop seasonality and growth stages* The growth stage of crops will have a significant influence on the relative damage and loss potential. It is often categorized between vegetative and reproductive stages, but can also reflect the crop fruiting and harvest period. Relative vulnerability between stages varies considerably with crop.
- *Mitigation and protection factors* Resilience factors such as the use of artificial hail covers, or the siting of tree wind breaks, can often result in a considerable reduction in crop damage. Livestock protection in barns can also improve loss potential. While understood at a best practice level (see for example, Changnon *et al.* 2009), it will be difficult to discriminate between hail loss mitigation impacts in most damage reports.

In short, development of hail vulnerability functions for agriculture risk requires consideration of the same challenges of data suitability, availability and provenance as with any other class of risk, but is exacerbated by unique issues related to the range of temporal, spatial and risk types inherent to agriculture. The hail example provides a clear example of the wider challenges facing agricultural risk modelling, including windstorm risks for forestry, and surge/flooding risk across all asset types. Model developers therefore have to be fully aware of the appropriateness of the data, and its structure when constructing vulnerability functions for agricultural risk.

### 4.5.7 Validating Vulnerability Models

As with hazard and exposure, validation methods for vulnerability functions will also tend to follow a generic approach, for example via the use of historical damage and claims data not included in the initial function construction and calibration ('out of sample' testing) to test the predictive nature of the resulting function.

As with all vulnerability functions the ability to validate the resulting loss relationships is limited by the degree of 'out of sample' (i.e. independent of the data used to produce the function) empirical and other data available, but is complicated by the fact that the modes of damage causality are often complex, and the driving factors inter-dependent. For example, modelled flood intensity is usually defined in terms of maximum water depth and vulnerability functions are therefore designed to estimate MDR as a function of water depth. The actual cause of damage, and potentially the level of damage experienced, will be due to a number of often localized hazard factors, including the flow velocity, duration of inundation, and whether the water is foul, saline or clean. In general, damage causation will be due to either hydrostatic (due to the in-situ body of water – depth and standing water pressure, pressure seepage, water quality related) or hydrodynamic (damages due to the flow of water, including pressure differentials due to flow around them, scouring around structures, and flood missiles). Equally, the response of a particular type of structure, and its contents, will vary greatly with individual characteristics. Depth is however, likely to be a primary determinant of overall damage, particularly in respect to floor levels (including basements) and contents exposure. Consequently, most flood validation exercises will attempt to test the depth-damage relationship for both direct (physical damage impact) and indirect (equivalent to business interruption) losses (see Nafari *et al.* 2016).

There tend to be two primary sources for validation data. First, damage surveys post event, either by claims adjusters, or third party engineering organizations (and increasingly catastrophe modelling teams) and second, claims data. Claims data may be based on insurance claims, or other sources for instance local or national authority data. Care must be taken to ensure that claims used are those related to the type of flooding being modelled, for example, for an inland flood model, claims due to wind-driven water ingress, or storm surge should be weeded out. Centralized databases of industry losses are increasingly used as validation data, for example PERILS AG (e.g. Peiris, 2015), and this is as true for flood as other perils. Data from ground surveys will also require careful interpretation in relation to intensity-damage, particularly as in many cases, the affected buildings will not be reachable until the inundation has receded. Wrack marks and damp levels may assist in determining water depths against individual structures.

Damage validation can take the form of a component based assessment, where detailed damage to key construction features or contents types is recorded and validated, or through total damage estimation as a percentage damage estimate at the property/structure level. Either way, it is important that the relative characteristics of the recorded damages are as well described as possible, for example in respect to both hazard intensity and resilience or vulnerability modifiers. For flood, an important modifier will be the recording of basement damages. A good example related to insurance based damage is that from the post-July 2013 Norwegian flooding damage survey undertaken (Berg *et al.*, 2015), including the spread of damage costs collected by water depth.

Quantification of the validation is often undertaken via model runs post event using the vulnerability functions created, either bottom up (per site and coverage) or top down (total event loss). This relies on the availability of reported losses. Results are often reported in terms of root mean squared error (RMSE) of the modelled-observed residuals. Bias between modelled and observed may highlight particular model failings, for example, inaccurate representation of structural or usage characteristics, or inadequate claims characterization, for instance, the incorrect coding of business interruption loss.

As an example, Simmons *et al.* (2016) applied a number of statistical validation tests on their regression model of FBC claims to building code, to assess model robustness. These included both sampling methods (e.g. removing ZIP codes where no claims had been recorded for a given year), as well as a 'hurdle' model to test whether the regression did not include any additional contributing factors. These tests were applied to an out-of-sample validation group to assess predictive skill. The results of their validation suggested that the regression model was skilful, except for very large losses.

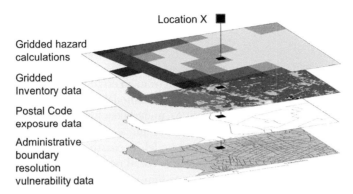

Location X

Gridded hazard
calculations

Gridded
Inventory data

Postal Code
exposure data

Administrative
boundary
resolution
vulnerability data

Figure 4.34 Schematic illustrating the geographical framework for a catastrophe model, illustrating the 'layer' concept, and various geospatial structures including vector (point, line, area) and raster (grid). *Source*: Reproduced with permission of RMS.

## 4.6  Integrating Model Components and the Geographical Framework

Catastrophe models can be considered as inherently spatial, in as much that one of the underlying concepts of loss correlation follows the concept of spatial dependency – that is, that local values of a given parameter, such as wind damage, will be similar (de Smith *et al.*, 2009: Chapter 2). In addition, the geo-spatial structure underpins the integration of hazard, exposure and vulnerability, as it provides the common reference between each component.

While this does provide a valuable method for integrating otherwise diverse datasets and model components, and also enables model user interaction via geocoding (see Chapter 1.9.1.1), there are a number of issues relate to spatially structured data which can determine overall model validity and applicability.

For example, model developers typically perform the hazard calculations on a regular or variable sized geographical grid or mesh. However, data available to the end user on the location and value of exposure do not always have precise geographical co-ordinates. In addition, the developer often uses data of varying precisions such as administrative region, postal code, census tract, etc., for defining the inventory of exposed properties at risk and their vulnerability to a particular peril.

Developers therefore use an underlying geographical framework first to relate the hazard and vulnerability elements of the model (see Figure 4.34). The final modelling product typically then includes a formal geocoding engine which facilitates the use of geographically coded data by the end user (see Chapter 1.9.1.1).

In this framework, developers use various geospatial analysis techniques to relate grid-based hazard calculations to other appropriate modelling attributes at a location. These analytics then become embedded into the hazard modelling platform that the end users employ for retrieving appropriate hazard and vulnerability information for the location being analysed. The details of uncertainties and potential biases created from spatial integration, especially in relation to the combination of data from various formats and resolutions, is beyond the scope of this book. The reader is directed to Zhang and Goodchild (2002), and particularly Chapters 5 and 6 for a detailed explanation of the spatial considerations of model construction. However, some important considerations of geospatial processing of multiple data sources are outlined below.

### 4.6.1  Relative Spatial Resolution/Nominal Scale of Source Data

Catastrophe modelling processes will make use of a range of spatially structured data, including continuous variable data (e.g. digital terrain models, or DTMs) and categorical variable data (e.g.

census tract data, land cover data, ZIP/postal code claim counts). These will be based on varying types of base spatial unit – whether a regularized grid of a given cell resolution (**raster** data), or a series of administrative or other digital boundary data. These can include discrete categorical irregular areas such as land cover described geometrically using digital points, lines and areas (**vector** data), which can sometimes be a derivation from an original raster grid data source such as satellite imagery. In many cases, the data used in the model development process will be the product of multiple derivations and modifications, undertaken for purposes other than the ones intended for the model. Of particular importance for hurricane and storm surge modelling will be the definition of coastline within the spatial model, as coastline is used as a key feature for determining landfall, surge risk, and as a distance to coastline measure. Given coastline features are inherently fractal and therefore self-similar, as shown in Mandelbrot's seminal paper on the British coastline (Mandelbrot, 1967), there will always be an element of uncertainty applied to the definition of the coast within the digital data framework. This will be determined by the initial scale of the data used, and any subsequent generalizations applied. What is important to the catastrophe modeller is that the relative scale and sources of each data set when combined are understood.

### 4.6.2   Geodetic and Coordinate Bases of Data Sources

The integration of data sets within the model may itself require additional data modification, for example the re-projection of raster datasets into a standard geodetic framework (e.g. WGS-84; see Chapter 1.9.1.1), allowing data produced from different map or other sources to be combined geographically. This is achieved by resampling the original grid via a cubic convolution method (see Mitchell and Netravali, 1988). This process, while enabling data from different sources and resolutions to be combined (overlaid) also creates new values on a spatially averaged basis in the new, combined grid structure. This method is often applied to land cover grid data to combine with other data, such as a DTM. Convolution modifications can have a significant influence on the resulting values, and in the case of flood modelling, can alter the depths generated by cell.

### 4.6.3   Relative Vintage of Source Data

All geospatial and geostatistical datasets can exhibit significant variability over time. In particular, postal geographies can change extensively over relatively short time periods (see, e.g. Raper *et al.*, 1992). Consequently, given many catastrophe models will utilize postal codes as a base spatial unit, the time-stamp between model version and current version (e.g. between one annual update of postal codes) may be significantly different. This can have significant implications on the proportion of exposure data which can be modelled, can affect the ability to undertake detailed validation analysis between claims date and current, as well as influencing disaggregation outputs, if used.

### 4.6.4   Point Representation Versus Cell, Area Geographies

Often, models will employ point vector representations of higher density locations, for example, high resolution postcodes, to reduce data storage and computational overheads. There are various methods for assigning the point location to represent an area (termed a **centroid**), with two common methods employed in catastrophe modelling being geometric centroids (the point located at the geometric centre of the area polygon) and weighted centroid (the point located within the area polygon, but weighted by a density value, such as population or building density). The choice of centroid can influence the relative hazard applied to given location, particularly in respect to flood plain or surge inundation as the spatial offset can be in the order of 100 s or 1,000 s of metres.

Some models will apply **disaggregation** methods, to enable exposure data captured at the area geography level (postal code/ZIP or county) to be distributed proportionally to a higher spatial resolution, based on an *a priori* representation of expected exposure density, inferred from available data such as land use, building stock, and in many simpler cases, population data. Again, the industry exposure database (Section 4.4.2) is often a key component of many disaggregation methods, enabling typological differentiation of key building types or land use to be included. In some cases, the centroid point distributions of higher resolution geographic units, such as postcodes, can be used as proxies for the building stock and population density within lower resolution units. The consideration of appropriate spatial resolution, currency and generalization as noted in Section 4.6.5, are key influence son the overall accuracy of any disaggregation method. Disaggregation is always considered a lower-quality approach to exposure modelling.

### 4.6.5 Data Generalization, Interpolation and Smoothing

As well as resampling, geographic data sets can be generalized, or interpolation methods applied to generate surfaces between point sample data (see Section 4.3.3.2). Generalization may be due to re-scaling of data within a vector based boundary dataset, to reduce data storage. Smoothing is often applied to DTMs to reduce noise and improve the structure of the surface model for better representation of the physical processes being modelled, for example, in hydraulic modelling of inundation, or when removing edge discontinuities from multiple sourced DTMs, with varying vertical resolutions. Model developers will often apply such generalizations and smoothing processes using standard GIS algorithms, but best practice is that the resulting surfaces are validated to ensure that no artefacts or discontinuities remain. Interpolation is a commonly applied procedure for creating hazard surfaces in catastrophe models. They are also applied in various forms by third party data providers. For example, the 2014 version of the USGS National Seismic Hazard Model's background historical seismicity is generated on a grid using two kernel-based smoothing algorithms, one a fixed radius, the other an adaptive radius based on the historical point density, and the results from both methods are weighted in the final background seismicity rate model (see Petersen *et al.*, 2014).

## 4.7 The Financial Model

Previous sections have focused on the hazard, exposure and vulnerability components of a catastrophe model. One component that is essential, but sometimes receives insufficient focus, is the financial model. This allows the (re)insurance policy terms (see Chapter 1.8.4) to be applied and enables (re)insurance companies to obtain an output from the model that is meaningful and relevant: a distribution of *insured loss*.

The fact that catastrophe models have a financial model is one of the reasons for the expansion of catastrophe model usage within the (re)insurance industry since the late 1980s. Without a financial model, the outputs of catastrophe models would be interesting but largely useless in helping a (re)insurance company make informed decisions about their catastrophe risk.

The following sub-sections explore why we need a financial model and the functionality that a financial model should provide as well as some case-studies based on one of the predominant cat models available on the market today.

### 4.7.1 Why We Need a Financial Model

The financial model enables the translation of the hazard, exposure and vulnerability approaches described in Sections 4.3, 4.4, and 4.5, into financial metrics of loss. The financial model translates the property damage into costs covered by insurance policies, and partitions these costs, incurred by the policyholders at each individual location, among all entities with a financial liability. To do this, it must consider model uncertainty (see Section 4.5.4 and Chapter 2.16.1), and ensure this is considered within the loss estimation calculation.

This complex calculation process can be considered one of the most challenging parts of the catastrophe model development process, not least due to the traditional focus of model teams towards the scientific and engineering specialisms. However, for most insurance policies, it is essential that loss estimates that reflect the impact of policy conditions, coinsurance, and risk transfer mechanisms such as facultative, risk and treaty reinsurance. The loss estimates must be hierarchically reconcilable, for instance, between location, policy and insurance portfolio levels.

Traditionally, catastrophe model developers have followed two approaches:

- large-scale generalizations of market conditions, and simplified assumptions related to the structure of exposure and vulnerability using industry databases and market share or other approaches (sometimes termed 'aggregate' models);
- a 'bottom-up' per risk approach, aggregating loss estimates from vulnerability functions across the range of hazard events, reflecting the assumed geographic and asset (e.g. similar risk type) correlations existing across multiple, spatially distributed assets – sometimes termed a 'detailed' model.

The former approach relies on the model and its data providing reasonable proxies for the business being written, the latter on the ability to properly represent exposure at the individual asset level, and enable proper interpretation within the loss calculation of insurance policy wording in respect to coverage, exclusion and other limits, often geographically. Both are important to the insurance risk assessment process, and will reflect the applicability of the data and knowledge available.

This section will focus on the model development considerations related to a detailed modelling approach, given that this is now the most common approach being taken to the development of catastrophe models.

Without a financial model, the output from a catastrophe model would comprise a set of losses for every location-coverage (see Chapters 1.8.4 and 1.9.1.2) for every stochastic event. These losses would not reflect the impact of any insurance or reinsurance policy conditions (such as deductibles or limits), but rather would be the overall loss ignoring any (re)insurance; this is termed the **ground-up loss** (see Chapter 1.8.4). These loss values for each location coverage would normally contain a measure of uncertainty around the location-coverage loss. This is termed secondary uncertainty (see Section 4.5.2 and Chapter 2.16.1).

This set of information represents the input to the financial model. From this starting point, the financial model performs a series of calculations to ensure that the model output is relevant and in a usable form for the (re)insurer.

The main functions that a financial engine should perform (see also Chapter 1.8.4) are:

- reflecting any insurance and reinsurance policy conditions, i.e. partitioning the loss between the different parties (insured, insurer, reinsurer);
- aggregating the location-coverage losses to higher levels, for example, to policy level or county level;
- back-allocating the impact of higher-level policy structure to lower levels so that impact of the higher-level structures can be understood and summarized at a more detailed level;
- calculating summary metrics such as average annual loss (AAL), occurrence exceedance probability and aggregate exceedance probability (AEP) curves (see Chapter 1.10).

To achieve the functions listed above, the financial model must handle the uncertainty specified in the input set of location-coverage losses, including any geographical correlations in this uncertainty. Model uncertainty is discussed in the Section 4.7.2; subsequent sub-sections then describe the four main functions listed above.

### 4.7.2  Uncertainty

A full description of uncertainty is provided in Chapter 2.16.1; the focus of this section is to summarize some of the basic concepts of uncertainty before providing a discussion of the aspects relevant to the financial model.

#### 4.7.2.1  Types of Uncertainty

As noted in Chapter 2.16.1, there are three broad classes of uncertainty:

- aleatory: the uncertainty *inherent* in the process that the model is attempting to describe; this cannot be reduced;
- epistemic: the *known* uncertainty in the model; this can be reduced through improved knowledge and capabilities;
- ontological: the *unknown* uncertainty in the model.

Since the sources of ontological uncertainty are by definition unknown, the following discussion focuses on the sources of aleatory and epistemic uncertainty present in the models.

As noted in Chapter 2.16.1, and in Figure 4.35, the uncertainties within catastrophe models are referred to as *Primary Uncertainty* and *Secondary Uncertainty*, both of which have elements of epistemic and aleatory uncertainty. The financial model will incorporate both primary and secondary uncertainty as propagated through the model component architecture and the loss calculation process. See Chapter 2.16 for a description and classification of the uncertainty sources in catastrophe models.

#### 4.7.2.2  Primary Uncertainty

Primary uncertainty represents both the epistemic and aleatory uncertainty included in the generation of the stochastic event catalogue, as described in Section 4.3.6. There is uncertainty in the following areas:

- the parameterization of the probability distributions used to build the stochastic event set: parameter uncertainty;

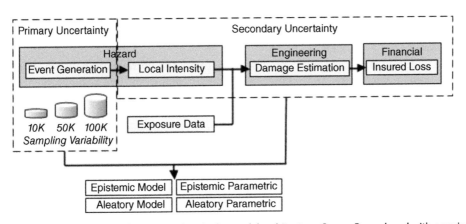

**Figure 4.35**  Primary and secondary uncertainty in the model architecture. *Source*: Reproduced with permission of AIR Worldwide Corporation.

- the choice of the model(s) used to represent the process under consideration: model uncertainty;
- whether the size of the stochastic event set wholly accounts for the uncertainties present in the expected realization of the modelled hazard: sampling uncertainty.

Taking each of these in turn, there is uncertainty in past data due to the *implicit deficiencies in the historical record*. Since there is under-reporting both of small events and of significant events in recorded history, both tails of the probability distribution, and thus the parameters that govern the distribution, are affected by this deficiency in the historical record. This uncertainty is called *parameter uncertainty*.

There is uncertainty in the choice of model used to represent the process, which is called *model uncertainty*. Modelling agencies often offer multiple stochastic event catalogues for certain models. For example, US hurricane models often offer stochastic event sets based on both the standard and climate-conditioned catalogues; the latter catalogue shows the effect of a warmer or cooler ocean on hurricane frequency. There are also time-dependent and time-independent earthquake catalogues for some earthquake models; the time-dependent earthquake catalogue accounts for a decreased probability of an earthquake at a seismic fault after the occurrence of an earthquake at that fault. In the absence of a clear consensus in the scientific community, a multiple-catalogue approach reflects some of this primary uncertainty and should be used by clients as a means of sensitivity testing the impact of the uncertainty on their loss estimates.

The *sampling uncertainty* is associated with catalogue size. A catalogue with more scenario years (e.g. 100,000 years vs. 10,000 years) has inherently less sampling variability than a smaller catalogue because it better reflects the full range of possible outcomes for the upcoming year. Although this source of variability can be reduced by using ever-larger samples of events, for the purposes of computational efficiency and workflow requirements (a larger catalogue results in longer analysis times), it is desirable to constrain the size of the catalogue.

Although the impacts of primary epistemic uncertainty can be stress-tested, for example, by using alternative event-sets as described above, this source of uncertainty is usually not explicitly represented in the current generation of catastrophe models and is thus not propagated through the financial model. It is also a significant source of uncertainty (Chapter 2.16.1). This may mean that some users of catastrophe models have a false impression as to the certainty in catastrophe model results.

Secondary uncertainty is, however, usually explicitly represented within catastrophe models through the uncertainty that is specified around the mean damage ratios within the vulnerability curves (see Section 4.5.7 for more details on secondary uncertainty).

### 4.7.2.3 Uncertainty Correlation

An important role of catastrophe models is to represent correlation between losses, in particular, geographical correlation. Losses from a particular event in a region are correlated due to the correlation between the hazard values from the event. For example, if an event has a high wind speed in location A and a high wind speed in location B, it is likely that both location A and B will experience high losses. This correlation due to driving hazard variables is sometimes called primary correlation (e.g. Lohmann, 2011).

The uncertainty in loss is also correlated. For example, the locations A and B may experience exactly the same wind-speed, and may be exactly the same type of property (as far as the model can tell), yet may experience different losses. This is called secondary uncertainty and is described in Section 4.5.2 and Chapter 2.16.1. This type of uncertainty is usually incorporated as a distribution about the mean damage ratio for each vulnerability curve. Sometimes models include the standard deviation and maximum exposure, which allows parameterization of a distribution; sometimes non-parametric distributions are included through providing damage ratios at different quantiles for a given level of hazard.

From the perspective of model construction, it is important to decide how to represent uncertainty correlation within the model. Since models usually aggregate loss from the location-coverage level upwards, the first consideration is whether or not the uncertainty in coverage level losses should be correlated. For example, consider an event that impacts a particular location with a specific hazard level such that the model mean damage ratio is 10% for the buildings coverage and 5% for the contents coverage. If the secondary uncertainty distribution is sampled from, the realization of the buildings coverage damage ratio may be 15% (e.g. a high percentile loss, perhaps 75th percentile, has been drawn from the simulation). The question in this case is whether the contents loss should also be higher than its 5% mean damage ratio. Most models assume full location-coverage correlation, and so in this case would also select the 75th percentile loss from the contents secondary uncertainty distribution: leading to an *above mean damage ratio* contents loss. In terms of the standard deviation of loss, full correlation implies that the standard deviations around the buildings and contents losses should be added to arrive at the standard deviation of combined (buildings + contents) loss.

A further consideration is whether the uncertainty correlation should vary by location separation: are two locations close together more likely to have damage ratios higher than the mean damage ratios, for the same level of hazard, than locations far apart?

The treatment of this effect in catastrophe models varies widely. Some models do not include geographical correlation in secondary uncertainty; some models do not include this, but allow the user to select a correlation parameter themselves. Some models do include this effect and vary the correlation level by the hazard, and the spread of an industry portfolio (see Section 4.4), but do not vary the effect by distance between locations in any given input portfolio.

The degree of correlation depends on how close together the locations are, and how dispersed the peril severity is. Perils such as earthquake, where the losses tend to be geographically concentrated, show more correlation than perils where the losses are geographically dispersed, such as windstorm.

Both Wang, Hofmann and Park (2016) and Lohman and Yue (2011) show that the impact of increased location correlation is significant; ground-up portfolio losses increase for high return periods and decrease at smaller return periods. Wang, Hofmann and Park (2016) also show that allowing the correlation to vary by distance between locations can results in very different results to assuming a fixed level of correlation based on an industry portfolio.

### 4.7.3 Case Study: Combining Distributions: Convolution

One of the functions of a financial model is to combine the distributions of loss to obtain a distribution at the required resolution. As discussed, the 'raw' output of a catastrophe model prior to the financial model is a set of ground-up location-coverage losses and associated secondary uncertainty information, for every stochastic event. Without considering uncertainty, a distribution of loss could easily be formed by ordering the mean event losses for each location-coverage (see Chapters 1.10.5 and 1.10.6). However, a catastrophe model user is likely to require a distribution of loss at a more aggregated level; for example, at location or policy level; this means the location-coverage distributions need to be combined. In addition, it is important that uncertainty is included in the process of combining the loss distributions. The financial model must address both of these tasks.

Accurately aggregating loss distributions becomes complicated when considering uncertainty. Doing so with relative transparency, while enabling the flexibility of additional user-defined assumptions and accomplishing it with minimum run-time, is a further challenge.

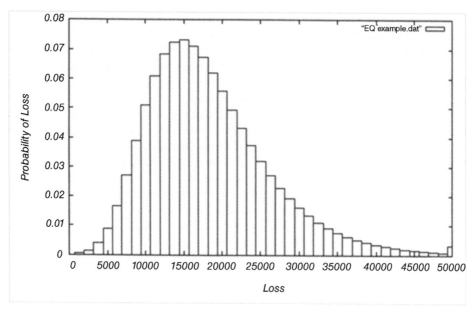

**Figure 4.36** Probability distribution example. *Source*: Reproduced with permission of AIR Worldwide Corporation.

Secondary uncertainty is captured using a probability distribution that represents the range of losses to a location as a result of being affected by a hazard of a specific intensity. This could be captured within a model framework through the specification of parameters such as the standard deviation of loss and maximum loss; which are used to specify a parametric distribution (e.g. Beta distribution; see Chapter 1.11.2.1 and Section 4.5.4). Alternatively the model framework may allow a generic distribution to be specified through allowing probability of loss to be captured at different discrete loss levels.

An example of this is illustrated in the secondary uncertainty distribution shown in Figure 4.36, where the x-axis represents the range of losses for a *single* location-coverage for a *single* event and the y-axis represents the probability of loss. The probability distribution is divided into discrete intervals. Each interval of loss is called a 'bin' and the amount on the y-axis represents the corresponding probability that the incurred loss lies within the range of the bin. For example, the last bin in the plot below ranges from $48,750 to $50,000, and the probability that the loss for this location and coverage occurs within this range is 0.005 for the event under consideration.

When *multiple* locations are affected by an event, their loss distributions need to be combined together. The process by which probability distributions are combined or added together is known as convolution.

This technique computes all possible loss distribution combinations $L_1 + L_2$, and their associated probabilities, given the probability distributions of $L_1$ and $L_2$ separately. In this case, $L_1$ and $L_2$ are the loss distributions for location 1 and location 2 respectively, for *the same event*. For independent distributions this is shown in the following equation, where $L$ represents the total loss for two locations; $P(L_1)$ is the probability distribution for location 1, and $P(L_2)$ the probability distribution for location 2.

$$P(L) = \sum_{L=L_1+L_2} P(L_1) \times P(L_2) \tag{4.3}$$

Figure 4.37, along with the associated Table 4.12 provides an example of the convolution process for location 1 and location 2.

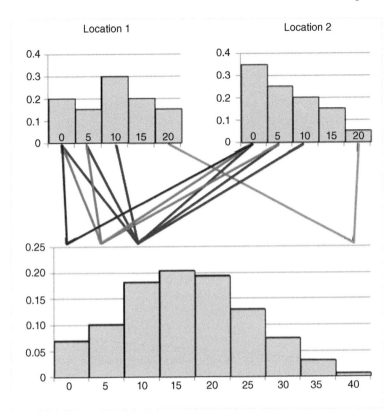

Figure 4.37 Example of the convolution process. *Source*: Reproduced with permission of AIR Worldwide Corporation.

As can be seen in Table 4.13, the convolution process is used to calculate the probability of each total loss. Notice that a total loss of 10 can be obtained by three possible combinations of losses from locations 1 and 2: (0, 10), (5, 5) and (10, 0).The individual probabilities for each component of the combinations are multiplied together and the three products summed; hence the probability of there being a total loss of 10 is 0.1825.

Convolution of loss distributions for two locations results in a loss distribution range that equals the sum of the ranges for the individual loss distributions. This process is transparent and can also be extended to calculate the impact of financial structures. When policy terms, such as deductibles and limits are applied, the loss values at each discrete point on the distribution are

Table 4.12   Convolution example: loss and probability.

| Loss for Location 1, $L_1$ | P(Loss = $L_1$) | Loss for Location 2, $L_2$ | P(Loss = $L_2$) |
|---|---|---|---|
| 0 | 0.2 | 0 | 0.35 |
| 5 | 0.15 | 5 | 0.25 |
| 10 | 0.3 | 10 | 0.2 |
| 15 | 0.2 | 15 | 0.15 |
| 20 | 0.15 | 20 | 0.05 |

**Table 4.13** Convolution example: total loss and probability

$$P(L) = \sum_{L=L_1+L_2} P_1(L_1) \times P_2(L_2)$$

$$P(10) = P_1(0) \times P_2(10) + P_1(5) \times P_2(5) + P_1(10) \times P_2(0)$$

$$P(10) = 0.2 \times 0.2 + 0.15 \times 0.25 + 0.3 \times 0.35 = 0.1825$$

| Total Loss, I | P(Total *Loss* = $l_1$) |
|---|---|
| 0 | $0.2 \times 0.35 = 0.07$ |
| 5 | $0.2 \times 0.25 + 0.15 \times 0.35 = 0.1025$ |
| 10 | $0.2 \times 0.2 + 0.15 \times 0.25 + 0.3 \times 0.35 = 0.1825$ |
| 15 | $0.2 \times 0.15 + 0.15 \times 0.2 + 0.3 \times 0.25 + 0.2 \times 0.35 = 0.205$ |
| 20 | $0.2 \times 0.05 + 0.15 \times 0.15 + 0.3 \times 0.2 + 0.2 \times 0.25 + 0.15 \times 0.35 = 0.195$ |
| 25 | $0.15 \times 0.05 + 0.3 \times 0.15 + 0.2 \times 0.2 + 0.15 \times 0.25 = 0.13$ |
| 30 | $0.3 \times 0.05 + 0.2 \times 0.15 + 0.15 \times 0.2 = 0.075$ |
| 35 | $0.2 \times 0.05 + 0.15 \times 0.15 = 0.0325$ |
| 40 | $0.15 \times 0.05 = 0.0075$ |

modified to reflect these terms, and the modified loss values are then used in the convolution process.

This process can be used to combine location-coverage values to location level, or location level distributions to policy level as shown in Figure 4.38.

However, without any methods to speed up the convolution process, a vector of probabilities of length $n$ (the number of discrete levels) would require $n^2$ calculations to arrive at the result for the convolved product of the two vectors (representing two locations). This should be intuitively reasonable since each one of the $n$ points needs to be multiplied by each of the other $n$ points a

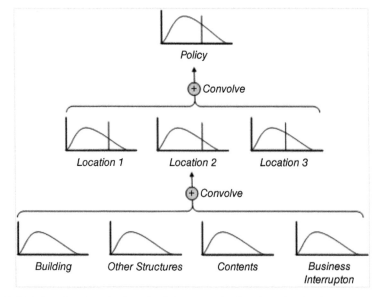

**Figure 4.38** Multiple locations in a single policy being convolved to the policy level. *Source*: Reproduced with permission of AIR Worldwide Corporation.

total of *n* times. Given that portfolios can contain millions of locations, this approach quickly becomes computationally unfeasible.

In practice, catastrophe model engines use various methods to speed up the convolution process so that analyses are completed within a reasonable amount of time. Three such methods are:

1) aggregating the mean and standard deviation of the individual distributions and then fitting a parametric distribution to these aggregated quantities;
2) using a numerical method such as Panjer or the Fast Fourier Transform (FFT) to speed up the convolution process (e.g. Embrechts and Frei, 2010);
3) using Monte Carlo simulation to sample from the individual distributions.

It is outwith the scope of this book to go into these methods in detail, however, they all have advantages and disadvantages.

- Method 1 is a quick method to use, however, there is little information about the shape of the distributions propagated. Instead an assumed form of parametric distribution is made and fitted to the aggregated mean and standard deviation. This assumed form is a large assumption; it is difficult to quantify the impact (or error) in this assumption.
- Method 2 is a fairly quick method to use (although not as fast as method 1). However, a discretization of the distributions is needed, and the transparency provided in the convolution example is lost once the numerical efficiencies are implemented.
- Method 3 is the most computationally expensive of the three approaches (although still quicker than full convolution). This method has the advantage of being very transparent. However, the stability of the results depends on the number of simulated samples and so there is a computational trade-off between accuracy and speed.

Given the increases in computational power, catastrophe modelling platforms are being released that are simulation based, and which represent uncertainty distributions more realistically.

As well as combining or aggregating distributions, the financial model must also apply the relevant insurance and reinsurance financial structures. This is discussed in Section 4.7.4.

### 4.7.4 Applying Financial Structures

Chapter 1.9.2 describes common insurance financial structures and Chapter 2.4.2 describes reinsurance financial structures. The financial model must partition or allocate the loss cost to the correct parties involved in the insurance or (re)insurance contract. This results in a variety of financial loss perspectives, described in Chapter 1.8.4.

For example, Figure 4.39 illustrates how a simple location-level property insurance policy may allocate the value at risk from a property from the point of view of an insurance company. The total rebuilding cost of the property is represented by the vertical axis. In the event of a loss, the policyholder pays the amount up to the value of the deductible. Beyond the deductible, the loss is paid by the primary insurance companies up to the blanket limit. In some cases, the risk may be shared on a proportional basis between multiple insurance companies (represented by 'other insurer' in Figure 4.39). Thus the loss to the insurance company (the gross loss) consists of the ground-up loss, minus the deductible, capped at the blanket limit, and reduced to allow for the portion of the loss ceded to the other insurers.

The way in which the financial model partitions loss will be closely related to the method used to convolve the loss distributions (Section 4.7.4):

- If method 1 is used to aggregate the distributions, then it is likely that an integration will be used to evaluate some of the financial structures. This is because if a parametric distribution

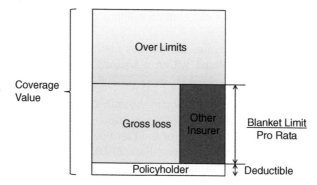

**Figure 4.39** Apportioning value at risk among simple location-level insurance policies. *Source*: Reproduced with permission of RMS.

form is already assumed, this form can be used to evaluate the integrals that represent the financial structures. Not all financial structures will be possible to evaluate in this way, but the evaluation is computationally quick.

- If a numerical method such as FFT is used, then the effect of the financial structures can be applied to the discretized distributions of loss to calculate the partitioning.
- If a simulation method is used, then the financial structures can be evaluated through simple arithmetic.

### 4.7.5 Case Study: Back-Allocation

Following on from the above case study, once the loss distributions have been convolved to the required levels, and the losses have been partitioned to the relevant parties (according to the financial structures in place), the financial model may be required to back-allocate the impact of any higher level financial structures to lower levels. For example, consider a policy level limit in place on a multi-location policy. The location level loss distributions will have been combined to the policy level. The policy level limit will have been evaluated and the loss partitioned accordingly, using the policy level loss distribution. However, the model user may wish to understand what the location level loss net of the policy limit looks like. For example they may wish to view this loss on a map or aggregate the location level loss at county level. To satisfy this need, the financial model may perform back allocation.

For example, one way of performing this is to pro-rate the policy level losses in proportion to the underlying location level losses prior to the policy financial terms. An alternative approach to conducting convolutions and back-allocations is demonstrated by the OASIS LMF Financial Module (OASIS, 2015).

### 4.7.6 Financial Model Output

The final function of the financial model is to output these aggregated, partitioned and back allocated distributions of loss. The output is normally in the form of event loss tables or year loss tables. See Chapters 1.10.5 and 1.10.6 for a description of these.

In addition, the financial model may also calculate summary metrics such as average annual loss (AAL) or the standard deviation of loss. These are described in Chapter 1.10.

A standard output from most catastrophe models is an exceedance probability (EP) curve. This measures the probability that a given level of loss is met or exceeded in a year. There are two types of exceedance probability curves commonly used: aggregate and occurrence. The aggregate exceedance probability (AEP) is created using total losses in a year and the occurrence

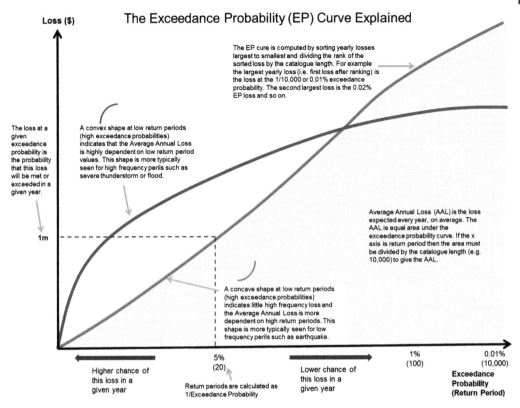

The Exceedance Probability (EP) Curve Explained

The EP cure is computed by sorting yearly losses largest to smallest and dividing the rank of the sorted loss by the catalogue length. For example the largest yearly loss (i.e. first loss after ranking) is the loss at the 1/10,000 or 0.01% exceedance probability. The second largest loss is the 0.02% EP loss and so on.

The loss at a given exceedance probability is the probability that this loss will be met or exceeded in a given year.

A convex shape at low return periods (high exceedance probabilities) indicates that the Average Annual Loss is highly dependent on low return period values. This shape is more typically seen for high frequency perils such as severe thunderstorm or flood.

Average Annual Loss (AAL) is the loss expected every year, on average. The AAL is equal area under the exceedance probability curve. If the x axis is return period then the area must be divided by the catalogue length (e.g. 10,000) to give the AAL.

A concave shape at low return periods (high exceedance probabilities) indicates little high frequency loss and the Average Annual Loss is more dependent on high return periods. This shape is more typically seen for low frequency perils such as earthquake.

Figure 4.40 Exceedance probability curve infographic. *Source*: Reproduced with permission of AIR Worldwide Corporation.

exceedance probability (OEP) uses the largest loss in a given year. EP curves are used to quantify the risk at different levels; for example to a portfolio, or an individual risk.

Figure 4.40 provides a visual explanation of the EP curve. The construction of an EP curve in the financial model will incorporate both the secondary uncertainty component of the event set, and the mean losses. More information about EP curves, and how they are constructed, can be found in Chapter 1.10.4.

## 4.8 Model Validation

Model validation refers to the testing of the model to assess its ability to produce results in line with expected reality (Wainwright and Mulligan, 2004) and is different to model calibration (for example the use of claims data in the construction of empirical vulnerability functions (Section 4.1.2). A publicly available description of a typical catastrophe model validation related to hurricane (RMS, 2012) illustrates the main approaches taken. Prior to releasing a catastrophe model, the development team and other internal stakeholders perform a series of validation checks that involve assessing the model's performance against benchmark observations. Ideally these benchmarks are independent of the data used to construct the model. In practice, however, due to the relatively sparse historical record of extreme events, the modellers are at least aware of these historical benchmarks, and in many cases are forced to incorporate them into the development process due to the scarcity of other relevant historical information.

Validation will be undertaken by the model developer, as well as independent organizations (such as regulators) and users developing their own view of risk (see Chapter 5 and Chapter 2).

An example of a comprehensive and independent validation exercise is that of the Florida Commission on Hurricane Loss Projection Methodology (see Chapter 2.11.5) which reviews hurricane catastrophe model methodologies used in residential insurance rate calculations. Model users will also undertake various levels of validation, including those related to regulatory governance requirements (such as Solvency II).

Model validation can be considered in two key areas - particularly from the perspective of a model developer:

- Component validation compares specific components of the model output to historical observations—both by event and across the whole population of events. Section 4.1.7 provides examples of vulnerability component validation. External expert reviews are often employed to validate methodologies applied in the model formulation, for example.
- Overall model validation or loss validation compares the overall output of the model to the historical experience of a portfolio of elements at risk—such as an insurance company portfolio, or an estimated industry exposure database—again, both by event and across the whole population of events. Limitations of data availability and appropriateness (e.g. the length of insurance claims data records, or the lines of business covered) will be important considerations in the validation process.

Depending on the availability of historical data benchmarks, additional complexities for considering model validity can be superimposed on this model framework,. For example, if data are available, modellers typically conduct validation tests across different spatial scales – local, regional, national and transnational – and over different averaging periods.

As the density of the available historical data increases, so does the potential complexity of the comparisons between modelled and observed outputs. This increase in complexity can provide more confidence in the model output. However, it can also add potential pitfalls for unwary end users, who are increasingly being asked to provide the results of such validation exercises on their own portfolios to interested parties such as regulatory bodies, by the inappropriate comparison of models against public databases, which may be constructed in ways that reduce their comparability to catastrophe model outputs.

For example, data vintage can be a major source of difference between modelled and reported losses, particularly for older events. Modellers typically employ sophisticated exposure- and loss-trending methodologies to account for changes in the inventory of exposed elements between the time of an event happening and the present day. Regardless of the sophistication of this process, model validation by its nature can only be applied to the lower end of the return-period curve, as it is limited by the availability of reliable historical data.

From an end-user perspective, the details of all the developer's model validation tests should be made available, along with clear explanations of the tests, the meaning of the results and the limitations of those tests. The model developer should also be fully transparent about parts of the model which cannot be validated, due to limitations in data, and explain their rationale for why they are comfortable with the model and its results in the absence of validation. It is also helpful for the end-user to understand what alternatives were considered by the model developer, and why they were rejected. In today's world, openness and transparency are as important as model sophistication, particularly where that sophistication adds volatility rather than any additional accuracy.

Additional, independent model validation tests relevant to their specific portfolio should always be conducted by the end-user. At longer return periods beyond the time period which can be validated, the end-user is reliant on the model developer and assumptions, but should evaluate the methodologies and assumptions used by the developer. It is of course important that the evaluator has correct technical expertise in order to do so and consults others who have useful and important insights, such as in-house risk engineers and underwriters. The relative paucity of available datasets to compare model outputs, particularly in respect to non-standard lines of business, adds to the challenges for model users to conduct effective validation. See Chapter 5.4.1 for a detailed discussion of end-user model validation considerations.

## 4.9 Conclusion

This chapter has the key considerations and decision points affecting the design and construction of component-based catastrophe models from the perspective of model developers. Every catastrophe model is an amalgam of various individual and integrated calculation blocks, based on varying assumptions and parameterizations. Common characteristics of model development processes can be identified, around the component structure, and the integration between these, via both geographical frameworks, and model parameters, are essential to the interpretation of a model's applicability to a particular portfolio of assets at risk. The relative uncertainty of any component will vary, and in the current formulation of catastrophe models, overall uncertainty due to sub-optimal data or methods, or parameterizations due to computational limitations, are not propagated through to the final model output. However, uncertainties due to parameter error, sampling error or data errors are increasingly captured within each individual component and the most recent models are now considering uncertainty within the calculation process.

Whether constructing the components of a catastrophe model, undertaking validation of a model and its underlying construction, or interpreting the outputs, it is important to be aware of the primary data sources, decision points and modifications, as well as the sources of uncertainty as they are derived and propagated through each component to the final loss estimates.

The complexities of the model development process make it difficult to isolate, or determine relative proportionality of each component's strengths or weakness, but it is possible to gain a general view of the relative confidence placed on each component.

## Note

1. http://www.europeanwindstorms.org/cgi-bin/storms/storms.cgi?storm1=Daria

## References

Abrahamsen, N. and Blanpied, M. (2003) Appendix G: Moment-balancing the fault rupture models. In *Earthquake Probabilities in the San Francisco Bay Region: 2002–2031.* Open File Report 03–214, Working Group on California Earthquake Probabilities, USGS.

Alavi, N., Warland, J. and Berg, N. (2006) Gap filling of evapotranspiration measurements for water budget studies: Evaluation of a Kalman filtering approach. *Agricultural and Forest Meteorology*, **141**, 57–66.

Andersson, J., Zehnder, A., Wehrli, B. and Yang, H. (2012) Improved SWAT model performance with time-dynamic Voronoi tessellation of climatic input data in Southern Africa. *Journal of the American Water Resources Association*, **48**, 480–493, DOI: 10.11111/j.1752-1688.2011.00627.x

Aspinall, W. and Cooke, R. (1998) Expert judgment and the Montserrat Volcano eruption. In *Proceedings, 4th International Conference on Probabilistic Safety Assessment and Management, PSAM4* (eds A. Mosleh and R. A. Bari), vol. 3, Springer, New York, pp. 2113–2118.

Atwater, B., Musumi-Rokkaku, S., Satake, K., et al. (2005) *The Orphan Tsunami of 1700: Japanese Clues to a Parent Earthquake in North America*, US Geological Survey, Department of the Interior, Reston, VA:

Barredo, J. (2010) No upward trend in normalised windstorm losses in Europe 1970–2008. *Natural Hazards and Earth System Sciences*, **10**, 97–104.

Bayliss, A. and Reed, D. (2001) The use of historical data in flood frequency estimation. *Report to Ministry of Agriculture, Fisheries and Food.* Centre of Ecology and Hydrology, http://nora.nerc. ac.uk/8060/1/BaylissRepN008060CR.pdf (accessed 3 Dec. 2015).

Berg, H., Ebeltoft, M. and Nielsen, J. (2015) Flood damage survey after major flood in Norway 2013 – cooperation between insurance business and a government agency. Presentation at the Fifth EU Loss Data Workshop, 30–31 March 2015, Joint Research Centre of the European Commission, Ispra, Italy, http://drr.jrc.ec.europa.eu/Portals/0/Loss/March2015_workshop/ additional%20material/Berg%20et%20al_Damage%20survey_FRIAR%202014.pdf (accessed 16 June 2016).

British Standards Institution (2011) NA to BS EN 1991-1-4-:2005+A1:2010. National Annex to Eurocode 1 – Actions on structures Part 1–4: General actions – Wind actions. British Standards Institution, London.

Cameron, D., Beven, K., Tawn, J., Blazkova, S. and Naden, P. (1999) Flood frequency estimation by continuous simulation for a gauged upland catchment (with uncertainty), *Journal of Hydrology*, **219** (3–4), 169–187.

Causevic, M. and Mitrovic, S. (2010) Comparison between non-linear dynamic and static seismic analysis of structures according to European and US provsions. *Bulletin of Earthquake Engineering.* Published online 22 July 2010. (accessed 16 June 2016).

Changnon, S., Changnon, D. and Hilberg, S. (2009) Hailstorms across the nation. Illinois State Water Survey, Contract report 2009-12, Champaign, Illinois.

Chen, L., Singh, V., Guo, Z., *et al.* (2012) Flood coincidence risk analysis using multivariate copula functions. *Journal of Hydrologic Engineering*, **17** (6), 742–755.

Chen, S., Chen, M., Zhao, N., *et al.* (2009) Florida Public Hurricane Loss Model: Research in multi-disciplinary system integration assisting government policy. *Government Information Quarterly*, **26**, 285–294.

Claesson, S. and Paulsson, J. (2005) Flyginventering av stormfälld skog – januari 2005. PM 2005-02-02. Skogsstyrelsen. 10.

Coleman, J. and LaVoie, S. (2012) 09 Paleotempestology: Reconstructing Atlantic tropical cyclone tracks in the pre-HURDAT era, *Modern Climatology.* Book 6. InTech. http://www.intechopen. com/books/modern-climatology (accessed 16 June 2016).

Coles, S. (2001) *An Introduction to Statistical Modeling of Extreme Values*, Springer, London.

Colette, A., Leith, N., Daniel, V., Bellone, E. and Nolan, D. (2010) Using mesoscale simulations to train statistical models of tropical cyclone intensity over land. *Monthly Weather Review*, **138**, 2058–2073.

Compo, G., Whitaker, J., Sardeshmukh, P., Matsui, N., and Allan, R.J. (2011) The twentieth century reanalysis project. *Quarterly Journal of the Meteorological Society*, **137**, 1–28.

Cook, N.J. (1985) *The Designer's Guide to Wind Loading of Building Structures*, Building Research Establishment Report, Butterworths, London.

Cook, N.J. (1997) The Deaves and Harris ABL model applied to a heterogeneous terrain. *Journal of Wind Engineering and Industrial Aerodynamics*, **66**, 197, 214,

Daniell, J., Wenzel, F., Khazhai, B., Santiago, J., and Schaefer A. (2014) A worldwide seismic code index, country-by-country global building practice factor and socioeconomic vulnerability indices for use in earthquake loss estimation. Proceedings of the Second European Conference on Earthquake Engineering, Istanbul.

Danish Hydraulic Institute (2015) MIKE 21 and MIKE 3 Flow Model. Hydrodynamic Module, Short Description, https://www.mikepoweredbydhi.com/download/product-documentation (accessed 16 June 2016).

D'Ayala, D., Meslem, A., Vamvatsikos, D. *et al.* (2013) Guidelines for Analytical Vulnerability Assessment, Vulnerability Global Component project (GEM). Version 3.2: www.nexus. globalquakemodel.org/GEM_VULNERABILITY_PHYSICAL/posts/(accessed 16 June 2016).

Dee, D., Uppala, S., Simmons, A., Berrisford, P. *et al.* (2011) The ERA-Interim reanalysis: Configuration and performance of the data assimilation system. *Quarterly Journal of the Royal Meteorological Society*, **137** (656), Part A, 553–597.

Della-Marta, P., Liniger, C. Appenzeller, C. *et al.* (2010) Improved estimates of the European winter wind storm climate and the risk of reinsurance loss using climate model data *Journal of Applied Meteorology and Climatology*, **49**, 2092–2120.

Della-Marta, P., Mathis, H., Frei, C. *et al.* (2009) The return period of wind storms over Europe. *International Journal of Climatology*, **29** (3), 437–459.

Deloitte (2015) Four years on: Insurance and the Canterbury earthquakes. https://www.vero.co.nz/sites/default/files/documents/Vero%20NZ%20Report_Four%20Years%20On%20-%20Insurance%20and%20the%20Canterbury%20Earthquakes.pdf (accessed 12 May 2016).

Demuth, J., DeMaria, P. and Knaff, J. (2006) Improvement of advanced microwave sounder unit tropical cyclone intensity and size estimation algorithms. *Journal of Applied Meteorology*, **45**, 1573–1581.

Donat, M.G., Leckebusch, G.C., Wild, S. and Ulbrich, U. (2011) Future changes in European winter storm losses and extreme wind speeds inferred from GCM and RCM multi-model simulations. *Natural Hazards and Earth System Sciences*, **11**, 1351–1370.

Dumedah, G. and Coulibaly, P. (2011) Evaluation of statistical methods for infilling missing values in high-resolution soil moisture data. *Journal of Hydrology*, **400**, 95–102.

Elsner, J. and Jagger, T. (2006) Prediction models for annual U. S. Hurricane counts. *Journal of Climate*, **19**, 2935–2952.

Flores-Estrella, H., Yussim, S. and Lomnitz, C. (2007) Seismic response of the Mexico City Basin: A review of twenty years of research, *Natural Hazards*, **40** (2), 357.

Fukutani, Y., Suppasri, A. and Imamura, F. (2014) Stochastic analysis and uncertainty assessment of tsunami wave height using a random source parameter model that targets a Tohoku-type earthquake fault. *Stochastic Environmental Research and Risk Assessment*, **29** (7), 1763–1779.

Gesch, D. (2009) Mapping and visualization of storm-surge dynamics for Hurricane Katrina and Hurricane Rita: U.S. Geological Survey Scientific Investigations Report 2009–5230.

Goldfinger, C., Nelson, C., Morey, A. *et al.* (2012) Turbidite event history—Methods and implications for Holocene paleoseismicity of the Cascadia subduction zone: U.S. Geological Survey Professional Paper 1661–F, http://pubs.usgs.gov/pp/pp1661f/ (accessed 10 May 2016).

Gregor, N., Silva, W., Wong, I. and Youngs, R. (2002) Ground-motion attenuation relationships for Cascadia subduction zone megathrust earthquakes based on a stochastic finite-fault model. *Bulletin of the Seismological Society of America*, **92**, 1923–1932.

Grossi, P. and Kunreuther, H. (2005) *Catastrophe Modeling: A New Approach To Managing Risk.* Springer, New York.

Gumbel, E.J. (1941) The return period of flood flows. *Annals of Mathematics and Statistics*, **12** (2), 163–190, http://projecteuclid.org/euclid.aoms/1177731747 (accessed 16 June 2016).

Gunasekera, R., Ishizawa, O., Aubrecht, C. *et al.* (2015) Developing an adaptive global exposure model to support the generation of country disaster risk pofiles. *Earth-Science Reviews*, **150**, 594–608.

Gutenberg, B. and Richter, C. (1954) *Seismicity of the Earth and Associated Phenomena* (2nd edn), Princeton University Press, Princeton, NJ.

Haas, R. and Pinto, J. (2012) A combined statistical and dynamical approach for downscaling large-scale footprints of European windstorms, *Geophysical Research Letters*, **39**, L23804.

Hall, T. M. and Jewson, S. (2007) Statistical modeling of North Atlantic tropical cyclone tracks. *Tellus*, **59A**, 486–498.

Hart, R. and Evans, J. (2001) A climatology of the extratropical transition of Atlantic tropical cyclones. *Journal of Climate*, **14**, 546–564.

Haylock, M. (2011) European extra-tropical storm damage risk from a multi-model ensemble of dynamically downscaled global climate models. *Natural Hazards and Earth System Sciences*, **11** (10), 2847–2857.

Henstock, T., McNeill, L., Dean, S. *et al.* (2010) Exploring structural control on Sumatran earthquakes. *Eos*, **91** (44) 405.

Hillier, J., Macdonald, N., Leckebusch, G., and Stavrinides, A. (2015) Interactions between apparently 'primary' weather-driven hazards and their cost. *Environmental Research. Letters*, **10**, 104003.

Hoffman, R. (2005) Neutral stability height correction for ocean winds, http://map.nasa.gov/data/ssw.old/reason_sample/doc/src/height-correct.pdf (accessed 10 May 2016).

Jain, V. and Davidson, R. (2007) Forecasting changes in the hurricane wind vulnerability of a regional inventory of wood-frame houses. *Journal of Infrastructure Systems*, **13**, 29–42.

Jain, V. and Guin, J. (2009) Modeling business interruption losses for insurance portfolios. Proceedings of the 11th Americas Conference on Wind Engineering, San Juan, Puerto Rico, June 22–26 2009, http://www.iawe.org/Proceedings/11ACWE/11ACWE-Jain.Vineet1.pdf (accessed 16 June 2016).

Jarrell, J., Herbert, P. and Mayfield, M. (1992) Hurricane experience levels of coastal county populations from Texas to Maine. *NOAA Tech. Memo. NWS NHC-46*.

Jarvinen, B., Neumann, C. and Davis, M. (1984) A tropical cyclone data tape for the North Atlantic Basin, 1886–1983: Contents, limitations, and uses. *NOAA Technical Memorandum NWS NHC 22*. Coral Gables, Florida.

Jayaram, N., Shome, N. and Rahnama, M. (2012) Development of earthquake damage functions for tall buildings, *Earthquake Engineering and Structural Dynamics*, **41** (11), 1495–1514.

Kappes, M.S., Keiler, M., von Elverfeldt, K. and Glade, T. (2012) Challenges of analyzing multi-hazard risk: a review. *Natural Hazards*, **64**, 1925–1958.

Kashani, M.H. and Dinpashoh, Y. (2012) Evaluation of efficiency of different estimation methods for missing climatological data. *Stochastic Environmental Resarch and Risk Assessment*, **26**, 59–71.

Keef, C., Tawn, J. and Svensson, C. (2009) Spatial risk assessment for extreme river flows. *Journal of Royal Statistical Society C*, **58** (5), 601–618.

Keller, J.L., Dailey, P.S. and Fischer, M.D. (2004) AMS ANNUAL MEETING-CD-ROM EDITION, 84; 5.1 Applied Climatology, 14th Conference.

Khanduri, A. and Morrow, G. (2003) Vulnerability of buildings to windstorms and insurance loss estimation. *Journal of Wind Engineering and Industrial Aerodynamics*, **91**, 455–467.

Kirchmeier, M., Lorenz, D. and Vimont, D. (2014) Statistical downscaling of daily wind speed variations. *Journal of Applied Meteorology and Climatology*, **53** (3), 660–675.

Kjedsen, T. and Jones, D. (2009) A formal statistical model for pooled analysis of extreme floods. *Hydrology Research*, **40** (5), 465–480.

Kramer, S. (1996) *Geotechnical Earthquake Engineering*, Prentice Hall. Englewood Cliffs, NJ.

Kunreuther, H. and Useem, M. (2010) *Learning from Catastrophes: Strategies for Reaction and Response*, Wharton School Publishing, Pennsylvania, PA.

Lamb, H. (1991) *Historic Storms of the North Sea, British Isles and Northwest Europe*, Cambridge: Cambridge University Press.

Lamb, R., Keef, C., Tawn, J.A. *et al.* (2010). A new method to assess the risk of local and widespread flooding on rivers and coasts. *Journal of Flood Risk Management*, **3**, 323–336.

Landsea, C., Franklin, J. and Bevan, J. (2015) The revised Atlantic hurricane database (HURDAT2), http://www.nhc.noaa.gov/data/hurdat/hurdat2-format-atlantic.pdf (accessed 16 June 2016).

Lhota, J. (2012) Statement from MTA chairman Joseph J. Lhota on Service Recovery, 30 October, http://live.reuters.com/Event/Tracking_Storm_Sandy/54277687 (accessed 10 May 2016).

Loveless, J. and Meade, B. (2011) Spatial correlation of interseismic coupling and coseismic rupture extent of the 2011 MW = 9.0 Tohokuoki earthquake. *Geophysical Research Letters*, **38**, L17306.

Lozowski, E. and Strong, G. (1978) On the calibration of hail pads. *Journal of Applied Meteorology*, **17**, 521–528.

Macdonald, N. (2013) Reassessing flood frequency for the River Trent, Central England, since AD 1320. *Hydrology Research*, **44** (2), 215–233.

Mandelbrot, B. (1967) How long is the coast of Britain? Statistical self-similarity and fractional dimension. *Science*, **156**, (3775), 636–638.

Minas, S., Galasso, C. and Rossetto, T. (2015) Assessing spectral shape-based intensity measures for simplified fragility analysis of mid-rise reinforced concrete buildings, paper presented at SECED 2015 Conference: Earthquake Risk and Engineering towards a Resilient World, 9–10 July 2015, Cambridge, UK, http://www.seced.org.uk/images/newsletters/MINAS,%20GALASSO, %20ROSSETTO.pdf (accessed 16 June 2016).

Mitchell, D. and Netravali, A. (1988) Reconstruction filters in computer graphics. *Computer Graphics*, **22**, 4.

Nafari, R., Ngo, T. and Lehman, W. (2016) Calibration and validation of FLFAs – a new flood loss function for Australian residential structures. *Natural Hazards and Earth System Sciences*, **16**, 15–27.

OASIS Loss Modelling Framework. (2015). OASIS Financial Module, White paper 20th July 2015, http://www.oasislmf.org/the-toolkit/documentation/download/669/ (accessed 28 February 2017).

Olsen, A. and Porter, K. (2010) What we know about demand surge. Structural Engineering and Structural Mechanics, Department of Civil Environmental and Architectural Engineering, University of Colorado, Boulder, CO, http://www.sparisk.com/pubs/Olsen-2010-CU-WWKADS. pdf (accessed 16 June 2016).

Parsons, T., Toda, S., Stein, R., Barka, A., and Dieterich, J. (2000) Heightened odds of large earthquakes near Istanbul: An interaction-based probability calculation. *Science*, **288**, 661–665.

Parsons, T. (2004) Recalculated probability of M>7 earthquakes beneath the Sea of Marmara, Turkey. *Journal of Geophysical Research*, **109**.

Peiris, N. (2015) Vulnerability modelling for insurance loss estimation – what are the challenges? Paper presented at SECED 2015 Conference: Earthquake Risk and Engineering towards a Resilient World, 9–10 July 2015, Cambridge, UK, http://www.seced.org.uk/images/newsletters/ PEIRIS.pdf (accessed 16 June 2016).

Petersen, M.D., Moschetti, M.P., Powers, P.M. *et al.* (2014) Documentation for the 2014 update of the United States national seismic hazard maps: *USGS Open File Report* 2014–1091.

Pielke, R., Jr. Gratz, J., Landsea, C. *et al.* (2008) Normalized hurricane damage in the United States: 1900–2005. *Natural Hazards Review*, **9**, 29–42.

Pinelli, J.P., Pita, G., Gurley, K. *et al.* (2011) Damage characterization: Application to the Florida Public Hurricane Loss Model. *Natural Hazards Review*, **12** (4), 190–195.

Pinelli, J.P., Subramanian, C.S., Artiles, A., Gurley, K., and Hamid, S. (2006) Validation of a probabilistic model for hurricane insurance loss projections in Florida. *Proceedings and Monographs in Engineering, Water and Earth Sciences*. **1–3**, 1377–1384.

Pinto, J., Bellenbaum, N., Karremann, M. and Della-Marta, P. (2013) Serial clustering of extratropical cyclones over the North Atlantic and Europe under recent and future climatic conditions. *Journal of Geophysical Research*, **118** (22), 12476–12485.

Pita, G., Pinelli, J., Cocke, S. *et al.* (2012) Assessment of hurricane-induced internal damage to low-rise buildings in the Florida Public Hurricane Loss Model. *Journal of Wind Engineering and Industrial Aerodynamics*, **104–106** (2012), 76–87.

Porter, K. (2016) *A Beginner's Guide to Fragility, Vulnerability, and Risk,* http://www.sparisk.com/pubs/Porter-beginners-guide.pdf (accessed 16 June 2016).

Porter, K., Farokhnia, K., Cho, I. *et al.* (2012) Global vulnerability estimation for the global earthquake model. Proceedings of the 15 WCEE, 24–28 Sept 2012. Lisbon. Paper No. 4504.

Porter, K., Kennedy, R. and Bachman, R. (2007) Creating fragility functions for performance-based earthquake engineering. *Earthquake Spectra*, **23** (2), 471–489.

Porter, K. and Ramer, K. (2012) Estimating earthquake-induced failure probability and downtime of critical facilities. *Journal of Business Continuity & Emergency Planning*, **5** (4), 352–364.

Powell, M.D., Murillo, S., Dodge, P. *et al.* (2010) Reconstruction of Hurricane Katrina's wind fields for storm surge and wave hindcasting. *Ocean Engineering*, **37**, 26–36.

Powell, M.D., Soukup, G., Cocke, S. *et al.* (2005) State of Florida hurricane loss projection model: Atmospheric science component. *Journal of Wind Engineering and Industrial Aerodynamics*, **93**, 651–674.

Ralph, D. (2013) Modelling the effect of catastrophes on international supply chains. Chapter 6, Supply Chain and Contingent Business Interruption (CBI). A perspective on Property and Casualty. SCOR, https://www.scor.com/images/focus_cbi.pdf (accessed 16 June 2016).

Raper, J., Rhind, D. and Shepherd J. (1992) *Postcodes: The New Geography*. Longman, Harlow.

Reid, H.F. (1910) *The Mechanics of the Earthquake: The California Earthquake of April 18, 1906,* Report of the State Investigation Commission, Vol. 2, Carnegie Institution of Washington, Washington, DC.

RMS (2012) Principles of Model Validation: United States Hurricane Model, http://forms2.rms.com/rs/729-DJX-565/images/tc_2012_principles_model_validation_us.pdf (accessed 16 June 2016).

RMS (2013) The 2012 UK Floods. RMS White Paper, http://forms2.rms.com/rs/729-DJX-565/images/fl_2012_uk_floods.pdf (accessed 16 June 2016).

RMS (2015) Tianjin is a wake-up call for the marine industry. http://www.rms.com/blog/2015/11/18/tianjin-is-a-wake-up-call-for-the-marine-industry/ (accessed 16/06 2016).

Roberts, J., Champion, A., Dawkins, L. *et al.* (2014) The XWS open access catalogue of extreme European windstorms from 1979–2012. *Natural Hazards and Earth System Sciences*, **14**, 2487–2501.

Rossetto, T., Ioannou, I., and Grant, D. (2013) Existing empirical vulnerability and fragility functions: Compendium and guide for selection. GEM Technical Report 2013-X, GEM Foundation, Pavia, Italy.

Sampson, C., Fewtrell, T., O'Loughlin, K., *et al.* (2014) The impact of uncertain precipitation data on insurance loss estimates using a flood catastrophe model. *Hydrology Earth System Sciences*, **18**, 2305–2324.

Schneider, T. (2001) Analysis of incomplete climate data: Estimation of mean values and covariance matrices and imputation of missing values. *Journal of Climate*, **14**, 853–871.

Schreck, C. (2013) The Climate Data Guide: IBTrACS: Tropical cyclone best track data. https://climatedataguide.ucar.edu/climate-data/ibtracs-tropical-cyclone-best-track-data (accessed 16 June 2016).

Schwierz, C., Koller-Heck, P., Mutter, E. *et al.* (2013) Modelling European winter wind storm losses in current and future climate. *Climate Change*, **101** (3–4), 485–514.

Shome, N., Jayaram, N. and Rahnama, M. (2012) Uncertainty and spatial correlation models for earthquake losses. Proceedings of the 15th WCEE, Lisbon, 2012.

Simmons, K., Czajkowski. J. and Done. J. (2016) Economic effectiveness of implementing a statewide building code: The case of Florida. Working Paper # 2016-01. Wharton School, Pennsylvania, PA, http://opim.wharton.upenn.edu/risk/library/WP201601_Simmons-Czajkowski-Done_Effectiveness-of-Florida-Building-Code.pdf (accessed 16 June 2016).

Skamarock, W., Klemp, J., Dudhia, J. *et al.* (2005) A description of the Advanced Research WRF version 2. NCAR Tech Note NCAR/TN-4681STR.

SMHI (2005) *Gudrun-Januaristormen 2005.* Swedish Meteorological and Hydrological Institute, http://www.smhi.se/kunskapsbanken/meteorologi/gudrun-januaristormen-2005-1.5300 (in Swedish) (accessed 10 May 2016).

Spence, R., So., E. and Scawthorn, C. (eds), (2011) Human casualties in earthquakes, in *Advances in Natural and Technical Hazards*, Springer Science & Business Media, Berlin.

Stedinger, J., Vogel, R. and Foufoula-Georgiou, E. (1993) Frequency analysis of extreme events, in *Handbook of Hydrology* (ed. D. Maidment), New York: McGraw-Hill.

Stein, T. (1997) Progressive failure on the North Anatolian fault since 1939 by earthquake stress triggering. *Geophysical Journal International*, **128**, 594–604.

Swiss Re (2013) Natural catastrophes and man-made disasters in 2012, *Sigma* No. 2/2013 http://www.swissre.com/media/news_releases/nr_20130327_sigma_natcat_2012.html (accessed 20 April 2016).

Taylor, A. and Glahn, R. (2008) Probabilistic guidance for hurricane storm surge. Preprints, 19th Conference on Probability and Statistics, New Orleans, LA, American Meteorological Society, 7 (4).

Thomson, M. and Page, B. (2011) Insured exposure development for estimating the financial consequences of windstorm risk in Europe. *Geophysical Research Abstracts*, **13**, EGU2011–7643.

Vickery, P. and Skerlj, P. (2005) Hurricane gust factors revisited, *Journal of Structural Engineering*, **131** (5), 825–832.

Vitolo, R., Stephenson, D., Cook, I. and Mitchell-Wallace, K. (2009) Serial clustering of intense European storms, *Meteorologische Zeitschrift*, **18**, 411–424.

Von Storch, H. and Zweiers, F. (1999) *Statistical Analysis in Climate Research*, Cambridge University Press, Cambridge.

Wainright, J. and Mulligan, M. (eds) (2013) *Finding Simplicity in Complexity* (2nd edn), John Wiley & Sons, Ltd, Chichester.

Wald, D., Jaiswal, K., Marano, K., Bausch, D., and Hearne, M. (2010) PAGER—Rapid assessment of an earthquake's impact. U.S. Geological Survey Fact Sheet 2010–3036.

Wald, D., Jaiswal, K., So, E. *et al.* (2011) The role of PAGER in improving global hazard building and loss inventories. Seismological Society of America Annual Meetings, Memphis, 13–15 April 2011. *Seismological Research Letters*, **82**, (2), 337.

Wang, Z., Hofmann, G. and Park, S. (2016) Distance dependent correlation in earthquake loss simulation, paper presented at the Reinsurance Association of America Conference, Orlando, FL, February 16–18, 2016.

Warner, T. (2011) *Numerical Weather and Climate Prediction*, Cambridge University Press, Cambridge.

Willoughby, H., Darling, R. and Rahn, M. (2006) Parametric representation of the primary hurricane vortex. Part II: A family of sectionally continuous profiles *Monthly Weather Review*, **134**, 1102–1120.

Winkel, G., Kaphengst, T., Herbert, S. *et al.* (2009) Final Report of EU policy options for the protection of European forests against harmful impacts: inc. Appendix 3, http://ec.europa.eu/environment/forests/pdf/Final_Report_Appendix_3.pdf (accessed 10 May 2016).

Woo, G. (2011) *Calculating Catastrophe*. World Scientific Press, Singapore.

Woodward, J., Tooth, S., Brewer, P. and Macklin, M. (2010). The 4th International Palaeoflood Workshop and Trends in Palaeoflood Science. *Global and Planetary Change* **70** (1–4), 1–4.

Wu, S., Jin, J. and Pan T. (2015) Empirical seismic vulnerability curve for mortality: case study of China. *Natural Hazards*, **77** (2), 645–662.

Yeo, G. and Cornell, C. (2005). Stochastic characterization and decision bases under time-dependent aftershock risk in performance-based earthquake engineering. PEER Report 2005/13 Pacific Earthquake Engineering Research Center, College of Engineering, University of California, Berkeley.

Zappa, G., Shaffrey, L. and Hodges, K. (2013) The ability of CMIP5 models to simulate North Atlantic extratropical cyclones. *Journal of Climate*, **26** (15), 5379–96.

Zhang, J. and Goodchild M. (2002) Uncertainty in geographic information, in *Research Monographs in Geographic Information Systems* (eds P. Fisher and J. Raper), Routledge, London.

# 5

# Developing a View of Risk

*Matthew Jones*

## 5.1 Overview

### 5.1.1    What Is Included

This chapter includes discussion of the reasons for developing a view of risk and the need to ensure suitable prioritization when assigning resource to this task. A conceptual model of the view of risk process is presented and discussed while also considering the importance of robust governance for this and other important processes. Details of how to develop a view of risk are provided, including documentation analysis, sensitivity testing, comparison to claims experience, assessment of multiple models and comparison to industry experience. The importance of understanding what is not in the model is emphasized, together with techniques for identifying and quantifying this non-modelled risk. Finally, practical considerations when implementing a view of risk are discussed, including the types of model adjustment and the topic of multiple model blending.

### 5.1.2    What Is Not Included

Precise details of model fusion, shoe-horning or morphing techniques are not included.

### 5.1.3    Why Read This Chapter?

Using catastrophe models without detailed understanding and appropriate adjustment can lead to wrong decisions. This chapter aims to provide the reader with a good depth of knowledge regarding the considerations and methods used to obtain a better understanding of such

*Natural Catastrophe Risk Management and Modelling: A Practitioner's Guide*, First Edition.
Edited by Kirsten Mitchell-Wallace, Matthew Jones, John Hillier and Matthew Foote.
© 2017 John Wiley & Sons Ltd. Published 2017 by John Wiley & Sons Ltd.

models. It also aims to equip the reader so that they will be able to implement any necessary adjustments, leading to better catastrophe risk-related decisions.

## 5.2 Introduction

This chapter focuses on the process of developing an appropriate understanding of the catastrophe risk faced by a company. Given that the main purpose of a (re)insurance company is to take on risk, at a suitable price, it is clear that understanding this risk is of vital importance. As discussed in Chapter 1, catastrophe risk is particularly difficult to understand because it cannot be assessed based on claims experience alone given its low-frequency nature. It is also important to understand this risk well because, unlike many other risk types, it can be difficult to diversify risk by writing more policies. The large spatial extent of catastrophes means that writing more business (even in a broad geographical area such as a country or state) can concentrate risk rather than diversify it. To help understand this risk better, specialist vendors producing catastrophe risk models, designed for (re)insurers to use, emerged in the 1980s (see Chapter 1.5 for more detail). However, for the reasons described in Section 5.2.1, it is not usually appropriate to use catastrophe model results without a proper investigation and evaluation exercise. There are also perils that are not covered by the existing models and so a way of including the risk from these non-modelled perils is needed. Additional work – beyond just licensing and running a model – is therefore required before a company can claim to understand its catastrophe risk.

There are various names given to the work of understanding whether models are suitable and adjusting them where necessary. In particular, **model validation**, **model evaluation** and developing a **view of risk** are terms often used interchangeably within the industry. In this chapter model evaluation is taken to mean the definition given in LMA/Lloyd's (2012a) as 'the process by which you determine whether the external catastrophe model provides a valid representation of catastrophe risk for your portfolio'.

The reason that the phrase 'model evaluation' is used in this chapter instead of 'model validation' is that in some contexts model validation may indicate a specific regulatory-focused purpose. For example, within the context of the **Solvency II** regulatory regime (EU-LEX 2015; Chapter 2.11.4), there is a requirement that an internal model validation team validate an entity's internal model and report up through the risk function to ensure independence. This does not mean, however, that all work of evaluating catastrophe models must be performed by the risk validation team, or that the catastrophe modelling team must sit within the risk function (see Section 5.3.10 for more details).

Evaluating a model is a prerequisite in understanding whether it properly represents the company's risk, however, it is usually then necessary to adjust and modify the model output, as well as considering items that are not within the model (**non-modelled risk**). In the context of this book, the term 'developing a view of risk' is taken to mean the overall process including adjustments to models, incorporation of non-modelled risk and implementation. Indeed, 'developing an *own* view of risk' is a phrase which seems to have entered the insurance industry's lexicon (at least within the area of catastrophe risk management) as it highlights the importance of a company owning their view of risk as a fundamental aspect of good management desirable in any risk-taking entity.

### 5.2.1 Why Develop a View of Risk?

The effort involved in building catastrophe models and the complexity and implied precision of model results can lead the uninformed user to the conclusion that the model output, in its raw form, may be relied on. Most of the time, this is not the case and the model output should only be used as a starting point in developing a view of risk that is appropriate for the entity and the intended usage. For any company in the business of assuming risk, that risk must be properly understood in the context of the *type of business written by the company*; unadjusted catastrophe model output is unlikely to be fit for this purpose. The reasons for this are as follows:

- *Models are incomplete.* Models include many aspects of the peril under consideration, but even the most sophisticated models will not contain all aspects of the real risk that the entity faces. Examples of the gaps in what is covered by the models (or non-modelled risk – see Section 5.4.7) include:
  - Omission of **allocated loss adjustment expenses (ALAE)**; these are insurance company expenses directly associated with settling a specific claim.
  - Missing **secondary perils** which are correlated with the primary peril; examples include a tsunami in an earthquake model (see Chapter 3.7.5) or riverine flooding induced by a land-falling hurricane (see Chapter 3.2.5).
  - An incomplete **event set**, for example, no magnitude 9.0 earthquakes along the fault line where the 2011 Japanese Tohoku event occurred.
  - In many cases there is no available model for an entire **peril-region**. These gaps must be systematically identified and quantified in some manner if the company's risk is to be represented completely.
- *Models should be used in context.* Models are built to serve many purposes and entities. They may, in unadjusted form, be unfit for the intended usage and company in question. For example, a model may be calibrated using claims data from a particular set of companies (see Chapter 4.8) whereas another company's claims experience could be different from that of the calibration set, for example, because the company writes business in a different niche or has a different claims handling process. It is extremely important that the individual characteristics of the company in question are reflected when using model output.
- *Different views of underlying model assumptions and parameters exist.* Models contain many assumptions about both the form of the underlying component models and the parameters within each of these models, all of which are subject to uncertainty (see Chapter 2.16). There may also be diverging scientific opinions (see Chapter 3), and different methods for analysing and summarizing the same data sets, leading to differences between different vendor models. It is useful as part of any model evaluation to compare and contrast modelling approaches and model results to help create a company's own view of risk. This evaluation must of course be done by individuals with sufficient scientific knowledge and understanding of the perils being modelled.
- *Evolution of knowledge.* As knowledge increases through scientific research and data from more recent events, model vendors update and change their models to incorporate this new information (see Chapter 4). For older models, the science underlying the model may be out-dated as research has progressed since the model was first built. Even for new models, scientific assumptions could still be out of date at point of release given the timescales involved in building, testing and releasing a model. The speed of model update depends upon the importance of the peril-region to the vendor's client base and the resource constraints and existing development plan of the vendor. If an insurance company's risk assessment is based only on the unadjusted output of the latest version of a particular model, then a model change (by the vendor) will flow through directly into the company's risk metrics (see Chapter 1.10). This should only happen if the company has made a conscious decision that it is appropriate. Ideally, the company's view of risk will also evolve with increased knowledge and so a new model version would contain few surprises. In practice, there will often be unexpected developments in a new model version. In addition, the modelling vendor may have invested in

new research not available prior to the release of that model, and thus not accessible to the model users. As new model versions are released, there should be an evaluation of the new version and a revised view of risk developed based on this new version before it is used operationally.

### 5.2.2   What Developing a View of Risk Involves

Developing a view of risk involves:

- gathering and analysing loss data relevant to the risk being modelled;
- investigating models that are available and evaluating them;
- modifying model output where necessary;
- ensuring non-modelled risk is considered;
- documenting the analyses;
- ensuring sign-off at the appropriate level;
- implementing the changes.

Evaluating the models is a critical component of developing a view of risk, involving using appropriately qualified people to investigate the model assumptions (by reading the technical model documentation, speaking to the model developers, keeping up to date with the relevant science, and being familiar with the model calculation framework), running quantitative tests on the model and assessing which elements may be missing in the model.

Once a view of risk has been developed, it should be justifiable to many stakeholders. A wide audience, from technical experts and management within the company to external regulators and rating agencies (see Chapter 2.11), will need to understand why a certain view of risk has been applied. This means that a comprehensive technical piece of work justifying a view of risk should be performed and that strong **governance**, including good documentation and a model change management process, are needed before a view of risk is changed.

Different levels of complexity are possible when applying a view of risk. Implementation can be as simple as multiplying all model output by a factor, for example, to scale up results for non-modelled items. However, it can also be as complex as adjusting specific vulnerability curves for certain occupancy classes within specific wind speed ranges for particular geographies. The approach adopted will depend on the risk profile and resource levels of the company, whether it is a reinsurer or insurer, the materiality of the risk in question, the granularity of the input data and the openness (transparency and adjustability) of the model being used.

It is clear that there is a range of possible effort in developing a view of risk. Given the resource constraints that all companies face, there needs to be a robust prioritization process around which perils need most focus.

### 5.2.3   Practical Considerations in a Resource-Constrained World

Developing a view of risk is a time-consuming process; there is an almost limitless depth to the work that can be done. In the real world, resources are constrained; developing a view of risk is just one step in the process of understanding and managing catastrophe risk, and catastrophe risk itself is just one risk among many a company might face.

In practice, it will be necessary to prioritize resource. If no resource is available to focus on this, and the company is taking on a significant level of catastrophe risk, then the appropriate resources should be put in place. The process of understanding and owning the risk is fundamental to the success or failure of a company when it assumes a meaningful amount of catastrophe risk.

Given resource constraints, two elements are necessary to prioritize the work: (i) a comparison of the significance of catastrophe risk to other risk types; and (ii) within the catastrophe risk type (if the catastrophe risk is significant), a comparison of the relative significance of the peril-regions to which a company is exposed. Some common methods to do this are as follows:

1) For each peril-region of interest, determine the premium written and rank in descending order.
2) As (1), except use the net exposed limit (or net sum insured).
3) If modelled metrics are available, use a tail metric such as 1:100 VaR or TVaR (see Chapter 1.10)

None of these methods are perfect. Method 1 will not yield an appropriate prioritization if the premium is driven by non-catastrophe risk types. Method 2 does not take into account the difference in hazard of each peril-region. Method 3 will not be available for all peril-regions (as modelled metrics will not be available for each peril-region). In practice, a combination of the above methods is often used. It is helpful then to categorize each peril-region into a significance (or materiality) risk-tier (say, low, medium and high) with a differential amount of effort applied to the view of risk development process for each.

The rest of this chapter will expand on how to develop a view of risk, together with implementation and governance considerations; it is mostly geared towards the type of process that could be carried out for a peril-region with a highly material level of risk. First, however, some differences in perspective between insurance and reinsurance companies are considered.

### 5.2.4 Insurance Versus Reinsurance

Both insurance and reinsurance companies, which are exposed to catastrophe risk, will need to develop an own view of risk. While the rationale, practical considerations, governance and methods are somewhat similar between an insurer and a reinsurer, there are some differences in approach related to forming a view of risk, in particular:

- An insurer should have more detailed knowledge of the original business they are writing and should be aware of any differences between their risk selection and claims settlement procedures, compared to the industry average, for example, specific underwriting criteria which mean certain high risk flood zones are avoided. An insurer will therefore be more interested in the modelled loss from their specific portfolio than the modelled loss for the industry as a whole (although this will still be of some interest to an insurer so that they can ensure they are not over-represented in specific areas).
- The translation of the original exposure data into the catastrophe model will be carried out by the insurer (or delegated to the reinsurance broker or outsourced provider; see Chapter 2.5.4.2). The checks, assumptions and translations that are carried out are specific to the insurer and may result in a view of risk being adjusted, for example, if specific financial structures cannot be coded into a particular model, the insurer will be aware of this and will need to adjust for the impact.
- An insurer may have access to detailed claims data and loss adjuster reports. If these data, along with the relevant exposure data, are suitably organized and available, they should be helpful in performing actual versus modelled claims analyses to test the appropriateness of the vulnerability within a catastrophe model (see Section 5.4.3).
- A reinsurer will have access to the portfolios of multiple insurers. They will therefore be more interested in the modelled loss across a wide variety of portfolios than the losses from any one particular portfolio (unless one portfolio drives their overall losses). A reinsurer is therefore

likely to be more interested in how modelled losses compare to overall industry losses than an insurer (see Section 5.4.5).

- A reinsurer will normally have more knowledge and interest across an entire peril-region, whereas an insurer may only be interested in specific geographical parts. For example, an insurance company may take on risk in specific states in the United States (or specific countries within Europe) while a reinsurance company probably has more exposure over the entire peril-region.
- A reinsurer will have information presented to them by the insurer and reinsurance broker in respect of the data checks and modelling assumptions (and sometimes view of risk assumptions) made by the insurer as part of the reinsurance submission (see Chapter 2.6.4.2). They will need to decide how much reliance they put on this information when forming their view of risk.
- A reinsurer will have their own view of risk which may require translation of multiple underlying portfolios, so may prefer to also receive unadjusted results.
- Although a reinsurer will not always have access to detailed claims data (sometimes this is possible through specific arrangements with an insurer or may be part of the claims payment process), they will know the aggregate losses for each event for each company they reinsure. They will therefore have a good insight into the variation in catastrophe losses between different insurer portfolios and which insurers perform better than others (e.g. in terms of a comparison between actual and modelled losses for different insurers). Reinsurers will also have a larger amount of overall loss information (albeit aggregated) than any single insurer to help refine their view of risk.

## 5.3 Governance and Model Change Management

### 5.3.1 Governance

Catastrophe models are used for many different purposes in an organization, as described in Chapter 2. The output from these models is used to make important decisions, for example, about the level of **outwards reinsurance** to purchase, about the **solvency** level of a company, about where to grow (and where not to), and about the **pricing** and **profitability** (Chapter 2.6) of contracts. Given this, it is important that the models are *selected and used in an appropriate and controlled way* through suitable governance. Any (re)insurance company will have governance around many aspects of conducting their business. However, the complexity of catastrophe models, the fact they are usually built by a company external to the (re)insurance company and the fact that they normally need to be adjusted or supplemented with additional information, make it appropriate to discuss and consider governance specifically in the context of catastrophe models.

As catastrophe models become more prevalent and embedded within (re)insurance companies, **regulators** and **rating agencies** (see Chapter 2.11) also require more assurance that models are being used appropriately. For example, the European Solvency II (SII) regulations (see Chapter 2.11.4) require that the (re)insurance entity is responsible for all aspects of its **internal model**, including external components (built by someone other than the (re)insurer, e.g. a catastrophe model). Many of the standards outlined in SII apply equally to these external components of the model. In particular, the company must be able to justify the use of an external model, such as a catastrophe model, to the regulator; it is not sufficient just to say that it is a market-leading model used by many entities. Ratings agency requirements (see Chapter 2.11.2) are also increasing through more detailed and comprehensive questionnaires and scrutiny of the company's **enterprise risk management** (see Chapter 1.2) processes.

Irrespective of regulator and rating agency focus, it is, in any case, good management to ensure that the risk that a company assumes is understood and to provide evidence of this understanding. This means that robust governance is an important aspect of the catastrophe risk management of a company.

Good governance will include the identification of key processes, documentation of these processes and appropriate and clearly identified decision-makers for the decision points within each process. It will also include an appropriate peer-review process. Examples of some catastrophe-related processes that should be governed appropriately for a typical (re)insurance company are:

- *Catastrophe event response process*: What happens when a catastrophic event is impending or has just happened? Who receives the modelled event loss information and how is this used? What sort of information can be expected within specific timescales, and how does this vary by peril? Often this information is very sensitive (e.g. it could impact the share price of a company) and so is usually limited to a need-to-know basis (see Chapter 2.9).
- *Reporting process*: What information is reported, at what frequency, for what purpose and to whom (see Chapter 2.7.2)?
- *Pricing process*: How are catastrophe models used in pricing? What leeway is there to adjust the model output for pricing purposes? Who is responsible for this in the company (see Chapter 2.6.2)?
- *Planning or business forecast process*: How is catastrophe model output included in the forward planning process to understand the impacts of growing or shrinking in certain areas due to varying underwriting strategies (see Chapter 2.8.1)?
- *Capital modelling process*: How is catastrophe model output used to calculate **regulatory** or **economic risk-based capital** (see Chapter 2.6.2.5)?
- *Risk tolerance*: What metrics are used for risk tolerance and what are the limits that these metrics are compared to? How often are these reviewed? What happens if a limit is breached? Who can change the limits (see Chapter 2.10.1)?
- *Model selection process*: How should a model, or models, be selected in the first place? How frequently should this be revisited? Does this vary according to the significance of the peril being modelled? Ideally the selection of a vendor should also include an assessment of their models' suitability and adequacy, if not full model evaluation.
- *View of risk process*: Is the model fit for purpose for the portfolio being modelled or does it need adjusting? This process could be combined with the model selection process, but traditionally a suite of models covering a variety of perils will be licensed from one vendor for operational and commercial reasons and so the process of deciding which vendor to select may well be different from the process of deciding whether a model is fit for purpose and the adjustments to apply. This should also include consideration of non-modelled risk (or *model completeness* in Lloyd's terminology; see Section 5.4.7). It should include the frequency of review for a particular view of risk, who should do the investigations and propose a particular view of risk, and who should sign off on a view of risk change. It should also consider how quickly a view of risk should be implemented in the different areas of catastrophe model usage once it has been signed off.

The remainder of Section 5.3 focuses only on the governance around the process of creating a view of risk (including model selection). Detail on how to *actually develop and implement* a view of risk is given in Sections 5.4 and 5.5.

### 5.3.2 The View of the Risk Process

Forming a view of risk is not a one-off piece of work but rather a continuous process. Given the resource constraints always present, it is also a process that lends itself to appropriate prioritization. Typical components within the process are shown in Figure 5.1.

The steps shown are discussed in more detail in the subsequent sub-sections.

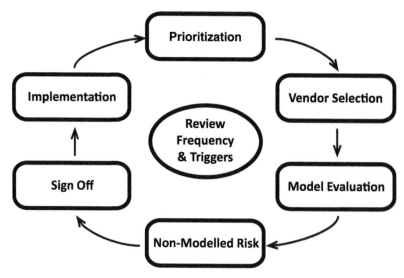

Figure 5.1 A conceptualization of the view of risk process.

### 5.3.3 Prioritization

Developing a view of risk is time-consuming and, as resources are always limited, it is important to prioritize this resource to focus on the peril-regions that are most important to the company. It can be helpful to lay out clearly the peril-regions in order of perceived risk to the company, using a suitable metric, before prioritizing them into different categories to help guide the resource allocated to each. Rankings are often based on premium written, net exposed limit, or on the basis of contribution to a tail-risk metric, as described in Section 5.2.3. This can then help drive how much budget is spent on model evaluation, how many people are assigned to it, how much expertise is built up in different knowledge areas, and the level of detail in the evaluation studies. Prioritization of perils can help when selecting models, as no vendor has complete coverage of all perils. A vendor with a better coverage of high-risk category perils for a particular company will be more useful to that company than one with a lower coverage (assuming the models are of the same quality). Developments in open, model-agnostic, platforms mean that in the future, the choice of model vendor could become independent of the platform vendor, and companies might be able to mix and match their model suite from different vendors.

Prioritization is a continuous process and as more is understood about a peril-region, it may increase or decrease in priority. Growth or reduction in business in particular areas could also cause different peril-regions to increase or decrease in risk and so in priority.

Priorities can sometimes be influenced by the development and model release agendas of vendors; models with new features that significantly change losses may be more desirable to evaluate compared to long-standing models that are not due to be updated.

If a company currently has no model coverage, then prioritization should take place based on knowledge of the exposure (e.g. **premium** or **sum insured/exposed limit**; see Chapter 1.2 and Chapter 1.9) and other information available on the level of hazard and vulnerability in each region. This may be, for example, from expert knowledge or published maps of natural peril hazards, or both. Clearly when (and if) the company acquires models, this initial view of risk can be refined as part of the continuous process.

In terms of governance, the prioritization methods, frequency and triggers for review should be clearly documented, and the appropriate people (or committees) responsible both for presenting the case for a change to prioritization and authorizing prioritization, identified.

### 5.3.4 Vendor Selection and High-Level Model Evaluation

In an ideal world, following the prioritization of perils and regions, a company would fully evaluate which model or models are best for the company and purchase accordingly; each model would be subject to a detailed model evaluation prior to purchase. In practice, model vendors often do not allow this sort of *evaluation* licence, although that is changing now with newer model vendors coming onto the market. Additionally, resource and cost constraints have meant that a company will traditionally not have been able to 'pick and choose' from different vendors; licensing a suite of models from one vendor usually attracts a substantial discount and there are steep increases in resource requirements for each additional vendor platform adopted. For this reason, it is possible that a company will run a model vendor selection exercise with the intent of licensing just one vendor's models. A typical exercise (which can also be used when picking and choosing different peril models from different vendors) will consider the following aspects:

- completeness of peril-region coverage compared to the company's prioritized perils;
- quality of models and appropriateness for the company's portfolio;
- number and quality of vendor development and quality assurance staff;
- quality of vendor documentation and other support (e.g. help desk, training, seminars, ability of staff to respond to questions, etc.);
- age of models, and frequency of update of models;
- ability of models to cope with the (re)insurance company's financial structures and lines of business;
- ability of platform to integrate with the (re)insurance company's systems;
- computational constraints and vendor software/hardware footprint.

It should be noted that there is always a barrier (including extra risk) to changing vendors. Long-standing use of a specific vendor means that processes and system integration will have evolved to accommodate this vendor's platform – these will all need to be revised. In addition, training of catastrophe risk analysts on the new platform will be needed, as will model evaluation of the new models. Model evaluation within a short time frame will prove problematic due to resource constraints (as the current vendor models will have been evaluated over a long period of time) and so it is likely that many of the new models adopted will essentially not be evaluated; with the attendant risks and surprise potential that this brings.

Governance on vendor selection needs to document the frequency of vendor reviews, the triggers for reviews, the criteria for selecting a vendor, and the decision-taking person or committee. Companies have faced criticism for changing vendors to maintain low model results and thus keep both the **technical price** (Chapter 2.6.2) and the **solvency capital requirement** (Chapter 2.10) low; known as 'model shopping'. The risk of a bias leading the company to select a vendor based on low model results should be explicitly addressed in the process design and documentation.

### 5.3.5 Detailed Model Evaluation

Once a model (or set of models) is purchased, a model evaluation study should be carried out for each peril-region. The level of detail this goes into will vary according to the priority assigned to the peril-region. Analyses (such as those described in this chapter) should be used to evaluate how well the model captures the catastrophe risk to the company's portfolio and the adjustments needed to ensure that the risk from the peril-region in question is appropriately represented. Any adjustments may differ by portfolio, return period or vulnerability type. If **model blending** (see Section 5.5.4) is being used, then the appropriate blending scheme should be selected. At the end of the evaluation work, a proposal should be made describing the extent of the work, its conclusions and the modifications to the pure model results required to provide a complete and appropriate view of risk for the company for that peril-region. A suitably senior individual or committee should authorize this.

A set of generic (i.e. not catastrophe risk-specific) validation tests or tools for internal models are described in Lloyd's (2014). These are summarized below:

1) **Sensitivity testing** means varying the inputs to the model and observing the changes to the outputs. Typically one input is varied at a time in order to quantify the sensitivity of the model output to that input.

2) **Scenario testing** is similar to sensitivity testing but means changing multiple inputs simultaneously (to reflect a particular scenario) to see how the model results change.

3) **Stress testing** is similar to sensitivity or scenario testing, however, it focuses on the most extreme or stressful events (or changes in inputs). For example, a stress test could consider multiple scenarios in the same year.

4) **Reverse stress testing** involves starting from the perspective of the scale of loss that could threaten the viability of a company and aiming to understand the type of scenarios (multiple or single) that could give rise to this scale of loss.

5) **Stability testing** focuses on understanding how sensitive model results are to different number of simulations.

6) **Back testing** means running models with historical data and comparing the results from the models to what actually happened.

7) **Profit and loss attribution** means understanding the actual causes of profit and loss each year and checking that the model is constructed in such a way that the causes of profit and loss are reflected in the model. It bears some similarity to back testing since in both cases historical experience is being used to understand whether a model is appropriate. (Profit and loss attribution in the context of Solvency II is discussed in Chapter 2.11.4.)

8) **Benchmarking** means comparing the results of a model to other benchmarks. For example, the results from the model currently used by a company could be compared to the results from other models.

9) **Analysis of change** means comparing how the assumptions and outputs of a model change from one version to the next.

10) **Model functioning tests** investigate whether a model has been constructed properly and therefore functions correctly. For example, a test set of model inputs that give rise to known outputs (calculated outside the model) could be developed. If these inputs to the model give outputs that are different from those expected, the model is not functioning correctly.

Although these are generic validation tests for any (or indeed all) types of risk, they are applicable to catastrophe model evaluation.

It is very important that this part of the process is well documented, as such evidence is frequently required in order to satisfy external stakeholders such as rating agencies and regulators. The documentation should include records of all meetings where information has been gathered to help inform a view of risk (e.g. meetings with model developers) because evidence of these meetings (and thus the level of knowledge transfer) may be required by regulators.

### 5.3.6 Non-Modelled Peril Evaluation

A complete view of risk process should always include an evaluation of non-modelled risk, as described in Section 5.4.7. In practice, this will usually result in either an adjustment to existing model output (to account for non-modelled risk which is correlated with modelled risk) or a set of simplified models in order to fill the gaps (e.g. for non-modelled perils).

From a governance perspective, it is important that the process of identification, evaluation and quantification of the non-modelled risk is well designed and documented. From a sign-off perspective the same principles outlined in the next section can be followed.

### 5.3.7 Sign-off

There is no *correct* sign-off process that will suit all companies; the best approach will depend upon the significance of the catastrophe risk compared to other risks within the company, as well as the company's existing governance structures. Some principles helpful to the process of signing off a view of risk are provided below.

A senior executive, or senior committee, should authorize each view of risk proposal. The level of seniority will depend upon the significance of the risk. Given a prioritization of perils it is common for the more senior committee (or person) to delegate sign-off to a lower level committee (or person) for the lower priority perils, or for changes that are relatively insignificant.

Given the technical and complex nature of the work, it is often useful to have a technical committee to review and challenge the initial work and suggest any changes prior to it progressing towards a more senior committee for sign-off. The technical committee could be comprised of catastrophe analysts not directly involved in the original evaluation. This extra level of review requires more resources but can provide a suitable level of challenge and can be helpful when it is unrealistic to expect a senior committee to grasp the full complexities of the work that has been performed. If the senior committee know that there has already been substantive technical challenge, this can be reassuring when providing sign-off. For a particularly significant peril-region, external technical review can be helpful to provide assurance.

A robust process can be developed by ensuring the person or committee responsible for sign-off is organizationally independent from the team performing and proposing the view of risk. For example, if the catastrophe modelling team is located within the underwriting function, a risk or actuarial function committee could authorize the view of risk. This helps ensure an independent review of the work and avoids potential conflicts of interest.

From a governance perspective, the sign-off process should be well documented (including any delegation of authority for the less significant peril-regions) and the minutes and presentations from sign-off meetings should be stored as evidence. These may well need to be provided to regulators at some future date to prove good governance around the view of risk.

### 5.3.8 Implementation

Once sign-off is complete, the view of risk should be implemented within the affected areas in the organization. Depending on the tools used in each area, it may be impossible to implement a view of risk exactly the same way in each area, and if this is the case, it should be noted as part of the sign-off process. The impacts of the implementation should be properly communicated, and again should form part of the proposal and sign-off process, so that stakeholders are informed as to the impact of the change in the view of risk.

Governance around implementation should include documentation about which areas the view of risk will flow into and the timescales in between sign-off and expected implementation. Also, any common differences between the signed-off view and practical implementation should be noted. It is usual that governance and documentation in other business areas can be referenced and should be consistent. For example, a technical pricing policy may specify one month as the timescale between completion of actuarial pricing analysis for a line of business and subsequent update of the pricing tool. In this case, it would make sense to align the period between the sign-off of a view of risk and its implementation within the pricing tool to the same one-month time scale.

### 5.3.9 Review Triggers and Frequency

The view of risk governance process should include documentation of how frequently view of risk analyses should take place and what events could trigger an analysis before the scheduled review date. It makes sense to vary the frequency of review according to the prioritization of perils, with perils of higher priority being reviewed more frequently. The frequency will depend upon the resource available for this kind of work. Triggers for an off-cycle review could include:

- a significant catastrophe event providing new insights and claims experience;
- a new model (or new version of an existing model) being released;
- new science emerging;
- new data sets relevant to the region of interest.

A log should be kept of the date of each review as a regulator or auditor may want to compare the actual review frequency to that documented in the governance policy.

### 5.3.10 Other Governance Aspects

Governance developed for a view of risk process should fit within the wider governance in a (re) insurance entity, and also with any regulatory imposed governance. For example, Solvency II specifies that the risk management function should be responsible for the internal model and for the validation of the internal model. In many companies under the Solvency II regime, this means there will be a separate validation team within the risk function with a remit to review the internal model and its components – including catastrophe risk. This does not mean that this risk validation team should do all the catastrophe risk model evaluation, nor does it mean that the catastrophe modelling team should be part of the risk function. Rather, it means that there should be an extra independent check (provided by the risk validation team) on the view of risk representation with the internal model.

**Enterprise risk management** has the fundamental concept of three *lines of defence* against unintended risk (Sweeting, 2011). The first line is the relevant business function or department, the second line is the risk function, and the third line is the audit function. The concept is that functions other than risk and audit (e.g. underwriting or claims) already have fundamental business as usual responsibilities for managing risk, and so they will have governance and controls in place to ensure that the risk they are responsible for is managed appropriately. The governance described in this chapter is part of this *first-line* governance. The risk and compliance function (*second-line*) is responsible for forming the risk management strategy, facilitating the determination of risk appetite (see Chapter 2.10.1), and providing oversight and reporting into the board risk committee. The audit function (*third-line*) provides independent assurance to the risk governance framework and typically reports directly into the board. Any governance related to the catastrophe view of risk process should be consistent with the wider organizational governance process and also take advantage of it. For example, the risk function may be able to provide reviews to provide assurance for significant peril-regions, perhaps support in terms of constructing a suitable process for non-modelled risk (see Section 5.4.7), and a compliance reporting system.

Peer reviews are a form of governance often used in a typical (re)insurance entity. For example, actuarial peer reviews within a reserving or pricing department are a relatively well-established form of governance within many companies. Underwriting reviews, where a particular segment of business is reviewed by underwriters from a different segment, are also reasonably common. Within the context of developing a view of risk, a technical peer review ahead of the sign-off process has already been mentioned, but this peer review concept could also be extended to the review of the catastrophe analysis of individual account pricing, for example.

Calder, Couper and Lo (2012) introduce the concept of a top-down and bottom-up framework for model blending (see Section 5.5.4.1). An essential part of the governance within this framework is the monitoring of the overall level of bottom-up adjustments; both to inform future R&D priorities and also to ensure that the adjustments are not being abused (e.g. by a systematic lowering of the catastrophe technical price in order to write more business).

Although governance can often be viewed as a resource drain, robust governance is a fundamental aspect of catastrophe risk management. Governance will be put to the test through scrutiny from the risk and audit functions as well as regulators and rating agencies, so it is important to ensure it is well designed and fit for purpose.

## 5.4   How to Develop a View of Risk

There are many aspects to a view of risk analysis and this section focuses on the considerations and types of analysis that can be helpful when forming a view of risk. It is unashamedly the longest section in this chapter.

A typical in-depth view of risk analysis for a high-priority peril-region model should contain the following elements:

- A thorough study of all the relevant vendor documentation, noting the methodologies employed, key assumptions and data used within the model (Section 5.4.1). This will normally include discussions with the model developers, and perhaps external experts, as well as reading other relevant literature on the most important areas for the model in question.
- Several sets of exposure (**portfolios**) run through the model (Section 5.4.2) with output at various resolutions, typically including overall level, line of business level and policy level. The following sets of exposure are typically used:
  - an industry portfolio (Chapter 4.4);
  - the (re)insurers' current portfolio, or proxy for it;
  - a specially constructed test portfolio to enable sensitivity testing.
- A consideration of the exposure data quality and any potential impact on model results (Section 5.4.2.1).
- The sensitivity of model output to variations in the input (sensitivity testing), including the impact of different model settings, primary and secondary modifiers (Section 5.4.2.8) and correlated loss components (Section 5.4.2.10).
- The impact of financial structures on model output (Section 5.4.2.9).
- A comparison of actual claims experience with modelled claims (Section 5.4.3), both at industry level and claims experience specific to the (re)insurers' portfolio (back-testing).
- A comparison with other models, from other vendors (Section 5.4.4) where possible, given restrictions imposed by the previously discussed licensing issues (benchmarking).
- A comparison with the previous version from the same vendor (analysis of change; Section 5.4.2.4).
- A consideration of what is not explicitly modelled, and must therefore be included in some way (Section 5.4.7).

These areas, and others, are discussed in more detail in the following sections.

### 5.4.1   Understanding What Is in the Model

The first step in developing a view of risk based on a particular model is to understand how each component in the model in question has been constructed. The starting point is to study the available model documentation in detail. This may sound obvious but this step can sometimes

| Category | AIR | CORELOGIC | IF | RMS |
|---|---|---|---|---|
| Model and Version | Extra-tropical cyclone model for Europe v13.0 | Eurowind v.15.1 | European Windstorm Model v1 | EUWS v15.0 |
| Last Update | summer 2011 | Jul-14 | Released March 2014 | Jul-05 |
| Event Frequency Distribution | No specific parametric distribution assumed. Based on lock bootstrapping of a large storm data base in order to capture clustering | Negative binomial | Based on output from Global Climate Model (GCM). Events are assigned a weather type and frequency compared against observed frequencies (NCEP) and later calibrated | Poisson |
| Stochastic Event Set | 53,046 events in 10K year catalogue | 23,107 events | 12,044 events/4,731 simulated yrs | Low Frequency Set: 31,029 events (Low+High: 40,969 events) |
| Stochastic Event Set Development | Hybrid method used. Large sethistorical storms are identified via the MM5 mesoscale model. Hourly time series for each storm are then perturbed and the new footprints are placed into the 10K catalog according to the output of a bootstrapping process | A hybrid approach is used. Perturbations of historic footprints from 1960-2013 and an ensemble of 1200-year simulations of an Atmospheric Ocean General Circulation Model (AOGCM). | Stochastic set based on GCM output, where 4,730 years are identified and ranked. The 12,044 strongest events are extracted and statistical downscaled to high resolution. These are calibrated and bias corrected against observed climatology (NCEP) | Hybrid modelling: 1. Global circulation model (CAM) with statistical model 2. Dynamical downscaling (WRF) 3. Statistical downscaling (calibration on wind observ) 4. Boiling Down |
| Alternative Event Set | No | Empirical - frequencies based on historic experience (1960-2013) | No | Climate variability view (based on past 25 years) |
| Geographical Coverage | Austria, Belgium, Czech Republic, Denmark, Estonia, Finland, France, Monaco, Germany, Ireland, Latvia, Lithuania, Luxembourg, Netherlands, Norway, Poland, Sweden, Switzerland, UK | Austria, Belgium, Czech Republic, Denmark, Estonia, Finland, France, Germany, Hungary, Ireland, Latvia, Lithuania, Luxembourg, Monaco, Netherlands, Slovakia, Sweden, Switzerland, UK, Norway, Romania, Poland, Spain, Portugal, North Sea, Irish Sea, and the Baltic Sea | UK, Ireland, France, Belgium, Netherlands, Luxembourg, Denmark and Germany | UK, Ireland, France, Germany, Switzerland, Austria, Belgium, Netherlands, Luxembourg, Denmark, Norway, Sweden, Czech Republic, Poland, Slovakia |
| Subperils | Wind only is modelled. Frozen precipitation and freezing temperatures are not explicitly modeled but implicitly captured via calibration against losses. Storm surge available separately for eastern and southern England | 1. Storm surge (available for the UK, France and Sweden) 2. Scandinavian forestry risk (Sweden and Finland) | No subperils are included in the current version | Storm surge on the UK East Coast |

**Figure 5.2** Subset of a sample model comparison exhibit for four different European windstorm models (Waisman, 2015). *Source*: Reproduced with permission of Federico Waisman.

be neglected which ultimately results in uneducated or needless questions directed to the model provider (although this may be viewed as part of the service the vendor provides and the client pays for). Regulators will often ask for evidence that the (re)insurance company has the appropriate model vendor documentation available to develop an informed view of risk. The aim in studying the documentation is to gain a good understanding of the source data, component model forms and key assumptions within the model. This should be done while considering what is new in this model, compared to previous versions, to help understand any changes in results between versions. At the same time, the similarities or differences between the model in question and other available models should be noted in order to try and understand the differences in results and to help decide which model to give more weight to if multiple models are available and a blending scheme is used (Sections 5.4.4 and 5.5.4). A subset of a sample model comparison exhibit from Waisman (2015) is shown in Figure 5.2.

When considering how a model is built, it is worth highlighting some points in particular:

- *What period of time is the underlying hazard data from?* This is important because it represents a gauge for judging how far the model can be extrapolated in time. A model based on over 100 years of hazard data (e.g. North Atlantic hurricane) is likely to be more robust than one based on 20 years of hazard data (although this depends on the peril; higher frequency perils being able to be represented by shorter time periods). However, it is not only the length of the period in question, but also the relevance of the historical period to the period for which the risk is being assessed. The activity rates of perils such as tropical and extra-tropical cyclones feature inter-seasonal variability, and if a model is based on an abnormally quiescent (or active) period, then this may not be appropriate for the risk period for which the model is being used. See Chapters 3.2.1 and 3.3.1 for further information on this topic.

- *How much detailed claims data are used in the model?* It is important to understand exactly how much claims data are used in the model, and of what granularity the claims data are.

Claims data will typically be used in developing vulnerability functions (bottom-up) as well as ensuring overall model results seem sensible (top-down). Chapter 4.5 expands on this topic in more detail. Claims data are often scarce and so it is common to derive vulnerability functions through transferring knowledge from one country to another, with adjustments based on building code comparisons, engineering knowledge of construction standards and other techniques. Knowing how much detailed claims data are used in the development of country-specific or line of business-specific vulnerability functions will help in assigning credibility to the model assumptions. Ideally this should be split by line of business and building character-istics, as well as by hazard value. If a large amount of risk is emanating from a line of business for which there are very few detailed claims data used in the model calibration, then the results from this part of the model should be assigned a high uncertainty, allowing more adjustment to the model results if they appear to disagree with experience. It is also important to understand the process used by the modelling company to match claims to the exposure data (in force at the time of the loss) and to cleanse the claims data of additional sources of loss not covered by the model.

- *Is the science in the model associated with a consensus in the scientific community or is it new science that is still subject to debate and possible future revisions?* When assessing new or revised models, care must be taken where the latest *cutting-edge* science has been incorpo-rated. In general, this is a positive thing and very often is the underlying reason for a model revision. However, scientific opinion and new scientific results or theories are prone to revisions and modifications as they get challenged subsequent to being initially published; indeed, this is the main point of the scientific process. In contrast, the current scientific consensus can be too readily accepted and not challenged, with detrimental results; for example, the possibility of a magnitude 9 earthquake on the Japan Trench was discounted by the consensus scientific view before it actually happened in 2011. Clearly independent scientific expertise is useful here: either from within the company or through engaging an appropriate external scientific resource.

Reading through the model documentation is only constructive up to a point. It is just as likely to lead to more questions as it is to provide answers. These questions then need to be directed to the model vendor. For the particularly important or complex questions, a conversation with the model developers is probably the best way to seek clarity.

A common criticism of catastrophe models is that they are 'black boxes'. While there is certainly a trend now to move towards more open modelling with consistent frameworks, some models are more open than others. For example, Impact Forecasting's *Elements* platform and JBA's *JCALF* platform are examples of more transparent modelling with full access to vulnerability functions and hazard maps, and the Oasis initiative is an example of an open and consistent development framework. Wherever possible, the analyst should seek access to the underlying hazard and vulnerability information. It is enlightening to find out that although model documentation describes a model as having hundreds of vulnerability curves to deal with many types of exposure, in fact, there are far fewer separate vulnerability functions in the model as many of the combinations map back to the same underlying curve.

If hazard information is obtained from the model, this can often be compared to publications or data in the scientific community. For example, North Atlantic hurricane activity rates can be compared to rates derived from the official North Atlantic hurricane database (HURDAT,[1] see Chapter 3.2.8 and Chapter 4.3.3.4).

A sample comparison of US hurricane landfall rates for all hurricane categories is given in Figure 5.3. This shows hurricane landfall rate by region for five different models as well as from actual experience (HURDAT). There are significant overall differences in landfall rates both between models (e.g. between models 4 and 5) and between models and history (e.g. between

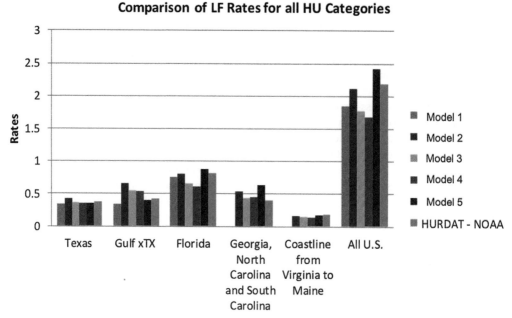

Figure 5.3 Comparison of US hurricane (HU) landfall (LF) rates between models and observation (HURDAT – NOAA) by region. *Source*: Reproduced with permission of SCOR Global P&C (SCOR, 2014).

models 1, 3 and 4 and HURDAT). The reasons for such differences should be investigated in order to understand whether a view of risk adjustment is needed for a particular model.

When performing such comparisons, it is important to understand detailed aspects such as:

1) What is the definition of landfall and does this vary between models and observed experience? For example. is landfall defined based on wind speed or central pressure? Is it the wind speed or central pressure *at landfall* or at the first observed point *before landfall*?
2) How are multiple landfalls, or bypassing storms (where the eye of the storm does not cross the coastline but damaging wind speeds still impact coastal locations), treated?
3) Are there likely to be biases in the observational data and can these be corrected for?

One must beware *not to expect an exact match between model hazard and available scientific data as the point of catastrophe models is to compensate for the lack of historical data, extrapolate beyond it, and fill in the gaps.* Thus, expertise is required to decide what is reasonable, acceptable or even preferred if trying to choose between two models. Nonetheless, any substantive differences between the model and observed data should be explained by the model vendor. For other perils, similar hazard metrics can be compared. For example, modelled versus observed earthquake size-frequency distributions are often compared using the parameters of the Gutenberg-Richter relationship $N = 10^{a-bM}$ (see Chapter 3.7.1). Flood frequency analyses use a measure of peril magnitude such as peak flow rate in a river in $m^3 s^{-1}$ (see Chapter 3.5.1).

If vulnerability information is obtained from the model, this can be compared to detailed claims data, if available, along with wind speed information and the associated exposure data (Section 5.4.3). The literature on vulnerability curves is much more limited than that for hazard and this is often an area where (re)insurance companies have more knowledge (and sometimes more claims data) than model vendors, leading to company-specific adjustments.

A reasonable amount of information can be found by reading documentation, talking to model developers, cross-checking with experts and scrutinizing the model tables; however, there is no substitute for running exposures through the model and analysing the results of these runs in order to understand how the model is behaving in practice. This process is described in Section 5.4.2.

### 5.4.2 Analysing Model Output (Including Sensitivity Testing)

Analysing the output from a catastrophe model is a key part of deciding whether the model is fit for the intended purpose. This includes running a set of exposures through the model and analysing a variety of *metrics* and *loss perspectives* across different *dimensions* (see Chapter 1.10 for examples of output metrics and loss perspectives). It also includes model sensitivity testing, which is the process of varying inputs to the model and analysing the model output in order to investigate the sensitivity of the model output metrics to changes in the inputs. This provides insight into how the model behaves which can be compared to the behaviour communicated by the model vendor; the two do not always match up and reasons for the differences should be understood. One can conduct many sensitivity tests to understand the model behaviour, such as which characteristics and secondary modifiers have the most impact on loss results. This in turn can be used to help the model user know where to focus most resources, both in terms of exposure data quality improvement (see Chapter 2.5.4.1) and understanding and evaluating the model.

The subsequent sections describe various aspects to consider when analysing model output.

#### 5.4.2.1 Data Quality and Exposure Coding

The aim of developing a view of risk is to understand how well a particular model represents the risk for a company's portfolio and, where necessary, to make adjustments to enable the model to better represent the risk. In this context, one of the reasons that a model may not be representing the risk appropriately is that of poor exposure data quality. Data quality is an important subject described in Chapter 2.5.4.1. For the purposes of this section it is sufficient to note that developing a robust view of risk is dependent on the quality of the data and, if the data quality substantially changes, the view of risk may well need to be revised.

#### 5.4.2.2 Defining the Input and Output Resolution

One of the first things to consider is the required resolution and choice of output, as this defines both the granularity of exposure input and which options are selected prior to running the model. The selected settings and resolution of output will depend upon the type of analysis required and the computing resources available (see Chapter 1.9).

A basic-level run will output a portfolio level **exceedance probability** (EP) curve and **event loss table** (ELT) or **year loss table** (YLT), while a more advanced (and computationally resource-intensive) run will output a location level ELT or YLT. See Chapter 1.10 for information on EP curves and ELT/YLTs.

The input exposures should include:

- An industry exposure portfolio in order to establish industry-level loss metrics. These industry metrics are often provided as part of the model documentation, but each vendor will have different industry portfolios and so the company evaluating the models may wish to run a bespoke and consistent industry portfolio through several different vendor models in order to gain a consistent comparison (see Section 5.4.5).
- The (re)insurance company's own portfolio (or proxy, reference or forecast portfolio) in order to ascertain the model metrics relevant to the company. These metrics will typically be output at the portfolio level as well as at class level, account level and some coarse geographical level (e.g. a US county).

- One or more sets of idealized portfolios in order to enable sensitivity testing and investigation of the geographical distribution of risk across key exposure classes. For example, a portfolio where each exposure is the same apart from an incremental change in the value of a specific modifier (**rating factor**) will give the sensitivity of the model output to this change in modifier. A file where every exposure is the same apart from a change in the location of the exposure (perhaps changing on a regular grid) will yield the geographical distribution of risk from the model. The output resolution for these types of analyses is typically the location level in order to properly quantify the model sensitivities.

### 5.4.2.3 Analysis Metrics, Perspectives and Dimensions

In any analysis it is important to consider which metrics (or values) will be studied, which loss perspectives are used, and across which dimensions they will be examined. Typical output metrics from catastrophe models are described in Chapter 1.10. They include the **average annual loss** (AAL), **standard deviation** (SD) of loss and **aggregate** and **occurrence exceedance probability** curves (AEP and OEP). ELTs or YLTs are fundamental forms of output from catastrophe models and these are described in Chapter 1.10.4; they can have a significant impact on the size of the output dataset and so the resolution at which event level output is turned on should be checked carefully against database size constraints or available disk space.

Loss perspective refers to which level of financial structures (deductibles, limits, coinsurance, reinsurance) have been applied to the losses. Financial structures are described in Chapter 1.9.2 (primary insurance) and Chapter 2.4.2 (reinsurance).

Typical dimensions upon which to investigate sensitivity in model results are:

- portfolio (a number of policies organized into a sensible grouping; for example, by business source, legal or business unit, or by country and line of business; Chapter 1.9.3);
- account/policy (Chapter 1.9.3);
- geography (e.g. by postcode, by county, on a regular grid);
- coverage (e.g. buildings, contents, business interruption; Chapter 1.9.1.2);
- peril and secondary peril (e.g. wind, surge; shake, tsunami; Chapter 3);
- primary modifiers (e.g. occupancy, construction type, number stories, year built; Chapter 1.9.1.3)
- line of business (e.g. property, construction, marine; Chapter 1.2).

The following sub-sections describe considerations when investigating result changes in the relevant metrics and loss perspectives across the above dimensions.

In practice, it is unlikely that there will be time available to use all of these analyses, and so often a top-down approach is necessary, first of all looking at the overall company portfolio changes, and then drilling down into the most significant drivers of change as necessary.

### 5.4.2.4 Model Version

Forming a view of risk is not a one-off process (see Section 5.3.2). Models are updated by vendors over time. The frequency of update will depend upon the importance of the model and emerging claims information and research. For a high importance peril-region like US Hurricane or European Windstorm, minor changes could be expected most years, with a complete model rebuild every three to five years or so.

Following a model version change, the same portfolio should be run through both the existing and new model versions to analyse changes in results for the key metrics and perspectives across the relevant dimensions. Even if overall portfolio changes are small, it is important to look into the detail (e.g. a breakdown by line of business and some geographical resolution) as changes at a more granular level can be large but offset each other. For example, overall model AAL changes could be small, whereas the AAL on a key policy could triple with the new model version; this large change is being masked by reductions in AAL from other policies. Ideally, both overall changes and policy-level changes will be investigated.

A consideration at this stage is whether the observed result changes make sense, given what has been learnt about the model from reading the documentation and discussing with the model developers. If they do not, detailed investigation should be undertaken to find out why.

It is important to note that IT limitations can complicate the comparison of different model versions; there is a need for significant investment in test environments as current generation platforms normally only allow the use of one single version of a given model.

Running, outputting and analysing the results of model change constitute a time-consuming process, but one that can be standardized. Sometimes model vendors or brokers will be able to provide this information. For example, they often provide studies comparing model versions for specific portfolios, drilling into certain features, together with some commentary on what is driving the changes.

### 5.4.2.5 Geographical Variations

A feature common to all catastrophic perils (and indeed implicit in the definition) is that of geographical correlation. Irrespective of peril, a certain event will affect a sizeable area leading to multiple locations being affected. This wide-ranging impact is what causes an accumulation of risk, which gives rise to the need for catastrophe models. It is therefore important to know how the risk from a catastrophe model varies by region, and to sense-check this against previous versions of the model, other models, other external data, claims experience or historical/third party risk assessment maps.

One way of understanding the geographical characteristics of a model is to construct a portfolio comprised of multiple exposures with exactly the same modifiers placed in a sensible geographical distribution (e.g. a regular grid) within the domain of interest. Running this through the model and obtaining location-level output for key metrics (e.g. for AAL) will result in a set of data that can be displayed as a risk map. Comparing this to risk maps built up in the same way from previous model versions or other vendor models is a useful way of identifying variation between results of different models and for different portfolios. Figure 5.4, from Waisman (2015), shows an example of such risk maps for four different European windstorm models.

For geographical analyses, care must be taken in differentiating between geographical change in hazard, geographical change in vulnerability and geographical changes in building stock.

*Spatial Variations Due to Building Stock*
Even with uniform hazard across a region, the risk can still vary geographically because of a regional variation in different types of building (building stock). For example, there may be a higher proportion of flats (multi-family dwellings) than houses in a city compared to a rural area. If flats are modelled as less vulnerable to a peril than houses, then there will be a decreasing effect in the modelled risk in cities compared to the rural areas due solely to the change in the building stock. This same effect could also apply to buildings of different ages, construction types, and heights. This, of course, is a desirable feature to have within a model – it is good to know that the risk is changing geographically because of geographical differences in the building stock. The reason that it can become an issue in geographical analyses is if we are aiming to better understand the model by eliminating changes in risk due to changes in the mix of building stock, in order to get better insight into the underlying hazard and vulnerability changes.

Models will usually define regions with the same assumed mix of building stock; in some models these are called *inventory regions*. A particular geographical area (e.g. postcode) within a model would be assigned a certain inventory region. This specific mix of buildings is then used where there is not enough information input by the model user. For example, the model user

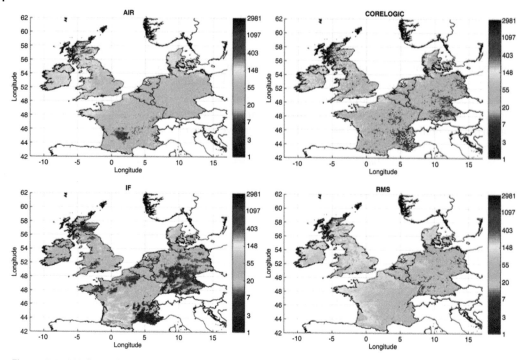

**Figure 5.4** AAL for uniform exposure on a regular grid from four different European windstorm models (Waisman, 2015): AIR (Touchstone v2 July 2011), Corelogic (Eurowind v15.1 July 2014), Impact Forecasting (IF; European Windstorm v1 March 2014), Risk Management Solutions (RMS; Risklink v15 March 2015). *Source*: Reproduced with permission of Federico Waisman.

may know that in a certain area there are residential buildings, but may not know whether these are houses or flats, nor the age of the buildings. In this case a model will typically look up the relevant inventory region for the area in question, and based on this will assign a **composite vulnerability curve** (i.e. one that is weighted across several different types of specific vulnerability) to try and reflect the building stock in the absence of full information.

Geographical analyses using a test portfolio comprising a number of standard risk locations on a standard grid must therefore have *all the primary modifiers specified explicitly* in the input exposure data for each location. This will ensure that composite vulnerability curves are not used by the model, which means that any changes in risk (e.g. changes in the model expected loss cost by location) are being driven by either hazard or vulnerability, and not by a variation in the building stock.

*Spatial Variations Due to Changes in Vulnerability*

Even with uniform hazard across a region, and with a uniform mix of building stock (or with this effect removed by specifying the explicit value combinations of all primary modifiers), the spatial risk can still change because the vulnerability in a model can change by region for the same type of building (e.g. a two-storey detached residential house built in 1950). These *vulnerability regions* try and reflect the variation in building regulations (building codes) or regional construction quality. Also, older properties, built prior to building regulations being put in place, can sometimes be more robust, as local builders developed building types to resist the harsher weather. These vulnerability regions try to account for the variations in vulnerability by region caused by construction changes that are too specific and detailed to be represented in the input exposure data, and therefore would not be taken into account in the model unless an explicit variation in vulnerability is specified by geography. For example, design wind loadings for UK wind risk are higher in the northern and western areas (because the hazard is higher) than other areas. A building in the north-west of the country would typically be constructed to these design wind speeds and should therefore be more resilient to wind damage than the same building in the south in order to mitigate the higher hazard in the north-west.

Conceptually one could imagine that, in a perfect world, the enhanced building regulations (being themselves based on some kind of hazard map) would cancel out the higher hazard, leading to a uniform risk. In the real world, there are many reasons why this is not the case, including:

- Hazard maps, upon which regulations are based, are not perfect and evolve with time; this means that an exact balancing off of increased hazard with increased resilience can never be achieved.
- Building regulations evolve with time, and so even if current regulations perfectly mitigated higher risk, houses built before regulations came into force will not be built to the current standards.
- Building regulations are typically focused on saving lives, not completely mitigating property damage. It is up to every country and region to determine their own risk tolerance: there are few global building regulations.
- Sometimes to save cost and time, regulations are not followed perfectly. This will depend upon the local culture and economy.

In order to understand the changes in risk due to this effect, it is important to establish, from the model vendor, the methodology for defining the vulnerability by region. If there is access to the exact vulnerability functions in the model, these can be compared (e.g. in the case of UK wind risk, comparing the vulnerability in north-west regions for a specific type of building to the vulnerability in southern regions for the same type of building). This modelled variation can be assessed by experts for appropriateness, or compared to detailed claims data if available (along with wind speed estimates or measurements).

If there is a large step-change in vulnerability across an artificially defined geographical line, this can have a number of undesirable consequences. For instance, a portfolio could become very sensitive to a small shift in the geographical distribution of exposures around the boundary region of the defined vulnerability areas. This is likely to be a particular issue if the step-change in vulnerability runs through an area of high exposure density. The second issue is that if the model is used for individual risk underwriting purposes, then a building on one side of the step-change could have a significantly different AAL (and thus catastrophe premium) when compared to a building on the other side of the line. This is unlikely to represent reality and may lead to loss of model credibility and adverse risk selection. Any such step-changes should be well tested with claims data from events if available.

### Spatial Variations in Hazard and Correlation

Once the impacts of inventory (building stock mix) and vulnerability regions have been understood, then the remaining focus should be on understanding the variation in the underlying hazard (also see Chapter 3).

If the model provider is sufficiently open and transparent in their approach, it may be possible to output hazard metrics directly. For example, there may exist the option to create an output of modelled peak gust wind speed against each location for each event, from which metrics such as the expected peak gust wind speed, or the 2% exceedance frequency peak gust wind speed, could be calculated. An example of this is shown in Figure 5.5.

These metrics could be compared to data from a third party provider independent from the vendor to help evaluate the hazard component of the model; ideally a separate data source to that used in the model building would be used, although this is not always possible.

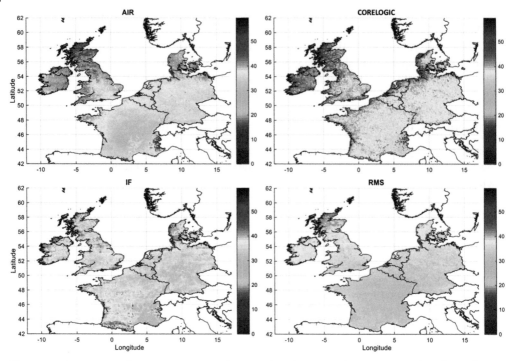

Figure 5.5 Comparison of 50 year return period 3-second peak gust wind speed (in m/s) from four different European windstorm models (Waisman, 2015): AIR (Touchstone v2 July 2011), Corelogic (Eurowind v15.1 July 2014), Impact Forecasting (IF; European Windstorm v1 March 2014), Risk Management Solutions (RMS; Risklink v15 March 2015). *Source*: Reproduced with permission of Federico Waisman.

If using external data for model evaluation, the evaluator must take steps to quality assure the data they are using, making any adjustments and corrections necessary for a true 'apples to apples' comparison. For example, model developers apply downscaling and roughness coefficients in their calculations of wind hazard at any location. Comparing model peak gust wind speeds at surface level to mean wind speeds collected from anemometers at 10 m height will therefore be misleading.

If functionality to output such hazard data directly does not exist, an appropriate financial metric, such as the ground-up AAL, can be generated at each location using a fixed risk type (to eliminate building stock regionality). If vulnerability regions exist and the effect of these has been understood, then these ground-up AAL data can also be normalized to eliminate the impact of vulnerability regions, leading to data that can be compared, at least in terms of overall geographical pattern, against other sourced hazard metrics.

Once the basic geographical variations in risk are understood and visualized, metrics can be calculated to quantify this geographical variation. For example, ELTs can be output at discrete geographical points for a standard risk, and the correlation (see Chapter 1.10.5.8) between these different points calculated to visualize the spatial correlation structure of the model. An alternative, but similar, approach is to calculate the probability of loss (over some loss-threshold) at one location given a loss (over the same threshold) at an initial location. Mapping the resulting output will help illustrate the spatial structure of models and can be very useful when comparing one model with another. An example showing how correlations between different countries can vary between models is shown in Figure 5.6.

Understanding the geographical variation in risk contained within the model is a key part of understanding the variation in model results. Indeed, the presence of spatial correlation, and

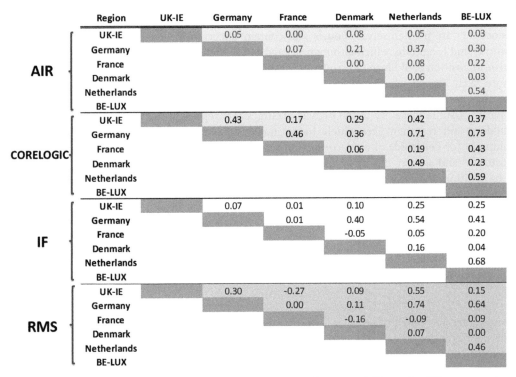

| | Region | UK-IE | Germany | France | Denmark | Netherlands | BE-LUX |
|---|---|---|---|---|---|---|---|
| **AIR** | UK-IE | | 0.05 | 0.00 | 0.08 | 0.05 | 0.03 |
| | Germany | | | 0.07 | 0.21 | 0.37 | 0.30 |
| | France | | | | 0.00 | 0.08 | 0.22 |
| | Denmark | | | | | 0.06 | 0.03 |
| | Netherlands | | | | | | 0.54 |
| | BE-LUX | | | | | | |
| **CORELOGIC** | UK-IE | | 0.43 | 0.17 | 0.29 | 0.42 | 0.37 |
| | Germany | | | 0.46 | 0.36 | 0.71 | 0.73 |
| | France | | | | 0.06 | 0.19 | 0.43 |
| | Denmark | | | | | 0.49 | 0.23 |
| | Netherlands | | | | | | 0.59 |
| | BE-LUX | | | | | | |
| **IF** | UK-IE | | 0.07 | 0.01 | 0.10 | 0.25 | 0.25 |
| | Germany | | | 0.01 | 0.40 | 0.54 | 0.41 |
| | France | | | | -0.05 | 0.05 | 0.20 |
| | Denmark | | | | | 0.16 | 0.04 |
| | Netherlands | | | | | | 0.68 |
| | BE-LUX | | | | | | |
| **RMS** | UK-IE | | 0.30 | -0.27 | 0.09 | 0.55 | 0.15 |
| | Germany | | | 0.00 | 0.11 | 0.74 | 0.64 |
| | France | | | | -0.16 | -0.09 | 0.09 |
| | Denmark | | | | | 0.07 | 0.00 |
| | Netherlands | | | | | | 0.46 |
| | BE-LUX | | | | | | |

Figure 5.6 Comparison of the correlations between European windstorm model losses for different countries, and how they vary for four different models (Waisman, 2015): AIR (Touchstone v2 July 2011), Corelogic (Eurowind v15.1 July 2014), Impact Forecasting (IF; European Windstorm v1 March 2014), Risk Management Solutions (RMS; Risklink v15 March 2015). *Source*: Reproduced with permission of Federico Waisman.

trying to understand the consequences of this, are important reasons why catastrophe models exist. However, this is not the only consideration. Models will typically produce variations in results due to differences in the following areas:

- Address/geocoding resolution
- Coverage type (Buildings, Contents, BI)
- Primary and secondary modifiers
- Financial structures
- Model settings and correlated loss components
- Data quality and exposure coding

These are now discussed in turn.

### 5.4.2.6 Geocoding

**Geocoding** is the process of assigning a geographical coordinate to an item (such as a building) from address information. Most models accept pre-geocoded exposures in the form of a latitude and longitude, or contain a geocoder which translates address information into a latitude and longitude together with an associated geocoding resolution. This geocoding resolution indicates the precision of geocoding that has been achieved with the supplied address information. For more details on geocoding, see Chapter 1.9.1.1.

Some aspects related to geocoding that should be considered when analysing model output, or changes in output, are described in the following sub-sections.

### Precise but Inaccurate Geocoding

Geocoders are in themselves models and contain uncertainty. This can lead to an assigned latitude and longitude, which although flagged as the highest resolution, is still incorrect. Any error in position could lead to erroneous model results. Other than mistakes in the geocoder, a more likely way in which precise but inaccurate geocodes can be entered is when the user supplies pre-geocoded exposures as latitude and longitude input to the model. A model will typically assume these are at the highest accuracy (i.e. represent the location of an individual building) and treat them accordingly. In fact, the latitudes and longitudes input could be just the midpoint (or **centroid**) of an accumulated set of exposure in a very large area (e.g. two-digit postcode) and so may not be at all accurate, leading to misleading model results.

In addition, the underlying geodetic system applied to generate the latitude and longitude coordinate may differ from that of the catastrophe model. Typically, most catastrophe models will employ a common global geodetic system, such as the World Geodetic System (WGS 84). Variations in geodetic parameters or conversion from national mapping systems to geodetic co-ordinates may result in spatial offsets at the ground scale in the order of kilometres. It is therefore important that the source and specification of any 'actual' co-ordinate data provided for modelling is understood.

### No Loss Produced for Coarse Geocoding

In some models (most often river flood or storm surge), if geocoding results are worse than a specific resolution, then no results are returned. In some models, post-import reports can be produced from the model tables which identify the level of geographic granularity achieved for a given risk location. This information may be used, when considered alongside exposure values, as a means to assess the proportion of non-modelled exposure and hence relative confidence in model result.

### Disaggregation of Exposures for Coarse Geocoding

In some models, if the geocoding resolution is too coarse, rather than returning model output at a specific point, the model will implicitly assume an underlying distribution of exposure throughout the region (consistent with the resolution of geocoding) and return the results accordingly. The intent of this is laudable, however, care is needed as the assumed underlying exposure distribution can contribute to misleading results. For example, if underwriting rules prohibit the writing of exposures in certain high-risk regions, the model may not allow for this in its disaggregation of exposure; this potentially leads to results that are too high. The converse could also be true; for example, if a commercial facility was located right next to a river, but the geocoding resolution was low, the disaggregation process could distribute some of the exposure far away from the river, leading to artificially low results. Some models average the hazard values for each event to match geocoding resolution. For example, if geocoding resolution is at postcode level, the model calls an event set in which the hazard has been averaged across all grid points within the postcode of interest (i.e. instead of disaggregating the exposure, the model is aggregating the hazard). This could also lead to artificially lower or higher modelled losses.

### Secondary Uncertainty Increased for Coarse Geocoding

**Secondary uncertainty** is the representation within catastrophe models of the fact that for the same hazard level, the same type of building (as represented in the model) can often experience a range of possible losses (see Chapter 2.16.1.4 and Chapter 4.5.1.1). In some models, the calculated secondary uncertainty may be increased as the geocoding resolution gets worse. The rationale is to allow for the increased uncertainty associated with poorer geocoding. While this makes sense, it will give a misleading view of the model uncertainty if coarse resolution exposure is imported with latitude/longitude coordinates derived from a previously matched aggregate unit, such as a postcode or city, effectively over-riding the model's own assessment of relative spatial accuracy.

*Change in Geocoding with Model Versions*

Just like other model components, geocoders are subject to periodic updates and improvements. Even if all the other model components remain the same, a change to the geocoding methodology or accuracy alone can lead to a change in model results, and should be considered in the model evaluation process.

*Understanding the Geocoding/Model Interactions*

It is important that for each peril-country combination, the four geocoding resolutions below are well understood, so that results, and the changes in these results, can be properly interpreted, and that view of risk adjustments can be made if necessary:

- the highest geographical resolution possible from the geocoder;
- the highest resolution used by the model;
- the geographical resolution at which exposure will be distributed over an area by the model (i.e. will be disaggregated);
- the geographical resolution at which no results are returned (if applicable).

In addition, it is possible that the model will make a choice between certain hazard modelling calculations depending on the resolution achieved; for example, an earthquake model may employ different calculations if locations are entered as coordinate versus non-coordinate. If this is the case, then the implications of this on model results should be understood.

### 5.4.2.7 Coverage Type

The basics of different coverage types are provided in Chapter 1.9.1.2. Some extra considerations are given in the following sub-sections.

*Building Sum Insured*

For commercial insurance the most likely complication is that of under-insurance (i.e. the insured value is less than the true value). If an **average clause** (a clause that reduces the amount of claim payable in proportion to the amount of under-insurance) is present in the insurance contract, this can mitigate the impact of under-insurance. For more details on under-insurance, see Chapter 2.6.3.

A further consideration for commercial insurance is whether or not a **day-one sum insured** is used. This is the sum insured with no explicit allowance for inflation during the policy and rebuilding period, but with a 'day-one uplift provision' in the insurance policy. An alternative is a **full reinstatement sum insured**, i.e. with an allowance for inflation already within the sum insured. Some policy wordings contain inflation protection provisions, expressed as a percentage of the day-one sum insured, which allows the insured to claim for more than the original day-one sum insured following a claim in order to account for inflation during the policy and rebuilding period. This period could be several years for a commercial claim. Care is needed to properly represent this day-one sum insured in the model, for example, how is the associated inflation protection provision represented? To some extent, the provision could already be implicit in the vulnerability functions due to the calibration process, as inflation will usually be operating while commercial claims are taking place, and so if this provision is common market practice, or at least common practice in the claims used to calibrate the model, then it will already be inherently included in the model to some extent. One way forward is to discuss with the model vendor and to seek agreement on a suitable average amount of inflation already inherent, and then apply exposure adjustments where the amount of inflation protection differs significantly from this average amount.

When modelling auto lines, the building sum insured field may well be used. In this case, it is vitally important to understand whether the model is based on depreciated car value, and what the exposures being provided are. This may vary depending on whether the exposures relate to personal auto or dealerships.

*Contents Sum Insured*

The main complication here is the potential presence of under-insurance; if this is different from the average level of under-insurance within the exposure data used in model calibration, then model results will be biased and an adjustment will be necessary. A further consideration for personal lines is the common practice of a limit being much higher than the real exposure value. This is common where the sum insured is calculated by the insurance company, i.e. a **notional sum insured**. If this limit is mistakenly used as the sum insured in a model, then the results will be biased high.

### 5.4.2.8   Primary and secondary modifiers

Variations in modifiers can have a large impact on results. The four main **primary modifiers** are usually: (i) occupancy class; (ii) construction type; (iii) building height; and (iv) year built. Other modifiers that are sometimes classed as primary are number of buildings (for aggregated data or within a campus scenario) and square footage. The impact of these does vary by peril, but they can be regarded as useful to obtain for most, if not all, perils.

**Secondary modifiers** are usually peril-specific and depend upon the model being used. For more details on primary and secondary modifiers see Chapter 1.9.1.3.

Vulnerability curves (Chapter 4.5) may be developed using an engineering-based model, empirically created using claims data, or (most common) a combination of both. As there are typically more vulnerability curves in a model than can be fully tested by claims data, the curves (and thus model results) are often based on engineering studies. The sensitivity of results to different vulnerabilities can be substantial.

It is therefore important to gain a good idea of how the model results change for different values of, and combinations of, primary modifiers. There are two main ways (other than reading model documentation) of understanding the impact.

The first is to obtain the vulnerability curves directly from the model. In some models this is available, while in other models it is not. With the trend towards more open modelling and an increased requirement for model evaluation, it is hoped that the model users will increasingly have direct access to vulnerability curves, but there is also an understandable caution from some vendors with respect to their intellectual property. If vulnerability curves can be obtained from the model, the curves can be compared in order to understand the differences in vulnerability between primary modifiers (and also between regions).

The second method, and possible in all cases irrespective of model openness, is to run synthetic portfolios through the catastrophe model, outputting the results (e.g. AAL) at location level in order to gain specific information about how the model results vary with different input modifier values (i.e. sensitivity testing). The synthetic portfolio should contain a number of identical locations, with only the value of one primary modifier changing at a time. The location-level model output then provides information on how the values within this specific modifier change the model results. This process can be repeated for each primary modifier until a comprehensive knowledge of how the primary modifiers affect model results is gained.

Although this sounds like a simple process, there are several complications. The main issue is that, since each vulnerability curve could be completely different to every other curve, there would in theory be an enormous amount of combinations to exhaust to gain complete knowledge of how the vulnerability works. For example, taking four primary modifiers with 10 levels each, there are $10^4 = 10,000$ potential vulnerability curves. If we consider that these could each vary by country (or state) and that each country could contain several vulnerability regions, then the amount of potential vulnerability curves increases further (e.g. 10 countries each with 2 vulnerability regions results in 200,000 unique combinations). Running a portfolio of 200,000 risks does not normally present a problem, but making sense of the resultant output can be a difficult task.

There are ways of coping with this complexity. These involve recognizing that pragmatically there will not be 200,000 or even 10,000 completely unique vulnerability curves – there simply is not the quantity of claims data to support such a variety. Typically, for a certain modifier, a particular form or shape of the curve will be specified based on engineering studies, which is then scaled up or down depending upon the value of the modifier; perhaps based on claims experience or expert opinion (Chapter 4.5). In this case, the relativities (i.e. the percentage amount that the model results change from one modifier level to another) can be fairly similar irrespective of the levels of the other modifiers. Once this is recognized, then several ways of progressing are as follows.

The first is to construct a portfolio which focuses on the levels within a particular modifier, but only allowing the most prevalent levels of the other modifiers to change so as to reduce complexity. For example, if occupancy type is the focus, the company could look at the other modifiers and pick two or three levels from each that best represent the exposure a company has so as to reduce the complexity and focus on what is relevant for that company.

Another method is to borrow from the skillset of most personal lines pricing actuaries (likely to be available within an insurance company) and use a Generalized Linear Modelling (GLM) approach. The output from the catastrophe model becomes the input to the GLM. Proprietary GLM packages (e.g. Willis Towers Watson's *EMBLEM*) can be used to investigate interactions between many combinations of factors and yield the relativities for each rating factor (or primary modifier).

In terms of secondary modifiers these are much more likely to be a straight scaling to the primary modifier vulnerability curve and this scaling is quite often documented by the model vendor. If it is not, then a run of a well-constructed synthetic portfolio can yield the secondary modifier scalings.

A final note with respect to the impact of secondary modifiers is that it can be very useful to run the whole company portfolio with and without secondary modifiers (in some models this is an option at model run-time) to identify the overall impact as well as the impact for specific lines or regions.

### 5.4.2.9 Financial Structures

The ability to apply financial structures to ground-up losses is one of the main reasons why catastrophe models are needed. Insurers or reinsurers can source information about hazard from academics and vulnerability from engineers (and their own claims data), but constructing this in a form which allows the impacts of the (re)insurance contracts to be recognized and allowed for is a difficult and complex task. At the time of writing, there is no catastrophe model available which completely represents *all* financial structures in use in the insurance market.

Most of the previous sensitivity testing is aimed at understanding how the model works and so will have taken place using the *ground-up* perspective (i.e. ignoring financial structures), as this is the best way to get an insight into the way in which the model hazard and vulnerability are working. In some cases, the impact of financial structures can be very large, and it is important to understand what the model can and cannot handle.

The model results (overall, portfolio-specific and at a range of return periods) should be studied using several relevant financial perspectives (typically ground up, gross and net loss pre-catastrophe reinsurance). The difference in the results for different financial perspectives can then be compared at various levels (e.g. overall portfolio, account-specific, line-specific) for a range of return periods to ensure the impact of the financial structures makes sense.

### 5.4.2.10 Model Settings and Correlated Loss Components

This section focuses on the different model settings, which can impact model results, such as the ability to turn demand surge, or correlated secondary perils, on and off and choose alternative events sets.

**Demand surge** or **post-event loss amplification** (**PLA**) refers to the increase (or inflation) in losses that is driven by the unusual scale of the event (Olson and Porter, 2011; see also Chapter 2.9.2). Demand surge itself refers to the increase in prices of materials and labour following an event, due to a limited supply and a *surge* in *demand*, and therefore cost, for these materials and labour. The term post-event loss amplification refers to the demand surge effect plus other impacts related to the scale of the event such as the very high number of claims causing challenges to loss adjusting, rioting, looting and civil commotion, lack of access (because of enforced evacuations) resulting in secondary damage after the event (e.g. mould, rain through broken roofs) and political pressure on insurers to provide coverage ('coverage expansion') even if coverage is not specified in the policy. In general, the impact of this demand surge or PLA is zero for small events and increases with event size. Usually model runs can take place with and without demand surge (or PLA) to gauge the impact of this component. One area of caution in applying demand surge or PLA is for commercial entities that have long-term building and renovation agreements in place with local suppliers at predetermined costs. In these cases the demand surge component should be reduced. Some models include demand surge (or PLA), some do not.

**Secondary perils** (e.g. fire following, sprinkler leakage, storm surge, inland flooding, tsunami) are perils which are not the main focus of a particular catastrophe model, but need to be included because they are caused by the primary peril and so will have a loss component that is correlated with the primary peril (see Chapters 2.6.3 and 4.3.7). Whether or not these should be applied depends upon whether the policy specifies the coverage or not, as it can quite often be the case that the policy covers the primary peril but not the secondary peril, for example, flood damage (caused by hurricanes) for residential policies in the United States. A caution here is that even if the policy is not intending to cover secondary perils, it may end up doing so due to poor policy wording that does not stand up to legal challenge, or due to political pressure to expand coverage. This is often called **policy leakage**. Also, a secondary peril loss may be covered partially due to the simple physical challenge of the loss adjustors to determine what proportion of the damage was caused by the primary (covered) peril, and what proportion by the (not covered) secondary peril. The following are common secondary perils or correlated loss components. If these are not explicitly represented in a model, they should be treated as non-modelled risk and evaluated and treated accordingly (Section 5.4.7).

- *Fire following* is extra damage from a fire that is caused by the primary peril (e.g. stemming from a gas explosion or electrical failure). Usually the primary peril is earthquake, but in Hurricane Sandy there was a small component of fire following. In the current generation of catastrophe models this is normally limited to some earthquake models (and in some models, to specific sub-regions of the model domain) and can sometimes be switched on or off at model runtime, enabling an evaluation of the impact of this loss component.
- *Sprinkler leakage* is caused by the primary peril damaging the sprinkler system and the resulting water damage increasing the loss beyond that which would have been incurred by the primary peril alone. Like fire following, this is typically seen in earthquake perils but may also be triggered by tropical cyclones.

- *Storm surge* is the impact of coastal flooding driven by the primary peril of extra tropical, or tropical, cyclone (see Chapters 3.2.5 and 3.3.5). Any flood-related peril is difficult to model, partly due to the high demands on location accuracy, and storm surge is no exception. Geocoding resolutions that may be acceptable for wind modelling (kilometres for extra tropical cyclone, hundreds of metres for tropical cyclone) are not acceptable for storm surge modelling where accuracy on the order of metres is required. Even if the peril is not directly covered, policies in storm surge-prone areas are often subject to policy leakage. For example, a tropical cyclone may partially destroy a building from wind damage, while a subsequent surge completely destroys that building, yet the insurance coverage is for wind and rain only, and not storm surge. However, by the time the insured (or loss adjustor) sees the building, it is very difficult to apportion the damage between the two perils, requiring significant judgement by the loss adjustors, or in some cases ruling by the courts. Thus the apportionment of loss driven by wind in the claims settlement may differ substantially to the modelled amount. Another example of policy leakage is where each peril (wind and surge) is covered to a certain peril-specific limit and the insured claims both limits against the loss (so long as this does not exceed the sum insured). Policies are generally not intended to work in this way, but poor wordings can result in this happening. Some models have inbuilt 'flat factors' to account for coverage leakage (e.g. apply 10%), which the user cannot change. Others have a sliding scale according to damage severity, which the user can change to a user-defined factor, and which varies by line of business in recognition of the different terms and conditions in place between residential and commercial policies. Further complications around modelling surge as a secondary peril are that the secondary modifiers collected are normally focused on the primary peril, not the secondary peril. So, for example, information on bespoke flood defences surrounding a property may not be captured in a portfolio focused on the wind peril. In terms of estimating the impact of storm surge, the models usually allow this loss component to be turned on or off, enabling an evaluation of the impact. For many portfolios the overall impact of the surge loss component will be lessened by locations outside of surge areas which feature no surge loss. However, it is generally worth focusing on specific model domain areas to fully test this component, because for some events it can be the dominant driver of loss; even though it is modelled as a secondary peril. The loss from Hurricane Sandy where the majority of the loss (65–70%) was caused by coastal flooding from the surge, not wind, demonstrated this (Swiss Re, 2013).
- *Inland flooding* is the flooding caused by precipitation, normally rainfall, resulting in inundation of areas that are normally dry (see Chapter 3.5). In the context of a secondary peril, it is the flooding caused by the rain that is produced by, or related to, the primary peril; usually tropical cyclone. The same comments relating to storm surge apply to inland flooding. Flooding can also accompany an extra-tropical cyclone. At the time of writing, inland flooding has not been explicitly included within an extra-tropical cyclone vendor model. This is not to say that the additional correlated risk of inland flooding does not exist for extra-tropical cyclones, but rather it should be treated as a non-modelled risk at this point in time, depending on coverage terms and policy conditions.
- *Tsunami* losses are caused by the large waves generated by underwater earthquakes or landslides into the sea (see Chapter 3.9). In the context of secondary perils, they are caused by earthquakes. Very often the earthquake causing the tsunami can take place a far distance from land and so there can be limited shake damage attributed to a tsunami-generating earthquake event. In the context of secondary perils, it is only the damage that is correlated with shake events that is included, and so the secondary peril tsunami risk will be a subset of the true overall tsunami risk. The Tohoku earthquake of 2011 is an example of an event with a large tsunami component (see Chapter 3.8). Coverage terms and conditions must be examined to understand whether the tsunami-related damage is considered 'earthquake', and whether it is covered or excluded.

Other items which may cause variability of loss estimates and whose impact can usually be investigated through changing the run-time model settings are as follows:

- *Secondary uncertainty* (see also Chapters 2.16.1.4 and 4.5.1.1). To allow for this, an uncertainty distribution is incorporated within each vulnerability function, and this is propagated throughout the loss perspectives. It is worth remembering that, in the absence of financial structures (or for the ground-up loss), the secondary uncertainty does not impact the mean loss cost. Secondary uncertainty will, however, impact the distribution of loss for the ground-up perspective and all metrics for perspectives other than ground up. In practice, this should always be turned on. The impact of this can be tested by comparing results with and without secondary uncertainty.

- *Uncertainty correlation* (see also Chapter 4.7.2.3). If secondary uncertainty is represented in a catastrophe model, then the model vendor must decide on how to treat any potential correlations in this uncertainty. For example, if a particular location, for a given level of hazard (e.g. a certain level of peak gust wind speed) has a higher than expected loss, should the location next to it also have a higher than expected loss? The same question can be asked for coverage type. If a location sustains a higher than expected loss for the buildings coverage, should it also sustain a higher than expected loss for contents? Most models assume 100% correlation between different coverages for a particular location. However, the treatment of location correlation is different between models; treatments range from a pre-set amount of location correlation that varies by peril-region, to an assumption of no correlation, to allowing the user to select the level. The sensitivity of results to the correlation can be explicitly tested if the parameter is user-defined. If it is not, then reference to the model documentation or model developers is needed in order to establish the results' sensitivity. With secondary uncertainty switched off, metrics will not be impacted by coverage or location correlation assumptions. The ground-up AAL (even with secondary uncertainty switched on) will not be impacted by correlation assumptions either. Other metrics will, in general, be impacted to different extents by these uncertainty correlation assumptions; typically as location uncertainty correlation increases, the model results increase at high return periods and reduce at low return periods.

- *Alternate event sets*. Diverging scientific views may lead model providers to implement multiple versions of a model in parallel: for some models, alternative event sets, or event rates, are provided by the model vendor. For example, for North Atlantic Hurricane, most model vendors provide either a medium-term (or warm sea surface temperature (SST)) set of events as well as the long-term event set. The medium-term or warm SST event sets take into account increased activity in North Atlantic hurricanes in recent years; albeit in very different ways. In one case, the model vendor recommends their medium-term event set be used as default, and another that the long-term view be used as the default. For European Windstorm, events sets with and without windstorm clustering (see Chapter 3.3.4) are available. A recent development has been one model vendor releasing a 'climate variability' view of European windstorm activity, to reflect that the last 20+ years have been less stormy than the average of the past 45 years. For earthquakes, some models allow the user to specify whether they want to use time-independent or time-dependent earthquake rate representations. All of these options around events sets give the model user more flexibility in their view of risk and are highlighting part of the primary epistemic uncertainty (see Chapter 2.16.1) in these models, which is often ignored. However, a decision about which event set to use should be substantiated with a rationale. All of the different event-set representations have their roots in science, but are often the subject of much debate. The most pragmatic way forward is first of all to understand which event rates form the chosen model vendor's recommended default, assess if that is appropriate, and then view the results with the alternative event rates. If the difference is material for a significant part of the company's risk, it is best to seek a scientific review in order to decide which view is most appropriate; this may come from external experts, together with any available internal or external data.

### 5.4.3 Actual Versus Modelled, Comparing to Own Company Experience

Comparing model output to the claims experience from an individual company, if available, is an important part of understanding if the models are fit for purpose for that company. It should be recognized that there will never be sufficient claims experience to test all aspects of the model, as if there were, there would be no need for exposure-based catastrophe models. In particular, catastrophe models are designed to provide an estimate of the risk for the low-frequency, high-severity events, where claims experience is insufficient.

However, claims experience is very important for two purposes. One is to compare actual events to model reconstructions of these events. This can be a useful test of some parts of the model (and also of the company exposure data). Findings from this type of analysis (e.g. leading to a change in vulnerability) can affect the model loss estimates across the whole model return period curve. The second is in comparing the low return period/high frequency part of the loss curve (see Chapter 2.6.4.1). If there is no agreement between actual and modelled loss in this part of the curve, then this provides a rationale for model adjustment in this range, perhaps extending to higher return periods if further investigations are carried out to establish why the model is different from experience.

Before focusing on both types of analysis, some general comments about comparing model output to claims data are required. While catastrophe modelling analysts will be experts in exposure data and catastrophe models, the actuaries and statisticians in a (re)insurance company will be the experts on claims data and associated analysis techniques. It is well worth engaging the relevant actuarial department's support and expertise for analyses involving claims data.

The focus of this section is on comparing claims data for a *particular company* to model results from that company's exposure data. Comparing *industry* claims to industry exposure model output is also useful in validating the model (discussed in Section 5.4.5). Ideally some level of industry analyses should have already been performed by the model vendor and should be an important part of the model documentation.

A common requirement for any analysis comparing catastrophe model output to claims data is that the exposure data (used to generate the model output) and the claims data are both referencing the same point in time. This can either be achieved by trending or 'on-levelling' the claims data (see Chapter 2.6.2.1), or by ensuring that the model output is derived from exposure data from the same period as the claims data being compared. The first of these options is particularly challenging if the nature of the portfolio has changed substantially (e.g. a change in geographical distribution, lines of business).

For detailed comparisons of model output to recent claims data, it is important to leave enough time to allow the claims to settle. This will vary according to the type of event, the line of business, and the country, but typically is in the order of six months (e.g. Europe wind personal lines claims) to several years (e.g. flood or earthquake large industrial claims). For cases where it is not possible to wait this long, claims should be projected to an estimate of their final amount (developed to ultimate) using standard actuarial techniques (e.g. Parodi, 2015).

Efforts must also be made to ensure that the claims data are 'cleansed' of sources of loss not covered by the model in question. For example, if validating a hurricane model which does not have a flood component, the claims data should be adjusted to reflect only the hurricane portion. Allocated Loss Adjustment Expenses (ALAE) must also be stripped out.

It is also important to ensure all corresponding exposures are captured and used in generating the model output, otherwise a misleading comparison will follow.

### 5.4.3.1   Event-Specific Comparisons

Comparison of actual event losses to model event representations can be used to investigate the model vulnerability and completeness of loss causes for those events. This can be used to inform a view on vulnerability. An important by-product of this type of analysis is that it often highlights incompleteness in the exposure data (e.g. when claims cannot be matched to exposure) as well as inadequacies in claims data (e.g. not knowing the split of the claim amount between wind and storm surge). Lessons in both of these areas can lead to process or system improvements to help in the future. Putting exposure data and claims improvements aside, there are several types of analysis possible when comparing actual to modelled events. The type chosen will depend upon the materiality of the peril, the available resource, and the quality of the claims and exposure data.

The most basic analysis possible, and the starting point for any claims analysis of this type, is to run the model vendor's 'best' footprint representing the event over the relevant exposure and compare the actual loss to the modelled loss. For these basic types of analyses, detailed claims data are not required; aggregated claims data (at the resolution required by the comparisons) are sufficient. Analysis should be undertaken at an overall company level, and across certain dimensions to the extent that the exposure and claims data allow. For example, the actual versus modelled loss by coverage type (buildings, contents, business interruption), by country, by class of business (e.g. property, motor) and by high-level occupancy (residential, commercial, industrial) should be compared. If there is a sufficient quantity and detail of claims data, then these types of actual versus modelled comparisons can be performed at higher granularities. The purpose of these analyses is always to identify segments where the model works well and segments where the model does not work as well for this event. The credibility of the underlying claims data (i.e. the quantity of claims), and the model uncertainty (see Chapter 2.16.1) must always be borne in mind so that conclusions are not made based purely on statistical fluctuation. It is helpful to repeat this type of analysis across several historical events; if the same pattern emerges across several independent events, it is a strong indicator that the pattern is valid, which further supports model adjustment.

If detailed, individual claims data are available that can be geocoded to a resolution commensurate with the event being analysed, then a more comprehensive type of analysis can be performed. In this case, it is extremely valuable to obtain the hazard event footprint from the model vendor. Ideally the footprint should be compared to observations of the event hazard, where possible, in order to validate it. Once this is done, hazard values can be assigned to each claim location. This allows for a graph of hazard value versus loss percentage to be derived, which can be directly compared to the model vulnerability curves. If the curves are not available directly from the model, the vendor may be able to assist in enabling this type of analysis in return for anonymized claims data. An example of this type of analysis for an earthquake event is shown in Figure 5.7, where distance from epicentre is used as a proxy for shaking intensity.

A further example using claims data from Hurricane Sandy is shown in Figure 5.8. A clear trend of increasing damage ratio with increasing wind speed is shown for all age bands. The trend is more distinct when the damage ratio is defined as the loss divided by the exposure for all properties affected, rather than the loss divided by the exposure from only the affected properties. This indicates that there is both an increase in probability of damage with increasing wind speed as well as in the severity of damage (given that a claim has happened) with increasing wind speed (see Chapters 4.5.1.2 and 4.5.2.3).

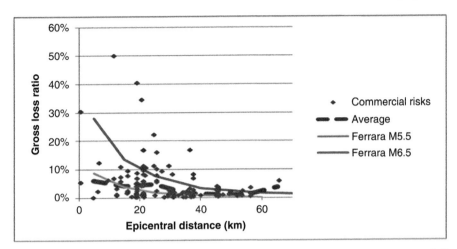

Figure 5.7 Claims experience compared to modelled event losses for Italy earthquake (SCOR, 2015). The blue diamonds represent the mean damage ratios (MDR) for individual risks (MDR = loss/exposure). The blue dashed line represents the damage ratios where the sum of the loss within a certain distance band from the epicentre (e.g. 20–21 km) is divided by the sum of the exposure (even for those locations with no loss) within the same distance band. The green and red lines show the predicted model loss from the portfolio for two different models events (magnitude 5.5 and 6.5 respectively). *Source*: Reproduced with permission of SCOR Global P&C.

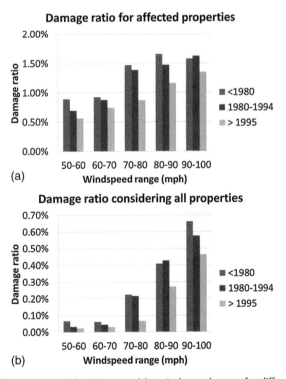

Figure 5.8 Analysis of damage ratio (= loss/exposure) by wind speed range for different age properties for Hurricane Sandy (SCOR, 2014). (a) the analysis including only those properties that had a claim; (b) includes all properties within the affected region irrespective of whether they gave rise to a claim. *Source*: Reproduced with permission of SCOR Global P&C.

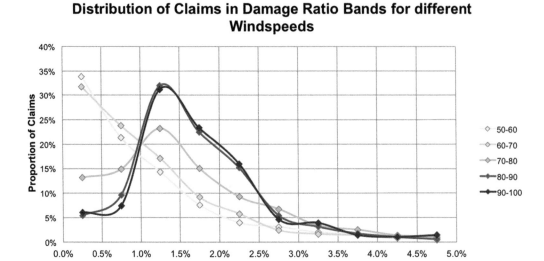

Figure 5.9 Comparison of damage ratio distribution from Hurricane Sandy for different wind speeds (in mph). *Source*: Reproduced with permission of SCOR Global P&C (SCOR, 2014).

The distribution of claims in different damage ratio bands can also be analysed, an example of which is shown for claims from Hurricane Sandy in Figure 5.9. As the wind speeds increase, the shape of the distribution changes and it shifts towards higher wind speeds.

The patterns shown in Figures 5.7, 5.8 and 5.9 can be compared to the model output for the same event to identify whether the model representation of claim distributions is similar to that observed. If the model representation is not similar to observations, then the reasons for this should be identified, which may lead to model adjustments.

An important aspect of these analyses is ensuring the loss amount assigned to each hazard level band is normalized by the correct exposure amount. What is correct will depend upon the form of the vulnerability curves, in particular, whether they are conditional vulnerabilities (i.e. given that a claim has occurred; part of a two-step vulnerability implementation) or unconditional (i.e. the mean losses include the chance that no claims occur; a one-step vulnerability construction). See Chapter 4.5.1.2 for a description of one-step and two-step vulnerability functions. If the curves are unconditional, then the claim amount should be divided by all of the exposure in the relevant hazard band, not just the exposure associated directly with the claims (average MDR including the non-affected risks in Figure 5.7). If, on the other hand, conditional vulnerability curves are used, then the exposure used to normalize the claim amount should be only that associated with the claim locations (localized MDR in Figure 5.7).

A further consideration for these analyses is ensuring that the claims data are comparable to the vulnerability functions in terms of financial perspective. For example, claims data may be net of primary insurance financial structures ('gross' perspective), whereas vulnerability functions are usually constructed prior to financial perspectives ('ground-up'). If this is the case, then adjustments will need to be made, normally to the claims data, in order to produce a like-for-like comparison. A particularly tricky aspect is adjusting for claims that were never reported as they were below the level of the deductible.

Geocoding claims can be problematic, especially for multi-location commercial policies. In some cases, the location provided is the location of the insured's head office and may not be the location of the claim.

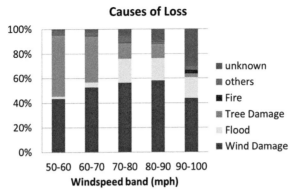

Figure 5.10 Distribution of different causes of loss for claim amounts from Hurricane Sandy by wind speed band. *Source*: Reproduced with permission of SCOR Global P&C (SCOR, 2014).

In conducting any types of comparison of individual events to actual claims, it is important to understand how the event footprint from the model is constructed, and in particular how independent it is from claims data. There is a tendency for modelling firms to release several footprints during the lifetime of a catastrophe event response process (see Chapter 2.9). An initial footprint is provided very quickly (usually within the first day or two), which may be sourced from real-time hazard information, and is sometimes the closest matching event in the model's stochastic catalogue. There is often a refinement of this when hazard information has increased in amount or quality. Finally, once loss information emerges, there is sometimes an attempt to select or modify a footprint based on closeness of fit to the industry loss.

Ideally, a comparison of an event to the modelled loss should be conducted with the footprint constructed from the best available hazard data, not selected based on closeness of fit to industry losses. The key purpose of such analyses is to test issues with exposure data, the vulnerability curves within a model, and identify loss sources not represented in a model but present in reality. If a footprint is selected based on closeness of fit to industry loss data, it negates the purposes above.

If individual claims data are available for an event, it can be very helpful to go through the details of some individual claims and ensure that all the loss causes giving rise to claims are covered by the model. For example, a distribution of cause of loss by wind speed band for Hurricane Sandy is shown in Figure 5.10.

If the causes are not all captured by the model, then an adjustment for non-modelled risk for these missing loss elements should be part of the view of risk proposal. When considering such an adjustment, care must be taken to consider loss causes that may be implicitly represented in the model; if vulnerability functions are developed with claims that already include the loss causes, the model loss estimates will also implicitly include them. The only way of establishing what is implicitly included, and what is not, is a detailed discussion with the model vendor about the type of claims information used in deriving the vulnerability curves. The evaluator needs to establish with the model vendor what efforts they have made to establish the cause of loss for the claims data used to develop the model. If possible, the cause of loss should be categorized to differentiate between explicitly modelled, implicitly modelled, and non-modelled types, as shown in Figure 5.11. A model adjustment is only needed for the non-modelled category.

**% of Total Loss by Cause of Loss**

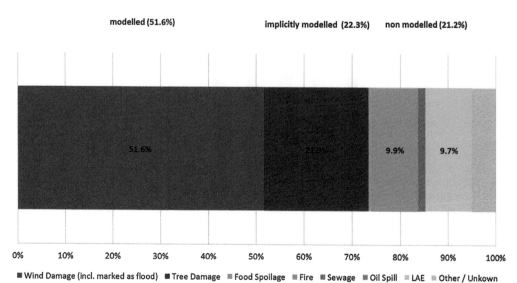

Figure 5.11 Categorization of loss causes. *Source*: Reproduced with permission of SCOR Global P&C. (SCOR, 2014)

### 5.4.3.2 Exceedance Frequency Curves

A systematic way of validating catastrophe models is to construct an empirically derived loss exceedance frequency distribution from actual event-level claims data to compare against the modelled loss exceedance frequency distribution from the catastrophe model. Some considerations for this type of analysis are given below.

Catastrophe models were originally designed for reinsurance purposes and so, by their nature, focus on large *catastrophe* events, with less focus given to smaller *attritional* events. Over the years there has been a tendency for some catastrophe models to include information on smaller events in order to better represent the entire frequency spectrum for a given peril. However, very often a primary insurer will have better information and knowledge about the impact of higher frequency events on their portfolio than that contained within a catastrophe model (see Chapter 2.6.4.1). An implication of this is that even if a catastrophe model does not fit experience well in the high frequency range, care must be taken in extrapolating this result into the lower frequency range. If the reason why there is a deviation between modelled and actual loss is known (e.g. a vulnerability effect identified through a detailed event-level comparison), then there is more rationale for extrapolating a result into the lower frequencies (higher return periods).

This type of analysis is mainly useful for the higher frequency catastrophe perils. These include perils such as hail, windstorm and flood. The earthquake peril is usually a lower frequency peril making these types of analysis less appropriate.

Comparisons must be *like for like* between model output and actual claims data. For example, it is important to understand what is meant by *storm* claims in a particular company's loss coding. For example, in the United Kingdom, *storm* claims will sometimes include claims from *flooding* that is caused by water that does not come directly from a watercourse, such as flash flood or pluvial flood; note, however, that the definition of flooding in the United Kingdom is changing to a more consistent and intuitive definition, partly due to the development of Flood Re (Chapter 2.13.6). From a catastrophe model perspective, these particular types of so-called storm claims will be represented in an inland flooding model, not a storm model, and so comparing storm claims to storm model output may not be a sensible comparison without some adjustment.

The best starting point for an empirical exceedance frequency analysis is a daily time series of claim number and amounts, over as long a period as possible. Claims should be, as far as possible, either settled or developed to ultimate with actuarial input.

**On-levelling** (i.e. bringing historic claims up to an appropriate current level) of the time series is an important, and usually difficult, part of this work (as with experience pricing in Chapter 2.6.2.1). The model output will represent the risk from the company's portfolio at a particular point in time. The claims time series will represent the claims from an evolving portfolio over the time period for which we have claims. The aim of on-levelling is to apply factors to the claim events to ensure, as far as possible, that they represent the current portfolio. These factors should compensate for inflation (e.g. in rebuilding cost) as well as any volume and composition (type of exposure) changes in the portfolio. Inflation indices are usually readily available to help with the inflation part of the on-levelling. Adjusting for volume and composition changes is more difficult. This will depend upon the extent to which historic information on relevant factors (e.g. sum insured, annual premium, number of locations, type of financial structures and type of business) is available. Actuarial input is again very important here. If the nature of the book has changed significantly, this will be very challenging to do robustly.

Defining an *event* is also an important aspect of this type of analysis. The claims time series should be checked to ensure it is consistent with the model peril definition and the claims should be developed to ultimate and on-levelled to represent the current portfolio. A definition of the event must then be applied so that event losses from the claims data can be isolated in order to construct an empirical event loss frequency distribution. The event definition in the catastrophe model is often a consequence of the science in the model, and based on geophysical parameters, rather than a time- or claims-based definition.

Several approaches can be used to categorize events from the claims data. The simplest, but still reasonable, approach is to consider the nature of the peril and pick a suitable time window within which an event normally runs its course. This is illustrated in Chapter 2.6.4.1. For windstorm, this may be a period of around three days. For flood in a small country, it could be around seven days. For flood in a large country, it could be more (14 or even 21 days). This time period can then be used to run a *moving window* through the time series, summing claim amount within the time window and defining events as the non-overlapping maximal claims amounts, on the condition that there is more than one claim in every event. This approach is fairly straightforward to implement, but does have the disadvantage that it could group together several claim events into one, or indeed cut off claims from an event that take place over a longer time period than the time window. A reinsurer may choose to pick a time window that matches the hours-clause in its contracts (see Chapter 2.4.2).

A more in-depth approach is to source hazard information from an external supplier (e.g. the country's meteorological office) and use this information to refine the definition. For example, a list of events, together with the event timestamp can be used to define the events directly from the hazard data rather than from claims data. The disadvantage of this approach is that hazard data are sometimes sparse and smaller events may be missed by the observations but present in the claims data.

Once events have been defined from the claims data, a frequency can be assigned to each event as $1/P$, where P is the time period for which we have claims data. The event losses are ordered in descending loss amount, and the exceedance frequency is defined as $1/P$ for the largest event, $2/P$ for the second largest event, and so on. This empirical exceedance frequency can be plotted against the exceedance frequency from the model event set (see Chapters 1.10.1 and 1.10.4) in order to see how well the model represents the claims experience.

It is important in this type of analysis that confidence intervals are placed around both the empirical and the modelled distributions so that it is clear where the model could be a valid representation of the actual experience. For example, if 15 years of loss history are available, the

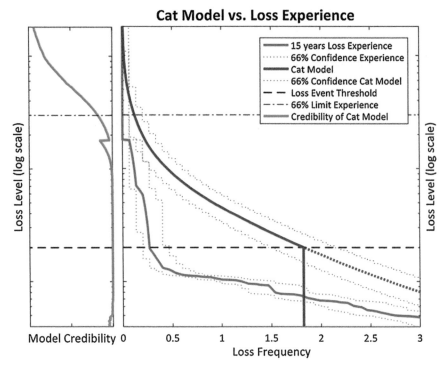

**Figure 5.12** Conceptual illustration of assessing the high frequency model range against observed loss experience using confidence intervals. The blue curve shows the exceedance frequency derived from a 15-year history of claim events. The red curve mimics the exceedance frequency curve from a catastrophe model. Both have 66% confidence intervals placed around them. The *66% limit experience* represents the loss threshold where the observation of loss frequency in the experience falls within the 66% confidence interval around the model. The green curve represents the model credibility, with the credibility increasing to the left. At high frequency levels the confidence intervals from experience and model do not overlap and so the model is assigned no credibility. At lower frequencies there is less confidence in the claims and more in the model and so the model credibility is higher. Below a certain *loss event threshold* the loss mechanisms of the actual claims are not consistent with the physical concepts of the model and actuarial methods are more applicable to assess that frequency and loss level range, provided that loss experience is available. *Source*: Reproduced with permission of Fortunat Kind.

loss level associated with the 15-year loss is simply the largest loss. This may or may not be representative of the true 15-year loss, but clearly there is low confidence in this figure, whereas the 5-year loss from 15 years of experience is a more credible figure to compare to the model output. An example of an analysis where confidence intervals are placed around both modelled and claims estimates, together with the credibility in modelled estimates, is shown in Figure 5.12. There are multiple statistical methods to determine confidence intervals with various degrees of sophistication, taking into account the information about the loss experience, the catastrophe model and the underlying frequency model assumptions. Ensuring confidence intervals are used in this type of analysis and giving the model more credibility for lower frequencies is the important theme. In Figure 5.12, a confidence interval of 66% is used, reflecting a 2-out-of-3 chance that the model should be aligned with the observations, which is quite restrictive on the model and gives a lot of weight to the observations. The choice of the interval size and the associated confidence should be reflective of the balance of trust in the reliability of the observations, on one hand, and the model's uncertainties and calibration flaws, on the other.

Unlike the comparison of individual events, this tests all aspects of the model (hazard and vulnerability), although is less likely to yield reasons why the model is different to reality.

### 5.4.4 Comparing Multiple Models

So far we have focused on understanding what is in the model and how well the model represents the company portfolio by:

- studying documentation and speaking to model developers;
- accessing external hazard data and expert scientific opinion;
- analysing model output, including sensitivity testing;
- comparing model results to individual company claims data.

For most peril-regions there are multiple vendor models available, and it is always useful and insightful to compare the differences between models and between model results. There will be different ways in which companies license and access these models, as shown in Table 5.1.

With no resource or cost constraints, companies with material catastrophe risk would license models from multiple vendors as well as developing their own. In practice, this option is only open to a handful of large companies, more usually reinsurers, whose brand or business model is focused around this type of insight and thought leadership, and whose nature of business warrants such an investment and approach, given the amount of catastrophe risk taken on.

All primary insurers should be able to access multiple vendors' model results through their reinsurance brokers (reinsurers will normally not be able to). Some will be able to access the results of reinsurers' proprietary models through knowledge and claims sharing arrangements. The issue with a primary insurer not licensing any models is that they will usually not have access to the model documentation nor the model development team, unless a specific arrangement is put in place with the vendor to gain access to these. A key part of managing catastrophe risk is understanding how appropriate the models are for the company's portfolio, and it is difficult to see how this can be done effectively without such access. Regulatory changes, such as Solvency II, mandate what should already be good practice; which is to understand the risk by understanding the models that are used. At the time of writing, some primary insurers with material catastrophe risk do not license catastrophe models, and some do not have a dedicated catastrophe risk management team (or even individual staff). It is likely that this state will change over the next few years, largely driven by Solvency II and enhancements to internal governance, and as the industry continues to grow and employ catastrophe risk experts.

Table 5.1 Model usage variants.

| Model usage (per peril-region) | Advantages | Disadvantages |
| --- | --- | --- |
| No licence Access all models through brokers | Cost (licence and resource) | More difficult to understand what model output represents. May prove hard to justify to regulators. |
| License one vendor's models Access other vendors' models through brokers | Intermediate cost, in-depth understanding of one model possible | Poorer understanding of non-licensed models – may end up optimising portfolio around one model |
| License multiple vendors' models | Good understanding of multiple vendors' models possible | Cost (licences and resource), and operational challenges, especially in respect of multiple platforms |
| License multiple vendors' models Develop own models | Very good understanding of multiple vendors' models. True thought leadership. | Cost (licences and resource). Substantial amount of resource is needed to develop high quality models, may be disproportionate to the amount of risk |

For the foreseeable future, many companies will choose to either license multiple vendors' models, or license one vendor's models and rely on brokers for results from the other vendors' models. The choice will depend upon the materiality of catastrophe risk to that company and the extent to which senior management want to focus on this risk; as evidenced by the budget and resource that are made available to the catastrophe risk team. It is a significant jump to license and utilize an additional vendor's models today, largely because of the extra resource and cost implications of adding another platform. This results in implications for training, systems integration, new hardware or hosting arrangements and new workflow arrangements; as well as additional licence fees. Given that many models use the same original data sources, some argue that the incremental benefit of adding an extra platform is not worth the additional cost. However, this is likely to change in the future given the development of model-agnostic platforms, such as Oasis. Indeed, it is only worth licensing multiple models if resource is available to make use of the output from the model and to follow the processes described earlier in this section for each model.

There is much debate about how best to use multiple models. Licensing at least one suite of vendor models and employing people who understand catastrophe models is recommended if the catastrophe risk is material for a company. It does not matter whether the company or an alternative provider (broker, vendor, or third party service provider) runs the exposures through the models and enables the type of investigations described in this chapter, as long as good processes and controls are in place.

From the perspective of developing a view of risk, at least one set of vendor models should be well understood. If the company chooses not to license other models (or not to develop their own models), then results of other models (run by reinsurance brokers) still can and should be used as a 'sense check' against the models that are licensed and better understood by the company.

If a company licenses multiple vendors' models, then the analyses described in this chapter can and should be applied to each model, in a proportionate way, depending on the materiality of the risk for each peril. Whether this is done in practice depends upon the available resources. An example of the type of comparison that can be made between models is shown in Figure 5.13.

In some cases, models seem to agree relatively closely at a high level (e.g. across all residential lines in Figure 5.13) perhaps due to similar data used to calibrate the different models. However, as the model output is segmented, the model comparisons often diverge. An example of this is shown in Figure 5.14 where the state-specific modelled losses for residential exposures show more

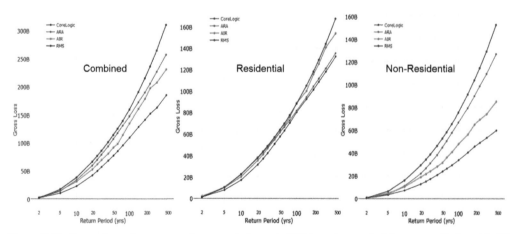

**Figure 5.13** Comparison of gross annual aggregate losses in USD by return period for a proxy Nationwide US portfolio for residential and non-residential exposures, representing the long-term view from four North Atlantic hurricane models (Waisman, 2016): AIR (July 2015), Corelogic (RQE v16 August 2015), ARA (HurLoss 6.9.2 October 2015) and RMS (March 2015). *Source*: Reproduced with permission of Federico Waisman.

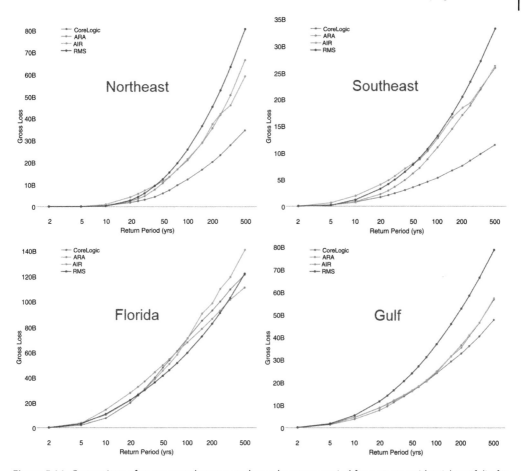

**Figure 5.14** Comparison of gross annual aggregate losses by return period for a proxy residential portfolio for four different regions, representing the long-term view from four North Atlantic hurricane models (Waisman 2016): AIR (July 2015), Corelogic (RQE v16 August 2015), ARA (HurLoss 6.9.2 October 2015) and RMS (March 2015). *Source*: Reproduced with permission of Federico Waisman.

divergence between models than the overall residential comparisons shown in Figure 5.13. An exception to this is for the state of Florida where the models seem very consistent at this level of output. A potential reason for this is the regulated model evaluation and authorization process conducted by the Florida Commission on Hurricane Loss Projection Methodology (FCHLPM; see Chapter 2.11.5). This is a largely open process and so each model provider has access to details and results from the other models seeking to be authorized by the Florida insurance commission. This potentially leads to model convergence which may not actually be indicative of a true reduction in the uncertainty for Florida compared to other states. Some further examples of comparisons that can be made using multiple models are discussed in Section 5.4.5.

If the outputs from multiple models are used in forming a view of risk, then there are various approaches to bringing these outputs together into one view (model blending). These are discussed in Section 5.5.4.1.

### 5.4.5 Using Industry Data

Section 5.4.3 describes comparing model loss output to an individual company's claims experience. These analyses are likely to be most useful to insurers who have detailed claims

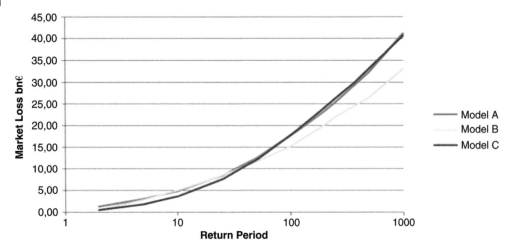

Figure 5.15 Comparison of occurrence losses for three European windstorm models using consistent industry exposure data (SCOR, 2015). *Source*: Reproduced with permission of SCOR Global P&C.

data available. Reinsurers, or insurers of business with high excesses (deductibles; see Chapter 1.9.2), may not have sufficient detailed claims experience to perform these types of analyses. Even if reinsurers do have access to enough detailed claims experience, they will also be interested in how the experience of different insurers compares to each other and to the wider industry.

Using a common set of industry exposure as input to the model (or models) is useful in understanding how models compare to each other and how model results compare to historical event losses. Although this is helpful background information to an insurer, since any single insurer is unlikely to have a portfolio well represented by an industry average, it is more directly relevant to a reinsurer, who is more likely to have a portfolio reflecting an industry average. A sample exhibit using consistent industry exposure data to compare industry losses between models is shown in Figure 5.15.

Using the same industry exposure across all models controls for exposure change and isolates all differences in results to those caused by the model components themselves. For the particular comparison shown in Figure 5.15, there is close agreement between the models for return periods less than 1 in 90 years, with one model diverging from the others above 1 in 90 years. This agreement between model results should not be interpreted as small uncertainty in the model output – as discussed in Chapter 2.16, model uncertainty will be significantly greater than may be inferred from Figure 5.15. Rather, this is evidence that each model provider has done a similar job in terms of their overall model calibration. Given that the same datasets for calibration are often used by each vendor, it is not unusual to see close agreement between models at an aggregated level such as this. It is common for model results to diverge as the results are compared in more granularity (e.g. by country rather than across Europe).

Figure 5.16 shows a comparison of historical event losses, derived from the same set of industry exposure, between models. This provides insight into the differences between the models either in terms of the way in which they have constructed the historical footprints (i.e. the variation in hazard between footprints) or in the vulnerability functions used in the models. Although the same observations of hazard data will often be used as input to footprint construction, in practice, the observational networks will be sparse and interpolation onto a regular grid will be necessary (see also Chapter 4.3.3). In some cases the observational data is so sparse that the model vendor uses their own hazard model to generate the footprint and then calibrate this to the few observations available. This will give a more realistic footprint than

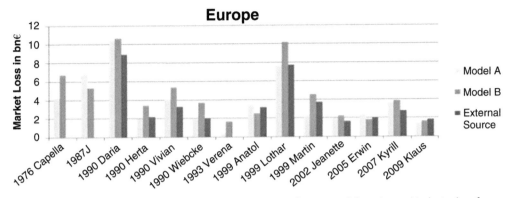

Figure 5.16 Comparison of losses between two European windstorm models and actual industry loss for historical events (SCOR 2015). *Source*: Reproduced with permission of SCOR Global P&C.

interpolation from a very small number of observations. The differences in interpolation techniques, and in the cleansing of the raw observational data, will lead to hazard footprint differences between the different models. These differences, coupled with differences in vulnerability give rise to the variations between models shown in Figure 5.16. This type of analysis alone is interesting but unlikely to be sufficient to result in a view of risk adjustment. However, when coupled with a more detailed investigation (e.g. a detailed claims analysis for an event or an analysis of an individual company's experience of historical events), it can provide useful context.

If an external source of market losses is available (e.g. PERILS or PCS), then these can be compared to the model losses, as shown in Figure 5.16. If using historical loss data, care must be taken to on-level the historical losses to the same point in time that the industry exposure represents. The method for doing this should also be peer-reviewed and double-checked because inadequate on-levelling will result in very misleading conclusions about model performance.

Figure 5.17 shows an analysis of historical event loss return period from each model based on the same consistent industry exposure dataset. This can be useful when comparing the model return periods to expert judgement of the return period for specific events (e.g. from a scientific analysis assigning a return period based on a classification of the hazard for the event). If a modelled return period for a specific event loss is very different from expectations, then the reason for the difference should be established. Variations in the return periods from different

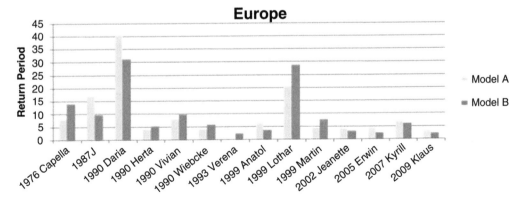

Figure 5.17 Comparison of return periods of historical European windstorm events between two different models (SCOR 2015). *Source*: Reproduced with permission of SCOR Global P&C.

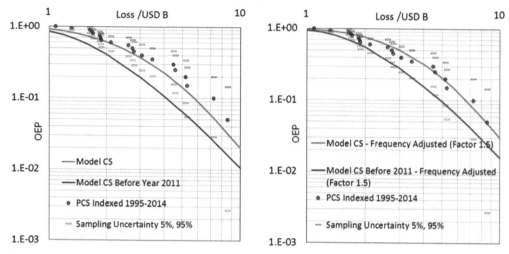

Figure 5.18 Comparison of model to industry claims experience for US convective storm (CS; SCOR, 2015). The blue and yellow curves represent the OEP curves for two different model vintages. The red dots represent the OEP curve from industry claims experience (PCS). The dashes represent the sampling uncertainty in the claims experience. The left-hand pane shows the analysis before any model adjustment. The right-hand pane shows the analysis where the model frequencies have been increased by 50%. *Source*: Reproduced with permission of SCOR Global P&C.

models could either be caused by differences in the model estimation of the historical loss (Figure 5.16) or by differences in the stochastic model EP curves. Knowledge of a return period for specific event losses is also useful in pricing analyses to decide whether a specific event loss has such a high return period that its impact in an experience analysis should be mitigated (see Chapter 2.6.4.2).

If sufficient industry claims data do exist, an industry empirical EP curve can be constructed and compared to the corresponding model-generated EP curve. An example of such an analysis is shown in Figure 5.18.

The industry claims data (PCS in this example) must be on-levelled to the same point as the industry exposure data vintage. It is also important to ensure consistency between the classes of business and types of loss present in the industry claims data and those represented in the industry exposure data, for example, if motor claims are included in industry losses, then motor exposures should also be included in the exposure data set. The uncertainty in the indexing means that this type of analysis is only really suitable for relatively short periods of time (a decade or two) compared to the much longer time period the catastrophe model is trying to represent (hundreds of years). From this type of analysis alone it is impossible to attribute the reason for any differences between modelled and actual losses; it could be due to on-levelling, hazard, vulnerability, or differences between modelled financial structures and those that actually applied when the events occurred. However, if there is a systematic difference between modelled and actual results, this can be the rationale for an adjustment. In the example above it seems that the model understates the risk for US severe convective storms and so a frequency scaling is applied (see Section 5.5.3.2) in order to increase the model output.

If historical event representations are available in the model, a *historical modelled* EP curve can be compared to the stochastic EP curve. Figure 5.19 shows a comparison of three different variants of modelled US hurricane losses (i.e. with and without secondary modifiers, and long-term view versus the current view) to modelled historical losses. An EP curve constructed from indexed historical actual losses is also shown (*PCS OEP*).

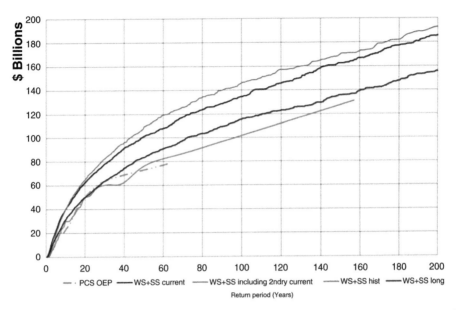

Figure 5.19 Comparison of US hurricane occurrence losses by return period for actual losses (PCS OEP), modelled historical windstorm (WS) and storm surge (SS) losses (WS + SS hist), modelled losses with a long-term view of hurricane frequency (WS + SS long), modelled losses with an increased view of hurricane frequency (WS + SS current) and losses with an adjustment for secondary perils and effects (WS + SS including secondary current; SCOR, 2014). *Source*: Reproduced with permission of SCOR Global P&C.

It is clear that the EP curve from the modelled historical losses (*WS + SS hist*) is fairly consistent with the EP curve from the indexed actual claims (*PCS OEP*). However, above a return period of around 30 years, the model results (*WS + SS current, WS + SS including 2ndry current*, and *WS + SS long*) are all higher; the biggest difference is that between the *long* and *current* views of hurricane activity. In this example, whether or not a (re)insurer chooses to scale the model losses down to match history more closely will depend largely on their view on multi-decadal variations in hurricane activity, and specifically on land-falling hurricane activity; this should be informed by scientific input.

If sufficient historical events are not available in a model, but there is a scientific record of some event parameters available to the (re)insurer, then SCOR (2015) suggests an interesting approach of matching the closest available stochastic event in the model to each historical event. To illustrate this, earthquake events could be matched on magnitude and distance between epicentres as depicted in Figure 5.20.

Once historical event parameters are matched to stochastic model events, the loss estimates for each historical event can be estimated from the model and a proxy historical EP curve can be constructed by putting together the loss estimates from the model with the historical event frequency. This *matched to historical* EP curve can then be compared to the model EP curve as shown in Figure 5.21.

This type of analysis does not provide the company with any information about vulnerability or hazard footprint accuracy. However, it does yield information on event frequency, so long as care is taken to ensure the historical record is complete. Unless there is a good explanation for the magnitude of difference shown in Figure 5.21, this would infer that a model adjustment is required.

In summary, industry data (both claims and exposure) can be used in a variety of ways to provide context and rationale for a view of risk.

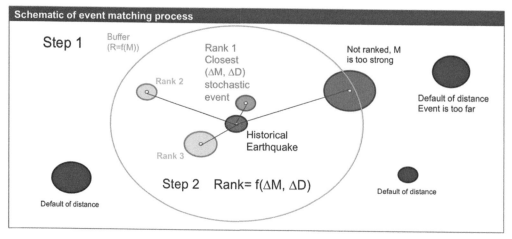

**Figure 5.20** Example of a process to match historical earthquake events with those within a model's stochastic event catalogue (SCOR, 2015). For each historical event, potential stochastic event matches are selected within a radius (R) from the location of the historical event epicentre. The buffer size varies with earthquake magnitude (R = f(M)). The potential stochastic events are then ranked based on difference in magnitude (ΔM) and difference in epicentre distance (ΔD) compared to the historical event. *Source*: Reproduced with permission of SCOR Global P&C.

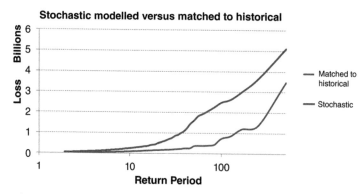

**Figure 5.21** Proxy historical OEP curve constructed by matching stochastic events to historical events compared to the pure model stochastic OEP curve (SCOR, 2015). *Source*: Reproduced with permission of SCOR Global P&C.

### 5.4.6 Considering the Time Period of Risk

In developing a view of risk, it is important to consider the relevant time period, or point in time, that the view of risk is meant to represent. This will depend on the intended use of the model results. If reporting key catastrophe risk metrics associated with the company's current exposure, then the relevant point in time is the *as at* date of the latest reported exposures. If pricing, then the relevant point in time will be halfway through the policy year being priced, and some inflationary factors will be necessary to project the exposure to this point in time (unless such inflation is already included within the sum insured). If the purpose is business planning, then the view of risk will need to be projected further into the future, typically one to three years, in line with forecast inflationary, exposure and business mix changes.

A second, and more complex, consideration is the extent to which climatic trends or fluctuations in risk are taken into account by the model. Is the model attempting to predict the historical risk level (and if so, based on what time period?), the risk level at the current point in time, the risk level over the next few years (perhaps to coincide with a typical business planning period) or the risk level over some longer future time period? For several perils these are interesting and relevant questions as there is evidence of fluctuating rates of activity in the hazard.

For example, at the time of writing, some consider that North Atlantic hurricane risk has a higher activity rate than the long-term average due to enhanced multi-decadal hurricane activity (and raised sea surface temperatures (SST)) in the North Atlantic since 1995. To reflect this, *medium-term rates* or *warm SST* event rates are provided as an alternative to the long-term historical rates by most vendors.

From an earthquake perspective, once an earthquake happens, this then may alter the chances of subsequent earthquakes within the vicinity (typically within a few tens to 100 km, depending on the fault line) of the original earthquake (see Chapter 3.7.1); this time dependence can sometimes be switched on or off within a model. Even if it is not an explicit option, event rates can be changed to reflect the redistribution of tectonic stresses following a large earthquake.

When reviewing a model, it is important to understand the base time period on which the model activity rates are calibrated (or targeted in the case of the forward-looking medium-term rates) and whether time dependence is taken into account for the earthquake peril (and if so, how). With this knowledge the model evaluator can judge, perhaps with the help of external scientific input if needed, whether the baseline used for the model is indeed appropriate. If not, it is often then possible to adjust model output to reflect the company's chosen view of risk.

### 5.4.7 Understanding What Is Not in the Model: Non-Modelled Risk

The previous sections have focused on understanding what is in the model. However, there are many sources of catastrophe risk that are not covered by vendor models, and so, in developing a view of risk, it is vital that these areas are also considered. Models (and model components) are typically developed in response to industry demand – usually for areas with high insurance take-up along with high hazard and exposed values. The first insurance-related vendor catastrophe model developed (in 1987) was for California earthquake, and the second for US hurricane. Today, these two perils continue to rank at, or near, the top of the peril list for most global (re) insurers. Given this, it might be tempting to think that the non-modelled risk would be relatively immaterial on a global scale, since where risk is material, demand will result in a supply of suitable catastrophe models, and the non-modelled risk will become modelled risk. However, for a regional or local insurer, materiality can have a very different definition. Generally, the materiality of non-modelled risk is indeed decreasing over time, and many peril-regions are now covered by vendor models. However, there are still several examples where, at the time of the event, there was either no model, or a large component of non-modelled loss. The 2011 Thailand floods gave rise to the largest freshwater insured loss ever (Swiss Re, 2012), and there was no vendor model available for this peril at the time of these floods. The losses from the 2011 Japanese earthquake also had a large non-modelled component (tsunami and contingent business interruption). Having said this, for most companies using a wide range of models (even from just one vendor), non-modelled risk is not likely to be as substantial as modelled risk, but it is certainly important to consider it when developing a view of risk. The requirement to consider non-modelled risk is implicit in the Solvency II requirements, and Lloyd's require each syndicate or managing agent to deliver information on 'model completeness'.

The material in this section is a summarized version of the 2014 paper written by a group of (re)insurance industry practitioners (ABI, 2014). The interested reader is directed towards this resource as a fuller exposition on this subject.

#### 5.4.7.1 Definition and Categorization of Non-Modelled Risk (NMR)

Although the definition of non-modelled risk (NMR) may seem obvious (i.e. *what is not in a model*), the question then arises – what is meant by a model? In many companies, a peril may not be modelled explicitly by a catastrophe model, but may be covered implicitly (e.g. within an internal model used for capital and solvency purposes). The definition given in ABI (2014) is as follows: 'Any potential source of non-life insurance loss that may arise as a result of existing catastrophe events, but which is not explicitly covered by a company's use of existing catastrophe models.'

It is useful to consider NMR by category as follows:

- peril-regions not covered by catastrophe models
- secondary perils and secondary effects not covered
- classes and lines of business (LOB) not covered
- coverages not considered.

Examples of each of these categories are given in Table 5.2.

Table 5.2  Examples of non-modelled risk, both in terms of categories and specific historical events. For more details of the events, see 'Key past events' sections in Chapter 3.

| | Peril-regions | Secondary perils | Classes/LOBs | Coverages |
|---|---|---|---|---|
| Examples of non-modelled risk types | Flood for most regions | Storm surge, fire following, tsunami, liquefaction, landslide, looting Demand surge, loss adjustment expenses, government intervention | Workers compensation, accident & health Offshore platforms, Wind farms, inland marine, engineering & infrastructure (motorways, bridges) Non static risks (auto, goods in transit, hull, cargo, pleasure craft, fine art and specie) | Contingent business interruption (CBI), freezer contents, additional living expenses, pollution Debris removal, day-one uplift, machinery breakdown |
| 2012 Tohoku EQ | | Tsunami | | CBI |
| 2012 Sandy HU | | Government intervention, fire following flood | Auto, marine | Pollution |
| 2012 Christchurch EQ | | Clustering/aftershocks, extreme liquefaction | | |
| 2012 Thailand FL | flood | | | CBI, 'interests abroad coverage' |
| 2005 Katrina HU | | Levee failure Loss amplification | Multiple non-property losses | Leakage, e.g. flood losses recovered from wind policy |
| 2008 Ike HU | | Inland damage, inland flooding | | |
| 2004 Ivan HU | | | Subsea pipelines | |

Key: EQ earthquake, FL flood and HU hurricane.

The following sections describe methods to identify and quantify NMR.

### 5.4.7.2  Identification of NMR

There are three main methods of identifying NMR:

- *Exposure-based identification*: this consists of analysing current or planned exposures and policy coverages and comparing to scientific datasets or other sources of hazard scoring.
- *Claims based identification*: this involves analysing claims from past events.
- *Identification using expert judgement*: namely gathering opinions and insight from relevant experts.

In practice, all three methods will need to be used as part of a structured process to identify NMR and also to help quantify the risk. Although materiality and the need for prioritization have already been mentioned in the context of developing a view of risk, it is worth emphasizing that this is especially the case for NMR. The risks are generally lower than those modelled, and identification and quantification of such risk can be time-consuming and resource-intensive.

*Exposure Analysis*

The first step in this approach is to estimate the catastrophe exposure (e.g. sum insured) at risk in each country (or other large geographical area, such as a state) for each of the main potentially catastrophe exposed line of business. Typically these are: personal lines (home & motor), commercial and industrial property, energy (onshore and offshore) and engineering (or construction) exposures. For multinational policies, care must be taken in distinguishing between the country of domicile (i.e. where the policy is written from) and the country where the exposure is located.

If exposure is material in a country, the next step is to draw up a list of the perils already covered by the models available to the company. Comparing this list to the list of exposures by country and line of business will enable a company to decide whether there are any material primary peril gaps that need to be quantified.

If there are any such gaps, the next step is to draw up a list of primary catastrophe perils for each country (or region) with some sort of risk or hazard grading. Given that we are considering NMR, this step will require more than a list of the currently available peril models. It will typically require expert opinion, external data and potentially scientific resources.

Having analysed the primary perils, secondary perils should also be considered for both the modelled and non-modelled primary perils in case any are likely to increase the primary peril loss significantly. Claims analysis and expert judgement will be helpful here. See Section 5.4.2.10, Chapter 3, and Chapter 4.3.7, for examples of secondary perils.

Care should be taken to ensure that such an analysis of exposure includes planned exposure growth, not just current exposure. Growth in a new region can often be a cause of NMR; the available models often lag behind expansion in business or insurance coverage.

*Claims Analysis*

One of the clearest ways of identifying non-modelled risk is through learning from claims analyses. Industry-wide studies and analysis of an organization's own claims are complementary ways of identifying NMR.

For the largest events a significant amount of information is published that can help identify areas of NMR. This ranges from descriptive media information to more detailed and relevant post-event reports from model vendors, reinsurance brokers or industry bodies such as PCS or PERILS. Some examples of the information that can be gleaned from industry-wide reports on events are as follows:

- *Past primary perils*: Are all these explicitly modelled? At the time of the 2012 event, the Thailand flood was not.

- *Secondary perils*: Are all the secondary loss causes described in the industry reports modelled currently? At the time of the 2012 Tohoku event, tsunami was not.
- *Lines of business*: Are there reports of claims from the event from lines or classes of business that are not currently modelled? Marine losses from Hurricane Sandy are an example of this.
- *Coverages*: Are all of the coverages giving rise to loss included within the current models? **Contingent business interruption**: a coverage that provides protection against lost profits due to an interruption of business at the premises of a *supplier* or *customer* of the insured in both the 2012 Tohoku earthquake and the 2012 Thailand flood are examples of a coverage that was not included.

A comparison of residential and commercial property claims to automobile physical damage claims from past US hurricanes is shown in Figure 5.22. Often automobile exposures are not modelled and so this could well be a non-modelled class of business for a (re)insurer. If so, an adjustment would be necessary to ensure that the company's overall view of risk is not understated. Figure 5.22 also shows that the level of automobile claims relative to property claims can vary by event, perhaps indicating the need for an adjustment that is more sophisticated than a simple flat increase across all property portfolios.

The advantage of industry reports is their scope; they may include aspects of NMR that would not have been identified by an individual company's claims experience. The disadvantage is that they are less directly relevant to the company in question. For example, although an NMR item is identified from an industry report, if it is not covered in the insurance policies issued by the individual company in question, then it is not relevant.

Internal claims analysis has the advantage of being directly relevant to the business written, but the disadvantage of a reduced credibility and scope. However, studying internal claims data normally yields a huge amount of insight on many topics associated with developing a view of

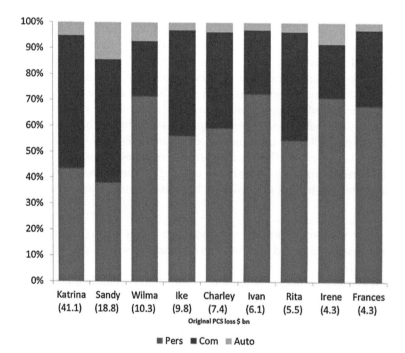

Figure 5.22 Comparison of the distribution of PCS event losses by claims class (Personal = Pers, Commercial = Com, Automobile = Auto) for past US hurricanes (SCOR, 2014). *Source*: Reproduced with permission of SCOR Global P&C.

risk (some examples are given in Section 5.4.3.1). From a NMR perspective the following information can often be derived:

- Claims without corresponding exposures. Is there a non-reported exposure issue?
- Comparison of sum insured recorded on the policy to the loss adjustor's estimate of sum insured. Is there an under-insurance (insured-to-value) issue?
- Comparison of the actual allocated loss adjustment expenses to the indemnity amount to obtain information on a suitable ALAE uplift.
- Review of loss causes (from secondary perils, lines of business or coverages) emerging in the claims that are not captured in the modelling.

### Expert Judgement

Exposure and claims analyses are an important part of NMR identification, however, these are potentially insufficient on their own; there is likely to be a need to seek expert opinion. This opinion should be captured in a structured enough way that it yields useful information and in an open enough way that it allows experts to think outside the normal range of risks. Expertise could be internal or external, academic or industry based. The following areas of expertise are useful in seeking to identify and quantify NMR:

- *Underwriters* take on risk in exchange for a commensurate premium (see Chapter 2.6.1 for a description of the underwriting process). They should therefore have a broad knowledge of the risks involved – including those beyond the scope of modelled risk. A structured process where the relevant underwriters quantify the various risks by region and line of business should yield a useful source of information to cross-check against the exposure analysis and the list of current modelled perils in order to identify NMR.
- *Claims handlers* and *loss adjusters* have detailed knowledge of the loss causes inherent in the claim and so can be a great help in identifying NMR. They are also key in identifying policy leakage. A communication process between claims, actuarial and underwriting should already exist in most companies in order to learn from experience and ensure reserving and pricing are appropriately informed. NMR identification could become part of this regular process.
- *Risk engineers* will be expert at providing a broad range of site-specific risk information. They should therefore be able to assist in the NMR identification process.
- *Local expertise* (i.e. country-specific knowledge) in any of the areas mentioned above can often be helpful.
- *Catastrophe model vendors* should have information on NMR, which could potentially be used within a company's NMR framework to help identify and quantify the risk.
- *Industry expert groups* can help share information on NMR and identify trends and patterns that would be hard to spot from an individual company's experience alone.
- *Hazard information*: There is a significant amount of information available on hazards, many of which are not included in conventional catastrophe models. This information ranges from flood maps to hazard indexes. Some of it is in the public domain, while some of it is available on a subscription basis (such as Swiss Re's *CatNet* or Munich Re's *Nathan*).

Although the process depicted seems fairly straightforward, in practice, there are several reasons why this can be more involved.

The first is that calculating the relevant sum insured by country for all catastrophe exposed lines of business for a global insurer with multinational business stored on multiple systems can be difficult. Very often an *aggregate* summary is produced and available for the purposes of outward reinsurance but this usually only represents certain perils, so for the purposes of NMR identification, it may not be complete. Establishing a complete data extract for all countries for all potential catastrophe perils and catastrophe-exposed lines of business can be a time-consuming task.

The second difficulty is in prioritizing the country perils (see Section 5.3.3). The sources of information already mentioned can help in assigning a Low/Medium/High rating to each country for each peril.

The third difficulty is that once a prioritized list is available, it can be difficult to establish whether peril coverage exists within the policies, as this may differ by line of business or even by customer segment by country.

The final challenge is in knowing which models are available. This should be straightforward for the suite of models used within a company but the model landscape is constantly changing and the pace of change will likely accelerate once open modelling frameworks, such as the Oasis Loss Modelling Framework (see Chapter 6.3.3), become more established. Knowing which models exist at any point in time will become an increasingly involved task.

The end result of these methods should be a prioritized list of non-modelled country-peril-LOB combinations.

### 5.4.7.3   Quantification of NMR

Once NMR has been identified and prioritized, it must be quantified using one or more of the four main methods described below:

1) expert judgement
2) geospatial (exposure/external data-based) methods
3) actuarial (claims-based) methods
4) catastrophe model modification.

The most appropriate one to use will depend upon:

• the purpose of the quantification
• the materiality of the NMR
• the data available (both internal and external)
• the resources available.

These four methods are described below.

#### Simple Expert Judgement

Before the existence of catastrophe models, most (re)insurers would use some form of expert judgement to underpin their catastrophe risk assessments. The expert judgements would typically be based on knowledge or experience of a line of business or peril, coupled with information about a key event in the relevant peril-region. Two uses of expert judgement in NMR quantification are the creation of simple deterministic **scenarios** (see Chapter 4.3.4) and the estimation of scaling factors for NMR.

A typical approach for non-modelled peril-regions is to take a percentage (the *damage factor* or *PML percentage*) of an exposure measure (typically the sum insured or premium) in a geographical region (e.g. country or region within a country) to represent a 'worst case' level of loss (i.e. create a very simplistic deterministic scenario). The damage (or PML) percentage would typically be established from studying a historical industry event, with adjustments where necessary based on expert knowledge of the company's portfolio. This is the same approach as advocated in Chapter 2.9.1.1.

A particular return period is sometimes assigned to this loss although this is, of course, solely judgement-based. The loss at this return period can be extrapolated to other return periods using either a parametric distribution (typically one-parameter Pareto, see Chapter 1.11.2.2), or a distribution based on the shape of a similar peril from another region – but with the overall level adjusted to match the reference loss and return period.

Another approach, often used for secondary perils, non-modelled lines of business or non-modelled coverages, is to use expert judgement to assign a simple adjustment to scale up the primary peril curve to account for the NMR in question. This adjustment may be based on information from past events, on industry reports or on the opinion of in-house or external experts. Care should be taken to consider how any non-modelled component scales with event severity when considering using a flat percentage.

These are simplistic approaches, but may well be the best method for the majority of secondary peril/line of business/coverage NMR, given the materiality and proportionality considerations discussed in Section 5.2.3. It is worth noting that the standard formula approach in Solvency II for catastrophe risk quantification (see Chapter 2.10.3) is an expert judgement damage factor approach.

For either approach, governance considerations (Section 5.3) require that the method and rationale be documented in a clear way.

### Geospatial Methods

Geospatial methods typically extend the damage factor approach described above by applying damage factors in a more refined, location-specific, manner. The starting point is usually exposure or an exposure proxy. The complexity of the geographical approach can vary from using political boundaries to custom-defined shapes to hazard layers. For example:

- CRESTA boundaries (see Chapter 1.9.1.1) can be used to summarize exposure within a country. Different damage factors (from expert judgement) can be applied by CRESTA to obtain an estimate of a reference loss by CRESTA (or, summed up, for the overall country).
- Custom shapes can be defined. For example, a boundary that is a certain distance inland from the coast for tsunami. Exposure can be calculated within these shapes and damage factors applied to derive a potential loss. The damage factors will be informed from historical events or modelling studies.
- Hazard layers, describing a granular variation in damage ratio, can be applied to geo-located exposure to calculate loss potential. This is the method typically used in the post-loss estimation process or in Realistic Disaster Scenario[2] (RDS) modelling (see Chapter 2.9.1 or Chapter 2.7.1 respectively). The damage ratios will usually be defined based on a past event, or modelling of a realistic but hypothetical event. In some cases, the footprint from a past event can be moved around to quantify what could have happened if the path or location of the event had been slightly different. Care needs to be taken when doing this so that the footprints are not moved so far that the revised event location would be physically implausible (e.g. moving a category 5 hurricane footprint from Florida over New York).

These methods also involve expert judgement but they are more complex than the simple expert judgement damage ratio approaches described above. They can be more appropriate where the peril is material and there is a significant hazard gradient to the risk. The geospatial and simple expert judgement-based approaches are exposure-based, unlike the actuarial approach described below.

### Actuarial Methods

The starting point for actuarial methods is a time series of claims data, either company-specific or industry-wide. The claims data are on-levelled using some measure of historical exposure and inflation to ensure they are relevant to today's exposure distribution and values (see Chapter 2.6.2). Finally, statistical models are fitted to these data to provide estimates of losses at different return periods. The considerations in developing claims data for such an analysis are described in Section 5.4.3.

Much has been written on loss modelling techniques and it is beyond the scope of this book to cover these. The reader is referred to a text such as Klugman, Panjer, and Willmot (1988) for further information.

It is worth noting that if claims-based techniques alone were suitable for estimating catastrophe risk, exposure-based catastrophe models would not have been developed. However, claims-based techniques are an important method of validating catastrophe models at low return periods (see Section 5.4.3.2). Regulatory regimes such as Solvency II will likely encourage more claims-based approaches as part of model evaluation.

In the context of NMR, actuarial techniques can prove useful in conjunction with catastrophe model modification in setting the appropriate level for a non-modelled peril-region.

*Catastrophe Modelling Modification*

Existing catastrophe model output can prove useful in helping to quantify NMR. For primary peril NMR, the model output from an existing peril-region can be scaled to approximate the risk from a similar, but non-modelled, peril-region. Adjustments would need to reflect differences in exposure, hazard, vulnerability and financial structure characteristics, and so this approach is only really suitable for the same peril in a similar geographical regime and market.

For secondary peril NMR, as well as line of business and coverage NMR, catastrophe model modification (e.g. by a flat percentage scaling) can also be a suitable implementation approach as it can incorporate the dependency between secondary and primary perils as well as accumulations between modelled and non-modelled lines of business and coverages.

The three previous methods (expert judgement, geospatial and actuarial) can all be used to derive adjustments that are *implemented* through catastrophe model adjustment. Implementation of a view of risk, including NMR, is the topic of Section 5.5.

## 5.5 Implementing a View of Risk

The main focus of this chapter has been on the need to develop a view of risk, the governance related to this and the required techniques, whether for modelled or non-modelled risk or for single or multiple models. Once a view of risk has been developed, it must be implemented within the many areas of a (re)insurance company using catastrophe models. This section highlights some of the practical considerations in developing a view of risk, including the important subject of model blending.

### 5.5.1 Different Uses in a Company

Catastrophe model output is used in the following areas within a (re)insurer:

- underwriting and pricing: expected loss cost and profit loading (see Chapter 2.6);
- accumulation, roll-up and capacity monitoring (including reporting and risk tolerance testing; see Chapter 2.7);
- portfolio management and optimization (including business planning; see Chapter 2.8);
- event response (see Chapter 2.9);
- capital modelling and management (see Chapter 2.10);
- reinsurance structuring and pricing (see Chapter 2.12).

The applications of catastrophe models are described fully in Chapter 2. The point of mentioning these areas here is two-fold. First, that in deriving or changing a view of risk, the downstream impact on each of these areas should be understood. The impact should not be used to inhibit or encourage a particular view of risk, but it should instead be identified and communicated to the relevant stakeholders as part of a complete process. This is discussed in Section 5.3.8.

Second, the practicalities of implementing a view of risk may well be different depending on the purpose for which the view is used. The view of risk should be consistent (with the caveats discussed in Section 5.5.2) in each area, but the implementation may vary a little depending on the tools used in each area. For example, the pricing of large and complex business may use the AAL directly from a catastrophe model. A view of risk implementation requiring an event-specific adjustment may be difficult to implement directly in the catastrophe model given the underlying model framework; it may, however, be more easily applied in a simulation package such as one of the asset-liability modelling (ALM) or dynamical financial analysis (DFA) systems on the market. A simpler scaling may be needed for the pricing implementation via a catastrophe model in this case.

### 5.5.2   Consistency in an Organization

It seems sensible that once a view of risk is developed, it should be used, in all the relevant areas in an organization, in a consistent way. Indeed, Solvency II regulation encapsulates this as part of the *use test* (see Table 2.7), where a company needs to demonstrate that the internal model (part of which should include catastrophe model output if the catastrophe risk is significant) is being used and plays an important role in the company's risk management system (including underwriting and reporting), their decision-making processes, and their risk-based capital assessment and allocation processes.

However, caution is required when propagating a single view of risk throughout an organization if models are used both to perform the risk selection (to price, underwrite and select terms) and to measure the outcome of the risk selection (reporting PMLs, comparing metrics to risk tolerance limits, calculating risk-based capital). The view of risk influences the prices charged, the underwriting criteria and the final limits and deductibles. The resulting portfolio therefore reflects this view of risk applied at the point of risk selection. If the company then uses the same model and the same view of risk to measure whether the underwriting process has been successful or not, there is a danger of the company fooling itself as to the efficacy of its catastrophe risk selection. In short, the company is optimizing around one model. This circularity issue will only become evident when a model is changed or when a catastrophe event happens and the company realizes the exposure to a peril is much greater than it expected.

This issue can be mitigated by using multiple models when building a view of risk (discussed in Sections 5.4.4 and 5.5.4) to ensure a portfolio does not just reflect one model. However, even using multiple models in constructing a view of risk will not fully mitigate the issue of using the exact same view for risk selection and for measurement of risk-selection efficacy.

There is no *correct* answer to this problem. However, for many perils, underwriting and pricing information do exist separate and independent from catastrophe models. This information is typically designed for assessing relative risk at any given location and so is useful for differentiating expected loss. It is not designed for assessing accumulations and the impact of large events; that is the role of catastrophe models. Usually this information will be in the form of flood or earthquake maps or a loss index by postcode. The information can be used for defining rating areas and as the basis for tariffs and, as it can be somewhat independent from the catastrophe model output, should help to mitigate the circularity issue of selecting risks and measuring success based on the same model.

### 5.5.3   Methods of Implementation: Single Model

This section focuses on different practical approaches to implementing a view of risk. There is no 'right' approach to this; the choice of method will be influenced by the materiality of the risk, the strategy around multiple models, the rationale for the view of risk adjustment and the transparency and ease of communicating and tracking both the adjustments and the rationale for the selected approach.

A generic consideration across all of the methods outlined here is the granularity at which the adjustments should be made. For example, are the adjustments appropriate to a whole peril-region, should they vary by country, should they vary by segment (e.g. personal lines, commercial lines) within a country, or should they be defined at some finer resolution? It is likely that some view of risk adjustments will apply across all granularities, while others will be specific to a particular subset of exposures. For example, a flat ALAE uplift (of 5%, say) may apply across all homeowners' portfolios, whereas an uplift for non-geocoded exposure will be different for each specific and granular portfolio according to the completeness of the address data and subsequent geocoding for the particular portfolio. An example of different adjustment ranges for different underlying causes is given in Figure 2.11 in Chapter 2.6.4.2.

In practice, there are different ways of implementing adjustments to models and the following techniques are all methods of implementing a view of risk. In today's platforms and modelling frameworks, the ability to apply these techniques is somewhat limited and constrained; this is likely to change as platforms evolve to enable the user to apply a view of risk adjustment directly within the model (rather than outside the model, as is currently the case for some platforms).

### 5.5.3.1 Adjusting the Input Exposure

In some circumstances the exposure can be scaled up or down prior to input into the catastrophe model. This is sometimes done as a workaround adjustment that ideally would be done directly within the model, but may not be possible given the limitations in some current platforms. This is an appropriate method of adjusting a view of risk under the following conditions:

- The effect is constant across return periods.
- The effect should occur *prior to* all policy conditions
- The effect is thought to be highly correlated with the input exposure being scaled.

This type of adjustment may therefore be appropriate for inflationary adjustments or adjustments for underinsurance. Care is needed when scaling the input exposure, as if it is done for the wrong reason it can lead to an unrealistic impact of financial structures. The relationship between the sum insured and the financial structures must be considered in order to avoid biasing the results. For example, consideration should be given to whether just the sum insured is scaled, or whether the deductibles and limits are also scaled. Adjustments can also be made to primary modifiers (such as floor area or occupancy type) to replicate claims behaviour (e.g. in a particular geographical area or a particular niche of business).

These kinds of adjustments are sometimes a necessary workaround, given the limited ability to apply adjustments in many current models and they need to be carefully documented, as downstream users of the model may not be aware of the adjustments and may misinterpret the adjusted exposure and results.

### 5.5.3.2 Adjusting the Outputs: Event Loss Table Scalings

The most important output from a catastrophe model is the event (or year) loss table. As discussed in Chapter 1.10.4, this contains mean event losses for each event together with other parameters that define the distribution of loss around the mean. The losses can be before financial structures ('ground-up'), after insurance financial structures ('gross'), after facultative reinsurance but before catastrophe reinsurance ('net loss pre-catastrophe') or after all financial structures and reinsurance ('net loss'). See Chapter 1.10.4 for more information on ELTs and YLTs, and Chapter 1.9.2 for details on financial perspectives. The model output, embedded in these tables, can be scaled to reflect a view of risk. The three options for adjusting output are: (1) severity scaling; (2) frequency scaling; and (3) uncertainty scaling. These are discussed below.

### Severity (Event Loss) Adjustment

The mean loss from each event can be scaled to reflect a company's view of risk for a particular peril. This scaling may also apply to the other parameters representing the uncertainty distribution, for example, standard deviation and maximum exposed value where a Beta distribution is used.

There are usually different loss perspectives available for each ELT/YLT and so the most appropriate, or pragmatic, perspective for scaling should be selected. Ideally scaling should be applied at the ground-up perspective and then flow through the other perspectives. In practice, this is difficult and not often done. It is more common to adjust the net loss pre-catastrophe reinsurance perspective, since this is the perspective most commonly needed within a company's asset liability model (treaty reinsurance structures are often represented within this type of model rather than the catastrophe model). The severity scaling will need to be appropriate for this financial perspective.

The scaling could be the same across all events or it could vary by event. For example, a view of risk analysis may conclude that only a subset of events need to be scaled, perhaps just those affecting a particular region. As a second example, an adjustment may be aiming to correct an identified weakness in the model hazard; perhaps the hazard is felt to be too high for large events. If scaling a hazard directly within the model is not practical, then a scale-down for larger events could be developed as a proxy for this effect.

Implementation can be achieved by scaling the mean loss (in which case, the scaled mean is used as a parameter within the uncertainty distributions) or by scaling the sampled representations of loss from the uncertainty distributions. If the view of risk adjustment is across all events, or specifically related to some event characteristic (e.g. event location), then it would make sense to scale the mean loss. If the adjustment is related to the size of loss of the event, then it could make sense to scale the sampled loss from the uncertainty distribution instead. A consideration when scaling the mean loss is whether or not also to scale the other uncertainty parameters (e.g. the standard deviation and the exposed value). Unless there is a good reason not to, in most situations it is sensible to scale these by the same amount in order to preserve a constant coefficient of variation; if a Beta distribution (see Chapter 1.11.2.1) is being used, this also preserves the values of alpha and beta.

Scaling the loss severity has the advantage of being intuitive, transparent and easy to explain. It can also be more directly related to the underlying need for adjustment (e.g. incorporating ALAE) than other methods. It is also possible to apply different adjustments to different size events *after* uncertainty sampling.

Note that when blending models, there are some disadvantages to blending on severity and a comprehensive paper on this topic (Calder, Couper and Lo, 2012) recommends blending on frequency. Nonetheless *scaling* or *adjusting* severity is still a useful technique and indeed is part of the recommended technical solution for the 'bottom-up' component in Calder, Couper and Lo (2012).

### Frequency (Event Rate) Adjustment

An alternative to severity scaling is adjusting the event frequency. This may be a preferred approach if the company gives credence to the modelled severity but not the modelled frequency. For example a company may want to correct a deficiency in the event rates, perhaps to implement their own medium-term view of risk. Unlike severity scaling, with frequency scaling, no consideration is needed on which financial perspective is adjusted. For the ELT model output, the scaling is straightforward; event rates can be modified directly as required. For the YLT model output, frequency scaling is less straightforward. Each event will have a rate of 1 divided by the number of years in the YLT simulation period. However, this rate is usually not contained or used on an event-by-event basis within a YLT implementation, rather, a specific year will be picked at random and the events from this year used to generate losses. One approach to change rates (for a constant frequency adjustment over the peril-region) is to randomly sample from all the events in the YLT, with the purpose of either removing events (decreasing frequency) or adding in duplicate events (increasing frequency). However, caution is needed in using this approach as, if there is dependency between events (e.g. Europe wind clustering), then such an approach could unwittingly alter the event dependency relationship.

*Uncertainty Adjustments*

ELT and YLTs usually include the mean event loss together with some representation of uncertainty. As discussed in Chapter 2.16.1, this representation is normally only a subset of the true uncertainty in the model. A view of risk analysis may conclude that while the mean event loss outputs from the model are fine, the representation of uncertainty is flawed and needs to be adjusted. In this circumstance the standard deviations in the ELT or YLT (if present) could be scaled.

### 5.5.3.3 Adjusting Specific Model Components

Implementing a view of risk by adjusting the inputs (exposure) or outputs (ELT/YLT) is possible for all catastrophe models. However, in some circumstances, a view of risk analysis will conclude that there is an issue with a specific model component. In this case it is desirable to adjust just this component to best represent the underlying issue. For example, an issue with the vulnerability curves for a particular type of business may have been identified. For insurers with a high market share in a particular type of business, it is possible that the insurer will have sufficient claims information to be better informed about the vulnerability than the modelling company. Although it is technically possible to adjust vulnerability in some models (if the user is advanced and knows exactly where to adjust database tables within the model database schema), this always carries an operational risk. Alternative options are either to adjust the exposure or to isolate this particular type of business into a specific portfolio and adjust the ELT or YLT for this tranche of business. Neither are particularly good solutions as they are roughly equivalent to a flat scaling across the entire vulnerability curve.

Recently, some model vendors have introduced platforms that are more open (e.g. Impact Forecasting's *Elements* or JBA's *JCALF*). In these platforms the vulnerability curves and event footprints can be viewed and modified by the user. Consequently, it is possible to implement a more granular and realistic view of risk adjustment. Despite this functionality, care must be taken when making such adjustments, and these should only really be made by an advanced practitioner with some model development skills in order to prevent unwanted effects. For example, an adjustment made to what the user believes is for a single combination of exposure characteristics could actually be applied to other classes of vulnerability curves (by mistake) as well. This type of adjustment also needs to be supported with strong governance to prevent multiple variations of a model existing and the wrong version being used unwittingly by other users within the (re)insurance company.

Other platforms (such as RMS's *RMS(one)* and AIR's *Touchstone*) have more granular model component adjustment on their development agenda and so it is anticipated that these types of adjustments will become a more common way of implementing view of risk.

### 5.5.4 Methods of Implementation: Multiple Models

Different strategies for using multiple models are discussed in Section 5.4.4, and this section focuses on practical implementation techniques when multiple models are used.

Section 5.5.3 focused on implementing a view of risk when a company is using one set of models as a base and adjusting this base according to the results of their view of risk analysis, which itself may well involve considering the results of other models. Another possibility is to blend output from multiple models together to form a composite view of risk, which is the subject of this section.

A comprehensive paper has been written on this topic (Calder, Couper and Lo, 2012). The interested reader is directed to this paper as it contains much more information than can be contained in this book section. Other helpful publications that consider this topic are Cook (2011), Guy Carpenter (2011) and ABI (2011).

Whether or not to blend catastrophe model output is itself a contentious topic within the catastrophe modelling community. Some believe that model blending may abdicate responsibility for understanding the constituent models, resulting in a naïve equal weighting of models; some of which might be very wrong and not at all appropriate for the portfolios being modelled. The alternative approach of deeply understanding and using one model as a basis is sometimes viewed as being preferable, since in this case the company understands one model well in the context of its portfolio and can calibrate it to its own view of risk. Another view is that blending several models, if they are well built and appropriate to the portfolio, reduces uncertainty to some extent and is a preferred approach to using one model (with the associated dangers of single model bias). *Educated* blending (which involves performing a full evaluation of each of the models to be blended and weighting them accordingly), as an alternative to naïve blending, is a comprehensive approach to understanding catastrophe risk, but of course is the most resource-intensive since the company must fully evaluate several models for a given peril-region and also document and be able to demonstrate this to their senior management and regulators. It is also the most costly option since a company must license a model to fully understand it.

One pragmatic solution is to relate the chosen approach to both the materiality of the peril (to the (re)insurance company) and the availability of resource. As discussed in Section 5.2.3, a company must prioritize its resource and should form a hierarchy of peril-regions, allocating more resource to those peril-regions which have the potential to cause most harm to the company. For example, a company could categorize each peril-region into Low, Medium and High (perhaps based on a TVaR metric, see Chapter 1.10.3). The view of risk implementation could then be varied as shown in Table 5.3.

Model blending in this section is very much focused on bringing together the output from different models in a reasonable way. However, there is also the concept of *model fusion* (Guy Carpenter, 2011) as a form of model blending. The idea is to combine different components from different models to create a fused or blended model. This idea is conceptually appealing. However, in order to take and combine the best parts of each model, in practice, each model component would have to be calibrated to work with the other components in the fused model. For example, many models will adjust vulnerability levels to ensure that the model meets overall industry-wide statistics (either return periods for key events, or levels of industry loss for key events). Models are built so that each component is individually calibrated and validated, as well

Table 5.3 Example of how the approach to multiple model usage could vary by peril-region priority.

| Peril category | Approach |
| --- | --- |
| Low | Equal weight blending |
| Medium | Educated adjustment of one model as a base |
| High | Educated blending |

as the total model output. However, limitations in data availability for some peril-regions mean that sometimes it is just not possible to do this to a precise degree and therefore one component of a model ends up compensating for another, although the total model output can be calibrated and validated. This of course complicates model fusion. There are also the technical difficulties of bolting together different model components. As frameworks such as Oasis take shape, and the concept of a wide variety of models and model components from different providers on the same platform evolves, it is likely that the technical difficulties will reduce. However, the overall calibration issues will remain, and it is likely that for a significant amount of time, model fusion will be the preserve of a very few, highly resourced and skilled entities. The remainder of this section focuses on model blending being the practice of combining the output from different models.

### 5.5.4.1 Different Methods for Blending Output

There are several different approaches for bringing together the outputs from multiple models and they all have their strengths and weaknesses. The operational implementation of the blend is always an important consideration. What may seem sensible and intuitive when considering blending EP curves from two different models may not be so straightforward when having to implement at the ELT/YLT resolution needed for many practical purposes.

#### Severity Blending

One way of blending model output is to perform a (weighted) average of the loss amounts at specific exceedance probabilities (or return periods). Since this is taking an average of the size (or severity) of the loss, it is often referred to as severity blending. Given that it is quite common to discuss and report loss sizes at specific return periods, this is perhaps the most common and intuitive method of model blending. It is also easy to calculate, for example with two models for the same peril run over the same portfolio:

Model A 200-year OEP loss = £200 million
Model B 200-year OEP loss = £400 million
50%/50% model blend 200-year OEP loss = £300 million

Some considerations with severity blending are as follows:

- Severity blending is financial perspective-specific, and so a decision will need to be made about which financial perspective it applies to, and how it propagates through to other financial perspectives in a consistent way.
- Practically, many implementations and uses of catastrophe model output require calculations at event (or year) loss table level. This means the blending must also be performed at this level. It is difficult to perform severity blending at ELT/YLT resolution in a way that conceptually makes sense and does not bias the outcome towards one model or another. This is discussed in more detail in Section 5.5.4.2.
- Conceptually, a 50%/50% severity blend implies that the practitioner thinks that given an event has happened, the size of the event loss has an equal chance of being of a magnitude of loss either from Model A or Model B.
- Related to the point above, the loss severities at a given return period for two different models may have underlying drivers that are very different to each other. For example, the 1-in-200-year loss from Model A might be driven by a large windstorm impacting the United Kingdom, while in model B the 1-in-200-year loss might be driven by a large windstorm impacting France. If this is the case, it is conceptually difficult to explain why it makes sense to average these two different quantities.

*Frequency Blending*

Frequency blending is a method suggested and strongly advocated by Cook (2011) and Calder, Couper and Lo (2012). It involves taking the average of exceedance frequencies at a given level of loss. For example:

Model A £200 million OEP = 0.5% (200 years)
Model B £200 million OEP = 1.0% (100 years)
50%/50% model blend OEP at £200 million = 0.75% (133.33 years)

Although straightforward to calculate, it does not easily yield (through manual calculation) the blended loss at a specific return period, and so it is perhaps slightly less intuitive than a severity blend.

Some considerations with frequency blending are as follows:

- A frequency blend is not financial perspective-specific.
- Practically, a frequency blend is easier to implement at ELT/YLT resolution. This is discussed in more detail later in Section 5.5.4.2.
- Conceptually, a 50%/50% frequency blend implies that the practitioner believes that Model A could be correct 50% of the time and Model B could be correct 50% of the time.
- A frequency blend allows for better consistency between accumulations and pricing purposes. The weighted combination of AALs from different models (as may be used as an input to pricing) is directly analogous and consistent with a frequency blend, but *not* with a severity blend.

*Rank Matching*

Rank matching (Maynard, 2011) is similar to the approach of using EP levels as a key to match events. It works as follows:

- A ranking is randomly distributed across a given number of years. For example, if 10,000 years is selected as a suitable base period, the numbers 1 to 10,000 are randomly assigned to each year.
- The metric to be blended (e.g. 200-year net AEP) is calculated from each model, for the same number of years as the base period and ordered so that the ranks for the metric are calculated for each model. For example, rank 1 would indicate the largest loss for a model, rank 2 the second largest loss for the model, and so on, until the smallest loss (which would be rank 10,000 in this example).
- These ranks are used to assign the losses for each model to the random distribution of ranks originally allocated to each year. For example, if year 1 in the base period has been assigned the rank of 50, the 50th largest losses from Model A and Model B will be selected for blending.

This is the way that Lloyd's bring together the results from the disparate model versions and model vendors used by each syndicate and managing agent (Maynard, 2011). An assumption embedded in this approach is one of perfect correlation between ranks for the different models. This is undoubtedly a pragmatically conservative assumption.

*Severity Scaling Versus Frequency Blending*

Model blends help combine outputs from models in order to help reduce uncertainty and avoid optimizing around any one model. However, even with blending there will always be a need to adjust model results for the features not explicitly captured by the models. As described in Section 5.4.7.3, these adjustments often take the form of an event loss scaling, as they are scaling up the loss in order to adjust for correlated elements not in the model. It is important to note that although frequency *blending* is the recommended method for blending multiple models together, severity *scaling* is a useful tool for adjusting model output for non-modelled risk. In fact Calder, Couper and Lo (2012) prescribe frequency blending adjustments from 'top down', and severity scaling adjustments from 'bottom up' as part of their technical solution (as described in Section 5.5.4.2).

*Other Considerations*

There are several other aspects to consider when model blending, as described here. In blending models there is always a need to average the results from the different models. Most practitioners would use an arithmetic mean to combine results. Using a geometric mean can give results that are optimistically low, as described in more detail in Calder, Couper and Lo (2012).

In some cases a company may want to use one model to replicate as closely as possible the result of another model; for example, perhaps they do not license the other model, or the other model works well for one financial perspective, but cannot cope with financial structures for an alternative financial perspective, or the accumulation platform is only based on one model format. In these cases there are techniques to *shoehorn* (ABI, 2011; Cook, 2011; Maynard, 2011) or *morph* (Guy Carpenter, 2011) one model into another model. These typically involve defining a transformation function to translate losses from one model into another and are beyond the scope of this chapter.

Calder, Couper and Lo (2012) advocate an approach to model blending which allows for both top-down and bottom-up adjustments. The top-down frequency blend is defined centrally and applied across all portfolios. However, in addition to this, severity adjustments for specific portfolios are allowed (e.g. reinsurance contracts in the context of Calder, Couper and Lo (2012)) to provide the individual underwriter with a way of allowing for items such as portfolio growth, data quality, non-modelled items, and experience and other pricing analyses. These adjustments need to be carefully governed and documented, but if done in a structured way, have the benefit of allowing underwriters to bring their experience to bear and also inform potential future R&D that can, at some future time, feed into the top-down blend.

As is the case for any view of risk adjustment, all model blending should be well documented and subject to robust governance, as described in Section 5.3.

### 5.5.4.2 Implementing Frequency and Severity Blends in ELT/YLTS

ELT and YLTs are the form of output most often used by (re)insurance companies. These can either be individual policy (or treaty) specific tables used for pricing purposes or tables at the level of a business unit of a company that need to be accumulated. A model blend will usually need to be applied at this ELT/YLT level so that the resulting blend flows through to the downstream applications of model results.

First, consider the potential ways of applying a severity blend. Assume there are two ELTs, one from Model A, one from Model B, each containing a number of events with event rates and losses particular to the portfolio being modelled. How should these event losses be combined to form a blended loss? Some options are:

- Match the events from Model A with the events in Model B in some way. For example, for a particular reference portfolio that the blend is based upon, each event could be assigned a precise exceedance probability level from each model, and these could be used as a key to match up events. There will be practical considerations, since the number of events in different models will usually be different, and there may be the need to combine an ELT from one model with a YLT from another. There is also the previously mentioned conceptual issue of combining very different types of events: this can be especially awkward when the blend is taken to granular levels and used in pricing analyses, for example.

- Derive an ELT-level adjustment to one model by considering the results of the severity-blended EP curve from both models for a particular reference portfolio. The scaling is applied to the event losses from just one of the models, using these as a base to try and reproduce the effect of blending the two models. An issue with this approach is that even with a 50%/50% blend, the portfolio will optimize towards the selected base model, as it will be impossible to fully reflect the complexities of the other model in such a scaling.

A significant issue with both approaches, and indeed any severity blending approach, is that the blend is sensitive to financial structures. A blend derived for one financial perspective will differ (often significantly) from that derived for another.

Frequency blending, on the other hand, is much easier to implement. The events from each model are simply appended into the same event loss table and the event rates for each model are scaled appropriately (e.g. halved for a 50%/50% blend). There is no need to try and match events in any way and the blend is not sensitive to choice of financial perspective. It is for these practical issues, as well as the conceptual issues with severity blending mentioned in Section 5.5.4.1, that Calder, Couper and Lo (2012) strongly advocate frequency blending.

This section has largely used the concept of an ELT to highlight considerations when blending model output. YLTs of course can also be used. For example, with two YLTs, a frequency blend can easily be constructed within a simulation by sampling (x) from a uniform distribution, and if $x < 0.5$, choosing one year's events from Model A, otherwise choosing that year's events from Model B.

For some purposes it may be necessary to construct a specific number of years (e.g. to fit in with a wider capital modelling framework). For a YLT, the number of years can be reduced by simply sampling a reduced number of years from the original number of years. Increasing the number of years is more problematic. Years can either be duplicated or the vendor can be contacted to see if they have a larger set available. For known frequency distributions, the intended number of years can be formed by simulation where the YLT is essentially treated as an ELT, with each event having frequency = 1/number of years in original YLT. However, some YLTs (e.g. those formed from the output of GCMs) may have an inherent (non-parametric) frequency distribution, which then precludes this kind of simulation approach to increase or decrease the number of years.

### 5.5.4.3 Choosing the Weights

Several methods for combining or blending model results have been described. However, so far in this chapter there has been no discussion of the value of the weights to be used in blending together model results: that is the subject of this section.

A naïve model blend implies an equal weighting, as nothing is known about either model. It is also a starting point that can be refined by a more educated model blending approach; an approach that gives most weight to the most suitable model for the portfolio in question. If a particular model is known to be unsuitable, then zero weight should be given; perhaps the geographical scope is not sufficient for the portfolio, or the model cannot cope with the complex financial structures in a particular portfolio. However, there will be many situations where several models appear equally valid, and in-depth work is needed in order to decide whether one model is more suitable than another and so assign appropriate non-equal weights.

The output from a view of risk analysis (Section 5.4) should lead to conclusions about model weightings. For example, Cook (2011) suggests a combination of technical factors and wider considerations when selecting weights. Technical factors could be compared using a check box or scoring approach to various categories relating to the model hazard, vulnerability and calibration. Wider considerations relate to the openness of the vendor, frequency of model revision, quality of documentation, and so on. A key question when assigning credibility to a model is the quantity of relevant and detailed claims experience used in the model calibration. This can often vary significantly between models. The quality of the match between historical actual event losses and the historical modelled losses for the company's portfolio can also be useful in deciding upon model blending weights.

## 5.6 Conclusion

Developing a view of risk to understand and manage the catastrophe risk that a company assumes is an important, but complex and resource-intensive process. In fact, the level of detail discussed in this chapter may lead the practitioner to despair that such a process is impossible due to resource constraints. However, as discussed at the outset of this chapter (Section 5.2.3), prioritization is extremely important and it will not be possible to apply many of the considerations and techniques in this chapter to every peril-region. Indeed, it may well be the case that, for many of the small peril-regions, a particular company uses the model results with no model evaluation, the rationale being that the level of modelled loss is small and it is better to have some model output (unevaluated) than no model output at all. So long as there is a rational process for defining when a model should be evaluated, with sensible definitions of what constitutes small, this seems a perfectly sensible approach.

Although the need for the analysis to be documented has been noted in this chapter, the format of how a view of risk analysis should be reported has not been discussed as this will be company-specific. However, the reader is referred to LMA/Lloyd's (2012b) as a good example of a model evaluation report for a high priority peril-region.

Model evaluation should always be focused on ensuring that the model is appropriate for the company in question. It is also important that the company retain both the understanding of why the model is appropriate for the company, and the decision-taking responsibility for whether and how to use the model output. This being said, there is much duplication and inefficiency in respect of model evaluation within the insurance industry. For example, in respect of US hurricane risk, HURDAT (Chapter 3.2.8) will have been downloaded and processed by many companies, and landfall frequency exhibits comparing industry data to model characteristics (such as in Figure 5.3) will have been produced many times as part of the evaluation process. Many companies will be running the same kind of sensitivity tests through the same models, producing similar exhibits. Many companies will be developing very similar tools to perform the type of analyses described in this chapter. It is hoped that in the future some aspects of developing a view of risk will be standardized to reduce the current market duplication and redundancy associated with this process. As the choice of models available continues to increase, efficiency in developing a view of risk will become more and more important.

## Notes

1. http://www.nhc.noaa.gov/data/#hurdat
2. https://www.lloyds.com/the-market/tools-and-resources/research/exposure-management/realistic-disaster-scenarios

## References

ABI (2011) *Industry Good Practice for Catastrophe Modelling: A Guide to Managing Catastrophe Models as Part of an Internal Model Under Solvency II.* Association of British Insurers, London.

ABI (2014) *Non-Modelled Risks: A More Complete Catastrophe Risk Assessment for (Re)Insurers.* Association of British Insurers, London.

Calder, A., Couper, A. and Lo, J. (2012) Catastrophe model blending: Techniques and governance. Paper presented at General Insurance Convention (GIRO), UK Actuarial Profession, Brussels.

Cook, I. (2011) Using multiple catastrophe models. In *The Latest Issues Surrounding Catastrophe Modelling.* UK Actuarial Profession, London.

EU-LEX (2015) Commission Delegated Regulation (EU) 2015/35 of 10 October 2014 supplementing Directive 2009/138/EC of the European Parliament and of the Council on the taking-up and pursuit of the business of Insurance and Reinsurance (Solvency II). *Official Journal of the European Union*, **58**.

Guy Carpenter (2011) *Managing Catastrophe Model Uncertainty: Issues and Challenges.* Guy Carpenter, London.

Klugman, S., Panjer, H. and Willmot, G. (1988) *Loss Models: From Data to Decisions.* John Wiley & Sons, Inc., New York.

Lloyd's (2014) *Solvency II Model Validation Guidance.* Lloyd's, London.

LMA/Lloyd's (2012a) *Validating External Catastrophe Models Under Solvency II.* LMA/Lloyd's, London.

LMA/Lloyd's (2012b) *External Catastrophe Model Validation: Illustrative Validation Document No.1 US Windstorm, High Materiality.* LMA/Lloyd's, London.

Maynard, T. (2011) *Lloyd's catastrophe modelling. The Latest Issues Surrounding Catastrophe Modelling*, UK Actuarial Profession, London.

Olson, A. and Porter, K. (2011) What we know about demand surge: Brief summary. *Natural Hazards Review*, **12** (2), 62–71.

Parodi, P. (2015) *Pricing in General Insurance.* CRC Press, Boca Raton, FL.

SCOR (2014) Hurricane risk: A reinsurer's perspective. *RAA*, Orlando.

SCOR (2015) What does it take to understand a model? *RMS Exceedance*, Miami.

Sweeting, P. (2011) *Financial Enterprise Risk Management*, Cambridge University Press, New York.

Swiss Re (2012) Natural catastrophes and man-made disasters in 2011. *Sigma, 2/2012.*

Swiss Re (2013) Natural catastrophes and man-made disasters in 2012, *Sigma, 2/2013.*

Waisman, F. (2015) *European Windstorm Vendor Model Comparison.* Aventedge CAT Risk Management and Modelling, London.

Waisman, F. (2016) *US Hurricane Vendor Model Comparison.* Aventedge CAT Risk Management and Modelling, London.

# 6

# Summary and the Future

*John Hillier, Kirsten Mitchell-Wallace, Matthew Jones, and Matthew Foote*

## 6.1 Overview

6.2    Key Themes in the Book
6.3    The Future: Progress, Challenges and Issues
References

The most important thing to remember about 'catastrophe models' is that they are just that, models. They are simplified mathematical approximations of reality. Thus, as with any model of any sort, they will never be correct. However, it is also critical to remember that this does not stop them being vitally useful; scribbling on the back of an envelope is better than guesswork, fitting statistical distributions to historical losses is better than the envelope, and catastrophe models add scientific insights into the multi-hazard Earth system making them better still. They are our current best guess.

## 6.2 Key Themes in the Chapters

### 6.2.1 Chapter1: Introduction

Most generally, a 'catastrophe' is any adverse event whose impacts exceed the capability of those affected to cope or absorb its effects. Thus, a catastrophe intrinsically requires it to have a human impact. Hazards generated by natural processes are a key source of risk and the subject of this book.

'Catastrophe models' are a type of probabilistic multi-hazard risk assessment tool for portfolios of exposed assets, and the key simplification that makes them practical is the use of an event set of many ($n > 10,000$) stochastic hazard 'footprints'. Usually, the aim is to manage the portfolio to determine efficient and effective ways of managing near-future risk (i.e. next year); this does not mean minimizing risk, rather, the management is to attempt to ensure that an organization's risk appetite and its risk-return profile (i.e. risk, cost, benefit requirements) are adhered to.

Exposure-based catastrophe models that run 'synthetic' realizations of next year over an organization's current exposure base to estimate risk within the next year are needed because:

- Catastrophes of a particular type in a particular location are rare (e.g. every 100 years), so our experience is typically insufficient to judge risk.

*Natural Catastrophe Risk Management and Modelling: A Practitioner's Guide*, First Edition.
Edited by Kirsten Mitchell-Wallace, Matthew Jones, John Hillier and Matthew Foote.
© 2017 John Wiley & Sons Ltd. Published 2017 by John Wiley & Sons Ltd.

- In any case, changing trends invalidate experience (e.g. climate change, demographic shifts, inflation, building standards, changes in business mix and original policy type).
- Hazard maps, even probabilistic ones, cannot assess the risk of more than one exposed asset; specifically, they do not factor in the spatial correlation between assets.

While catastrophe models are currently configured for financial losses, they provide a framework that requires no conceptual alteration to be applied to other 'costs' (e.g. lives, additional journey time). Indeed, there is no reason why multiple types of cost cannot be included, summed and weighted to reflect the views of different stakeholders. Thus, methods refined in insurance have wider potential uses in fields such as disaster risk reduction; see Section 6.3.4.

### 6.2.2 Chapter 2: Applications of Catastrophe Modelling

This chapter is aimed at users of models and model developers, since it attempts to encompass the diverse applications of the models in the current environment. The details of the topics covered should be familiar to those who work in a particular area but the collection of topics is intended to give a broad overview to provide basic context across a spectrum of uses rather than detailed actuarial or commercial insights.

The crux of this chapter is that catastrophe models are not an end in themselves, but are a tool to enable catastrophe risk management, including risk transfer. The traditional home of catastrophe models has been risk quantification to support underwriting and risk management in insurance and reinsurance, but their application is growing: not just to insurance-linked securities (ILS) but also into disaster risk reduction (DRR) and government initiatives.

Treatment of uncertainty will ultimately become more complex, and indeed this can already be seen in newer models, which attempt to provide more transparency around model development choices and the impact to test these. However, at present, there is no clear route to propagating this into the essential applications of catastrophe modelling and it still remains a relatively academic exercise.

### 6.2.3 Chapter 3: The Perils in Brief

This chapter is mainly aimed at model users, but should also be informative for those designing models. Model developers will have an in-depth familiarity with the perils that they are modelling, but there are insights that can be drawn by comparing practice between perils. The style is concise but highly referenced; the intention is for it to be a launch pad for investigations of any level of depth, from a brief familiarization to an in-depth study. Three cross-cutting points run across the perils.

First, knowledge of the frequency-severity distribution, location, and expected magnitude of losses (e.g. with respect to other perils or regions), is a vital tool to enable users to reality-check model outputs. This is necessary as it is inherent in the nature of sophisticated, or at least complex, numerical models that results are sometimes produced that are either 'obviously' wrong or contradict a user's view of how the world works either scientifically (i.e. physical processes) or in terms of risk (i.e. evidence from loss data).

Second, do not be afraid to challenge models or model vendors; all models are necessarily imperfect, and developing a view of a peril is a key part of developing a view of risk (Chapter 5), perhaps by accounting for non-modelled factors. In addition, if an unexpected high-impact event happens, as they do (e.g. the Tohoku tsunami), a good grounding in that peril permits an agile response without waiting for model revisions and potential over-reaction can be avoided. Answering no to questions such as those below might indicate that more investigation is needed.

- Can I explain how a hazard footprint is created for this particular model? Really? Could you put the key points in a 3-minute presentation to your boss?
- Do I understand the hazard severity metric used? What is it based on, scientific theory, historic losses or engineering?
- Could I prepare a 5-minute presentation on the subtleties of the processes that *are not* included in the model, and why each matters or doesn't?

Third, be aware of multi-hazard links and dependencies; better, including these will likely be one of the forces driving the development of future catastrophe models. Inter-relationships between perils can have material impacts on risk estimates (e.g. 250-year AEP loss) and are only just starting to be addressed in catastrophe models. Early models dealt in sets of independent (i.e. Poisson) primary hazard events (e.g. a hurricane), but there are simultaneous secondary hazards (e.g. coastal storm surges, tsunami, landslides) whose inclusion in models is of variable sophistication. There are also dependencies in impacted systems (e.g. cascades or have a mutual underlying cause), which may include sometimes substantial (e.g. months) time-lags, and apparent primary hazards may actually be linked; see Section 6.3.2.

### 6.2.4   Chapter 4: Building a Catastrophe Model

This chapter is designed to highlight the key decisions on data, methods and validation commonly used by modelling teams to construct both deterministic and probabilistic models; its structure explores sequentially each component (i.e. hazard, exposure, vulnerability, financial) that enables the estimation of a natural hazard risk from necessarily incomplete views of process and impact (see Figure 4.1 on p. 300). It does not attempt to detail all approaches and methods used in model development from the last twenty years or so. Similarly, it does not focus on the upcoming alternative methods (e.g. **Monte Carlo, 'plug and play'**), computational capability (e.g. cloud approaches) and data sources that are expected to become more common in the coming years; while these have the potential to facilitate more granular, locally tuned, and sophisticated models, the fundamental decisions described in the chapter will remain.

There are concepts, common to most models, that are used to form an overall model view from inevitably sub-optimal data; these are interpolation, extrapolation, and calibration. They are all covered, illustrated using examples related to a number of perils. This is done in a context where it can be argued that except at the highest (i.e. overview) level of model structure, no model is comparable in terms of its construction; namely, each is a bespoke combination of approaches and assumptions. This reflects the reality of model development, where each model will be constructed to reflect a range of specific aims, and in respect to the limitations inherent in the sources and appropriateness of the data used. As such, the chapter uses examples from a number of the specialist modelling organizations, for each component, but has not attempted to propose a unified step-by-step modelling methodology.

A final key point is that model development needs a team, incorporating inputs from diverse specialisms and external expertise. This makes overall model validation difficult, even for a model's developers; specifically, understanding the ultimate impact of decisions made at the component (e.g. hazard or vulnerability) level, is complex and made non-trivial by non-linearities in relationships between individual components.

### 6.2.5   Chapter 5: Developing a View of Risk

To manage risk properly, a company or organization must first understand the risk. Thus, developing its own view of catastrophe risk is important to any risk-bearing entity. Since a catastrophe model is only one representation of reality, the challenge is to ascertain how appropriate a particular model is for the company's portfolio. This is a time-consuming process and so it is important to prioritize which risks should receive the most attention according to the resources available and the importance of the peril region (e.g. US Hurricane) that is being modelled.

A typical process to establish and implement a view of risk includes the following aspects:

* reading model documentation and speaking to developers;
* using scientific data and experts to test and validate the hazard component;
* using both a company's own and industry-wide claims data to test the vulnerability component and the overall level of losses where this is possible (i.e. for low return periods);
* running sensitivity tests to check whether the model behaves as it should (i.e. as the documentation claims);
* comparing the model to, and blending it with, other models where appropriate;
* adjusting the model for non-modelled risk and any other areas where the model can be demonstrated to be deficient;
* documenting and communicating the work undertaken to establish the view of risk and recommendations this leads to;
* implementing recommendations following sign-off.

These elements are covered in the chapter along with discussion of appropriate governance and many sample exhibits of the process designed to help the catastrophe risk practitioner.


## 6.3   The Future: Progress, Challenges and Issues

The business of predicting, understanding and mitigating future risk is a dynamic and changing one. There are, naturally, many aspects of this that cannot be foreseen. However, the list below highlights some of the topics that at present seem likely to be important, discussed in the sections that follow.

1) Future changes in climate
2) Modelling dependency between weather-driven perils
3) Open modelling and open architectures
4) The role of modelling in disaster risk financing
5) Changing global demographics and growing insurance penetration


### 6.3.1   Future Changes in Climate

*Claire Souch*

The possible ramifications of future changes in climate have recently been at the forefront of discussions in the organizations that manage catastrophe risk. Questions include: What are the implications on extreme event occurrence? Are the impacts of climate change adequately reflected in catastrophe models? At a more macro level, how may climate change impact the demand for and availability of catastrophe risk (re)insurance in the future?

Climate changes on many different time scales, driven by both natural and anthropogenic (human) influences. In addition, answers to the questions above must consider each user's timeline of interest. For (re)insurance underwriting, the time horizon is relatively short: strictly speaking, next year, or perhaps three years if writing a multi-year policy. Strategically, many in the industry have an emerging risk task force with a 10-year time horizon. This, however, still contrasts with assessments of the anthropogenic contribution to climate change in projections for 50–100 years time, which is the focus of public policy and dominates over shorter-term variability in climate models. These time scales form the foundation of the Intergovernmental Panel on Climate Change (IPCC) analysis, including the special report on the Risks of Extreme Events and Disasters to Advance Climate Change Adaptation (SREX) (IPCC, 2012).

The dominant scientific view links anthropogenic emissions of some gases (e.g. $CO_2$) with a recent increase in global mean temperatures (IPCC, 2014). Questions remain as to how much greenhouse gas will continue to be emitted, and what magnitude the induced temperature rises will be. However, not all extremes are projected to increase with a warming planet. In addition, changes in extremes can happen without any changes in mean climate. Some extremes are modelled to be affected by warming (e.g. heatwaves, heavy precipitation), but the links for others are not established (e.g. floods, storms). Climate models are still very much in their infancy when it comes to identifying extremes – both their spatial distribution and actual magnitude. A summary of the current state of understanding of the impact on extreme events is presented in Table 6.1, including an assessment of the level of likelihood and uncertainty.

The underlying assumption of most catastrophe models in use today is that event frequency is stationary (i.e. unchanging) through time, and thus calibrating to the average of the available historical record is an adequate representation of expected activity in the coming few years of interest to the catastrophe risk management industry. Many studies propose natural multi-decadal climate-driven variability in the genesis and impact of some perils (e.g. hurricanes, extra-tropical cyclones), but this often differs by region, is ascribed various causes (e.g. AMO, NAO), and is a topic of active debate highlighting the uncertainties involved (see Chapter 3). A general caution is that a trend visible in historical data may be one phase of a multi-decadal cycle if the historical record is not long enough rather than a long-term trend. A few studies have considered longer-term (i.e. 2050–2100) consequences of a changing climate on catastrophe risk, focusing on Europe Windstorm, UK Flood, US Severe Thunderstorms, South Pacific Tropical Cyclones and US Hurricane (ABI, 2009; Lloyd's, 2014; Risky Business Project, 2014).

Table 6.1 Observed and projected changes in extremes.

| Extremes | Observed | Projections |
|---|---|---|
| Hot days frequency and magnitude increasing | Very likely ↑ | Virtually certain ↑ |
| Cold days frequency and magnitude | Very likely ↓ | Virtually certain ↓ |
| Heavy precipitation extremes | Likely more regions with ↑ | Likely ↑ in many regions. Very likely in mid-latitudes and tropics |
| Droughts | Medium confidence in some regional trends ↑↓ | Medium confidence in ↑ in some regions. Likely ↑ in some currently dry regions. |
| Storms | Low confidence | Low confidence in deailed regional projections |
| Floods | Low confidence (because of human and vegetation water use) | Low confidence in regional projections. Medium confidence related to ↑ in heavy percipitation events |
| Tropical Cyclones | Low confidence | More likely than not ↑ in intensity in some basins |

Notes: Arrows represent the change. 'Observations' describe whether the peril has already undergone some change as a result of human-driven influences, and 'projection' indicates what the future prediction is, along with confidence levels.
Source: Summarized from the IPCC's Fifth Assessment Report (AR5) and the IPCC Special Report on Managing the Effects of Extreme Events and Disasters (SREX) (IPCC, 2012, 2014); original summary by Seneviratne (2015).

To summarize, when considering climate change, we first need to be clear about what time horizon are we interested in. Then, we need to consider any evidence that natural or anthropogenic influences may cause event frequency to be different from the historical average over that time horizon. Regardless of the time horizon and driver, there is still a lot we do not know about the influence of a changing climate on extreme events.

### 6.3.2 Modelling Dependency between Perils

*Rick Thomas*

Catastrophe models now, arguably, provide a reasonable set of tools to model many of the key perils independently. Some account is also taken of directly connected 'secondary' perils (e.g. flooding associated with tropical storms or a tsunami following an earthquake), see Chapter 3. However, evidence is accumulating of other correlations between regions or perils that are still being ignored. An illustration is the co-variation of the El Niño Southern Oscillation (ENSO) with a number of global weather patterns. For example, during La Niña, hurricane activity in the Atlantic is enhanced (e.g. Bove *et al.*, 1998; Jagger *et al.*, 2011) as is cyclone activity in Australia (e.g. Kuleshov *et al.*, 2008) along with tornado activity in some parts of the United States (Allen *et al.*, 2015). Finally, there may be non-modelled links relating geophysical perils to either weather (e.g. earthquakes (Costain and Bollinger, 2010), volcanoes (Rampino *et al.*, 1988; Christiansen, 2007)) or each other (e.g. earthquake and volcano (Lupi and Miller, 2014)): see Woo (1999), for an overview.

As loss data are often sub-optimal, correlations involving losses remain more tentative but have been suggested for US Hurricane and ENSO (Pielke and Landsea, 1999), and combined losses (e.g. sum of flood losses and wind losses) may also be influenced by links between distinct weather-driven perils within a region (e.g. Hillier *et al.*, 2015).

Modelling these types of variability globally is not easy, and there are challenges from both the science and (re)insurance sides. The reinsurance complications are generally related to the interaction between annual contracts periods (e.g. January to January) and the seasonality of the perils (e.g. storms typically in October to March). Furthermore, the multi-year time scales of some of the modes of climate variability (e.g. AMO) can potentially create inter-annual clustering of bad results. Some initial efforts to adjust the rates of Atlantic Hurricanes have been made (e.g. Hunter, 2014).

To model dependency there are generally two end member approaches that could potentially be taken: purely statistical and purely physical. The statistical approach would introduce correlations between the statistical distributions that govern the frequency and severity of events. A choice of joint distribution needs to be made, perhaps introducing tail dependency, e.g. using t-copula. This approach requires careful calibration of correlation, and particularly of the overall joint distributions for loss severity. Unfortunately not nearly enough data exist to create models of this type at the moment. The physical approach is to use a General Circulation Model (GCM) to simulate all the key aspects of the climate system and produce a full set of realistically connected events. However, the current GCMs are too low resolution to realistically simulate hurricanes and would have prohibitively long run times to simulation of sufficient years of events to drive a catastrophe model.

Considering these challenges, the industry approach has been to focus on the effects of climate variability on the rates and severity of storms, and try and develop models that are a hybrid of the physical and statistical approaches. There are generally three steps in this approach:

1) Establish the physical mechanisms behind the connection, or at least demonstrate that it is statistically significant (e.g. ENSO and Atlantic hurricanes).

2) Quantify the variation in event frequency.
3) Quantify the changes in event geography and severity, ultimately loss severity.

Unfortunately the variability in the key perils is clearly not related to just one mode of climate variability. Hence, although pioneering modelling companies like KatRisk have started to produce global SST-driven global event sets, these are only the first step on the way to true global event sets. Nevertheless, these initial models should allow us to start quantifying the basics of global dependency, and, in future, focused GCM research should allow us to improve our understanding of the climate system to refine the models.

For work to progress, however, the willingness must be there to overcome one other important additional factor; most global portfolios are dominated by US event risk, thus even the most extreme losses in almost all other territories are a fraction of the 1-in-100-year loss estimated for US Hurricane (e.g. >100bn US$ for an 'as if' repeat (Karen Clark and Company, 2012)). Hence it would require extreme aggregation of events to match a major US loss.

### 6.3.3 Open Modelling and Open Architectures

*Dicke Whitaker*

In the past few decades, the use of catastrophe models in the (re)insurance industry has grown significantly. Many models have been developed for a significant number of perils and regions and the complexity and detail of the models, as well as the computing power required to run them, have increased many times. Today the (re)insurance market expends significant resources on licensing and operating catastrophe models. These are predominantly sourced from the two market-leading firms; their proprietary software largely is installed as an in-house platform. Typically, public information is limited on these products, leading to a perception of them being 'black boxes' even if significant information is provided in (proprietary) user documentation. This forces comparisons to be made by organizations holding multiple licences (e.g. WillisRe, 2007; Waisman, 2015).

Non-proprietary risk assessment software and publicly documented risk assessment software have existed for a number of years, such as HAZUS (i.e. http://www.fema.gov/hazus). It is not uncommon for academics to develop entry-level catastrophe models or publish methodologies (e.g. Powell *et al.*, 2005; Varghese and Rau-Chaplin, 2013; Punge *et al.*, 2014), but they have either tended to be subsumed into larger products (e.g. Hailcalc Europe into RMS) or the research is not implemented in the first place (e.g. the MATRIX project (Mignan, 2013)).

2011 was a pivotal year for the catastrophe modelling market due to three unrelated occurrences:

- The insured losses in 2011 of circa US$130 billion (see Table 3.3 on p. 190) were the largest of any historical year and were dominated by perils other than the peak perils, including substantial contribution from a peril-region where no commercially available model existed (Thailand flood).
- Model version changes from one of the market leaders for two peak peril-regions resulted in substantial, and largely unanticipated, model loss result changes for many (re)insurance companies.
- Solvency II requirements (e.g. Eling *et al.*, 2007; EU, 2009), and the impact on the way catastrophe models should be used under this regime were emerging (e.g. ABI, 2011).

This led to a recognized need for more openness and transparency in catastrophe models, so that (re)insurance companies could better understand the models they are using and develop their own view of risk; both to manage their risk in a better manner (e.g. the impact of model changes) and to be able to justify their understanding of models to regulators. An additional impact of 2011 was the increase in new models and model vendors, partly to try and fill the model gaps that 2011 had highlighted. These new vendors tended to build models with an ethos of openness, for instance, with KatRisk claiming to allow model subscribers to see all underlying code, or with JBA and Impact Forecasting allowing subscribers to see all the vulnerability curves and hazard maps. Free online demonstrations are also becoming common (e.g. KatRisk, RiskScape). Furthermore, feeding into a landscape of over 80 open source or freely available models (see Chapter 2.14.2), catastrophe models are freely available to download from a number of initiatives. For example:

- Global Earthquake Model's (GEM) OpenQuake model (http://www.globalquakemodel.org/openquake/);
- NORSAR's SELENA seismic risk package (http://www.norsar.no/seismology/engineering/SELENA-RISe/);
- CPRA's probabilistic multi-hazard risk assessment program (http://www.ecapra.org/software);
- Climada multi-hazard probabilistic risk tool (http://www.preventionweb.net/educational/view/42020).

However, such non-proprietary models are usually not fit for purpose for a typical (re)insurer, largely because the way that they deal with (re)insurance financial structures (Chapter 1.9.2) is too simplistic. In order to address this, the Oasis initiative aims to build and offer a new open source loss modelling framework that is owned and supported by the (re)insurance industry. The aspirations are that a platform built using the Oasis frameworks will be able to import various exposure data formats, run on a wide range of computers and mobile devices, and able to take on any catastrophe loss model into any user organization in a scalable way.

As of 2016, Oasis and other emerging open approaches have yet to significantly change how end-users use catastrophe models, but as and when a change occurs, the new paradigm could be significant in impact and may include the following characteristics:

- Components of selected models will, by necessity and due to the need for inter-comparison, become de facto industry standards; this includes exposure data schemas, hazard event sets, and financial module methodology. This may extend as far as core methodologies that underpin models.
- Similarly, approaches to verify the quality of a model will likely become standardized.
- Interoperability of components between software packages and architectures.
- Models that can be run in the cloud (private or public), with licences for one-off analysis or regular use.
- User interactivity built on open source software (e.g. R Shiny).
- A catalogue of models from many vendors readily available through an online marketplace.

What could a shift to open modelling and open architectures mean for organizations (e.g. governments, HGOs, insurers) assessing risk?

- Access to multiple views of risk would allow organizations to better evaluate models and make more informed model selection, potentially leading to more resilience.
- With access to standards and open source modelling components, independent software providers and technology firms would be able to build software and integration services leading to platform innovation.
- Creating a market in models and associated software could make catastrophe modelling licences and operating costs more economic.

- Standardization in the core modelling approaches could allow innovation, such as developing robust approaches to uncertainty and correlation.

### 6.3.4    The Role of Modelling in Disaster Risk Financing

*Rashmin Gunasekera*

Within the post 2015 Framework for Disaster Risk Reduction (HFA2), as discussed in Chapter 2.14.1, there is likely to be a surge in the uptake of probabilistic catastrophe models for disaster risk financing. However, it will still take time for policy and organizational demands to cause probabilistic catastrophe risk analyses to become common in the public sector; key challenges remain. First, stakeholders' technical understanding of catastrophe models, and particularly probabilistic modelling, remains low. Second, agreement is required from key public sector stakeholders on evaluating how fit models are for their intended purpose and on what the models should and should not be used for. Some aspects of these are currently being addressed by international development organizations and stakeholders through Technical Assistance Programs (TAPs). A key message has to be not only the need for construction and adaptation of catastrophe models for applications in the public sector, but also the importance of calibration of results for the numerous applications. It is important for the public sector to consider that even the best available exposure, vulnerability and hazard model components designed for insurance do not necessarily provide a robust answer for public sector financial protection or decision-making without the proper calibration of models.

Upcoming initiatives, such as RISE (http://www.preventionweb.net/rise/home), Geneva Association (https://www.genevaassociation.org/research/topics/climate-risk/), and the 1 in 100 initiative (http://climateaction.unfccc.int/), are seeking to build the quantification and management of risk more directly into the financial systems and regulatory environments.

Many governments and regulatory bodies in various sectors, including finance, are increasingly looking at how to make our cities, populations and economies more resilient to natural catastrophe disasters. New initiatives in open modelling (see Section 6.3.3) are helping to widen the availability of catastrophe models and to broaden their use. A number of disaster relief agencies are now beginning to explore the use of models and platforms to assist with planning and post-event response.

Under the Resilience, Adaptation and Disaster Risk Reduction (DRR) Action Area, the United Nations and the World Economic Forum have convened a unique coalition of accounting organizations, asset managers, central banks, credit ratings agencies, risk modellers, financial regulators and science leaders to confirm these principles and support the goals of securing commitment to accelerated and ambitious action in Climate Summits 2015 (http://climateaction.unfccc.int/).

Further, there are broader questions to answer. Probabilistic risk modelling quantifies structural measures well, however, the impact of non-structural measures is a significant driver of socio-economic loss (GFDRR, 2010) that is not necessarily captured in these models and raises the question: 'Are probabilistic catastrophe models in the public sector really useful for Disaster Risk Reduction purposes other than for financial protection?' Looking to the future, the answer is likely to be a resounding 'yes'. Public sector open data initiatives such as OpenCities, Geonodes, and InaSafe, discussed in Chapter 2.14.2, are already providing detailed exposure data that have been a missing ingredient in quantifying socioeconomic data and loss. Yamin *et al.* (2013) have also already highlighted examples from Bogota where public sector integration of quantitative probabilistic catastrophe modelled results with non-structural measure risk indicators have provided a roadmap for multi-purpose effective disaster risk reduction in the public sector.

### 6.3.5 Changing Global Demographics and Growing Insurance Penetration

*John Hillier*

Insurance is moving into new geographic areas and the characteristics of better-established regions are altering as a result of changes in demographics. Changes that have the ability to significantly affect losses include an increase in the value of assets at risk, an increase in the fraction of these assets covered by insurance (i.e. increased insurance penetration), the relocation of the assets by processes such as increased urbanization, and population growth (e.g. Crompton and McAneney, 2008). For instance, the US states most impacted by severe convective storms (SCS) (see Chapter 3.4) have seen a population growth of 15–48% since 1990 that contributes to a rise in SCS losses (Swiss Re, 2015). Similarly, the rise in potential tornado losses in the United States is ascribed to increasing urbanization (Swiss Re, 2013a). In general, these trends (e.g. more high-value technology at risk) have been attributed to economic development (Swiss Re, 2015; Impact Forecasting, 2015). There are, however, cultural differences between regions. Notably, in Italy, 44% of domestic properties have fire cover, but only 0.4% have policies including earthquake due to a conviction that government post-disaster intervention will suffice (Swiss Re, 2013a). In general, increasing urbanization and insurance penetration offer a challenge that generates accumulations of risk, but also presents opportunities (Swiss Re, 2013b).

## References

ABI (2009) The Financial Risk of Climate Change. Association of British Insurers Research Paper No. 19. A report by AIR Worldwide Corp. and the Met Office, London.

ABI (2011) Industry Good Practice for Catastrophe Modelling. https://www.abi.org.uk/~/media/Files/Documents/Publications/Public/Migrated/Solvency II/(accessed 31 March 2016).

Allen, J. T., Tippett, M. K. and Sobel, A. H. (2015) Influence of the El Niño/Southern Oscillation on tornado and hail frequency in the United States. *Nature Geoscience*, **8**, 278–283.

Bove, M. C., O'Brien, J., Elsner, J. B., Landsea, C. W. and Niu, X. (1998) Effect of El Niño on U.S. landfalling hurricanes, revisited. *Bulletin of the American Meteorological Society*, **79**, 2477–2482.

Christiansen, B. (2007), Volcanic eruptions, large-scale modes in the northern hemisphere, and the El Niño–Southern Oscillation. *Journal of Climate*, **21**, 910–922.

Costain, J. K. and Bollinger G. A. (2010), Review: Research results in hydroseismicity from 1987 to 2009. *Bulletin of the Seismology Society of America*, **100** (5A), 1841–1858.

Crompton, R. and McAneney, J. (2008) The cost of natural disasters in Australia: The case for disaster risk reduction. *The Australian Journal of Emergency Management*, **23** (4), 43–46.

Eling, M., Schmeiser, H. and Schmit, J.T. (2007) The Solvency II Process: Overview and critical analysis. *Risk Management Insurance Review*, **10** (1), 69–85.

EU (2009) Directive 2009/138/EC of the European Parliament and of the Council of 25 November 2009 on the taking-up and pursuit of the business of insurance and reinsurance (Solvency II). *Official Journal of the European Union*, **155**.

Global Facility for Disaster Reduction and Recovery (GFDRR). (2010) *Damage, Loss and Needs Assessment (DaLA): Guidance Notes*. World Bank, Washington, DC.

Hillier, J. K., Macdonald, N., Leckebusch, G. and Stavrinides (2015) Interactions between apparently 'primary' weather-driven hazards and their cost. *Environmental Research Letters*, **10**, 104003.

Hunter, A. (2014) Quantifying and understanding the aggregate risk of natural hazards. PhD thesis, University of Exeter. http://ethos.bl.uk

Impact Forecasting (2015) *2014* Annual Global Climate and Catastrophe Report. http://thoughtleadership.aonbenfield.com/Documents/20150113_ab_if_annual_climate_catastrophe_report.pdf (accessed 23 March 2016).

IPCC (2012) *Managing the Risks of Extreme Events and Disasters to Advance Climate Change Adaptation: A Special Report of Working Groups I and II of the Intergovernmental Panel on Climate Change* (eds C.B., Field, V. Barros, T.F. Stocker, *et al.*). Cambridge University Press, Cambridge.

IPCC (2014) *Climate Change 2014: Synthesis Report. Contribution of Working Groups I, II and III to the Fifth Assessment Report of the Intergovernmental Panel on Climate Change* (eds. R.K., Pachauri and L.A., Meyer,). IPCC, Geneva, Switzerland.

Jagger, T. H., Elsner, J. B. and Burch, R.K. (2011) Climate and solar signals in property damage losses from hurricanes affecting the United States. *Natural Hazards*, **58**, 541–557.

Karen Clark and Company (2012) Historical hurricanes that would cause $10 billion or more of insured losses today. http://www.karenclarkandco.com/pdf/HistoricalHurricanes_Brochure.pdf (accessed 2 Dec. 2014).

Kuleshov, Y., Qi, L., Fawcett, R. and Jones, D. (2008) On tropical cyclone activity in the Southern Hemisphere: Trends and the ENSO connection. *Geophysical Research Letters*, **35**, L14S08.

Lloyd's (2014) Catastrophe Modelling and Climate Change. https://www.lloyds.com/~/media/lloyds/reports/emerging%20risk%20reports/cc%20and%20modelling%20template%20v6.pdf.

Lupi, M. and Millier, A. A. (2014) Short-lived tectonic switch mechanism for long-term pulses of volcanic activity after mega-thrust earthquakes. *Solid Earth*, **5**, 13–24. DOI: 10.5194/se-5-13-2014.

Mignan, A. (2013) MATRIX Common IT sYstem (MATRIX CITY) Generic multi-hazard and multi-risk framework – the concept of Virtual City – IT considerations. *MATRIX* (Deliverable D8.4). http://matrix.gpi.kit.edu/Deliverables.php (accessed 31 March 2016).

Pielke, R. A. and Landsea, C. N. (1999) La Niña, El Niño, and Atlantic hurricane damages in the United States. *Bulletin of the American Meteorological Society*, **80** (10), 2027–2033.

Powell, M., Soukup, G., Cocke, S. *et al.* (2005) State of Florida hurricane loss projection model: Atmospheric science component. *Journal of Wind Engineering and Industrial Aerodynamics*, **93**, 651–674.

Punge, H., Werner, A., Bedka, K. and Kunz, M. (2014) A new physically based stochastic event catalogue for hail in Europe. *Natural Hazards*, **73**, 1625–1645.

Rampino, M. R., Self, S. and Stothers, R. B. (1988) Volcanic winters. *Annual Review of Earth and Planetary Sciences*, **16**, 73–99.

Risky Business Project (2014) A Climate Change Risk Assessment for the United States. http://riskybusiness.org/site/assets/uploads/2015/09/RiskyBusiness_Report_WEB_09_08_14.pdf

Seneviratne, S. (2015) Paper presented at the SCOR Foundation seminar on climate risks, 9 June 2015. Available from: http://scor-climaterisks-2015.com/seminar-content.

Swiss Re (2013a) Natural catastrophes and man-made disasters in 2012. *Sigma*, No. 2.

Swiss Re (2013b) Urbanisation in emerging markets: boon and bane for insurers. *Sigma*, No. 5.

Swiss Re (2015) Natural catastrophes and man-made disasters in 2014. *Sigma*, No. 2.

Varghese, B. and Rau-Chaplin, A. (2013) Accounting for secondary uncertainty: efficient computation of portfolio risk measures on multi and many core architectures. In *Proceedings of the 6th Workshop on High Performance Computational Finance*. DOI: 10.1145/2535557.2535562

Waisman, F. (2015) European windstorm vendor model comparison. Paper presented at International Underwriting Association of London (IUA) Conference, available at www.iua.co.uk (accessed 23 March 2016).

WillisRe (2007) Windstorm Kyrill, 18th Jan 2007. https://www.yumpu.com/en/document/view/ 35214333/download-file-willis-research-network/21(accessed 31 March 2016).

Woo, G. (1999) *The Mathematics of Natural Catastrophes.* Imperial College Press, London.

Yamin, L. E., Ghesquiere, F., Cardona, O. D. and Ordaz, M. G. (2013) *Modelación probabilista para la gestión del riesgo de desastre.* World Bank, Washington, DC. http://documents .worldbank.org/curated/en/2013/07/18100020/colombia-probabilistic-modeling-disaster-risk-management-modelacion-probabilista-para-la-gestion-del-riesgo-de-desastre

# Glossary

**Account** Another term for an insurance policy. Sometimes used to denote an overall grouping of policies to a single customer; i.e. a single customer may have several different policies.

**Accumulation** An aggregation of exposure or loss potential. Often used as shorthand for the process of combining multiple portfolios into a time specific snapshot of exposure and modelled results. See *aggregation* and *roll-up*.

**Accumulation zones** Geographical regions used for aggregation, which typically correspond to a pre-existing administrative definition such as country, city or postcode.

**Actual cash value (ACV)** A type of policy where the insurance payment is the replacement cost less depreciation and obsolescence.

**Actuary** Applied mathematician with a deep understanding of financial systems, typically working in insurance, who has passed the exams and courses, required by the relevant professional body, to become an actuary.

**Additional living expenses (ALE)** The cost of a home-owner being housed during repair work. See *alternative living expenses*.

**Adverse risk selection** The tendency for more demand for insurance from high-risk compared to low-risk individuals, coupled with pricing that does not fully reflect the difference in risk between low and high individuals, leads to a portfolio of business where the premium does not adequately cover the risk.

**Aggregate exceedance probability (AEP)** The probability of the sum of event losses in a year exceeding a certain level.

**Aggregate exposure** Exposure added up across a geographical area or by a class of business often, but not always, provided net of deductibles and limits.

**Aggregates** (Re)insurance term for the sums insured grouped by some factor like geography or type of insured activity/asset.

**Aggregation** The process of combining multiple portfolios into a time specific snapshot of exposure and modelled results. See *accumulation* and *roll-up*.

**Aleatory uncertainty** Uncertainty associated with the inherent, irreducible, randomness in a process.

**Allocated loss adjustment expenses (ALAE)** Insurance company expenses directly associated with settling a claim. Not normally included in catastrophe model output.

*Natural Catastrophe Risk Management and Modelling: A Practitioner's Guide*, First Edition.
Edited by Kirsten Mitchell-Wallace, Matthew Jones, John Hillier and Matthew Foote.
© 2017 John Wiley & Sons Ltd. Published 2017 by John Wiley & Sons Ltd.

**Alternative capital**  A loose term generally used to include cat bonds and collateralized reinsurance. It is an alternative to traditional reinsurance.

**Alternative living expenses (ALE)**  The cost of a home-owner being housed during repair work. See *additional living expenses*.

**American Society of Civil Engineers (ASCE)**  Professional body of American civil engineers, responsible for building codes.

**Amplitude**  The size or magnitude of a wave as measured from trough to crest (e.g. seismic waves).

**Analogue event**  An event which resembles an event of interest (i.e. past or present event that is the subject of an analysis) in terms of location and intensity and which is expected to have a similar impact; analogues can be chosen from either history or a synthetic event set.

**Analysis of change**  Comparing how the model assumptions and results change from one version to the next.

**Annual mean loss (AML)**  The expected loss cost over a one-year time period. See *expected loss*, *pure premium* and *average annual loss*.

**Antecedent conditions**  The state of the environment in a relevant time period before a hazard event.

**Area source**  Region where EQs are distributed uniformly as they are considered by scientists as seismically homogeneous.

**'As at' date**  The date at which the exposure snapshot has been taken.

**Aseismic**  A movement of tectonic plates past each other that does not cause earthquakes.

**As-if**  Bringing historical losses to a level that reflects today's risk. See *on-levelling*.

**'As if' losses**  The value of a historical loss as if it were to happen again today, i.e. reflecting current values and conditions. See *on-levelling* and *trending*.

**Asset-liability modeling (ALM)**  Another term for *Dynamic Financial Analysis*.

**Atlantic meridional mode (AMM)**  A climate mode characterized by an anomalous meridional gradient of sea surface temperature centred near the latitude of the thermal equator ($\sim5°N$).

**Atlantic multi-decadal oscillation (AMO)**  A climate mode usually defined from the patterns of sea surface temperature variability in the North Atlantic once any linear trend has been removed. Thought to be associated with tropical cyclone activity.

**Atmospheric instability**  A measure of the tendency of vertical motion in the atmosphere to be encouraged, e.g. warm, moist air is unstable with the tendency to rise and is associated with formation of tropical cyclones.

**Attachment point**  The monetary value at which a (re)insurance cover starts.

**Attenuation functions**  Equations used to describe the way ground shaking decreases with distance away from the epicentre of an earthquake. Also see *ground motion prediction equations*.

**Auto insurance**  Insurance to cover the physical car and sometimes also liability to others. See also *motor insurance*.

**Average** An insurance concept where underinsurance (insurance purchased for a sum which is lower than the value of the insured property) will lead to a proportional scaling down of the loss paid.

**Average annual loss** or **Annual average loss (AAL)** The expected loss cost over a one-year time period. See *expected loss*, *pure premium* and *annual mean loss*.

**Aviation insurance** Insurance for aircraft, including third party liability coverage.

**Back wash** The return flow of a wave (e.g. of a tsunami) to the sea under gravity.

**Back testing** Comparing the results from the models to what actually happened (e.g. actual versus modelled loss comparisons for specific historical events).

**Basin activity** The number of storms occurring within an ocean basin.

**Basis risk** The uncertainty that remains when there is no exact correspondence between the insured quantity and the true exposure.

**Bedrock** Geological term for rock, commonly underlying more significantly broken or reworked deposits (e.g. soils).

**Benchmarking** Comparing the results of a model to other benchmarks (e.g. to the results from other models).

**Beta distribution** A continuous statistical distribution sometimes used in catastrophe models to represent the uncertainty around the mean damage ratio.

**Book** A (re)insurance term for a portfolio.

**Bottom-up** A method for capturing location-level building characteristics for every insurable property and aggregates by geographical region.

**Bow echo** A bow-shaped line of convective thunderstorm cells.

**Building codes** Regulations stipulating minimum standards for the construction of buildings (e.g. houses). Stringency and enforcement are variable by territory.

**Building regularity** How structural homogeneous a building is. With irregularity comes increasing sensitivity to earthquake damage.

**Bulk coding** Where a set of standard assumptions has been applied to a class of exposures in lieu of detailed information.

**Burning cost** The burning cost is the pure loss to the programme from 'as if' historical losses over a period of time; typically per year.

**Business interruption (BI)** A coverage category that provides protection against lost profit or revenue due to damage which results in the insured's business not being able to operate. Business interruption exposure is the value insured.

**Business unit** A common term for an organizational component of a company.

**Calibration** A model development process which is used to choose values of model parameters, using independent data, such that apparently correct outputs are produced.

**Capacity** Amount of reinsurance cover or limit. Sometimes used as a generic term to indicate a constraint on the amount of business that can be written in a specific product line or region.

**Capital** A company's assets minus its liabilities. Working capital is the capital plus long term debt plus retained earnings. See *equity* and *own funds*.

**Capital allocation** The distribution of risk-based capital to different parts of a company according to the relative riskiness of each part.

**Capital intensity ratio (CIR)** The capital required per unit of technical premium.

**Capital models** Models which attempt to quantify and allocate the risk-based capital of a company.

**Cash in transit** Insurance that covers the movement of cash.

**Casualty insurance** Insurance to cover claims from third parties. See also *liability insurance*.

**Catalogue (of events)** A set of hazard events, with sufficient information to represent each individual event, e.g. probability, intensity distributions (i.e. colloquially 'maps'). The catalogue could be a representation of a historical set of events or could represent a synthetic history using stochastic events.

**Catastrophe** A catastrophe is something that exceeds the capability of those affected to cope with, or absorb, its effects; in the context of natural hazards the driver is an extreme event causing widespread and, usually sudden, damage or suffering.

**Catastrophe bond** An insurance linked security where the issuer (the cedant) pays premium coupons (interest payments), usually in excess of a floating benchmark yield (such as three-month LIBOR or EURIBOR). The principal (notional amount) of the bond serves as collateral to pay for potential losses from catastrophes.

**Catastrophe excess of loss (CAT XL)** Excess of loss insurance designed to cover the aggregation of losses caused by a catastrophe event.

**Catastrophe load** Catastrophe component of price in a (re)insurance policy.

**Catastrophe model** A computerized system that generates a robust set of simulated events and estimates the magnitude, intensity, and location of the event to determine the amount of damage and calculate the insured loss as a result of a catastrophic event such as a hurricane or an earthquake.

**Catastrophe risk** The uncertainty in the impact of catastrophes, be they natural or man-made.

**Catastrophe risk analysts** The primary users of the models; responsible for data entry, model operation, analysis, outputs and reporting (and sometimes model selection).

**Catastrophe risk management** The discipline of managing the uncertainty caused by catastrophes, or the function that does this within an organization. 'Exposure management' is often used interchangeably with this term.

**Catchment area** The area draining into a river above any given point.

**Cedant** The reinsurance policyholder.

**Ceding commission** Commission paid by the reinsurer to the cedant to compensate it for the costs of acquiring and maintain the original business.

**Centroid** A point reference used in geographic information systems to represent the general location of an area in relation to its plan geometry. Usually a centroid will be defined either as the geometric centre of gravity or alternatively, a location weighted by some other representative measure within the area, such as the relative density of built up area or population.

**Cession** Transfer of risk via a reinsurance (or retrocession) contract.

**Characteristic earthquake model** Hypothesis wherein fault segments dictate a maximum size of earthquake in a particular area.

**Claims experience** The insurance losses that the insurance company has had.

**Claims inflation** The increase in the value of a claim, which may be due increased costs associated with shortage of material or labour or due to other factors such as wage or price inflation.

**Clash potential** The possibility that multiple different business segments are affected by the same catastrophe event. This can refer to geographies (e.g. the US and Caribbean), or lines of business (e.g. marine and property).

**Clay minerals** Types of crystal that tend to form clay soils.

**Climate modes** A mode of climate variability is a pattern with quantifiable characteristics, specific regional effects, and typically oscillatory behaviour. These patterns in the atmosphere and oceans are commonly associated with variations in some weather-driven hazards.

**Clustering** Grouping in time or space such that more hazard events (e.g. earthquakes) occur in a given period and region than is expected by chance for independent events (i.e. Poisson model).

**Coastal flooding** Flooding associated with sea water.

**Coefficient of variation (CoV or CV)** The standard deviation divided by the expected loss cost.

**Coherent risk measure** A metric which satisfies the criteria of sub-additivity, positive homogeneity, monotonicity and translational invariance.

**Coinsurance** Sharing of insurance between the insured and the insurer or between different insurers. The same term is used for reinsurance.

**Cold conveyor belt** Flow of cold, dry air into and around the back of the central low pressure of an extra-tropical cyclone.

**Collateralized reinsurance** A reinsurance contract where the limit is covered by money set aside by investors (i.e. the collateral) purely for this contract. No credit rating is needed; the investors receive their investment return in the form of interest on the collateral and reinsurance premiums.

**Commercial lines** Insurance for corporations.

**Component** A sub-model that is one of the building blocks of a catastrophe model (e.g. hazard, vulnerability).

**Composite insurer** An insurance company providing both life and general insurance.

**Composite vulnerability curve** A vulnerability curve that is formed by weighting several constituent vulnerability curves together in order to form an approximate view of vulnerability in the absence of full exposure information.

**Compound distribution** A distribution resulting from combining the individual distributions of the frequency of losses and the size of losses.

**Construction lines** Insurance for building projects. See also *engineering lines*.

**Construction type** What buildings are made of (e.g. reinforced concrete), and sometimes how they are made. An input (i.e. characteristic of exposure) to catastrophe models.

**Contingent business interruption (CBI)** Losses due to a disruption of trade caused by damage or interruption of business by a third party; such as supplier or key partner. A type of indirect loss.

**Convergent** A tectonic setting where plates are moving together.

**Copula** A statistical term for a mathematical structure that enables a dependency between variables to be specified in a flexible way (e.g. if one variable is higher than average another variable may also be likely to be higher than average and this relationship may strengthen for larger values of one of the variables).

**Cost benefit analysis** An analysis of the financial losses (costs) and income streams (benefits) in order to ascertain whether a project is worth investment.

**Correlation** A statistical term for the extent to which variables fluctuate together. If variables generally increase or decrease together, this is a positive correlation. If one variable tends to decrease when another increases, this is a negative correlation.

**Cost of capital** The minimum return that a company should earn on its capital in order to satisfy its providers of capital. In the context of cat bonds, the return that the market expects from a cat bond with zero expected loss.

**County weighted industry loss (CWIL)** An insurance linked security. County weighted industry loss contracts pay when industry losses, multiplied by weights assigned to each county, cross specified thresholds.

**Coverage** Generally means the scope of insurance provision provided by the policy. In the context of catastrophe modelling, it is also used as a description of a type of exposure for vulnerability and loss modelling, typically categorized between buildings, contents and business interruption.

**Coverage expansion** A more liberal view of the interpretation of policy language/exclusions so that cover is more than originally anticipated.

**CRESTA (Catastrophe Risk Evaluating and Standardizing Target Accumulations)** An international body established to develop standard approaches for accumulation and exposure data: www.cresta.org

**Crown** Top of a landslide scarp.

**Cyclogenesis** The process of initiating a storm (e.g. tropical or extra-tropical cyclone).

**Damage ratio (DR)** The estimated repair cost of an asset at risk divided by the replacement cost of the asset.

**Day-one sum insured** The sum insured calculated with no explicit allowance for inflation during the policy and rebuilding period. A commercial property insurance term. A contrast to full reinstatement sum insured. See *day-one uplift.*

**Day-one uplift** A provision in some commercial property insurance policies allowing for extra payment to be made to the insured if losses have increased due to inflation. Usually specified as a percentage of the day-one sum insured.

**Debris avalanche** A type of mass movement where the material behaves as a fluid. A term used in volcanology.

**Debris flow** A type of mass movement where the material behaves as a fluid.

**Decay** When a tropical cyclone loses vigour.

**Deductible** The amount of loss a policyholder has to pay before they can reclaim from the policy. See also *excess*.

**Deep moist convection** Latent heat released by the condensation of rising air that drives thunderstorms.

**Demand surge** Increased prices due to a shortage of supply after an event; mostly in labour or materials.

**Derechos** The most severe sort of straight-line winds associated with severe convective storms.

**Derivative** A security that derives its price from an underlying quantity.

**Design code** An international, national, or regional (often legal) set of building and planning requirements designed to increase resilience to natural and artificial risks.

**Digital elevation model DEM** Any spatially structured set of elevation values (usually based on a grid or other tessellation), referenced to a vertical datum. Terrain models (DTMs) and surface models (DSMs) are types of DEM.

**Digital surface model (DSM)** A digital elevation model representing the highest, upper surface of all visible objects on the landscape, as recorded using a particular measurement technique. May contain natural and man-made features that are unwanted such as buildings (e.g. houses), and trees. When derived from an airborne data (e.g. radar), a DSM will be created from 'first returns'.

**Digital terrain model (DTM)** Typically thought of as a 'bare earth' digital elevation model representing the Earth's surface free from anthropogenic features. Can, however, be designed to explicitly include key linear terrain features such as levees (i.e. inclusion or not of earthworks is an area of debate).

**Direct & fac (D&F)** Usually 'Property D&F'. Property insurance arranged either on a primary or facultative reinsurance basis.

**Direct loss** The cost of repairing an insured asset.

**Disaggregation**

**Disaster risk financing (DRF)** Financial mechanisms for providing funds to countries following a disaster. Essentially insurance at a sovereign (country) or sub-national (e.g. state) level.

**Disaster risk reduction (DRR)** A method for identifying, assessing and mitigating the effects of disasters, via both hazard and socio-economic vulnerability reduction.

**Discharge** The volume of water flowing through a river in a unit of time.

**Diversification** The reduction in risk caused by aggregating together exposure from a number of insurance policies in order to reduce uncertainty in the outcome. See *pooling of risk*.

**Domain** Scope of a model as in 'modelled domain', often referring to the geographic extent of a catastrophe model.

**Downbursts** Severe downdrafts.

**Downdrafts** Air flowing downwards (e.g. in a severe convective storm).

**Downscaling** A statistical or modelling technique to estimate metrics at smaller scales than the raw output resolution from a model (e.g. catastrophe model or Global Climate Model); vital to correctly estimate extremes.

**Downside (risk)** The risk quantified by the set of outcomes that are worse than expected (e.g. scenarios where a company's profit is worse than that planned for).

**Drainage systems** Engineered conduits for water flow such as pipes, culverts or channels.

**Driving forces** Forces acting to drive geo-hazards, such as tectonic for earthquake or gravity for mass movement.

**Dynamic financial analysis (DFA)** The process by which a (re)insurance company uses internal and external data to calculate statistical distributions for key economic variables on which Monte Carlo simulation techniques are applied to estimate the possible range of economic outcomes.

**Earthquake (EQ)** A sudden release of energy stored by the tectonic plates being temporarily and locally prevented from moving past each other.

**Earthquake cycle** The idea that earthquakes repeat after some, potentially predictable, processes occur; e.g. building up sufficient stored energy after the release of the last quake.

**Economic capital** The risk-based capital calculated according to the company's own definitions and assumptions. Contrast to *regulatory capital*.

**Economic demand surge (EDS)** See *demand surge*.

**Economic exposure database (EED)** A database of exposed assets defined by the total economic value.

**Efficient frontier** The set of solutions maximizing and minimizing the metrics chosen for an optimization.

**Effusive** Type of volcanic eruption where lava is extruded.

**El Niño** A phase of the ENSO climate mode.

**El Niño Southern Oscillation (ENSO)** Periodic variations in sea surface temperatures within the equatorial Pacific and a key driver of global weather patterns.

**Emerging risk** An issue that is perceived to be potentially significant but which may not be fully understood or allowed for in insurance terms and conditions, pricing, reserving or capital setting.

**Empirical distribution** A statistical distribution that is not specified by a particular formula and set of parameters.

**Engineering lines** Insurance for building projects. See also *construction lines*.

**Ensemble forecasts** Probabilistic forecasts defining the spectrum of what could happen, often using a number of different models.

**Enterprise risk management** Risk management on a holistic basis; assessing all risks together, allowing for diversifications and concentrations and risks that are both difficult and easy to quantify.

**Epicentre** Point on the Earth's surface directly above the hypocentre.

**Epistemic uncertainty** Uncertainty that comes from a lack of knowledge associated with a process.

**Equity** A company's assets minus its liabilities. See *capital*.

**Eurocode** European building standards codes.

**Evapotranspiration** The process combining evaporation of water and transpiration (e.g., by trees) of it; both extract water from soil.

**Event** A distinctly definable occurrence of a hazard (e.g. hurricane), implicitly with a start and end point in time.

**Event footprint** The maximum intensity of the event hazard (e.g. flood depth) over the defined time period of the event.

**Event limit** A cap to the losses which can be paid out in response to a single event.

**Event loss table (ELT)** The output from a catastrophe model, comprising loss statistics for every loss generating event in the catastrophe model.

**Event response** The process of reacting to a catastrophe event that is imminent or has just happened. This could include a logistical 'on the ground' response as well as providing company management with information on the likely scale of impacts and financial losses.

**Event set** A group of a large number of events (e.g. $n > 10,000$) simulated, with associated probabilities, to reflect the range of what is possible in reality.

**Exceedance frequency (EF)** The annual frequency of events with losses greater than a certain level. Unlike exceedance probability, EF can be greater than 1.0.

**Exceedance probability (EP)** The probability of loss from an event exceeding a certain level. Usually refers to either the occurrence exceedance probability or the aggregate exceedance probability.

**Excess** The amount of loss a policyholder has to pay before they can reclaim from the policy. See also *deductible*.

**Excess average annual loss (XSAAL)** The expected value of loss above a specific level. Unlike tail value at risk and tail conditional expectation this is not a conditional measure, and so the XSAAL will always be less than or equal to the average annual loss (AAL).

**Excess of loss (XoL or XL)** Reinsurance which operates in layers.

**Excess tail value at risk (xTVaR)** The same as tail value at risk, but with the mean value subtracted.

**Excess value at risk (xVaR)** The same as value at risk, but with the mean subtracted.

**Ex-gratia payment** Claim payments falling outside the terms and conditions of the contract/policy.

**Exhaustion point** The monetary value at which a (re)insurance cover ends.

**Expected loss (EL)** The statistical mean loss of a quantity. In this context, usually over a period of one year. See also *pure premium*, or *average annual loss*.

**Experience models** Models which use the claims history from the policy (or locations) being modelled to estimate the risk from the policy (or locations). Contrast to *exposure models*.

**Experience rating** A pricing method based on the policy-specific historical losses.

**Explosive** Violent type of volcanic eruption.

**Exposed limit** The maximum amount of limit that can be lost by the (re)insurer.

**Exposure** In the context of catastrophe modelling, see *exposure data*. More generally exposure is an asset's or entity's susceptibility to loss.

**Exposure data** The data representing the assets to be modelled. Typically building values and characteristics, location information, and details of insurance financial structure (such as limits and deductibles).

**Exposure management** See *catastrophe risk management*.

**Exposure models** Models which take exposure data as input and thus provide an estimate of the risk without using only the historic claims data from the specific locations or policies being modelled. Contrast to *experience models*.

**Exposure rating** Pricing method based on the characteristics of the assets being priced, but not considering claims experience only from the assets being priced.

**Extrapolation** A process to extend data beyond the observed range.

**Extra-tropical cyclone (ETC)** Mid-latitude cyclonic weather systems (i.e. ±30-80°N/S).

**Extra-tropical transition** A term used to describe the movement of some tropical cyclones outside the tropics (i.e. ±30°N/S).

**Extreme value analysis (EVA)** Statistical methods to assess and describe rare (infrequent) events.

**Eye** A calm area in the middle of a tropical cyclone delimited by a wall of strong winds and rain known as the eye-wall.

**Eye-wall** A region of strong winds and rain delimiting a calm area in the middle of a tropical cyclone known as the eye.

**Eye-wall replacement cycle** Decay and replacement of an eye-wall in a tropical cyclone.

**Facultative reinsurance (Fac)** Reinsurance which is negotiated to cover specific risks; contrasts with treaty reinsurance.

**Fault** A crack in the outer brittle layer of the Earth.

**Fault plane** The surface on which slip occurs during an earthquake.

**Fault segment** Section of a fault thought to have a tendency to rupture as a unit, i.e. entirely and in isolation from areas around it.

**FEMA (Federal Emergency Management Agency)** US state body responsible for disaster mitigation.

**Financial calculation or financial engine** The component of a model that performs the statistical calculations, including the application of insurance and reinsurance policy structures.

**Financial structures** Insurance and reinsurance policy conditions such as deductibles, limits, and coinsurance shares. See also *policy terms*.

**Fire following earthquake (FFEQ)** Loss due to earthquake-induced fires. Significant secondary source of loss.

**First loss scales (or curves)** An exposure rating technique used particularly in the pricing of per risk reinsurance treaties.

**Flash flooding** Flooding whose onset is rapid.

**Flooding (FL)** The appearance of water on land where it is not usually present.

**Fluvial flooding** Flooding associated with rivers.

**Focusing** Focusing of waves (e.g. seismic or tsunami) concentrates energy in a given spatial location. Analogy: what a magnifying glass does to sunlight.

**Follow the fortunes** A reinsurance doctrine where the reinsurer must accept the insurer's reasonable claims decisions provided there is coverage under the terms of the original insurance contract.

**Foot** Bottom of a landslide scarp.

**Footprint** The area expected to be impacted for each event (e.g. by earthquake shaking, or inundation by flood water).

**Forecast** An assessment of the likelihood of a hazardous event occurring for a given area, size range and time frame; these are normally probabilistic.

**Fragility curves** Also known as fragility function. Typically used in engineering. Relates hazard intensity to damage state of building. See also *vulnerability curves*.

**Franchise deductible** A threshold amount that prevents a payment to the policyholder until the level of loss reaches the amount of the franchise. Once the amount is reached, the deductible vanishes and the full amount is payable to the policyholder.

**Frequency** (1) Speed of oscillations that make up a wave, e.g. pitch of a musical note. (2) Number of times that an event occurs in a given time period; typically annual.

**Frequency analysis** An assessment of the probability of an event (e.g. flood) of a given magnitude range occurring within a time-frame (e.g. next year).

**Fujita scale** A scale measuring tornado severity.

**Full reinstatement sum insured** The sum insured calculated including an allowance for inflation during the policy and rebuilding period. Contrast to *day-one sum insured*.

**General insurance (GI)** The insurance of everything other than the contingency of death.

**Geocoding** A computational method to convert address information into spatially structured data, allowing address position to be defined in a co-ordinate system (e.g. latitude/longitude).

**Generalized Pareto distribution (GPD)** A continuous statistical distribution commonly used to model extreme loss values.

**Geographical information system (GIS)** A digital, computerized system for the capture, storage, integration and display of information related to the Earth using a common geographic reference base and data structures. GIS and geographic data underpin catastrophe model integration.

**Global climate (circulation) model (GCM)** A computer model that incorporates the physical principles that govern the Earth's meteorological system, simplified to a series of equations and run at a global scale.

**Global positioning system (GPS)** A system for identifying real-world position on the Earth's surface via a constellation of dedicated near-Earth orbiting navigation satellites and ground based receivers.

**Governance** The processes, controls and oversight put in place to ensure that catastrophe risk (in this context) is appropriately managed.

**Graupel** Also called soft hail or snow pellets, this is precipitation that forms when super-cooled droplets of water are collected and freeze on falling ice particles.

**Gross loss** The loss to the insurer after limits and deductibles and co-insurance are applied, but before any forms of reinsurance.

**Ground motion prediction equation (GMPE)** Equations to predict the intensity of ground shaking due and its associated uncertainty at any given location due to the earthquake hazard; factors with predictive skill include earthquake magnitude, source-to-site distance, local soil conditions and fault mechanism. Also see *attenuation* functions.

**Ground-up loss (GUL)** The modelled loss before application of any insurance or reinsurance financial structures.

**Groundwater flooding** Flooding caused by rising water tables, such that flow appears close to or above the ground.

**Gust front** High wind speeds bordering a severe convective storm generated by downdrafts.

**Hailstreak** The area affected by a hail event.

**Hazard** The danger resulting from a peril. The hazard component of a catastrophe model reflects the extent and intensity of set of events for a given peril.

**Head of claim** Cause of loss.

**Heave** The distribution of shrinking or swelling beneath the foundations of buildings causing shrink swell damage.

**Hours clause** A time window (e.g. 72 hours) that is used to identify discrete loss-generating events for (re)insurance purposes. This length of this window may be specified in the loss occurrence clause in a reinsurance contract.

**HURDAT** Atlantic hurricane track and landfall database, provided by the US National Hurricane Centre.

**Hurdle price** Price below which the (re)insurance product will not be written, irrespective of the market.

**Hurricane**  Name given to tropical cyclones in North Atlantic and Central and East Pacific oceans.

**Hydraulic models**  Models that, for specified flows of water, calculate how water spreads out across a terrain in order to determine inundation extent or depth.

**Hydrological models**  Statistical or physical models that relate precipitation amounts and patterns to flows in rivers; used as an input for hydraulic models.

**Hydrological systems**  The pathways water takes such as soils, aquifers, hill-slopes, or rivers.

**Hydro-meteorological triggering**  Rainfall leading to wetter, heavier slopes becoming unstable causing mass movement.

**Hypocentre**  Centre point of the part of a fault that has slipped in an earthquake.

**Indemnity**  An obligation to provide compensation following a loss.

**Indemnity-based contract**  A kind of insurance-linked security contract where the policyholder is put in the same financial position as before the loss – this is the usual form of insurance.

**Indirect loss**  Loss caused by the wider impacts of an event (e.g. infrastructure ceasing to work, or business interruption).

**Induced seismicity**  Earthquake activity, raised above the expected natural level, by human causes.

**Industry/insured exposure database (IED)**  An economic exposure database calibrated to reflect insurance take up rate/penetration rate. Alternatively, database may be restricted to insured assets.

**Industry loss warranty (ILW)**  An insurance linked security contract which pays when the loss to the entire industry crosses specified thresholds.

**Inland filling rate**  The rate at which the central pressure anomaly of a tropical cyclone 'fills', and the storm decays.

**Insurance**  An arrangement whereby one party (insurer) promises to pay another party (the policyholder) a sum of money in the event of a loss due to a specific cause.

**Insurance linked security (ILS)**  A financial instrument whose price is determined by insurance related losses.

**Insurance-to-value (ITV)**  Where the reported insured value is less than the actual value. See *under-insurance.*

**Insurer**  A company that provides insurance.

**Intensity**  Strength of a hazard. For instance, severity of shaking as experienced at a geographic location during an earthquake.

**Interaction**  A statistical dependency or 'link' between observations, which may then be interpreted in terms of physical processes, i.e. the behaviours or occurrence of hazards may influence each other or be driven by some mutual underlying cause.

**Internal model**  A model used by an insurance company to understand its risk and assess its solvency level. See also *capital models.*

**International building code (IBC)**  International building code standard.

**Interplate earthquakes**  Those that are generated at boundaries between tectonic plates.

**Interpolation**  A range of statistical/geospatial data conversion methods used to 'fill in gaps'. For example, to convert point sample data to a complete extent over the study area or period.

**Intraplate earthquakes**  Those that happen away from the boundaries of tectonic plates.

**Inundation**  Flooding.

**Inundation depth**  The vertical amount by which the inundation is above the land surface.

**Inundation height**  For tsunami or storm surge, the maximum elevation above mean sea level, at any given point, that gets submerged during the wave or surge's passage.

**Inuring**  A reinsurance that applies before another is said to inure to the benefit of the other.

**Jökulhlaup**  Glacial outburst flood.

**Key risk indicator (KRI)**  A term used in enterprise risk management for a metric that quantifies the main risk exposures for a business.

**Lahar**  Mud flow comprised of volcanic material.

**Lake outburst flood**  Flood resulting from the breach of natural geomorphological features that act as dams.

**Landslide**  See *mass movement*.

**Landslide inventory**  Catalogue of landslide events.

**La Niña**  The cold phase of the ENSO climate mode.

**Lava flow**  Flow of molten rock, on the Earth's surface. A volcanic hazard.

**Layer**  A slab of non-proportional reinsurance cover which starts at an attachment point and ends at an exhaustion point.

**Liability**  Liability can mean the amount of exposure and is commonly used this way. It can also refer to a type of (re)insurance covering the cost of compensation to third parties

**Liability insurance**  Insurance to cover claims from third parties. See also *casualty insurance*.

**Life insurer**  An insurance company providing insurance contingent on whether an individual dies.

**Lightning**  Electrical discharge during a thunderstorm.

**Limit**  The maximum amount a policy will pay out.

**Line**  One line is the amount of the retention in a surplus treaty. Taking a line on a treaty, may simply mean underwriting a portion of it.

**Line of business**  The type of insurance, for example, motor, property, marine, aviation, construction or liability.

**Linear exposures**  An asset which takes the form of a line e.g. pipelines or roads.

**Liquefaction**  Shaking-induced change to water-saturated soils causing fluidity and loss of supporting strength.

**Loss amplification** The increase in losses due to loss inflating effects from large scale catastrophic events, including but not limited to: labour and materials supply-demand, claims inflation, or claims expansion.

**Loss occurrence clause** See *hours clause*.

**Loss on line (LoL)** Loss divided by limit.

**Loss perspectives** Terminology describing how the modelled loss is shared among different parties, usually: ground-up, retained (or client), gross, net-pre-cat, or net-post-cat.

**Loss ratio** The ratio of loss to premium, used often in communicating (re)insurance company results.

**Magma** Molten rock when still inside the Earth.

**Magnitude** The representative size of an event (e.g. earthquake).

**Man-made disasters** Perils that are entirely of human construction.

**Mapped fault** Fault with a surface expression (i.e. trace) that has been observed such that it can be recorded on a map.

**Marginal impact analysis** An analysis that compares the incremental impact to key metrics (such as AAL or AEP or XSAAL) of adding a unit of business (e.g. an individual policy, or a portfolio of many policies) to the company's current portfolio of business.

**Marine insurance** Insurance for ships ('hull') and the cargo they carry and liability to others.

**Market price** Price the customer pays (which may be different to the technical price).

**Market pricing curves** Curves which are used to compare catastrophe excess of loss pricing across a market, typically with RoL on the *y*-axis and the geometrical mid-point of the layer against some measure of exposure on the *x*-axis.

**Mass movements (MM)** Downslope movements of rock, earth or debris. See also *landslide*.

**Materiality** The relative significance of a peril, risk or exposure to the anticipated overall risk to the organisation. Materiality is defined within Solvency II as "[Information] is material if omission or misstatement could influence the [economic] decisions of users taken on the basis of the financial statements"- CEIOPS-DOC-50/9.

**Mesocyclone** A single rotating updraft at the core of a supercell.

**Mesoscale convective systems (MCS)** Large-scale assemblages of thunderstorms.

**Metarisk** Metarisk is a 'type of risk'. Florida Hurricane and California Earthquake are two examples of metarisks.

**Microbursts** Small downbursts.

**Mitigation** Actions taken to reduce the impact of a hazard; can be behavioural (e.g. education) or engineering (e.g. flood barrier).

**Model blending** The process of combining model outputs (or in some cases model components) in order to gain benefit from multiple models and thereby reduce the uncertainty in a company's view of risk.

**Model developers** The people or companies who build catastrophe models, model components or model platforms.

**Model evaluation** The process of determining whether a catastrophe model provides a valid representation of catastrophe risk for the portfolio in question. See *model validation*.

**Model functioning tests** These tests investigate whether a model has been constructed properly and functions correctly. For example, test inputs (with known results) can be run through the model to test whether the model generates the required set of results.

**Model validation** The process of determining whether a catastrophe model provides a valid representation of catastrophe risk for the portfolio in question. Sometimes carries regulatory connotations. See *model evaluation*.

**Model vendors** The specialist companies who develop and license catastrophe models.

**Modified Mercalli intensity (MMI)** A measure of the severity of shaking as experienced at a geographic location.

**Modifier** A variable that can help predict changes in loss with changes in variable values. See also *rating factor*.

**Modularization** An approach for reducing complexity in computer programs by creation of discrete modules which may be integrated into a complete system.

**Mono-line insurance** Provided by an insurance company that specializes in a particular type of insurance.

**Monotonicity** In the context of coherent risk measures in catastrophe modelling: if the modelled losses increase, the risk measure should increase.

**Monte Carlo simulation** A class of computational algorithms that rely on random sampling to solve mathematical problems.

**Moral hazard** A potential change in behaviour when the policyholder no longer needs to face the consequences of the loss, e.g. when the risk is insured.

**Motor insurance** Insurance to cover the physical car and sometimes to cover liability to others. See also *auto insurance*.

**Multi-cell thunderstorms** Thunderstorms consisting of more than one convective cell.

**Multi-line insurance** Provided by an insurer that offers multiple types of insurance (e.g. property, marine, accident).

**Multiple** The ratio of dollars of premium for each dollar of expected loss.

**Mutual insurance company** An insurance company owned by its policyholders.

**Negative binomial distribution** A discrete statistical distribution commonly used to represent the probability of a number of events occurring in a specified time where there is some known dependence between events.

**Net-post-cat** A modelled loss perspective representing the net-pre-cat loss with catastrophe treaties applied.

**Net-pre-cat** A modelled loss perspective representing the gross loss with facultative and per-risk reinsurance applied, but not catastrophe treaties.

**Non-life insurer** A company providing general insurance (i.e. not life insurance).

**Non-modelled risk (NMR)** A term used to describe the process of identification and quantification of risk not captured by the models employed. Lloyd's of London's definition is: 'Any potential source of non-life insurance loss that may arise as a result of catastrophe events, but which is not explicitly covered by a company's use of existing catastrophe models.'

**Non-proportional (NP)** A type of reinsurance where the losses paid are not proportional to the premium.

**Non-stationary** A process which is not *stationary*.

**North Atlantic Oscillation (NAO)** A climate mode. Pressure difference between Iceland and the Azores. Thought to be associated with tropical and extra-tropical cyclone activity.

**Notional (sum insured)** Sum insured that is estimated by the insurer instead of being defined by the policyholder. The 'notional' may also refer to the full 100% liability of a reinsurance contract.

**Numerical weather prediction model** (**NWP**) A mathematical model which forecasts the weather based on current conditions. Like a Global Climate Model, but typically run at a higher spatial and temporal resolution.

**Occupancy** The purpose for which the insured building is being used. At a high level, typically residential, commercial, industrial or agricultural.

**Occurrence exceedance probability (OEP)** The probability of the maximum event loss in a year exceeding a certain level.

**Offering circular** The cat bond equivalent to a reinsurance submission, but may not contain the exposure data.

**On-levelling** Adjusting historical claims to a level which reflects the current (or future) risk. See *as-if* and *trending*.

**Ontological uncertainty** Uncertainty associated with unknown unknowns.

**Outer-rise earthquakes** Those that occur on the bulge that results from forcing one tectonic plate beneath another.

**Outwards reinsurance** Insurance purchased to mitigate the insurance risk taken on by an insurance company.

**Over-dispersion** A measure of the variability in measured data over and above that expected from independent randomness; the ratio of the variance to the mean.

**Overspill** The loss which occurs after the exhaustion of the programme's or policy's limit and falls back to the policyholder, similar to retention but at the top of the programme or policy.

**Own funds** Solvency II terminology for *capital*.

**Parametric contract** A type of catastrophe bond where contract losses are calculated indirectly using a parameter (such as wind speed).

**Pareto distribution** A continuous statistical distribution commonly used to model extreme values.

**Payback period** The length of time in years it would take to pay for the limit with the current premium.

**Peak ground acceleration (PGA)** A measure of severity of shaking due to an earthquake as experienced at a geographic location.

**Peak gust windspeed** The maximum windspeed over a short defined time period; usually 3 seconds.

**Peak zone aggregate** The sum insured in a major zone considered for aggregation.

**Penetration rate** The ratio of the value of insured properties to the total value of exposed properties.

**Per event** A policy which protects against the accumulation of multiple losses caused by a catastrophe event.

**Per risk** A policy which protects specific locations.

**Peril** Insurance name for a natural phenomenon with the potential to cause loss or damage.

**Peril-region** A peril modelled for a specific area (e.g. UK flood). A term driven by tackling the modelling problem from the science of particular hazard processes, then limiting the spatial scope of the model to make it tractable.

**Period of indemnity** The period of time for which loss of profits or revenue can be calculated for a business interruption claim.

**Personal lines (PL)** Insurance for individuals.

**Platform** A computer system that ensures the model components interact together in the correct way and enables data input, model running, and results output.

**Plinian** A type of large magnitude explosive volcanic eruption.

**Plug and play** A system or platform built in such a way that a new model or model component (even developed by someone other than the platform developer) can be used within the platform easily and quickly.

**Pluvial** Flooding caused directly by rainfall (i.e. not from rivers). Sometimes called 'off-flood-plain' flooding.

**Point observation** Measurements of the intensity of a hazard at a single geographic location.

**Poisson distribution** A discrete statistical distribution commonly used to represent the probability of a number of independent events occurring in a specified time.

**Policy** The legal contract between the insurer and insured. May also be referred to as 'cover'.

**Policyholder** The individual who takes out an insurance policy.

**Policy leakage** Claims (or components of claims) paid out when not originally intended to be covered by the policy, often due to poor policy wording followed by legal challenge, political pressure, or operational errors.

**Policy terms** Features of the (re)insurance policy, typically designed to mitigate or control the loss to the (re)insurer. See also *financial structures*.

**Ponding** A type of shallow flooding where water accumulates on relatively flat ground.

**Pooling of risk** Aggregating together risk from a number of insurance policies, usually in order to reduce uncertainty in the outcome.

**Portfolio** A grouping of a number of (re)insurance policies for the purposes of summarizing loss statistics at a useful level. For example, a grouping of policies for a specific product line or territory.

**Positive homogeneity** In the context of coherent risk measures in catastrophe modelling: multiplying modelled losses by a constant factor should result in the risk measure changing by the same factor.

**Post event loss amplification (PLA)** Demand surge plus other impacts that increase the loss from an event and relate to the scale of the event such as: loss adjustment challenges caused by a high number of claims, civil commotion, lack of access leading to secondary damage, such as mould and rain ingress and political pressure on insurers to provide coverage.

**Precipitation** Falling $H_2O$ in various forms (e.g. rain, snow, hail).

**Prediction** Stating in advance an exact location, date and time of future hazardous events; contrast to *forecast*.

**Premium** The amount paid for the (re)insurance policy.

**Present value** The equivalent value at the current time of some money to be received in the future.

**Pricing** The process of ascertaining the correct price to charge for a (re)insurance policy in order to cover losses, expenses and profit requirements whilst allowing for investment income.

**Primary hazard (or peril)** The main hazard or peril covered by a model. Typically treated as independent, unrelated to and not triggered by other hazards.

**Primary modifier** The most significant variables that can help predict changes in loss according to different variable values for a range of perils. Typically primary modifiers are occupancy type, year built, property height, property age, and (occasionally) square footage. See *secondary modifier* and *rating factor*.

**Probable Maximum Loss (PML)** The value of the largest loss that is considered likely to result from an event. It is typically the loss assuming the normal functioning of protective features (e.g. firewalls, static flood defences) including the proper functioning of most, but perhaps not all, active responses (e.g. sprinklers, temporary flood barriers erected, timely flood alerts). Since probabilistic catastrophe models have been in use, Probable Maximum Loss is also commonly used to refer to exceedance probability losses at particular thresholds e.g. commonly the 1-in-100 OEP loss for wind perils or the 1-in 250 OEP loss for earthquake perils. However, confusingly it can be used for be any exceedance probability loss. The term is used inconsistently, but is in wide circulation in the underwriting community where it is often used as a metric for catastrophe exposure monitoring against agreed risk appetites and appears in this context of exceedance probability loss in this book.

**Property insurance** Insurance that covers properties, their contents, and loss resulting from not being able to use the building due to an insured peril. See also *Direct and Fac.*

**Pro-rata** Another name for proportional treaty.

**Profit** The amount by which the (re)insurance premium exceeds losses and expenses, allowing for investment income.

**Profitability** An assessment of the amount of profit a policy, or grouping of policies is expected to make, or has made.

**Profit and loss attribution** A model validation term which means understanding the actual causes of profit and loss each year and checking that the model reflects the underlying causes of profit and loss.

**Profit commission (PC)** Commission returning money to the insurer if the loss results of the insurer are low and the treaty is profitable for the reinsurer.

**Propagation (or translation) speed** Speed at which a storm (e.g. extra-tropical cyclone) taken as a whole entity moves over the land surface.

**Proportional reinsurance** A type of reinsurance where the reinsurer takes a share of the original premium in return for a fixed share of the losses. Also known as *pro rata.*

**Pure premium** The *expected loss* cost to the insurance policy, i.e. the premium without taking into account expenses, investment income or other factors.

**Pyroclastic density currents** Flow of hot (100s of °C) rock fragments of a wide range of sizes (e.g., ash to 10s of m) downhill. Also known as pyroclastic flows and surges, or nuée ardente.

**Quota share (QS)** A type of proportional reinsurance treaty.

**Rainfall-runoff models** A type of *hydrological model* that uses the physical properties of known processes within the river catchment to determine rate and magnitude of water movement through the system.

**Rapid intensification** For tropical cyclones, an increase in maximum sustained winds of at least 55 km/h in 24 hours.

**Raster** A form of geographic data structure where data (attributes) are represented by discrete regular and contiguous grids, usually in a Cartesian two-dimensional coordinate system.

**Rate of event occurrence** The number of times an event is expected to occur divided by the time period represented by the catalogue of events.

**Rate on line (RoL)** Premium divided by limit. Can be technical or market.

**Rating agencies** Corporations whose primary function is to assess and summarize the financial strength of governments or corporations (primarily their ability to pay back debt), through assigned credit ratings.

**Rating factor** A variable that can help predict changes in loss according to different variable values. See also *modifier.*

**Realistic disaster scenario** Catastrophe scenarios used for exposure management by Lloyd's.

**Reanalysis data** Past meteorological observations unified and gridded in a physically consistent way using a Global Climate Model that has been forced to be consistent with the observations.

**Recurrence interval** Another name for *return period*.

**Regulators** Supervisory authorities, usually at country or state level, who put in place and govern the rules and directives relating to the (re)insurance entities within their jurisdictions.

**Regulatory capital** The amount of capital required by a company according to regulatory solvency rules. Same as solvency capital requirement. Contrast to *economic capital*.

**Reinstatement** A feature where cover is automatically reset (or reinstated) to its original level after a loss, often for an extra premium.

**Reinstatement premium** Amount to be paid to reinstate the cover to its original level. Reinstatements are not usually optional but are specified as part of the reinsurance contract.

**Reinsurance** Reinsurance is the insurance of insurance companies.

**Reinsurance panel** Group of reinsurers who participate on the same business.

**Reinsurance programme** The combined set of reinsurance contracts that a reinsurer has, designed to meet its specific needs.

**Reserving** The process (usually carried out by actuaries) of estimating the final (ultimate) claim amount for a particular segment or line of business for a particular time period.

**Resolution** A metric which describes the level of detail captured by spatial data (e.g. a digital terrain model), often described in terms of 'real-world' distance.

**Retained (or client) loss** The loss that the policyholder retains.

**Retention** The amount of loss retained by the policyholder (usually used for the part below the attachment of the policy).

**Retrocession** Retrocession is reinsurance of reinsurers. Sometimes this may simply be called reinsurance.

**Return period (RP)** An event that has a 1/T frequency of occurring in any year is said to have a T-year return period. In other words, the reciprocal of the exceedance frequency. Commonly used as terminology for 1/OEP or 1/AEP.

**Reverse faults** A type of fault in which rocks of lower stratigraphic position are pushed up and over higher (often younger) strata. They accommodate tectonic forces compressing an area.

**Reverse stress testing** Considers the scale of loss that could threaten the viability of the company, and then attempts to understand the type of scenarios (single or multiple) that could give rise to this scale of loss.

**Risk** Uncertainty leading to potential adverse outcomes. Also used as shorthand for an insured object both in this book and widely in the insurance industry.

**Risk-adjusted premium** Adjustment of the premium so that it reflects the amount of exposure (and thus loss potential).

**Risk appetite** The level of downside risk the company is comfortable accepting relative to the level of return desired.

**Risk appetite statement (RAS)** A statement which defines the level of downside risk the company is comfortable accepting; typically relative to the level of return desired and the capacity of the company to absorb the risk.

**Risk-based capital (RBC)** A modelled estimate of the amount of capital a company needs to avoid insolvency at a specified probability level over a specified time frame.

**Risk-free rate** What a capital provider can earn without taking on any type of risk.

**Risk management** The process setting a risk appetite and identifying, quantifying and monitoring the level of risk against this risk appetite.

**Risk metrics** Statistical quantities used to provide information about risk.

**Risk tolerance** Quantifiable constraints that limit the amount of risk that the firm is willing to assume.

**Risk transfer** The process of moving risk from one party to another e.g. policyholder to insurer, insurer to reinsurer, or reinsurer to retrocessionaire, via a contract and in exchange for premium.

**Roll-up** The process of combining multiple portfolios into a time-specific snapshot of exposure and modelled results. See *accumulation* and *aggregation*.

**Runout** Process moving material in a mass movement outside the area of the original scarp.

**Runup height** The maximum height above sea level reached by waves, including tsunami. See *inundation height*.

**Rupture** The breaking of a fault during an earthquake.

**Scarp** A steep surface above a landslide caused by the downwards movement of mass.

**Scenario** A representation of a possible (catastrophic) event based on scientific, industry and other expert knowledge.

**Scenario testing** Similar to sensitivity testing but varying multiple inputs (to reflect a particular scenario) to see how model results change.

**Schedule of risks (or values)** List of what assets are where; often the basis for exposure data.

**Secondary modifier** A variable that can help predict changes in loss according to different variable values, typically for a specific peril. See *primary modifier* and *rating factor*.

**Secondary peril (or hazard)** A peril that is seen as being caused or triggered by another 'primary' peril.

**Secondary uncertainty** The uncertainty in the loss given that there has been a specific, defined event (e.g. particular intensity and location). Usually, characterized through uncertainty in the vulnerability curves.

**Securitization** The process of taking an illiquid asset and using financial engineering (i.e. mathematical methods applied to financial problems) to transform it into an instrument that can be traded.

**Seeding** The presence of existing meteorological features that can help initiate storms, for instance, by extracting energy from the environment in the case of extra tropical cyclones.

**Seismic gap**  A part of a fault plane that has slipped an unusually small amount in the past.

**Sensitivity testing**  Varying the inputs to the model (typically one at a time) and observing changes to the outputs.

**Severe convective storms (SCS)**  A type of potentially damaging thunderstorm.

**Shallow flooding**  The name for surface water flooding in FEMA flood zones.

**Shallow-water waves**  Those with a wavelength that is much larger than the water depth.

**Shear strength**  Ability of a fault plane or shear surface to resist movement on it. Consists of friction and cohesion.

**Sheet flow**  Shallow flooding on slopes.

**Short tail**  Business where the loss does not take a long time to develop, so that the final claim size will be known and paid in a reasonably short time after the event.

**Shrink-swell subsidence**  Ground movements induced by volumetric change in soils rich in some silicate clay minerals.

**Sidecar**  A sidecar is a collateralized vehicle that allows alternative capital to invest in an existing book of business; in effect, sharing the risk and (some of) the return of the existing book.

**Signing profile**  How the reinsurers' participation on layers changes with the layer height in a reinsurance programme, e.g. a preference for top or bottom layers, or a similar share for every layer.

**Single-cell thunderstorms**  Thunderstorms consisting of only one convective cell.

**Site amplification**  An increase in shaking severity at a particular location during an earthquake due to local conditions (e.g. soil type).

**Skill**  A formal measure of the accuracy of predictions that relates the forecast accuracy of a particular forecast to some reference method or model.

**Sliding scale commission**  The payment to an insurer by their reinsurer (to compensate for the cost of acquiring the original business which is covered by a proportional treaty); the level of which varies depending on the profitability of that business.

**Slip**  Movement on a fault, either related to an earthquake or mass movement.

**Slip surface**  The surface on which slip occurs permitting a mass movement.

**Soil amplification**  Greater shaking by earthquakes due to the presence of unconsolidated sediment (e.g. soils in sedimentary basins).

**Soil moisture deficit**  The extent to which soil is drier than usual.

**Solvency**  The degree to which the assets of a company exceed its liabilities; the ability of a company to meet its long-term obligations.

**Solvency II**  A pan-European regulatory regime for insurers.

**Solvency capital ratio**  The amount of capital (or own funds) divided by the solvency capital requirement.

**Solvency capital requirement (SCR)** The amount of capital to be held by an insurer to meet the solvency requirements of the regulator. The same as *regulatory capital*. Contrast to *economic capital*.

**Spectral acceleration (SA)** A measure of earthquake shaking based on the acceleration of the ground for particular frequencies.

**Sponsor** The organization or entity commissioning a catastrophe model development project.

**Squall lines** Linear bands of convective thunderstorm cells.

**Stability testing** Understanding how sensitive model results are to different numbers of simulations.

**Standard deviation (SD)** A statistical measure of the uncertainty or variation in a quantity. The square root of the variance.

**Standard formula** A standard approach in which capital requirement is calculated using pre-defined industry-calibrated factors applied to the company's exposure.

**Stationary** Statistical term meaning the distribution of the process under consideration does not change with time (i.e. statistical metrics such as the mean and variance of the process do not change with time).

**Step policy** A policy that pays out a set number of specific amounts. For example, a discrete number of levels of loss depending on an assessed building damage state.

**Stochastic** Random. In this context of this book, a stochastic model is one which assigns rates or probabilities to a set of events to enable estimation of statistical metrics from a given peril.

**Stock insurance company** An insurance company with shareholders.

**Storm surge** Where the combined force of the wind and the low pressure from a storm acts to lift and push water onshore, potentially leading to coastal flooding.

**Straight-line winds** In severe convective storms, severe ground-level winds other than tornadoes. See *derechos*.

**Stress testing** Similar to sensitivity or scenario testing, but focusing on the most extreme (or stressful) scenarios (e.g. considering multiple scenarios occurring in the same year).

**Stretch** The value between attachment and exhaustion for an excess of loss cover.

**Sub-additivity** In the context of coherent risk measures in catastrophe modelling: combining risk measures from multiple locations or portfolios should not lead to risk measures that are higher than the sum of the risk measures for the individual locations or portfolios.

**Sub-limit** A limit that applies to a subset of exposure within the policy. For example, a limit for a specific peril or a specific geographical region.

**Submission** Application for (re)insurance cover and associated information.

**Sum insured** The overall replacement cost of the insured asset without including deductions for policy deductibles and limits. See *exposure data*.

**Supercell thunderstorms** The most dangerous type of convective storms (i.e. SCS).

**Supercooled** Water that remains liquid at sub-zero temperatures because it cannot freeze homogeneously (i.e. with no nucleus).

**Surface water flooding** Water moving or ponding outside established flood plains, or yet to enter a watercourse; see also *shallow flooding, sheet flow,* and *pluvial flooding.*

**Surficial deposits** Geological term for deposits near the Earth's surface (e.g. soil) to distinguish them from those that extend to depth (i.e. bedrock).

**Susceptibility** The likelihood that a hazard will occur in a region.

**Tail conditional expectation (TCE)** The expected value of loss above a specific level, given that the level has been exceeded. See *tail value at risk.*

**Tail value at risk (TVaR)** The expected value of loss above a specific level, given that the level has been exceeded. See *tail conditional expectation.*

**Takaful insurance** A particular form of insurance which is used to overcome cultural and religious objections to insurance in the Islamic world.

**Take-up rate** The ratio of the number of insured properties to the total number of exposed properties.

**Technical price** The price required to cover the loss cost, expenses and company's profit requirement over some medium term, allowing for investment income.

**Tephra** Fragments of any size, including ash, explosively erupted from a volcano.

**Terrain data** See *Digital Terrain Model.*

**Thrust fault** A special class of reverse fault that has a low dip angle, less than 20° from the horizontal.

**Time value of money** The way that the value of money changes over time reflecting interest rates.

**Toe** The furthest extent down-slope of a mass movement.

**Top-down** An approach that first estimates country-wide exposures and then allocates these to higher resolution geographical regions with distribution factors. See *disaggregation.*

**Topographic amplification** Greater shaking/ground motion due to earthquakes on high mountain peaks.

**Tornadoes** Violently rotating columns of air in contact with the ground, associated with severe convective storms.

**Track** Route of a storm usually specified by the location of the central pressure or vorticity minimum.

**Transform margin** A tectonic setting where plates are moving past each other.

**Translational invariance** In the context of coherent risk measures in catastrophe modelling: if the amount of modelled loss is increased by a fixed amount, the risk measure should increase by the same amount.

**Treaty reinsurance** An overall umbrella-type agreement that all risks written in a given year/period of a certain type (property, marine, risks in France, homeowners', earthquake, etc.) are reinsured and the treaty will automatically cover all the risks within a particular portfolio.

**Trending** Adjusting historical claims to a level which reflects the current (or future) risk. See *as-if* and *on-levelling*.

**Trigger** A change or fluctuation that, although perhaps small in itself, starts a hazard event (e.g. landslide or tropical cyclone). Can also refer to the threshold used for payout for a parametric contract.

**Tropical cyclone (TC)** (a.k.a. *hurricane* or *typhoon*). A type of severe, large-scale storm.

**Tsunami (TS)** A series of long-duration waves that are generated by an impulsive, rapidly occurring disturbance of a large water mass.

**Tsunami earthquakes** Earthquakes that cause tsunami waves larger than expected for their magnitude.

**Tsunamigenic earthquakes** Earthquakes that cause tsunami.

**Two-risk warranty** A contract clause found primarily in catastrophe excess of loss reinsurance which states that at least two risks must be affected for the contract (and so cover) to be activated.

**Typhoon** Name given to tropical cyclones in the western North Pacific Ocean.

**Ultimate net loss (UNL)** The sum of applicable claims to the treaty reduced for any potential gains e.g. from other reinsurances or salvages.

**Under-insurance** Where the amount insured is less than the amount that should have been insured. See *insurance-to-value*.

**Underpinning** An engineering solution to heave or shrink-swell.

**Underwriter** A person who performs the job of underwriting.

**Underwriting** The process of defining which risks are acceptable, determining the premium to be charged and the terms and conditions of the insurance contract and monitoring each of these decisions.

**Underwriting authority** The maximum amount and type of risk that can be taken on by a specific underwriter.

**United States Geological Survey (USGS)** US geological science agency, responsible for US/worldwide earthquake and other hazard data.

**Validation** Part of a model development process comprising testing of the model results against expected independent values to ensure the model is appropriate. For validation in a Solvency II context. See *model validation*.

**Value at risk (VaR)** The loss value at a specific quantile of the relevant loss distribution (e.g. specific values of the OEP or AEP distributions).

**Variance** A statistical measure of the uncertainty or variation in a quantity. The square of the standard deviation.

**Vector** A geographical data representation where features are stored as points, lines or areas, often in geographical information systems (GIS).

**Vendors**  In the context of this book, the companies who develop and sell catastrophe models.

**View of risk**  Developing a view of risk is the overall process of model evaluation, consideration of items not within the model (non-modelled risk) and subsequent model adjustment where necessary.

**Volcanic aerosol**  Small particles or droplets (e.g. ash, sulphuric acid droplets $H_2SO_4.nH_2O$).

**Volcanic bomb**  Rock ejected from a volcano of size exceeding 64 mm in diameter.

**Volcanic explosivity index (VEI)**  Measure of the size of explosive volcanic eruptions.

**Volcanic flooding**  Flooding resulting from heat causing glaciers to melt.

**Volcanic gas**  Gases emitted by a volcano (e.g. carbon dioxide or hydrogen sulphide).

**Volcano**  Outlet of hot and often dangerous material from the Earth's interior.

**Volcano observatory**  Scientific installation to monitor a volcano.

**Vulnerability**  A function relating a hazard metric (e.g. water depth) to damage ratio in order to calculate ground-up loss.

**Warm conveyor belt**  The flow of warm, moist air ahead of an extra-tropical cyclone.

**Wavelength**  Distance between successive waves.

**Wave period**  Time between successive waves (e.g. for tsunami).

**Wind shear**  A difference between winds at low and high altitudes; affects severe convective storm and tropical cyclone.

**Windstorm**  Generic name for any event where the predominant hazard is strong wind. Sometimes used to specifically refer to extra-tropical cyclones that generate strong surface winds.

**Writing across the board**  Taking equal or similar shares across every layer of a reinsurance programme (e.g. 5% of every layer).

**Wording**  The terms of a (re)insurance contract.

**Year loss table (YLT)**  The output from a catastrophe model, comprising loss statistics for every loss generating year in the catastrophe model. Also known as a Period Loss Table if considering other lengths of time.

**Zoning map**  A method for representing hazard across geographical regions, often for planning or disaster management purposes.

# Index

*Natural Catastrophe Risk Management and Modelling: A Practitioner's Guide*, First Edition.
Edited by Kirsten Mitchell-Wallace, Matthew Jones, John Hillier and Matthew Foote.
© 2017 John Wiley & Sons Ltd. Published 2017 by John Wiley & Sons Ltd.

Printed and bound by CPI Group (UK) Ltd, Croydon, CR0 4YY

16/04/2025

14658553-0007